ADVANCES IN VEHICULAR NETWORKING TECHNOLOGIES

Edited by **Miguel Almeida**

Advances in Vehicular Networking Technologies
Edited by Miguel Almeida

Published by InTech
Janeza Trdine 9, 51000 Rijeka, Croatia

Publishing Process Manager Katarina Lovrecic

Technical Editor Teodora Smiljanic

Cover Designer InTech Design Team

Additional hard copies can be obtained from orders@intechopen.com

Advances in Vehicular Networking Technologies, Edited by Miguel Almeida
p. cm.
ISBN 978-953-307-241-8

Meet the editor

Miguel Almeida is both experienced in publishing papers for relevant conferences, and in working at the Telecommunication industry. His research background allowed him to participate in several European research projects, while his industry related experience has allowed him to manage advanced network performance solutions and software architecture definitions.
With an Engineering Degree and a Master Degree in Telecommunications, he is currently pursuing his PhD under the topics of performance management of remote devices. His main areas of interest are: Cross Layer Performance Analysis, Vehicular and Mobile Networks, 4G Architectures, End User Behavior Tracking, the Internet of Things Paradigm and Remote Device Management solutions.

Contents

Preface

It's fair to say that the array of commercially available vehicles is beginning to catch up with the technological advances made available by science during the past decade. With a lot of effort being employed by manufactures to provide cars with advanced networking capabilities, we can mainly distinguish the connectivity topologies in two very different groups: the Vehicular Ad-hoc Networks (VANETs), which make use of the scientific contributions provided by the Mobile Ad-Hoc Networks (MANETs), and the Vehicle Infrastructure Integration (VII) based networks. VII or Vehicle-to-Infrastructure (V2I) deserved special attention during the last couple of years, in part, due to the increase of interest on Cloud based communications, but also given the current highlight over the paradigm shift towards the Internet of Things. Nevertheless, the scope of technological challenges that have an immediate impact on the design and performance of such specific networks is extremely wide and particularly diverse.

This book provides an insight on both, the challenges and the technological solutions of several approaches, which allow connecting vehicles between each other and with the network. It underlines the trends on networking capabilities and their issues, further focusing on the MAC and Physical layer challenges. Mobile oriented technologies set up the basic requirements for high mobility scenarios. Having this in mind, particular attention was paid to the propagation issues and channel characterization models.

We tried to cover a vast multitude of topics, which reflect the current state of the art concerning Vehicular Networking Technologies, some of which include dealing with connectivity issues, networking topologies (VANETs, VII/V2I), MAC solutions, data forwarding, network/vehicle performance management, link layer coding techniques, mobile/radio oriented technologies, channel characterization and channel coding amongst others.

We are thankful to all of those who contributed to this book and who made it possible. We hope others can enjoy it as much as we do.

Miguel Almeida
University of Aveiro
Portugal

Part 1

Wireless Networks

1

Seamless Connectivity Techniques in Vehicular Ad-hoc Networks

Anna Maria Vegni[1], Tiziano Inzerilli[2] and Roberto Cusani[2]
[1]*University of Roma Tre, Department of Applied Electronics, Rome,*
[2]*University of Rome Sapienza, Department of Information Engineering, Electronics and Telecommunications, Rome, Italy*

1. Introduction

Emerging Vehicular Ad-hoc NETworks (VANETs) are representing the preferred network design for Intelligent Transportation Systems (ITS), mainly based on Dedicated Short-Range Communications (DSRC) for Vehicle-to-Vehicle (V2V) communications (Held, 2007). Future vehicles will be fully networked, equipped with *on-board* computers with multiple Network Interface Cards (NICs) (*e.g.*, Wi-Fi, HSDPA, GPS), and emerging wireless technologies (*e.g.*, IEEE 802.11p, WiMax, LTE).

Although V2V is potentially the most viable approach to low-latency short-range vehicular networks, connectivity in VANETs is often not available due to quick topology network changes, random vehicle speed, and traffic density (*i.e.* sparse, dense, and totally disconnected neighbourhoods) (Chiara *et al.*, 2009). As an alternative, longer-range vehicular connectivity are supported by a Vehicle-to-Infrastructure (V2I) protocol (Held, 2007), which exploits a pre-existing network infrastructure, for communications between vehicles and wireless/cellular access points (referred to Road Side Units, RSUs). As a further benefit, V2I protocols allow access to the Internet and delivery of traditional applications in addition to dedicated applications for ITS, thus making vehicle communications more versatile.

Both the paradigms —V2V and V2I— exhibit connectivity problems. Different speed and traffic densities result in low vehicular contact rate, and limit communications via V2V protocol, while V2I communications are reduced, especially in highway scenarios, by the low number of RSUs displaced on the roads. Moreover, V2I limitations are due to particular vehicular applications, and performance is also strictly dependent on the specific wireless technology for the RSUs. It can then become advantageous to adopt hybrid schemes combining V2V and V2I communication into a single protocol and allowing fast migration from V2I to V2V connectivity depending on the operation context (high/low density, high/low velocity). Such mechanism is the so-called Vertical Handover (Pollini, 1996).

In this chapter we shall describe the traditional techniques —Vertical Handover algorithms— used for seamless connectivity in heterogeneous wireless network environments, and in particular adopt them in VANETs, where V2V and V2I represent the main communication protocols. Section 2 deals with the basic features of Vertical Handover (VHO) in the general context of a hybrid wireless network environment, and it discusses how decision metrics can affect handover performance (*i.e.* number of handover occurrences, and throughput). Instead, Section 3 briefly introduces two proposed techniques achieving seamless connectivity in VANETs. The first technique is a vertical handover mechanism applied to V2I-only communication environments; it is presented in Section 4 via an analytical model, and main simulated results are shown in Subsection 4.1. The second approach is described in Section 5. It addresses a hybrid vehicular communication protocol (*i.e.* called as Vehicle-to-X) performing handover between V2V and V2I communications, and vice versa. Subsection 5 illustrates how messages can be propagating via V2X, while Subsection 6 shows the main phases of V2X algorithm and the simulation results. Finally, we conclude this chapter in Section 6.

2. Vertical Handover mechanism

Next generation wireless networks adopt a heterogeneous broadband technology model in order to guarantee seamless connectivity in mobile communications. Different network characteristics are basically expected for different multimedia applications, and ubiquitous access through a single network technology is not always guaranteed because of limitation of geographical coverage. Moreover, since mobile applications require Quality-of-Service (QoS) continuity in a ubiquitous fashion, cooperation of access networks in heterogeneous environments is an important feature to assure.

A Vertical Handover (VHO) is a process preserving user's connectivity *on-the-move*, and following changes of network (Pollini, 1996). In this context, VHO techniques can be applied when network switching is needed (*i*) to preserve host connectivity, (*ii*) optimize QoS as perceived by the end user, and (*iii*) limit the number of unnecessary vertical handover occurrences (*i.e.* the well-known *ping-pong effect*) (Inzerilli & Vegni, 2008).

VHO schemes can be classified on the basis of the criteria and parameters adopted for initiating a handover from a Serving Network (SN) to a new Candidate Network (CN). Namely, we can enlist the following main metrics whose monitoring can drive handover decisions:

- *Received Signal Strength* (RSS)-based VHO algorithms: when measured RSS drops below receiver sensitivity it denotes lack of connectivity which requires necessarily a VHO (Ayyappan & Dananjayan, 2008); (Inzerilli & Vegni, 2008);
- *Signal-to-Noise and Interference ratio* (SINR)-based VHO algorithms: SINR directly impacts achievable goodput in a wireless access network. A modulation scheme can sometimes adapt transmission rate a channel coding scheme to measured SINR (Yang *et al.*, 2007); (Vegni *et al.*, 2009);
- *Multi-parameter QoS*-based VHO algorithms: VHO algorithms can be based on the overall quality assessment for the available networks obtained balancing various parameters (Vegni *et al.*, 2007); (Jesus *et al.*, 2007);
- *Location*-based VHO algorithms: they estimate network QoS on the basis of the MT location relatively to the serving access point (Kibria *et al.*, 2005); (Wang *et al.* 2001); (Kim *et al.*, 2007).

In an RSS-based VHO approach, when the measured RSS of the SN drops below a predefined threshold, the RSS of the monitored set of CNs is evaluated in order to select the best network to migrate to. The authors in (Ayyappan & Dananjayan, 2008) adopt this basic approach for VHO and evaluate the performance of a vertical handover mechanism that is based on the RSS measurements. This approach represents the traditional handover and simplest mechanism, which however does not aim to optimize communication performance.

Differently, the SINR-based approach (Yang *et al.*, 2007) compares the received power against the noise and the interference levels in order to obtain a more accurate performance assessment, which brings about a slight increase of computational cost. SINR factor is considered for VHO decisions, as it directly affects the maximum data rate compatible with a given Bit Error Rate (BER). Therefore, when the SINR of the serving network decreases, the data rate and the QoS level decrease too. As a consequence, a SINR-based VHO (Yang *et al.*, 2007) approach is more suitable to meet QoS requirements, and can be used to implement an adaptive data rate procedure. RSS-based and SINR-based schemes are both reactive approaches, which means that they aim to compensate for performance degradation when this occurs, that is whenever either the RSS or the SINR drops below a guard threshold. Moreover, we expect that a combination of different VHO decision metrics (*i.e.* location information, RSS/SINR measurements, QoS requirements or monetary cost) can generate most effective and correct VHO decisions (Vegni *et al.*, 2009).

A Multi-parameter QoS-based VHO scheme has been illustrated in (Vegni, *et al.* 2007), which is instead representative of a proactive approach performing regular assessment of the QoS level offered by the current SN, as well as by other CNs. The proposed method attempts to select the best CN at any time thus preventing performance degradation, and sudden lack of connectivity. It can be based on the simultaneous estimation of a set of parameters such as RSS, throughput and BER and in the subsequent evaluation of an objective QoS metric, which is a function of such parameters. Its effectiveness is directly dependent on the ability of the objective QoS metric to mimic subjective Quality-of-Experience of the end-users, and on the accuracy of the assessment of the parameters on which the metric is based. QoS-based VHO is well suited for multimedia applications like real-time video streaming. As a drawback, preventive approaches may lead to high handover frequency and hence lead to algorithmic instability, *i.e.* the so-called *ping-pong* effect. A hysteresis cycle or a hard limitation in maximum handover frequency in the VHO algorithm can help reducing this phenomenon (Kim *et al.*, 2007).

In location-based VHO solutions, the knowledge of location information is exploited to assess the quality of the link between SN and the MT, and to predict its future evolution to some extent on the basis of the MT estimated path. User position can be determined in several ways (Kibria *et al.*, 2005), including Time of Arrival, Direction of Arrival, RSS, and assisted GPS (Global Positioning System) techniques. Examples of location-based VHO are discussed in (Wang *et al.* 2001), though the proposed technique shows a computational complexity of the handover decision that is rather high, and establishing and updating a lookup table to support a handover margin decision turns out to be time-consuming. In (Kibria *et al.*, 2005) the authors develop a predictive framework based on the assumption that the random nature of user mobility implies an uncertainty on his/her future location, increasing with the extension of the prediction interval.

Location-based VHO solutions are the most commonly-used techniques in the VANET context, where high-mobility of nodes makes it difficult to promptly react to performance degradation purely basing on RSS measurements.

3. Seamless connectivity in vehicular ad hoc networks

In this section we are introducing two different approaches for seamless connectivity in VANET scenarios.

The first approach is a Vertical Handover technique based on vehicle speed, where a vehicle switches from a Serving Network (SN) to a Candidate Network (CN) if its speed is lower than a fixed threshold (Vegni & Esposito, 2010); (Esposito *et al.*, 2010). Such technique *i.e.*, called as Speed-based Vertical Handover (S−VHO) mainly addresses necessary connectivity switching in vehicular environments for real-time applications, which require high QoS levels and seamless connectivity. S−VHO does not consider traditional vehicular protocol (*i.e.* V2V and V2I), but it simply guarantees a seamless connectivity when vehicles are crossing an area with overlapping wireless networks.

The problem of a seamless connectivity becomes more challenging in VANETs, because vehicles move across overlapping heterogeneous wireless cells environments. In such scenarios, frequent and not always necessary switches from a SN to a CN may occur, often degrading network performance. Vertical Handover techniques are able to fix seamless connectivity according to a well-defined decision criteria (*i.e.* the type of RSU technology, the RSS indicator, QoS metrics, and so forth (McNair & Fang, 2004)).

Traditional VHO decision metrics cannot be applied in vehicular environments, and may fail due to the speed and the time that the vehicle is going to spend in a wireless network. The problem of a seamless connectivity becomes even more challenging as vehicles move at high speed across overlapping heterogeneous wireless cells environments.

Therefore, in VANETs handovers should be performed on the basis of specific factors, such as vehicle mobility pattern, and locality information, rather than standalone QoS requirements. Past solutions have partially but not fully considered these aspects. In (Chen *et al.*, 2009) the authors deal with a novel network mobility protocol for VANETs, to reduce both handover delay and packet loss rate. In (Yan *et al.*, 2008) a vertical handover technique focuses on an adaptive handover mechanism between WLAN and UMTS, based on the evaluation of a handover probability, obtained from power measurements. In this case, the handover decision is taken by comparing the handover probability with a fixed probability threshold, which depends on the vehicle speed and on handover latency.

The second technique is a hybrid vehicular communication protocol, called as Vehicle-to-X (V2X), which achieves the advantages of both traditional V2V and V2I protocols (Vegni & Little, 2010). V2X supports VANET scenarios with a heterogeneous network environment, and aims for vehicles (*i*) to communicate between them (via V2V), and (*ii*) to connect to the Internet (via V2I). V2X permits hybrid vehicular communications, and each vehicle can switch from V2V to V2I, and vice versa, on the basis of a *protocol switching decision metric.*

We will illustrate the behaviour of V2X protocol, and analyze how information is propagated in VANETs with heterogeneous network infrastructure nearby. In the simulated scenarios, a data push communication model has been assumed, in which information messages are propagated via localized (limited range) broadcast. Effectiveness of V2X will be validated via a performance comparison —in terms of *message dissemination*— with traditional opportunistic networking technique (*i.e.*, V2V).

V2X results in a novel opportunistic forwarding technique that is the main approach to achieve connectivity between vehicles, and to disseminate information. In traditional opportunistic networking V2V communications exploit connectivity from other neighbouring vehicles by a *bridging* technique, where message propagation occurs through connectivity links which are built dynamically (Agrawal & Little, 2008). Each vehicle acts as next hop and subsequent hops form a path from a source vehicle to a destination vehicle. Because V2X does not rely only on V2V but also exploits the potentiality of V2I, it should be considered as an opportunistic forwarding technique, where messages are propagated along dynamically generated paths whose links are vehicles, as well as road side units.

4. Speed-based VHO technique

A Speed-based VHO technique (S-VHO) is now described in each single step of VHO decision (Vegni & Esposito, 2010). We recall that, as all the VHO techniques are mainly focused on maximization of network performance, and limitation of vertical handover occurrences, even for S-VHO the aim is to both minimize VHO frequency—the number of executed VHOs— and maximize the throughput measured at the vehicle.

Differently from traditional RSS-based vertical handover approaches (Inzerilli & Vegni, 2008), the proposed S-VHO method does not consider any signal strength parameters: such information might be out of date, unreliable and its variance may fluctuate significantly, especially in VANET scenarios, causing unnecessary and unwanted vertical handovers, as well as throughput degradation. With respect to traditional VHO algorithms, S-VHO takes a vertical handover decision on the basis of the estimation of the *cell crossing time* parameter, which represents the time spent in crossing a wireless cell by the vehicle. No RSS indicator is considered in such approach, and S-VHO decides to switch from a SN to a CN on the basis of throughput experienced at the vehicle receiver, and the time spent by the vehicle in a CN.

It is noticeable that S-VHO method approaches to the technique proposed in (Yan *et al.*, 2008), except that S-VHO focuses on a vehicle-controlled VHO, due to smart *on-board* computer equipped with GPS connectivity, and handover decisions are based on both vehicle speed, and handover latency (Vegni & Esposito, 2010); (Esposito *et al.*, 2010). Moreover, S-VHO differs from previous vertical handover techniques because its usefulness is extended to real time applications for vehicular scenarios (*i.e.*, video streaming, on-line gaming, Internet browsing, and so on).

We shall describe an analytical framework of S-VHO technique. First, we depict how throughput in both serving and candidate networks can be estimated; then we show analytically how a vertical handover decision is taken by the vehicle.

Figure 1 illustrates the vehicular scenario we are dealing with. We assume that a vehicle is driving at constant speed v [m/s], following a typical Manhattan mobility model, suitable for a urban area and where the vehicle's trajectory is constrained by a road grid topology, with straight paths (*i.e.* a highway composed of straight lanes). Each vehicle is equipped with a GPS receiver, and then the vehicle's location is continuously updated and tracked.

Let us denote as t_{in} and t_{out} the time instants when a vehicle enters and exits a wireless cell (*i.e.* an UMTS network), respectively. The distance the vehicle will cross inside the wireless cell during the time interval $\Delta T = t_{out} - t_{in}$ is:

$$\Delta x = \frac{v}{\Delta T}. \quad (1)$$

From (1) we can introduce the *Cell Crossing Time* [s] parameter, according to the following assessment, such as:

Definition *(Cell Crossing Time).* *Given a vehicle V, traversing an area covered by a wireless cell C at constant speed $\vec{v}$, the cell crossing time of V in C, denoted as ΔT [s], is the overall time that V can spend under C's coverage.*

According to Figure 1, the cell crossing time lasts since the vehicle enters the wireless cell in P_{in} at t_{in}, and then exits the wireless cell in P_{out} at t_{out}, respectively. During ΔT interval, the vehicle is crossing the Δx distance. Assuming an omni-directional radio coverage C for the wireless network, and denoting (*i*) $R \in \Re^+ \setminus \{0\}$ as the radius of the wireless cell, (*ii*) $\phi \in [0, \pi]$

as the angle between the vehicle's line-of-sight with the RSU and the direction of the vehicle, we can use classic Euclidean geometry to obtain

$$\Delta x = 2R\cos\phi. \tag{2}$$

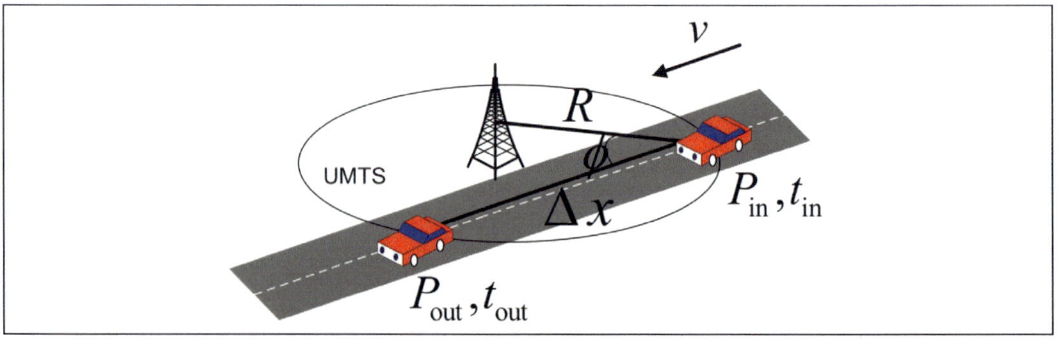

Fig. 1. VANET scenario where a vehicle moves at constant speed v, following a Manhattan mobility model. P_{in} and P_{out} are the wireless cell entrance and exit points, respectively

Since we have assumed that each vehicle is equipped with a GPS receiver, the coordinates $P_{in} \equiv (x_{in}; y_{in})$ of the entrance, and $P_{out} \equiv (x_{out}; y_{out})$ of the exit point of the wireless cell, with respect to a coordinate system centered in the cell centre, are easily calculated so that Δx is known. It follows that the *cell crossing time* can be expressed as

$$\Delta T = \frac{\Delta x}{|\vec{v}|} = \frac{2R}{|\vec{v}|}\cos\left[\arctan\left(\frac{y_{out} - y_{in}}{x_{out} - x_{in}}\right)\right]. \tag{3}$$

Notice that the computation in (3) is directly performed by the vehicle by assuming constant vehicle speed, and knowledge of the wireless cell radio coverage.

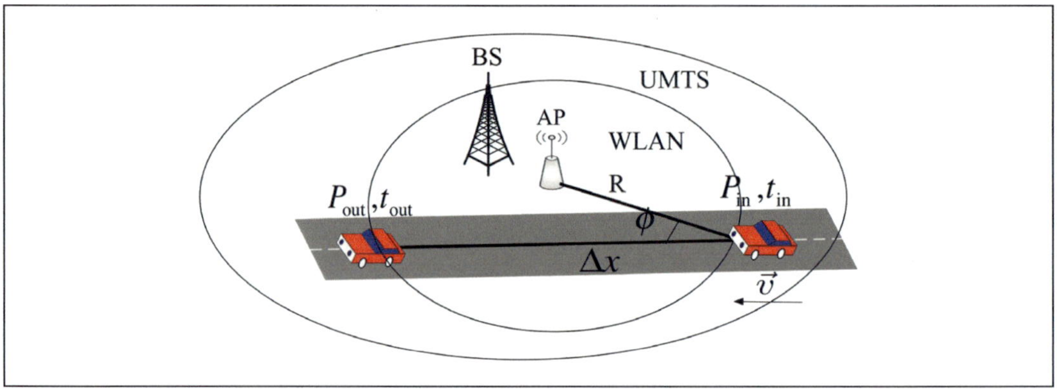

Fig. 2. VANET scenario with heterogeneous overlapping wireless networks

After defining geometric parameters, we can introduce the throughput Θ (t) [Bits], evaluated for $t = \Delta T$ [s], as:

$$\Theta(t) = B_{SN}\Delta T, \tag{4}$$

where B_{SN} [Bit/s] is the bandwidth of the actual wireless cell (*i.e.* Serving Network, SN), assumed to be constant during ΔT. Equation (4) defines the throughput Θ that a vehicle would experience by remaining connected with the actual wireless cell, during the *cell crossing time* ΔT.

Note that so far we have only modelled the throughput in a VANET network where a vehicle is crossing a single wireless network, and no overlapping cells are considered. We now extend the model to heterogeneous environments by capturing vertical handovers across different access networks as well. In this case, a vehicle is entering an area where two or more different wireless cells are co-existing and overlapped.

Figure 2 shows this vehicular scenario with heterogeneous overlapping wireless networks (*i.e.* UMTS, and WLAN).

Namely, the vehicle drawn in Figure 2 is actually connected to an UMTS network, and is entering a WLAN network in P_{in} at t_{in}. Then, UMTS represents a Serving Network (SN) while WLAN a Candidate Network (CN), respectively.

Since S-VHO aims to maximize throughput, system initiates a handover from the SN (*i.e.* UMTS) to the CN (*i.e.* WLAN), if and only if the estimated throughput measured in the CN is higher than the throughput in the current SN, such as:

$$\Theta_{CN} \geq \Theta_{SN} \,. \tag{5}$$

Let a vehicle be connected to a SN, entering the wireless range of a Candidate Network (CN). In this heterogeneous scenario, we model the data delivered between the two time instants t_{in} and t_{out}, as a positive range function $\gamma: \Re \rightarrow \Re^{+}$ defined as

$$\gamma = \alpha \cdot \left(B_{CN} - \delta\right)\left(\Delta T - L\right) + \left(1 - \alpha\right) B_{SN} \Delta T, \tag{6}$$

where α is an indicator function, such that $\alpha = 1$ when a vertical handover is executed, and zero otherwise. Note that when $\alpha = 1$, γ is equivalent to the throughput obtained in the CN, while for $\alpha = 0$, γ is the throughput in the SN. The parameter $\delta \in \Re^{+}$ is a hysteresis factor introduced to avoid vertical handover occurrence when the two competing networks have negligible bandwidth difference.

Function in (6) captures the data loss due to the vertical handover latency L [s], that is the time interval during which a vehicle, traversing an area covered by at least two wireless cells, does not receive any data due to control plane (socket switching) signalling messages exchange.

Since during the cell crossing time of a vehicle it is desirable to maximize throughput, S-VHO technique initiates a handover only when it represents a valid handover, that is when

$$\gamma\big|\left(\alpha = 1\right) > \gamma\big|\left(\alpha = 0\right), \tag{7}$$

from which it follows:

$$B_{CN} > \frac{B_{SN}}{1 - \dfrac{L}{\Delta T}} + \delta \,, \tag{8}$$

The inequality in (8) shows that switching decisions may not be necessary even though the bandwidth B_{CN} is higher than B_{SN}. Switching becomes necessary only if the time that the vehicle will spend in the wireless cell with higher bandwidth is long enough to compensate for the data loss due to the switching overhead, namely only if

$$1-\frac{L}{\Delta T}>0 \quad \Rightarrow \quad L<\Delta T. \tag{9}$$

This observation leads to the conclusion that the throughput Θ is influenced not only by the bandwidth of the considered technologies, but by a larger set of parameters, such as (*i*) the cell crossing time, (*ii*) the vehicle speed, and (*iii*) the overhead of the control-plane protocols adopted (*handover latency*). This consideration represents not only a novel definition for throughput by three previous parameters (*i.e.* cell crossing time, vehicle speed, and handover latency), but also gives an important result for vertical handover management and connectivity switching decisions in vehicular networks. By monitoring and limiting vehicle speed to a fixed upper bound, it is always possible to perform a valid handover. This result represents the following *Theorem of Speed Upper Bound* for valid VHO decisions (Esposito *et al.*, 2010).

Theorem (*Speed Upper Bound*). *Given a vehicle V, travelling with an average constraint speed* $\vec{v}$ *in a heterogeneous vehicular environment for a distance* Δx*, a valid handover for V occurs if* $|\vec{v}|$ *is bounded as follows:*

$$|\vec{v}|<\frac{\Delta x\left(B_{CN}-B_{SN}-\delta\right)}{\left(B_{CN}-\delta\right)L}. \tag{10}$$

Proof: the claim follows from (8), where we highlighted the term ΔT, such as

$$\begin{aligned}
&\left(B_{CN}-\delta\right)\left(1-\frac{L}{\Delta T}\right)>B_{SN},\\
&\left(B_{CN}-\delta\right)\left(\Delta T-L\right)>B_{SN}\Delta T,\\
&\left(B_{CN}-B_{SN}-\delta\right)\Delta T>\left(B_{CN}-\delta\right)L,\\
&\Delta T>\frac{\left(B_{CN}-\delta\right)L}{\left(B_{CN}-B_{SN}-\delta\right)}.
\end{aligned} \tag{11}$$

By replacing the term ΔT from (11) in the following definition of average vehicle speed, *i.e.*

$$|\vec{v}|=\Delta\vec{x}\,/\,\Delta T, \tag{12}$$

we obtain the result expressed in (10).

The Theorem of Speed Upper Bound is useful not only for VHO management, but also in designing V2I protocols, as well as to promote vehicle safety applications. Network providers may in fact offer lower data rate in those areas where the speed limit is lower, and so induce vehicles to maintain lower speeds, in order to experience acceptable QoS levels — low jitter and high throughput— throughout valid handovers. Users should be more motivated to respect speed limits, rewarded by QoS enhancement.

S-VHO technique is based on Theorem 1 in order to manage valid handovers for fast users driving in an heterogeneous vehicular network environment. This approach is acted by the vehicle itself each time is crossing a wireless network and needs to be connected with it. S-

VHO rules according to the following algorithm drawn in Figure 3. S-VHO accepts three inputs, such as (*i*) the vehicle speed $\vec{v}$, (*ii*) the ingress time t_{in} of the vehicle into a wireless cell, and (*iii*) the GPS location information P_{in}, respectively. The algorithm then returns the handover decision variable $\alpha \in \{0, 1\}$, whose value means if a vertical handover is executed or not.

Let a vehicle be connected to a SN, driving in an area with heterogeneous overlapping wireless networks (*i.e.* UMTS, and WLAN). Due to multi-network interface cards equipping the vehicle, it is able to recognize one or more available CNs to access. A vertical handover can occur whenever the inequality in (8) is verified.

After each handover execution, the algorithm enters in *idle mode* for an inter-switch waiting time period (*i.e.* T_w [s]), that means no vertical decision is taken. For example, if a vehicle travels at 15 m/s, a 10 seconds inter-switch waiting time results in 150 meters covered by the vehicle, before the algorithm is reactivated. The idle mode approach is a well-know solution for limitation of vertical handover frequency (Inzerilli & Vegni, 2008); (Inzerilli *et al.*, 2008).

Input: $\vec{v}, t_{in}, P_{in}$
Output: α
while *inside area with at least two overlapped cells,* **do**
  if $\left[B_{CN} > \dfrac{B_{SN}}{\left(1 - \dfrac{L}{\Delta T}\right)} + \delta \right]$
  then
    evaluate ΔT
    if $\Delta T > \Delta T^*$ **then**
      $\alpha \leftarrow 1$ (VHO executed)
      set a decreasing counter to $T_w[s]$.
      while $T_w > 0$ **do**
        idle mode
      end
    else
      $\alpha \leftarrow 0$ (no VHO executed)
    end
  end
end

Fig. 3. Speed-based Vertical Handover Algorithm

Many handover algorithms incorporate a hysteresis cycle within handover decisions, in order to prevent a mobile terminal moving along the boundary of a wireless cell to trigger handover attempts continuously. This phenomenon is well known in the literature as *ping-pong effect* (Kim *et al.*, 2007), and hysteresis is largely adopted in practical implementations.

A high number of vertical handover executions can lead to excessive network resource consumption and also affects mobile terminal's performance (*i.e.* battery life, and energy consumption).

Ping-pong effect occurs in vehicular environments, specially when vehicles travel on a border line between two wireless cells, and make frequent, often unnecessary or unwanted handovers. Limitation of handover frequency is acted by imposing a minimum interval of time between two consecutive handovers. Namely, the grater is the waiting time the smaller will be the number of vertical handovers.

S-VHO algorithm first measures the data rate from CN (*i.e.* B_{CN}), and then if inequality (8) holds, the cell crossing time ΔT is computed as shown in (3). After the cell crossing time evaluation, S-VHO decides whether the handover would be valid or not by comparing ΔT with the threshold for valid handover, (*i.e.* ΔT^*).

We can express the threshold for valid handover as

$$\Delta T^* = \frac{\Delta x}{v^*}. \tag{13}$$

The term v^* is the speed upper bound, obtained from the inequality (10) and formally expressed as

$$v^* = \frac{\Delta x\left(B_{CN} - B_{SN} - \delta\right)}{\left(B_{CN} - \delta\right)L} - \varepsilon, \tag{14}$$

where ε is a positive quantity (*i.e.* $\varepsilon \to 0$). Equation (13) becomes

$$\Delta T^* \simeq \frac{\left(B_{CN} - \delta\right)L}{B_{CN} - B_{SN} - \delta}. \tag{15}$$

The Theorem of Speed Upper Bound shows that the vehicle's speed is strictly limited by handover latency L, and the hysteresis factor δ. The impact of both two parameters and the effects on the speed upper bound are described by the following analytical results.

In Figure 4 (*a*) we show the impact of the handover latency L [s] on the speed upper bound, for a given bandwidth ratio of two available wireless technologies (*e.g.* UMTS, and WLAN[1]). The hysteresis factor δ was set to zero to isolate only the impact of handover latency, and the range of speed was bounded by 35 m/s (typical highway speed limit). As we can notice, for higher values of handover latency, the speed upper bound (*i.e.* the maximum speed at which vehicles experience valid handovers) decreases, and approaches to zero:

$$\lim_{L \to \infty} v^* = 0. \tag{16}$$

This result is reasonable since vehicles travelling at higher speed may not spend enough time under higher data-rate wireless networks to justify the degraded performance introduced by the handover overhead of the signalling messages.

For a bandwidth ratio equal to one (*i.e.* $B_{CN} = B_{SN}$), the speed bound is null as the hysteresis factor has been set to zero; in general (*i.e.* for $\delta \neq 0$), the speed bound is still approaching to zero for L approaching to infinity:

---

[1] The bandwidth ranges were chosen according to WLAN, and UMTS (Laiho *et al.*, 2005) requirements.

$$\lim_{L\to\infty} v^* = \lim_{L\to\infty} \frac{-\Delta x \delta}{(B_{CN} - \delta)L}. \tag{17}$$

The second result comes by observing the epigraph and hypograph —the set of points above, and below the drawn curves, respectively. Any point belonging to the epigraph represents no performance gain in initiating handovers, even if the CN has higher bandwidth than the SN. In contrast, for any point in the hypograph valid handovers occur. We can notice that the curve with zero bandwidth gap (*i.e.* $B_{CN} = B_{SN}$) has empty hypograph. This follows directly from the definition of valid handover, that means a vertical handover cannot be valid when the data rates from CN and SN are equal.

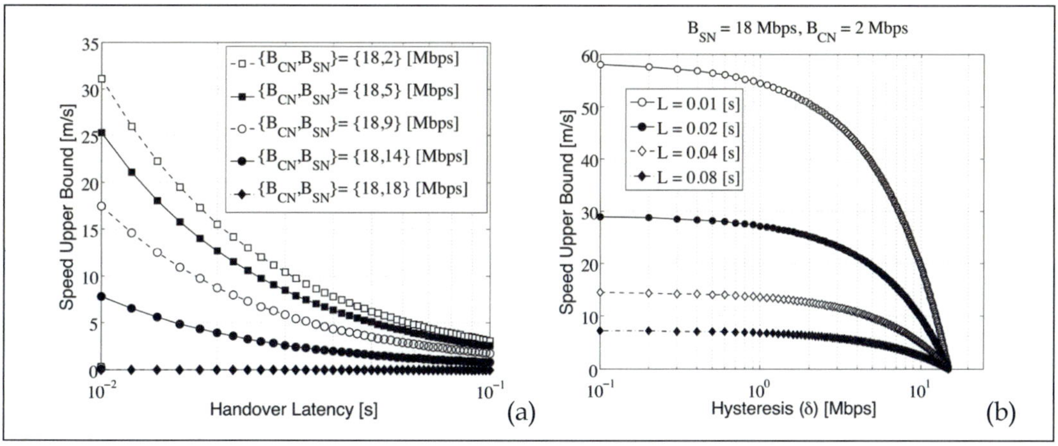

Fig. 4. Speed Upper Bound behaviour. Impact of (*a*) handover latency L, and (*b*) hysteresis δ

In Figure 4 (*b*) we show the impact of the hysteresis δ [Mbps] on the speed bound, for different values of the handover latency (*i.e.* L = {0.01, 0.02, 0.04, 0.08} [s]). The hysteresis range is $[0, B_{CN} - B_{SN}]$. It is useful to note that $B_{CN} - B_{SN}$ is the maximum value of δ after which no valid handover would occur. We have simulated the case $B_{CN} - B_{SN}$ = 16 [Mbps], which represents a typical gap in data rates between UMTS and WLAN.

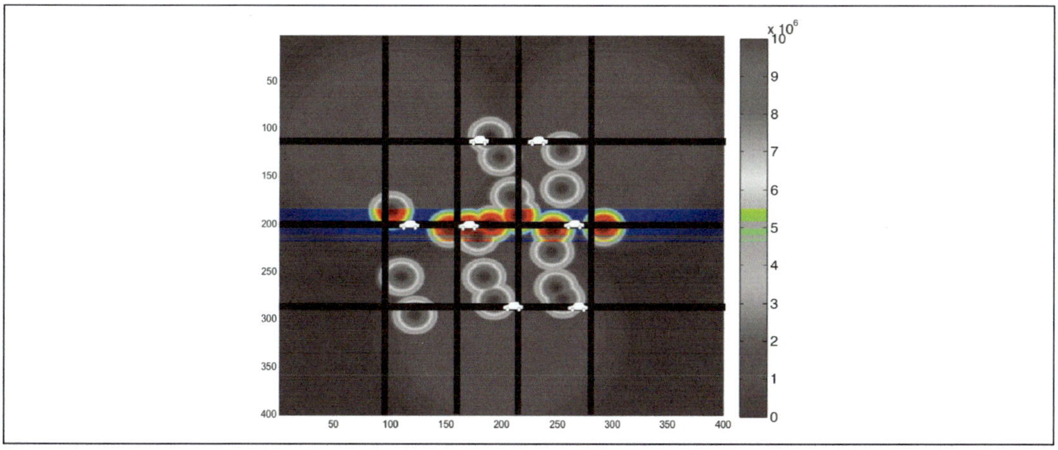

Fig. 5. Simulated VANET scenario with overlapping WLAN and UMTS cells

Notice how the hypographic area changes for different values of L: the lower the handover latency, the greater the hypograph. For example, when the handover latency increases (*i.e.* $L = 0.08$ [s]), the hypographic area significantly reduces. From this observation, it follows that *handover latency should be taken into account when designing protocols for seamless connectivity in VANETs*, and not only focusing on physical parameters or vehicle speed.

4.1 S-VHO simulation results

In this section we describe some simulation results for S-VHO mainly expressed in terms of throughput maximization, and limitation of *ping-pong* effect. They will show how S-VHO works for valid vertical handovers in VANETs.
Network performance, *i.e.* throughput, delay, and jitter, as well as the number of vertical handovers, have been obtained with an event-driven simulator. More details of the simulator can be found in (Vegni, 2010). The simulation scenario depicts a vehicle entering from a random location, restricted to travel along a grid of streets and intersections. The vehicle follows a random path inside a grid, according to the Manhattan mobility model, as shown in Figure 5. The event-driven simulator generates different scenarios, whose characteristics are similar to previous heterogeneous scenarios described in (Vegni & Esposito, 2010). Figure 5 depicts one of the simulated scenarios, in terms of data rate distribution from a set of three UMTS base stations, and twenty WLAN access points, modelling a 2 km^2 area.
The location of each wireless cell has been generated uniformly at random, and a vehicle moves in this area with speed in the range [5, 35] [m/s], capturing urban environment as well as highway scenarios. During its journey, a vehicle requires to download a series of video frames. For example, we could figure out that passengers are enjoying their travelling time by means of real time applications, *e.g.* video streaming and online gaming.
For the cell setup, we have considered typical values of WLAN and UMTS radio parameters, as described in (Vegni & Esposito, 2010). The cell radius was set to 120 [m] for IEEE 802.11/a outdoor environment, and 600 [m] for an UMTS microcell; the transmitted power in the middle of UMTS and WLAN cell has been chosen to be between 43 and 30 dBm, respectively. Finally, the average vertical handover latencies from UMTS to WLAN, and from WLAN to UMTS, have been set respectively to $L_{U\rightarrow W} = 2$ [s], and $L_{W\rightarrow U} = 3$ [s], while the data rate of UMTS and WLAN equal to 384 [kbps], and 11 [Mbps], respectively.
S-VHO network performance have been validated in terms of (*i*) throughput, (*ii*) jitter, and (*iii*) handover frequency. Moreover, a comparison between S-VHO and a previous vertical handover technique (Yan *et al.*, 2008), based on traditional power measurements, has been carried out. In both algorithms, the speed of the vehicle is used as handover assessment criterion. Hereafter we will name the technique described in (Yan *et al.*, 2008) as Speed Probability Based VHO (SPB). Simulated results[2] will prove the effectiveness of S-VHO for valid handovers.
Figure 6 shows the throughput expressed as cumulative received bits in a downlink connection for both S-VHO and SPB techniques, versus the inter-switch waiting time in the range [0, 50] [s]. The cumulative received bits represent the amount of received bits by a vehicle moving inside an heterogeneous wireless environment for the simulated period of time.

---

[2] Results obtained by averaging 100 heterogeneous network scenarios.

The throughput performance of S-VHO outperforms at every speed the SPB algorithm. Our performance improvement are justified by the absence of received signal strength dependence in S-VHO handover assessment criteria. Moreover, the effectiveness of S-VHO is clear when vehicle speed is below a given limit (*e.g.*, 20 [m/s]).
On the other hand, SPB does not appear sensitive to either speed or inter-switch waiting time, and its throughput is limited. However, the S-VHO throughput drops when the vehicle speed exceeds the desired limit. This is justified by the Theorem of Speed Upper Bound for valid VHOs.

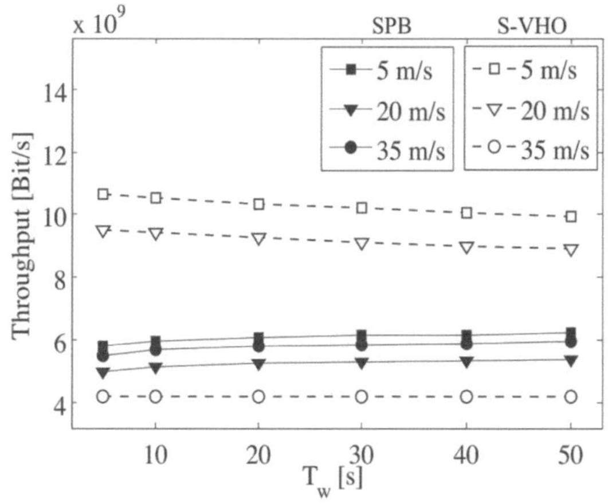

Fig. 6. Comparison of throughput between S-VHO (white markers), and SPB (black markers)

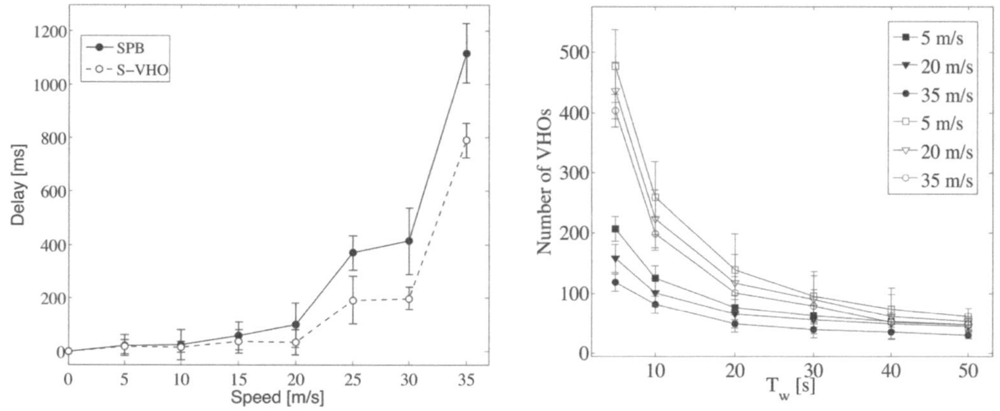

Fig. 7. Comparison of (*left*) delay, and (*right*) number of vertical handover occurrences, between S-VHO (white markers), and SPB (black markers) algorithms

The average frame delay for both S-VHO and SPB is shown in Figure 7 (*left*). The 95% confidence intervals express how the average frame delay increases for higher speeds. This is because there is not enough time to download the next frame before the signal from the SN gets too weak. Moreover, S-VHO experiences lower delays compared to SPB, since on average it performs less handovers.

Figure 7 (*right*) depicts, with 95% confidence intervals, the average number of handovers for different values of inter-switch waiting time. As expected, the number of vertical handovers decreases when the system is idle for longer periods (*i.e.* when T_w increases). Since our simulations count all the handovers (valid or invalid), the gap between the S-VHO and SPB curves represents the number of invalid handovers, executed not taking into account the handover latency L.

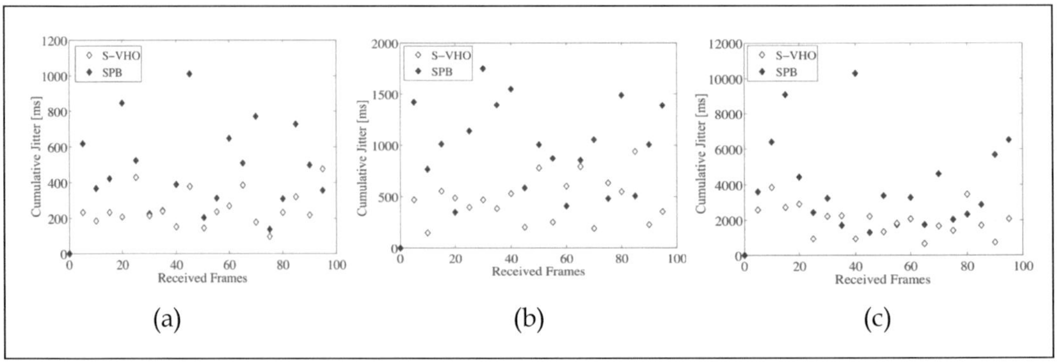

Fig. 8. Cumulative jitter experienced by a vehicle, for different speeds (*a*) $|\vec{v}| = 15$ [m/s], (*b*) $|\vec{v}| = 25$ [m/s], and (*c*) $|\vec{v}| = 35$ [m/s]

Jitter performance from S-VHO and SPB have been compared, and are shown in Figure 8 for different values of speed, and a fixed inter-switch waiting time value (*i.e.* T_w = 10 [s]). Each point represents the cumulative jitter, defined as the difference between maximum and minimum frame delay. Jitter increases with speed —note the scale difference among different graphs— since two frames may be more often coming from different wireless networks, and also because the cell crossing time decreases when the speed increases. For S-VHO higher speed implies higher jitter, unless the number of unnecessary handovers is reduced.

To summarize, S-VHO technique distinguishes from previous approaches by using both handover latency and cell crossing time estimation to simultaneously improve throughput and delay. It is driven by the vehicle speed, so that vehicles are required to maintain a given speed limit to maintain acceptable levels of throughput, delay and jitter. VHOs are limited to effective and necessary connectivity switching. A handover towards a CN with higher data rate does not necessary result in a throughput improvement.

S-VHO represents not only a novel VANET protocols for real-time applications, but also it could be useful for urban safety. As a matter, service providers assisted by vehicular networks could enforce speed limits and safety, while delivering real-time services as video-streaming or online gaming.

5. Hybrid vehicular communication protocols

As known (Held, 2007), one of the main characteristics of VANETs is that vehicles move in clusters —interconnected blocks of vehicles— due to different traffic scenarios (*i.e.* dense, sparse or totally disconnected traffic neighborhoods). For this reason, vehicular connectivity is not always available, and messages propagation in VANETs is still an open issue.

Many authors have addressed how to improve message forwarding by opportunistic techniques, and multi-hop approaches. In (Resta *et al.*, 2007) a multi-hop emergency message dissemination is described by means of a probabilistic approach. The authors derive lower bounds on the probability that a vehicle correctly receives a message within a fixed time interval. Similarly, in (Jiang *et al.*, 2008) an efficient alarm message broadcast routing protocol is presented. This technique estimates the receipt probability of alarm messages sent to vehicles.

In other works the message propagation model is based on the main VANET characteristics such as number of hops, vehicle position, mobility, etc. In (Yousefi *et al.*, 2007) the authors consider a single-hop dissemination protocol based on Quality-of-Service metrics, while in (Chen *et al.*, 2008) a robust message dissemination technique is illustrated, based on the vehicles position. Finally, the authors in (Nadeem *et al.*, 2006) present a data dissemination model based on bidirectional mobility of paths between a couple for vehicles.

Common aspect in all previous works is that data traffic is disseminated only through vehicles communicating via V2V; no network infrastructure nor V2I protocol have been considered.

The use of the vehicular grid together with a network infrastructure has been discussed in (Gerla *et al.*, 2006); (Marfia *et al.*, 2007). The benefits of using the opportunistic infrastructure placed on the roads result in an enhancement of message propagation. In fact, a Road-Side Unit represents a fixed node which is able to forward message information to vehicles driving inside a wireless cell.

V2X relies on the network scenario depicted in (Gerla *et al.*, 2006), but it represents a novel protocol providing switching from V2V to V2I, and vice versa. It enables vehicles to communicate via V2V or V2I on the basis of a protocol switching decision. The message propagation via V2X is then improved by a correct use of vehicular communication protocols (*i.e.*, V2V and V2I).

The heterogeneous vehicular scenario we are referring to is depicted in Figure 9. Let us consider a cluster C comprised of a set S of vehicles (*i.e.*, $S = \{1,2,\dots,n\}$). Then, m Road-Side Units (RSUs) (*i.e.*, $m < n$) are displaced in the network scenario. Each vehicle is able to communicate via V2V on the basis of a fixed transmission range radio model (Vegni & Little, 2010). We assume that only a limited subset of vehicles in the cluster C, (*i.e.*, $S' = \{1,2,\dots,l\} \subset S$, with $l < n$), is able to connect to an RSU via V2I. For example, not all the vehicles might have an appropriate network interface card, and/or are not in the range of connectivity of an RSU. Analogously, we assume that only k RSUs (*i.e.*, $k = \{1,2,\dots,h\}$ with $h < m$) are available to V2I communications.

In such scenario, we are now introducing the *Protocol Switching Decision* as follows:

Definition *(Protocol Switching Decision).* *A source vehicle V_s, sending via V2V a message of length L to a destination vehicle V_d, will switch to V2I if there exists an optimal path linking V_s with an RSU. Analogously, a source vehicle V_s, sending via V2I a message of length L to a destination vehicle V_d, will switch to V2V if there exists an optimal path linking V_s with neighbouring vehicles.*

The *optimal path* will be defined hereafter.

For the connectivity link from the *i*-th to the *j*-th vehicle we define as *link utilization time* $q_{(i,j)}$ [s] the time needed to transmit a message of length L [bit] from the *i*-th to the *j*-th vehicle, at an actual data rate $f_{(i,j)}$ [Mbit/s], such as

$$q_{(i,j)} = \frac{L}{f_{(i,j)}}. \tag{18}$$

For a link between the i-th vehicle and the k-th RSU, the data rate $f_{(i,j)}$ is obtained by the nominal data rate $\tilde{f}_{(i,k)}$ by applying a *Data Rate Reduction* (*DRR*) factor (*i.e.*, $\rho_{(i,k)}$ [s]) that depends on the distance from the vehicle to the RSU, such as

$$f_{(i,k)} = \rho_{(i,k)} \tilde{f}_{(i,k)}. \tag{19}$$

The *DRR* factor increases when a vehicle is laying within the bound of a wireless cell.

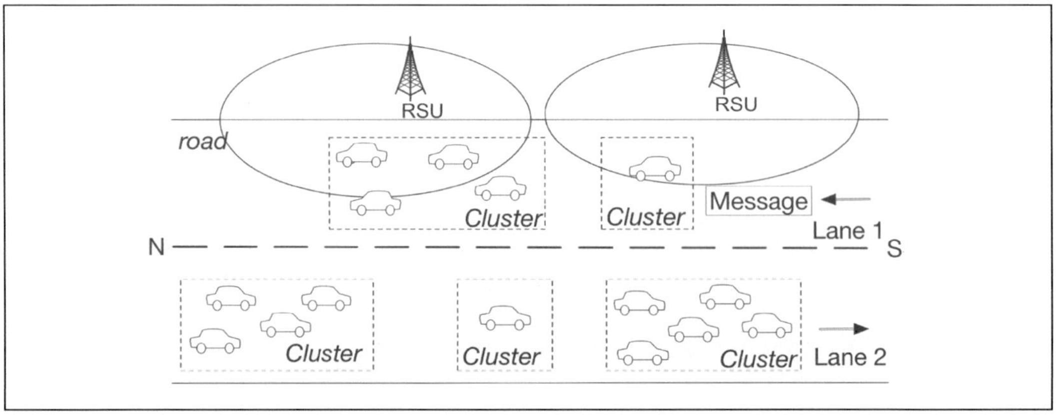

Fig. 9. VANET scenario with heterogeneous wireless network infrastructure partially covering the grid

Let us now define a *Path* in a vehicular network as:

Definition (*Path*). *Given the i-th vehicle and the k-th RSU, a path is a sequence of M hops connecting the i-th vehicle with the k-th RSU, where a single hop represents a link between two neighbouring vehicles. The path length represents the number of hops M for a single path.*

It follows that the maximum number of directed links from a vehicle to an RSU is $\alpha = l \cdot h$, while the maximum number of different paths that can connect the i-th vehicle to the k-th RSU is $n \cdot \alpha$.

From the definition of path, we define the *path utilization time* $Q_{(i,k)}$ [s] from the i-th vehicle to the k-th RSU as the sum of single link utilization time parameters (*i.e.*, $q_{(i,j)}$), for each hop that comprises the path, as

$$Q_{(i,k)} = q_{(i,j)} + q_{(j,x)} + \ldots + q_{(x,k)} = L \sum_{\substack{i=1 \\ x \in S}}^{n} \left[f_{(j,x)}^{-1} \right]. \tag{20}$$

The *optimal path* will be the one, among all the paths $n\alpha$, with the minimized path utilization time, such as

$$\min_{s=1,2,\ldots,n\alpha} Q_{(i,k)}^{(s)} = L \cdot \min_{s=1,2,\ldots,n\alpha} \sum_{\substack{i=1 \\ x \in S}}^{n} \left[f_{(j,x)}^{(s)} \right]^{-1}. \tag{21}$$

Equation (21) is compared with the link utilization times in V2V communications in order to detect the most appropriate vehicular protocol. It represents our criterion for the *optimal path detection technique* in VANETs where vehicles are V2X enable.

5.1 Data propagation rates

In this section we illustrate how a message is propagated in a VANET incorporating a heterogeneous network infrastructure, where vehicles are communicating via V2X. For this purpose, we shall give several definitions of message dissemination rates for different cases. The network scenario we are referring to is that one as depicted in Figure 9. Vehicles move in clusters in two separated lanes (*i.e.*, lane 1, and 2), where north (*i.e.*, N), and south (*i.e.*, S) represent the directions of lane 1, and 2, respectively. The message propagation direction is assumed to be N.

Each vehicle in the grid is able to know about its local connectivity through broadcast "hello" messages forwarded in the network. Local connectivity information received by each vehicle establishes if a vehicle is within a cluster or is traveling alone on the road. In contrast, a vehicle will know if there are neighbouring wireless networks to access, on the basis of broadcast signaling messages sent by the RSUs.

Let the vehicles be traveling at a constant speed c [m/s]. The *message propagation rate within a cluster* (*i.e.*, v [m/s]) is defined as

$$v = \frac{x}{t}, \tag{22}$$

where x [m] is the transmission range distance between two consecutive and connected vehicles communicating via V2V, and t [s] is the time necessary for a successful transmission, which depends on the single link of connected vehicles. Typical value of transmission range distance is $0 < x \leq 125$ m, as shown in (Agarwal & Little, 2008).

Equation (22) represents the average message propagation rate within a cluster, because it consists of each single contribution due to each single link (i, j) in the cluster, such as

$$v = \frac{1}{h}\sum_{i,j} v_{(i,j)} = \frac{1}{h}\sum_{i,j}\frac{x_{(i,j)}}{q_{(i,j)}} = \frac{1}{hL}\sum_{i,j} x_{(i,j)} \cdot f_{(i,j)}, \tag{23}$$

where $v_{(i,j)}$ [m/s] is the message propagation rate for the link (i, j), and h is the number of hops occurred within a cluster. The message propagation rate within a cluster v [m/s] depends on the average message propagation rate for each single hop, and increases for a low number of hops h.

Now, let us consider v_{RSU} [m/s] as the message propagation rate within the network infrastructure, as

$$v_{RSU} = \frac{d}{T_{RSU}}, \tag{24}$$

where d is the distance between two consecutive RSUs, and T_{RSU} is the time necessary to forward a message between two consecutive RSUs. T_{RSU} is defined as the ratio between the message length L [bit], and the effective data rate B [bit/s], for the link between the m-th and $(m + 1)$-th RSU,

$$T_{\text{RSU}} = \frac{L}{B}. \tag{25}$$

Notice that v_{RSU} is strictly dependent on the message propagation direction: a message is forwarded to an RSU if it is placed along the same message propagation direction. The potential for communications between RSUs aims to avoid connectivity interruptions caused by low traffic densities, and that the V2V protocol cannot always solve.

In Figure 10 we show the data propagation rates for the considered VANET scenario. Notice that each message forwarded by an RSU to the next RSU has been previously sent by a vehicle driving inside the wireless cell of the RSU. Moreover, each time an RSU receives a message from another RSU, it will send the message (*i*) to the destination vehicle if it is driving inside the actual RSU coverage, or (*ii*) to forward the message to next RSU.

In the first case, the message propagation rate will depend on a downlink connection from RSU to a vehicle, while in the second case the message propagation rate will be equal to v_{RSU}. By leveraging these considerations, we define the message propagation rate in *uplink* (*downlink*), when a vehicle sends a message to an RSU (and vice versa), as:

$$v_{\text{UP}} = \frac{x_r}{L} \cdot \tilde{f}_{(i,m)}, \quad v_{\text{DOWN}} = \frac{x_r}{L} \cdot \tilde{f}_{(m,i)}, \tag{26}$$

where x_r is the distance that separates the *i*-th vehicle and the *m*-th RSU, while $\tilde{f}$ is the effective transmission data rate for the link (*i*, *m*) (*uplink*), and (*m*, *i*) (*downlink*), respectively.

From (24) and (26), it follows that the message propagation rate v_{V2I} [m/s] for communications between vehicles and RSUs via V2I depends on the effective transmission data rates in uplink and downlink, and on the effective data rate for intra-RSU communications, such as:

$$\begin{aligned} v_{\text{V2I}} &= v_{\text{UP}} + v_{\text{RSU}} + v_{\text{DOWN}} = \\ &= \frac{1}{L}\left[d \cdot B + x_r \cdot \left(\tilde{f}_{(i,m)} + \tilde{f}_{(m,i)}\right)\right]. \end{aligned} \tag{27}$$

After defining *message propagation rates for communications via V2I*, we introduce the *message propagation rate for communications via V2V* (*i.e.*, v_{V2V} [m/s]), as

$$v_{\text{V2V}}^{(\pm)} = \pm (v + c), \tag{28}$$

which depends on the constant velocity *c* of vehicles and on the effective transmission data rates within a cluster *C*, according to (23). The positive or negative sign of v_{V2V} is due by the message propagation direction.

Finally, when no connectivity occurs (*i.e.*, a vehicle is traveling alone in the grid), the message propagation rate is equal to $\pm c$, which depends on message propagation direction.

To sum up, we characterize the behaviour of the whole system in terms of six transition states as follows:

1. Messages are traveling along on a vehicle in the N direction at speed *c* [m/s];
2. Messages are propagating multi-hop within a cluster in the N direction at speed $v_{\text{V2V}}^{(+)}$ [m/s];

3. Messages are traveling along a vehicle in the S direction at speed $-c$ [m/s];
4. Messages are propagating multi-hop within a cluster in the S direction at speed $v_{V2V}^{(-)}$ [m/s];
5. Messages are transmitted via radio by an RSU in the N direction at speed v_{V2I} [m/s];
6. Messages are transmitted via radio by an RSU in the S direction at speed $-v_{V2I}$ [m/s].

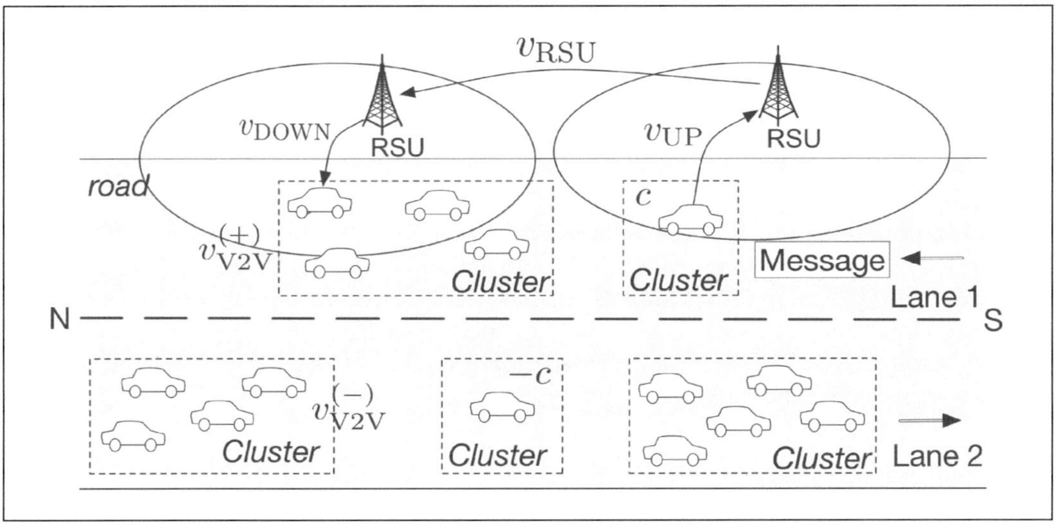

Fig. 10. Data propagation rates in VANET scenario with network infrastructure

States (1−4) are typical for data propagation with opportunistic networking technique for vehicles communicating only via V2V, while states 5, and 6, have been added for vehicles communicating via V2I. All the six states can occur when vehicles communicate via V2X.

In our assumptions, we considered two message propagation directions (*i.e., forward* and *reverse propagation*).

In *forward message propagation,* each vehicle is assumed to travel in the N direction at speed c [m/s], and the message is propagated in the N direction as well. The message propagation rate has a minimum value due to the speed of the vehicle (*i.e.,* c [m/s]) since the message is traveling along the vehicle. When a connection between two consecutive vehicles traveling in the N direction is available, the message will be propagated via V2V at a rate $v_{V2V}^{(+)}$. Moreover, if no vehicle connection is available, the *bridging technique* can attempt to forward a message to clusters along the S (opposite) direction, whenever they are overlapping with the cluster along the N direction.

Analogously, it is easy to evince that in *reverse message propagation,* each vehicle is assumed to travel in the S direction at speed $-c$ [m/s], and the message is propagated in the S direction as well. The message propagation rate will have a minimum value due to the speed of the vehicle (*i.e.,* $-c$ [m/s]), and a maximum bound when a message is propagating via V2V at a rate $v_{V2V}^{(-)}$. Again, if no vehicle connection is available, a message will be forwarded via *bridging* to clusters along the N (opposite) direction, whenever they are overlapping with the cluster along the S direction.

Such considerations occur for vehicles communicating via V2V and when opportunistic networking is available. In contrast, when vehicles are communicating via V2I, the *forward*

message propagation will have a maximum bound equal to v_{V2I}, while for *reverse message propagation* range the maximum bound is $-v_{V2I}$.
The definitions for *forward* and *reverse message propagation* rates are given below, respectively.
Definition (*Forward Message Propagation rate*): *the forward message propagation rate, when a vehicle is communicating via V2V, is in the range* $[c,\ v_{V2V}^{(+)}]$. *In contrast, when a vehicle communicates via V2I, the forward message propagation rate is in the range* $[c,\ v_{V2I}]$.

Definition (*Reverse Message Propagation rate*): *the reverse message propagation rate, when a vehicle communicates via V2V, is in the range* $[-c,\ v_{V2V}^{(-)}]$, *while for vehicles communicating via V2I, the range of reverse message propagation rate is* $[-c,\ -v_{V2I}]$.

5.2 V2X algorithm

This section illustrates how V2X takes a protocol switching decision.
The algorithm for handing over from V2V to V2I, and vice versa, is described by its pseudo-code in Figure 11. It is mainly based on (*i*) the *Infrastructure Connectivity* (*IC*) parameter, which gives information if a vehicle is able to connect to an RSU, and on (*ii*) the *optimal path detection technique*. The algorithm accepts one input (*i.e.*, the vehicle's *IC*), and returns the actual message propagation rate (*i.e.*, $\{v_{V2V}, v_{V2I}\}$).

```
Input:  IC
Output: { v_V2V,  if a vehicle communicates via V2V
        { v_V2I,  if a vehicle communicates via V2I
while IC = 0 do
|  A vehicle is connected via V2V, ← v_V2V
end
else
|  if IC = 1 then
|  |  Optimal path detection, ← v_V2I or v_V2V
|  end
end
if A vehicle communicates with an RSU via V2I then
|  the RSU tracks the destination's position,
|  if Destination vehicle is inside the actual RSUs coverage then
|  |  Direct link from RSU to destination vehicle
|  |  else
|  |  |  The actual RSU will forward the message to next RSU
|  |  end
|  end
end
```

Fig. 11. Algorithm for protocol switching decisions in V2X

Let us consider the following VANET scenario. A source vehicle is communicating with other vehicles (*relay*) via V2V in a sparsely connected neighbourhood, where the transmission range distance between two consecutive vehicles is under a connectivity bound, *i.e.* $x \leq 125$ m.
The source vehicle is driving inside any wireless cell, and is receiving "hello" broadcast messages from other vehicles nearby. Local connectivity information will notify the vehicle the availability of vehicles to communicate with via V2V; no RSU presence will be notify to the vehicle. In this case (*i.e.*, V2V availability, and no V2I) the *IC* parameter for vehicle *A* will be set to 0. Otherwise, when a vehicle enters a wireless network, the presence of an available RSU to access will be directly sent to the vehicle by means of its associated *IC* parameter set to 1.
Finally, a destination vehicle is driving far away from *A*, and other vehicles (*relay*) are available to communicate each other.
In such scenario, the algorithm works according to two main tasks, such as (*i*) checking *IC* parameter, and (*ii*) tracking the destination vehicle(s). Every time a vehicle forwards a message it checks its *IC* value. When $IC = 1$, the vehicle calculates the *optimal path* according to (21) in order to send the message directly to the selected RSU via V2I. Otherwise, the vehicle forwards the message to neighbouring vehicles via V2V.
By supposing the RSU knows the destination vehicle's position (*i.e.* by A-GPS), if the destination vehicle is traveling within the RSU's wireless coverage, the RSU will send the message directly to the destination vehicle. Otherwise, the RSU will be simply forwarding the message to the RSU that is actually managing the vehicle's connectivity. Finally, the message will be received by the destination vehicle.
Some simulation results are now shown in order to verify the effectiveness of V2X approach as compared with traditional opportunistic networking scheme in VANET. As a measure of performance, we calculate the *average message displacement* (*i.e.* X [m]) in VANETs via V2X. The *message displacement* is a linear function, depending on time, and varying for different traffic scenarios, message propagation speeds, and network conditions. It follows that in each of the six states listed in Section 5.1, the message displacement $X(t)$ will be as follows:

1. $X(t) = c \cdot t$, for messages traveling along on a vehicle in the N direction at speed c [m/s];
2. $X(t) = v_{\text{V2V}}^{(+)} \cdot t$, for messages propagating multi-hop within a cluster in the N direction at speed $v_{\text{V2V}}^{(+)}$ [m/s];
3. $X(t) = -c \cdot t$, for messages traveling along a vehicle in the S direction at speed $-c$ [m/s];
4. $X(t) = v_{\text{V2V}}^{(-)} \cdot t$, for messages propagating multi-hop within a cluster in the S direction at speed $v_{\text{V2V}}^{(-)}$ [m/s];
5. $X(t) = v_{\text{V2I}} \cdot t$, for messages transmitted via radio by an RSU in the N direction at speed v_{V2I} [m/s];
6. $X(t) = -v_{\text{V2I}} \cdot t$, for messages transmitted via radio by an RSU in the S direction at speed $-v_{\text{V2I}}$ [m/s].

States 1, 2, and 5 refer on a *forward message propagation*, while stated 3, 4, and 6 on a *reverse message propagation*, respectively.

We simulated a typical vehicular network scenario by the following events:

i. at $t = 0$ s a source vehicle is traveling in the N direction and sends a message along on the same direction, (*state* 1);

ii. at $t = 2$ s the message is propagated multi-hop within a cluster in the N direction, (*state* 2);

iii. at $t = 6$ s a relay vehicle enters an RSU's radio coverage, and the message is transmitted via V2I to the RSU. Finally, it will be received by other vehicles at $t = 10$ s, (*state* 5).

We compared this scenario with traditional opportunistic networking technique in VANETs, where the following events occur:

i. at $t = 0$ s a source vehicle traveling in the N direction sends a message along on the same direction, (*state* 1);

ii. at $t = 4$ s the message is forwarded to a vehicle in the S direction, (*state* 3);

iii. at $t = 6$ s the message propagates via multi-hop within a cluster in the N direction, (*state* 2). The transmission stops at $t = 10$ s.

For comparative purposes, main simulation parameters has been set according to (Wu *et al.*, 2004), including $c = 20$ m/s, $d = 500$ m, typical message size $L = 300$ bit, data rate transmission $B = 10$ Mbit/s (*e.g.*, for WiMax connectivity), and $x_r = 400$ m. The transmission rates in DSRC have been assumed equal to 6 Mbit/s (Held, 2007). We assumed a cluster size equal to $h = 5$, and different distances between couples of vehicles (*i.e.*, 100, 75, 50, 40, and 30 m). For each hop the transmission range has been hold (*i.e.* < 125 m).

Figure 12 (*left*) depicts the maximum and minimum message propagation bounds for V2X in *forward message propagation mode*. Notice a strong increase in the message propagation with respect to other forms of opportunistic networking: after $t = 10$ s, the message has been propagating for approximately 30 km in V2X (Figure 12 (*left*)), while only 1.5 km in traditional V2V (Figure 12 (*right*)). The high performance gap is mainly due to the protocol switching decision of V2X, which exploits high data rates from wireless network infrastructure. In contrast, opportunistic networking with V2V is limited to use only DSRC protocol.

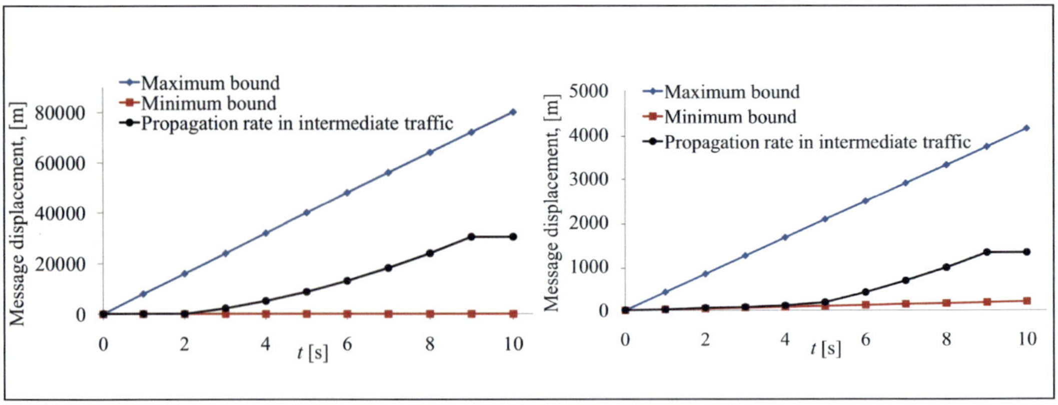

Fig. 12. Forward message propagation for (*left*) V2X protocol, (*right*) traditional opportunistic networking

Analogously, we simulated how a message is forwarded in *reverse message propagation mode*, where vehicles are traveling in an opposite direction (Figure 13). In this case, the message

propagation rates are in the range $[-c;\ -v_{V2I}]$ and $[-c;\ v^{(-)}_{V2V}]$ [m/s], for V2X and traditional opportunistic networking scheme, respectively. Once again, while V2X assures high values for message displacement (*i.e.*, at $t = 10$ s, a message has been propagated up to around 70 km, as shown in Figure 13 (*left*)), traditional V2V can achieve low values (*i.e.*, at $t = 10$ s, messages have reached 1.3 km far away from the source vehicle (see Figure 13 (*right*)).
Notice the fluctuations of message displacement in *forward* and *reverse* cases with V2X (*i.e.* 50, and 70 km, respectively). They are mainly due to traffic density, and RSUs' positions (*i.e.* inter-RSU distance). In general, high performance are obtained with V2X, while low message propagation distance with traditional V2V.

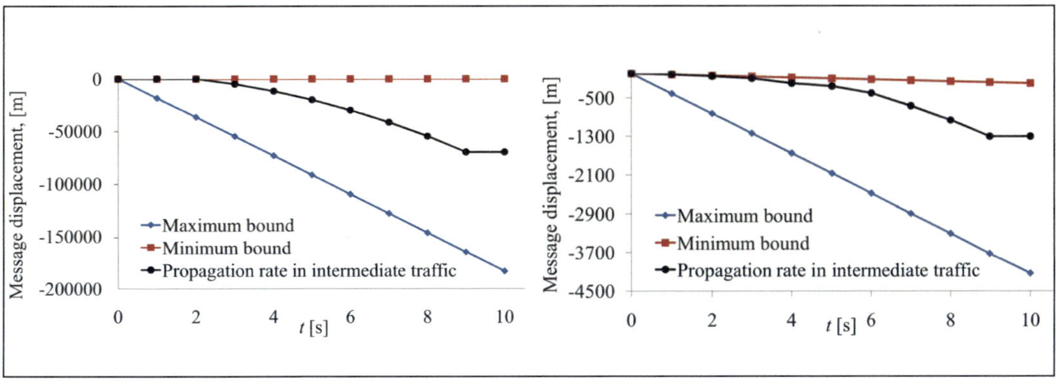

Fig. 13. Reverse message propagation for (*left*) V2X protocol, (*right*) traditional opportunistic networking

6. Conclusions

In this chapter we have discussed application of VHO in the context of VANETs in order to optimize application delivery through a mixed V2V/V2I infrastructure. Vertical handover strategies can be applied to assure VANET connectivity *context-aware*, and *content-aware*.
Various metrics can be adopted to trigger handover decisions including RSS measurements, QoS parameters, and mobile terminal location information. This last represents the most common parameter used to drive VHO decisions.
Hence, a geometrical model has been presented where GPS-equipped mobile terminals exploit their location information to pilot handover and maximize communication throughput taking into account mobile speed. The proposed technique has been described via both analytical and simulated results, and validation of its effectiveness has been supported by a comparison with a traditional vertical handover method for VANETs (Yan *et al.*, 2008).
Moreover, we have described a hybrid vehicular communication protocol V2X and the mechanism by which a message is propagated under this technique. V2X differs from traditional V2V protocol by exploiting both V2V and V2I techniques, through the use of a fixed network infrastructure along with the mobile ad-hoc network. In this heterogeneous scenario, we have characterized the upper and lower bounds for message propagation rates. Validation of V2X has been carried out via simulation results, showing how V2X protocol

improves network performance, with respect to traditional opportunistic networking technique applied in VANETs.

7. References

Held, G. (2007). Inter- and intra-vehicle communications, CRC Press.

Chiara, B.D.; Deflorio, F. & Diwan, S. (2009). Assessing the effects of inter-vehicle communication systems on road safety, *Intelligent Transport Systems, IET*, Vol. 3, No. 2, June 2009, pp. 225–235.

Pollini, G.P. (1996). Trends in handoff design, *IEEE Communication Magazine*, Vol. 34, No. 3, March 1996, pp. 82–90.

Inzerilli, T. & Vegni, A.M. (2008). A reactive vertical handover approach for WiFi-UMTS dual-mode terminals, *Proceeding of 12th Annual IEEE International Symposium on Consumer Electronics*, April 2008, Vilamoura (Portugal).

Vegni, A.M.; Tamea, G.; Inzerilli, T. & Cusani, R. (2009). A Combined Vertical Handover Decision Metric for QoS Enhancement in Next Generation Networks, *Proceedings of IEEE International Conference on Wireless and Mobile Computing, Networking and Communications* 2009, pp. 233–238, October 2009, Marrakech (Morocco).

Vegni, A.M. & Esposito, F. (2010). A Speed-based Vertical Handover Algorithm for VANET, *Proceedings of 7th International Workshop on Intelligent Transportation*, March 2010, Hamburg (Germany).

Vegni, A.M. & Little, T.D.C. (2010). A Message Propagation Model for Hybrid Vehicular Communication Protocols, *Proceeding of 2nd International Workshop on Communication Technologies for Vehicles*, July 2010, Newcastle (UK).

McNair, J. & Fang, Z. (2004). Vertical handoffs in fourth-generation multinetwork environments, *IEEE Wireless Communications*, Vol. 11, No.3, (June, 2004), pp. 8–15.

Esposito, F.; Vegni, A.M.; Matta, I. & Neri, A. (2010). On Modeling Speed-Based Vertical Handovers in Vehicular Networks —Dad, slow down, I am watching the movie—, at *IEEE Globecom 2010 Workshop on Seamless Wireless Mobility* 2010, December 2010, Miami (USA).

Inzerilli, T.; Vegni, A.M.; Neri, A. & Cusani, R. (2008). A Location-based Vertical Handover algorithm for limitation of the ping-pong effect, *Proceedings on 4th IEEE International Conference on Wireless and Mobile Computing, Networking and Communications*, October 2008, Avignon (France).

Kim, W.I.; Lee, B.J.; Song, J.S.; Shin, Y.S. & Kim, Y.J. (2007). Ping-Pong Avoidance Algorithm for Vertical Handover in Wireless Overlay Networks, *Proceeding of IEEE 66th Vehicular Technology Conference*, pp. 1509-1512, September 2007.

Chen, Y.S.; Cheng, C.H.; Hsu, C.S. & Chiu, G.M. (2009). Network Mobility Protocol for Vehicular Ad Hoc Network, *Proceeding of IEEE Wireless Communication and Networking Conference*, April 2009, Budapest (Hungary).

Yan, Z.; Zhou, H.; Zhang, H. & Zhang, S. (2008). Speed-Based Probability-Driven Seamless Handover Scheme between WLAN and UMTS, *Proceeding of 4th International Conference on Mobile Ad-hoc and Sensor Networks*, December 2008, Wuhan (China).

Laiho, J.; Wacker, A. & Novosad, T. (2005). Radio Network Planning and Optimisation for UMTS, 2nd edition, Chapter 6.

Resta, G.; Santi, P. & Simon, J. (2007). Analysis of multihop emergency message propagation in vehicular ad hoc networks, *Proceeding of the 8th ACM International Symposium on Mobile Ad Hoc Networking and Computing*, pp. 140–149, September 2007, Montreal (Canada).

Jiang, H.; Guo, H. & Chen, L. (2008). Reliable and Efficient Alarm Message Routing in VANET, *Proceeding of the 28th International Conference on Distributed Computing Systems Workshops*, pp. 186–191.

Yousefi, S.; Fathy, M. & Benslimane, A. (2007). Performance of beacon safety message dissemination in Vehicular Ad hoc NETworks (VANETs), *Journal of Zhejiang University Science A*.

Chen, W.; Guha, R.; Kwon, T.; Lee, J. & Hsu, Y. (2008). A survey and challenges in routing and data dissemination in vehicular ad hoc networks, *Proceeding of IEEE International Conference on Vehicular Electronics and Safety*, Columbus (USA), September 2008.

Nadeem, T., Shankar, P. & Iftode, L. (2006). A comparative study of data dissemination models for VANETs, *Proceeding of the 3rd Annual International Conference on Mobile and Ubiquitous Systems*, pp. 1–10, San Jose (USA), July 2006.

Gerla, M.; Zhou, B.; Leey, Y.-Z.; Soldo, F.; Leey, U. & Marfia, G. (2006). Vehicular Grid Communications: The Role of the Internet Infrastructure, *Proceeding of Wireless Internet Conference*, Boston (USA), August 2006.

Marfia, G.; Pau, G.; Sena, E.D.; Giordano, E. & Gerla, M. (2007). Evaluating Vehicle Network Strategies for Downtown Portland: Opportunistic Infrastructure and Importance of Realistic Mobility Models, *Proceeding of MobiOpp 2007*, Porto Rico, June 2007.

Agarwal, A. & Little, T.D.C. (2008). Access Point Placement in Vehicular Networking, *Proceeding of 1st International Conference on Wireless Access in Vehicular Environments*, Dearborn (USA), December 2008.

Wu, H.; Fujimoto, R. & Riley, G. (2004). Analytical Models for Information Propagation in Vehicle-to-Vehicle Networks, *Proceeding of ACM VANET*, Philadelphia (USA), October 2004.

Ayyappan, K. & Dananjayan, P. (2008). RSS Measurement for Vertical Handoff in Heterogeneous Network, *Journal of Theoretical and Applied Information Technology*, Vol. 4, Issue 10, October 2008.

Yang, K.; Gondal, I.; Qiu, B. & Dooley, L.S. (2007). Combined SINR based vertical handover algorithm for next generation heterogeneous wireless networks, *Proceeding on IEEE GLOBECOM 2007*, November 2007, Washinton (USA).

Vegni, A.M.; Carli, M.; Neri, A. & Ragosa, G. (2007). QoS-based Vertical Handover in heterogeneous networks, *Proceeding on 10th International Wireless Personal Multimedia Communications*, CD-ROM no. of pages: 4, December 2007, Jaipur (India).

Jesus, V.; Sargento, S.; Corujo, D.; Senica, N.; Almeida, M. & Aguiar, R.L. (2007). Mobility with QoS support for multi-interface terminals: combined user and network approach, *Proceeding on 12th IEEE Symposium on Computers and Communications*, pp. 325–332, July 2007, Aveiro (Portugal).

Kibria, M.R.; Jamalipour, A. & Mirchandani, V. (2005). A location aware three-step vertical handover scheme for 4G/B3G networks, *Proceeding on IEEE GLOBECOM 2005*, Vol. 5, pp. 2752–2756, November 2005, St. Louis (USA).

Wang, S.S.; Green, M. & Malkawi, M. (2001). Adaptive handover method using mobile location information, *Proceeding on IEEE Emerging Technology Symposium on Broadband Comm. for the Internet Era Symposium*, pp. 97–101, September 2001, Richardson (USA).

Vegni, A.M. (2010). Multimedia Mobile Communications in Heterogeneous Wireless Networks -Part 2, *PhD thesis*, University of Roma Tre, March 2010, available online at http://www.comlab.uniroma3.it/vegni.htm.

2

Asynchronous Cooperative Protocols for Inter-vehicle Communications

Sarmad Sohaib[1] and Daniel K. C. So[2]
[1]*University of Engineering and Technology, Taxila*
[2]*The University of Manchester*
[1]*Pakistan*
[2]*United Kingdom*

1. Introduction

Inter-vehicle communication is envisioned to play a very important role in the future, improving road safety and capacity. This can be achieved by utilizing cooperative relaying techniques where the communicating nodes exploit spatial diversity by cooperating with each other (Laneman et al., 2004). This alleviates the detrimental effects of fading and offers reliable data transfer. The source node broadcasts the signal to the destination node directly, and also through the relay nodes. Both the direct and relayed signals are combined at the destination. However, conventional cooperative communication systems require frame or symbol level synchronization between the cooperating nodes. The lack of synchronization results in inter-symbol interference (ISI) and degrades the system performance. This problem will be more severe in inter-vehicle communication as maintaining synchronization in fast moving nodes is very difficult. In this chapter, we present the major asynchronous cooperative communication protocols that can be employed for inter-vehicle communications. These are the asynchronous delay diversity technique (Wei et al., 2006), asynchronous space-time block code (STBC) cooperative system (Wang & Fu, 2007), and asynchronous polarized cooperative (APC) system (Sohaib & So, 2009; 2010).

2. Conventional cooperative communication system model

A three node cooperative network containing the source (S), relay (R) and destination (D) nodes is shown in the Fig. 1. The information will be transmitted from the source node to the destination node directly and also through the relay node. Both the direct and relay signals are combined at the destination using combiners (Brennan, Feb 2003). In general, there are two kinds of relaying modes; *amplify-and-forward (ANF)*, where the relay simply amplifies the noisy version of the signal transmitted by source, and *decode-and-forward (DNF)*, where relay decodes, re-encodes and re-transmits the signal.
The conventional ANF channel model is characterized by transmitting and receiving in orthogonal frequency bands or time slots (Laneman et al., 2004; Sohaib et al., 2009). Here we consider the ANF scheme with the relay node transmitting at the same frequency band as the source node, but in subsequent time-slot.

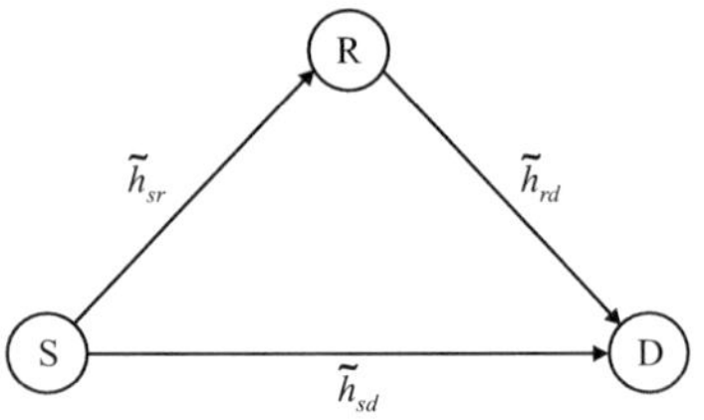

Fig. 1. Cooperative communication netwrok.

The channel $\tilde{h}_{ij}$ between the i-th transmit and j-th receive antenna is given by

$$\tilde{h}_{ij} = \frac{\sum_{u=0}^{U-1} h_{ij}(u)}{\sqrt{PL_{ij}}} \tag{1}$$

where, $h_{ij}(u)$ is the normalized channel gain, which is an independent and identically distributed (i.i.d.) complex Weibull random variable with zero mean. This describes the random fading effect of multipath channels, and is assumed to be frequenct selective fading with U the total number of frequency selective channel taps. Weibull distribution is used for the analysis of APC in vehicle-to-vehicle communication as it fits best (Matolak et al., 2006). The path loss factor PL_{ij} models the signal attenuation over distance, and is given by (Haykin & Moher, 2004)

$$PL_{ij} = \frac{(4\pi)^2}{G_t G_r \lambda^2} \left(d_{ij}\right)^{\alpha} = PL_0 \left(d_{ij}\right)^{\alpha} \tag{2}$$

where PL_0 is the reference path loss factor, d_{ij} is the distance between i-th transmitter and j-th receiver, α is the path loss exponent depending on the propagation environment which is assumed to be the same over all links, λ is the wavelength, and G_t and G_r are the transmitter and receiver antenna gains respectively.

In a typical three node system, single transmission is normally divided into two timeslots (Peters & Heath, 2008; Tang & Hua, 2007). In the first timeslot, the source node broadcasts the signal to the destination and the relay node. The received signal at the destination node directly from the source node is

$$y_{sd}(t) = \sqrt{\frac{E^s}{PL_{sd}}} \sum_{u=0}^{U-1} h_{sd}(u)x(t-u) + n_d(t) \tag{3}$$

where x is the transmitted signal from the source with unit energy, E^s is the transmitted signal energy from the source, h_{sd} is the normalized channel gain from the source to the destination with a corresponding path loss of PL_{sd}, and $n_d(t)$ captures the effect of AWGN at the destination. Similarly, at the same timeslot the relay node receives the same signal from the source, given by

$$y_{sr}(t) = \sqrt{\frac{E^s}{PL_{sr}}} \sum_{u=0}^{U-1} h_{sr}(u)x(t-u) + n_r(t) \tag{4}$$

where h_{sr} is the normalized channel gain from the source to the relay with a corresponding path loss of PL_{sr}, and $n_r(t)$ is the AWGN at the relay.

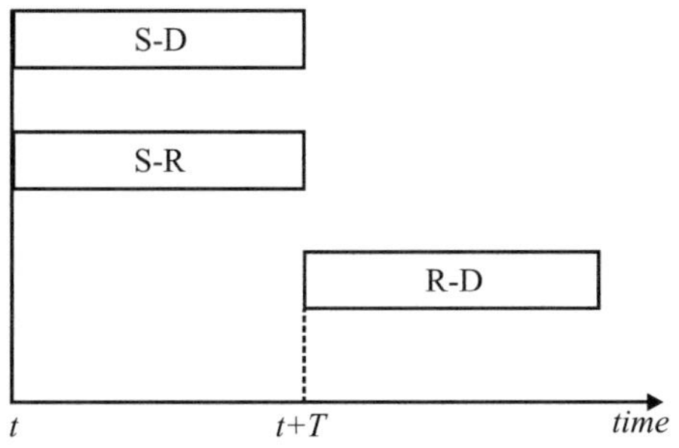

Fig. 2. Timing diagram of ANF cooperative scheme.

In the second timeslot the signal received at the relay node is amplified by a factor k_r' and forwarded to the destination given by

$$y_{rd}(t+T) = \frac{k_r'}{\sqrt{PL_{rd}}} \sum_{u=0}^{U-1} h_{rd}(u) y_{sr}(t-u) + n_d(t+T) \tag{5}$$

where $T = LT_s$ is the timeslot or frame duration with L being the total number of symbols per frame and T_s the symbol period, h_{rd} is the normalized channel gain from the relay to destination node having a corresponding path loss of PL_{rd}, and $n_d(t+T)$ is the AWGN at the destination node. The transmitter estimates path loss through the reverse link and is assumed to be perfectly estimated. On the other hand, instantaneous channel fading gain is not assumed to be known at the transmitter, as it requires feedback information. Therefore, setting identical received signal energy from the direct and relayed link, the amplification factor k_r' is given by

$$k_r' = \sqrt{\frac{E^s\,\mathbb{E}\left[\left|\tilde{h}_{sd}\right|^2\right]}{E^r\,\mathbb{E}\left[\left|\tilde{h}_{rd}\right|^2\right]}} = \sqrt{\frac{E^s/PL_{sd}}{\left(E^s/PL_{sr} + N_0\right)/PL_{rd}}} \tag{6}$$

where E^r is the received signal energy at the relay node. All AWGN noises are modeled as zero mean mutually independent circular symmetric complex Gaussian random sequences with power spectral density (PSD) N_0. Exact channel state information (CSI) is assumed to be available at the receiver only, and not at the transmitter.

For conventional ANF system, the signal in (3) and (5) are combined at the destination node using diversity combiners, e.g. Maximal Ratio Combiner (MRC). The diversity gain achieved through cooperation can compensate the additional noise in the relay (Laneman et al., 2004). Hence, cooperative diversity schemes achieve better performance than non-cooperative schemes.

Fig. 2 illustrates the timing diagram of ANF cooperative system, where, t is the time when the source node starts transmitting the data to the destination and relay nodes. The relay node will start transmitting after a duration of T. Therefore it takes two orthogonal channels for one complete transmission, thus decreases the spectral efficiency of the system. Also frame level synchronization is required in conventional ANF, which is not always achievable in wireless communication. The diversity gain achieved through cooperation can compensate for the additional noise in the relay (Laneman et al., 2004). Hence, the cooperative diversity schemes achieve better performance than non-cooperative schemes.

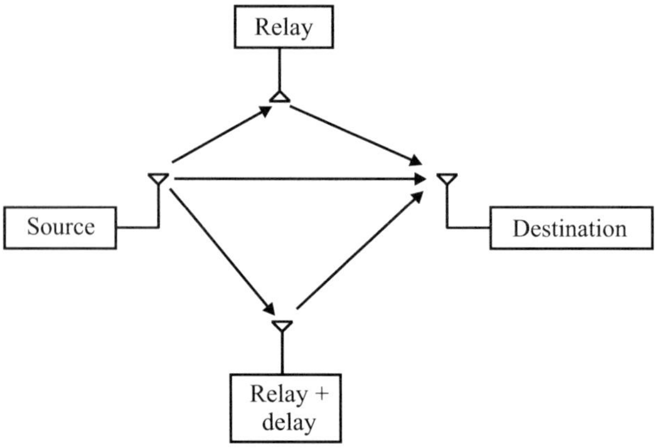

Fig. 3. System structure of cooperative communications.

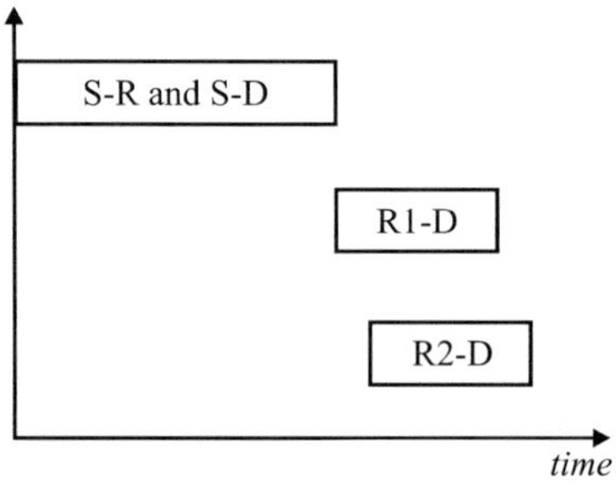

Fig. 4. Timing diagram of asynchronous delay diversity cooperative scheme.

3. Asynchronous cooperative systems

In this section we present a brief summary of the three major inter-vehicle asynchronous cooperative communication systems.

3.1 Asynchronous delay diversity technique

In (Wei et al., 2006), a distributed delay diversity approach is proposed in the Relay-Destination (R-D) link to achieve spatial diversity as shown in Fig. 3. Error detection schemes such as cyclic redundancy check (CRC) is employed at the relay nodes to determine whether the received packet is error free or not. If the received packet is error-free, the relay node will then forward the information packet to the destination, after an additional artificial delay. On the contrary if the packed is in error, it will be dropped at the relay node. Assuming the CRC code can perfectly detect any packet error the forwarded signal from the relay is thus a delayed version of the transmitted symbols. Hence, the destination node will see an equivalent frequency selective fading channel in the form of artificially introduced delays. Fig. 4 illustrates the timing diagram of this scheme.

To equalize the frequency selectivity, a decision feedback equalizer (DFE) is employed at the destination node. It also combines the inputs from the direct link channel, and relay link ones. Although this scheme can mitigate the synchronization problem, it uses half duplex relay node which reduces the spectral efficiency due to the bandwidth expansion or extended time duration. Constellation size has to be increased to maintain the spectral efficiency which then reduces the performance gain over non-cooperative single-input single-output (SISO) scheme.

3.2 Asynchronous space-time block code cooperative system

Instead of using the simple delay diversity code in the R-D link, the asynchronous STBC is proposed in (Wang & Fu, 2007) to achieve distributed cooperative diversity. The system and timing diagram for this scheme is identical to that of the asynchronous delay diversity scheme in Fig. 3 and Fig. 4. At the relay, the detected symbols are mapped into the orthogonal STBC matrix. Each relay then randomly select one row from this matrix for transmission. The random cyclic delay diversity technique is then applied to make the equivalent channels frequency selective. At the destination node the frequency domain equalizer (FDE) is employed to combine and equalize the received signal.

The scheme has a disadvantage that it could suffer performance degradation due to diversity loss by random row selection. Similar to the previous scheme, this system also assumes the relay to be half duplex which results in low spectral efficiency.

3.3 Asynchronous polarized cooperative system

Most cooperative communication systems, including (Wang & Fu, 2007; Wei et al., 2006), employ half duplex relays. This is because full duplex relay that uses the same time and frequency for transmission and reception is difficult to implement. The transmitted signal will overwhelm the received signal. In view of this, the asynchronous polarized cooperative (APC) system is proposed in (Sohaib & So, 2009; 2010), and is illustrated in Fig. 5. It allows full duplex relay operation, and does not require frame of symbol level synchronization. In this scheme every vehicle is equipped with dual polarized antennas that can auto-configure itself to be the source, relay and destination node. The vehicle working as a source only activates the vertical polarized antenna for transmission, whereas the destination vehicle configures the dual polarized antennas for reception. The vehicle working as a relay uses dual polarized antennas for transmission and reception at the same time and at the same frequency thereby achieving the full duplex ANF communication and effectively reducing the transmission duration and increasing the throughput rate. The solid lines represent transmission and reception on the same polarization, also known as co-polarization. On the other hand, the dotted lines represent transmission in one polarization but reception in the other polarization, also known as cross-polarization. The effect of cross-polarization is considered as it is impossible to maintain the same polarization between the transmitter and the receiver due to the complex propagation environment in terrestrial wireless communications. For more practical consideration, path loss is also included in the analysis.

For a relay to operate in full duplex mode the transmission and reception channels must be orthogonal either in time-domain or in frequency domain, otherwise the transmitted signal will interfere with the received signal. In theory, it is possible for relay to cancel out interferences as it has the knowledge of transmitted signal. In practice, however, the transmitted signal is 100-150dB stronger than the received signal and any error in the interference cancellation can potentially be disastrous (Fitzek & Katz, 2006). Due to this reason, the installation of co-polarized antennas at the relay node in place of dual-polarized antennas is not feasible for full duplex relay. However, with dual-polarized antenna the transmitted signal on one polarization is orthogonal to the received signal at another polarization, thereby, enabling the relay to communicate in full duplex mode, not the overall system.

The source node will broadcast using vertical polarization. The vertically polarized received signal at the relay node is the same as (4).

The received signal at the relay node is amplified by a factor k_r, and transmitted immediately to the destination node through horizontal polarization. Radio propagation and signal

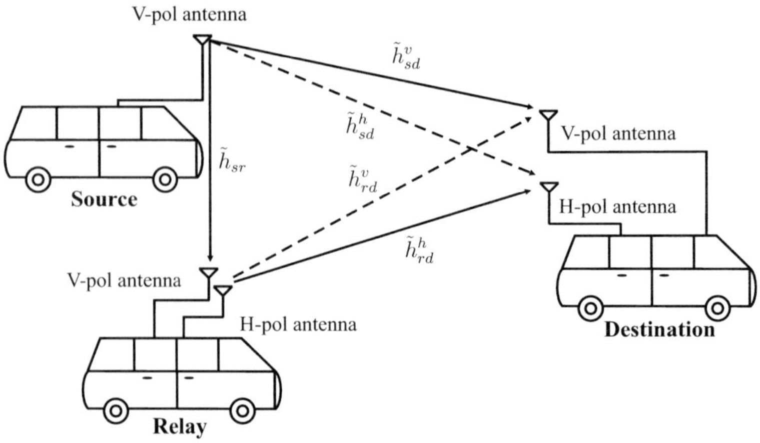

Fig. 5. Asynchronous polarized cooperative system for inter-vehicular communication.

processing at the relay node will cause some additional time delay τ, which could be a few symbols duration and is much shorter than the frame duration T. It must be noted that the APC system does not require symbol level synchronization, between the source and relay, and thus τ can be any positive real number. Fig. 6 illustrates the timing diagram of this scheme. The vertically and horizontally polarized signal received at the destination, denoted as y_{d_v} and y_{d_h} respectively, are given by

$$y_{d_v}(t) = \sqrt{E^s}\,\tilde{h}^v_{sd}\,x(t-u) + k_r\,\tilde{h}^v_{rd}\,y_{sr}\,(t-\tau-u) + n_{d_v}(t) \tag{7}$$

and

$$y_{d_h}(t) = \sqrt{E^s}\,\tilde{h}^h_{sd}\,x(t-u) + k_r\,\tilde{h}^h_{rd}\,y_{sr}\,(t-\tau-u) + n_{d_h}(t). \tag{8}$$

The received signals of the above equations can therefore be written in matrix form as

$$\underbrace{\begin{bmatrix} y_{d_v}(t) \\ y_{d_h}(t) \end{bmatrix}}_{\mathbf{y}_d} = \underbrace{\begin{bmatrix} \tilde{h}^v_{sd} & \tilde{h}^v_{rd} \\ \tilde{h}^h_{sd} & \tilde{h}^h_{rd} \end{bmatrix}}_{\mathbf{H}} \begin{bmatrix} \sqrt{E^s}x(t-u) \\ k_r y_{sr}(t-\tau-u) \end{bmatrix} + \underbrace{\begin{bmatrix} n_{d_v}(t) \\ n_{d_h}(t) \end{bmatrix}}_{\mathbf{n}} \tag{9}$$

where $\mathbf{n}$ is the 2×1 i.i.d. zero mean complex AWGN vector with variance $\mathbb{E}\left[\mathbf{n}\,\mathbf{n}^H\right] = N_0\,\mathbf{I}$, and $\mathbf{I}$ is an identity matrix. The diagonal elements of $\mathbf{H}$ correspond to co-polarization, while the off-diagonal elements correspond to cross-polarization. The relay amplification factor k_r is

$$k_r = \sqrt{\frac{E^s\left(\mathbb{E}\left[\left|\tilde{h}^v_{sd}\right|^2\right] + \mathbb{E}\left[\left|\tilde{h}^h_{sd}\right|^2\right]\right)}{E^r\left(\mathbb{E}\left[\left|\tilde{h}^v_{rd}\right|^2\right] + \mathbb{E}\left[\left|\tilde{h}^h_{rd}\right|^2\right]\right)}} \tag{10}$$

where E^r is the received signal energy at the relay node given by

$$E^r = E^s\,\mathbb{E}\left[\left|\tilde{h}_{sr}\right|^2\right] + N_0. \tag{11}$$

Since the source and relay node are spatially separated apart, we can assume the channel from the source to the destination is not correlated with the channel from the relay to the

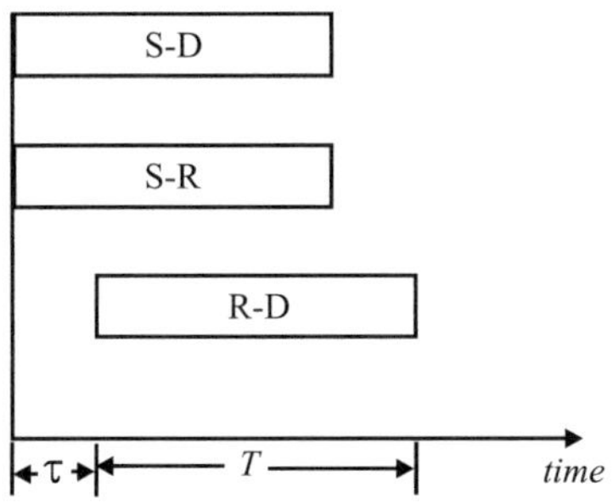

Fig. 6. Timing diagram of APC scheme.

destination. In other words, the co-polarization elements of the channel h^v_{sd} and h^h_{rd} and the cross-polarization elements h^h_{sd} and h^v_{rd} are assumed to be completely un-correlated. Therefore

$$\mathbb{E}\left[h^v_{sd}h^{h*}_{rd}\right] = \mathbb{E}\left[h^h_{sd}h^{v*}_{rd}\right] = 0 \tag{12}$$

and

$$\mathbb{E}\left[h^v_{sd}h^{v*}_{rd}\right] = \mathbb{E}\left[h^h_{sd}h^{h*}_{rd}\right] = 0. \tag{13}$$

We define the receive correlation coefficient as

$$\rho_r = \frac{\mathbb{E}\left[h^v_{sd}h^{h*}_{sd}\right]}{\sqrt{\chi}} = \frac{\mathbb{E}\left[h^v_{rd}h^{h*}_{rd}\right]}{\sqrt{\chi}}. \tag{14}$$

At the destination node, the vertical and horizontal polarized signals are received at different time due to the signal processing and additional propagation delay τ caused by the relay. Because of cross polarization, the delayed signal from the relay becomes an ISI. Therefore equalization for each polarization is required. As there are two branches from the vertical and horizontal polarization, diversity combiner is needed. The frequency domain diversity combiner and equalizer (FDE-MRC) is therefore used and is shown in Fig. 7. Assuming that

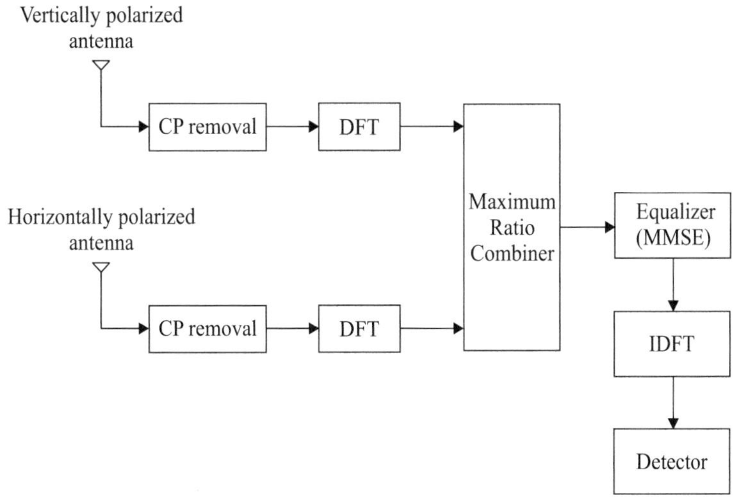

Fig. 7. Receiver structure of the APC MIMO system.

cyclic prefix (CP) with duration longer than delay τ is inserted before transmission from the source node, and removed at the destination node, the signals received at the destination node from the source and relay nodes are transformed into frequency domain by taking L points discrete Fourier transform (DFT). The resulting signal spectras at the k-th subcarrier from vertical and horizontal polarized branches are respectively given by

$$\begin{aligned} Y_{d_v}(k) &= \sqrt{E^s}\; X(k)\left[\tilde{h}^v_{sd} + k_r \tilde{h}^v_{rd}\tilde{h}_{sr}e^{-j2\pi\frac{k}{L}\tau}\right] + k_r\tilde{h}^v_{rd}N_r(k)e^{-j2\pi\frac{k}{L}\tau} + N_{d_v}(k) \\ &\triangleq \sqrt{E^s}X(k)\; H_v(k) + N_v(k) \end{aligned} \tag{15}$$

and

$$\begin{aligned} Y_{d_h}(k) &= \sqrt{E^s}\; X(k)\left[\tilde{h}^h_{sd} + k_r \tilde{h}^h_{rd}\tilde{h}_{sr}e^{-j2\pi\frac{k}{L}\tau}\right] + k_r\tilde{h}^h_{rd}N_r(k)e^{-j2\pi\frac{k}{L}\tau} + N_{d_h}(k) \\ &\triangleq \sqrt{E^s}X(k)\; H_h(k) + N_h(k) \end{aligned} \tag{16}$$

where $k = \{1, 2, \ldots, L\}$, $X(k)$ is the transmitted signal in frequency domain, $N_r(k)$ is the relay noise in frequency domain, and $N_v(k)$ and $N_h(k)$ are the effective noises at the vertical and horizontal antennas respectively at the destination node, $H_v(k)$ and $H_h(k)$ are the effective channels at vertical and horizontal antennas respectively at the destination node given by

$$H_v(k) = \tilde{h}^v_{sd} + k_r\tilde{h}^v_{rd}\tilde{h}_{sr}e^{-j2\pi\frac{k}{L}\tau} \tag{17}$$

and

$$H_h(k) = \tilde{h}^h_{sd} + k_r\tilde{h}^h_{rd}\tilde{h}_{sr}e^{-j2\pi\frac{k}{L}\tau}. \tag{18}$$

The polarized frequency domain signals $Y_{d_v}(k)$ and $Y_{d_h}(k)$ are combined through MRC at the destination node and the resultant signal spectrum $Y(k)$ is

$$Y(k) = Y_{d_v}(k)H^*_v(k) + Y_{d_h}(k)H^*_h(k). \tag{19}$$

The combined signal $Y(k)$ is input to MMSE equalizer given by

$$W(k) = \arg\min_{W}\; \mathbb{E}_h\left[\left|W(k)\; Y(k) - \sqrt{E^s}X(k)\right|^2\right] \tag{20}$$

where $\mathbb{E}_h\,[.]$ denotes the expectation conditioned on the channel gains. For ease of notation and without loss of generality, we drop the index k in the following derivation. Substituting the value of Y from (15), (16), and (19) into the objective function of (20)

$$\begin{aligned} J =& \mathbb{E}_h\left[\left|W\left(\sqrt{E^s}X\,|H_v|^2 + N_vH^*_v + \sqrt{E^s}X\,|H_h|^2 + N_hH^*_h\right) - \sqrt{E^s}X\right|^2\right] \\ =& \mathbb{E}_h\left[\left|\left(W\,|H_v|^2 + W\,|H_h|^2 - 1\right)\sqrt{E^s}X + WN_vH^*_v + WN_hH^*_h\right|^2\right]. \end{aligned} \tag{21}$$

Solving the above equation for minimum value of W, we take the derivate of J w.r.t. W and set it to 0, i.e. $\frac{dJ}{dW} = 0$

$$\Rightarrow E^s \left(|H_v|^4 W^* + |H_h|^4 W^* + 2 |H_v H_h|^2 W^* - |H_v|^2 - |H_h|^2 \right) + N_0^v |H_v|^2 W^* + N_0^h |H_h|^2 W^* = 0$$

$$\Rightarrow \left(E^s |H_v|^4 + E^s |H_h|^4 + 2 |H_v H_h|^2 + |H_v|^2 N_0^v + |H_h|^2 N_0^h \right) W^* = E^s \left(|H_v|^2 + |H_h|^2 \right) \quad (22)$$

Rearranging (22) we obtain,

$$W^* = \frac{E^s \left(|H_v|^2 + |H_h|^2 \right)}{E^s \left(|H_v|^4 + |H_h|^4 + 2 |H_v H_h|^2 \right) + |H_v|^2 N_0^v + |H_h|^2 N_0^h} \quad (23)$$

Assuming $H = |H_v|^2 + |H_h|^2$, (23) becomes,

$$W^* = \frac{H}{|H|^2 + |H_v|^2 \frac{N_0^v}{E^s} + |H_h|^2 \frac{N_0^h}{E^s}}. \quad (24)$$

Taking the conjugate on both side and adding the index k, we obtain the final form

$$W(k) = \frac{H^*(k)}{|H(k)|^2 + |H_v(k)|^2 \frac{N_0^v}{E^s} + |H_h(k)|^2 \frac{N_0^h}{E^s}} \quad (25)$$

where

$$H(k) = |H_v(k)|^2 + |H_h(k)|^2, \quad (26)$$

$$N_0^v = N_0 \left(1 + k_r^2 \, \mathbb{E} \left[\left| \tilde{h}_{rd}^v \right|^2 \right] \right) = N_0 \left(1 + \frac{k_r^2 \chi}{PL_{rd}^v} \right) \quad (27)$$

and

$$N_0^h = N_0 \left(1 + k_r^2 \, \mathbb{E} \left[\left| \tilde{h}_{rd}^h \right|^2 \right] \right) = N_0 \left(1 + \frac{k_r^2}{PL_{rd}^h} \right). \quad (28)$$

As the dual polarized antennas at the destination node are closely spaced, we can assume the distance for the cross-polarized channels from the same node are the same, i.e., $d_{sd}^v = d_{sd}^h = d_{sd}$ and $d_{rd}^v = d_{rd}^h = d_{rd}$. Therefore (27) and (28) becomes

$$N_0^v = N_0 \left(1 + \frac{k_r^2 \chi}{PL_{rd}} \right) \quad (29)$$

and

$$N_0^h = N_0 \left(1 + \frac{k_r^2}{PL_{rd}}\right). \tag{30}$$

The detected data in frequency domain is then transformed back to time domain by using inverse discrete Fourier transform (IDFT). Due to the full duplex nature of the relay, the transmission time is reduced, which in turn increases the data rate as compared to the conventional ANF protocol. Also no frame or symbol synchronization is required at the relay node because of the use of FDE-MRC at the destination node.

3.4 Capacity analysis of asynchronous polarized cooperative system

In this section, the capacity of the APC scheme with one relay node will be presented. For fairer comparison, we also present the capacity of ANF cooperative system which employs dual polarized antenna at the destination node, where polarization diversity is also exploited.

3.4.1 Asynchronous polarized cooperative scheme

Given the channel information at the receiver, the ergodic capacity of the system in (15) and (16) can be computed as

$$\begin{aligned} C &= \max_{p(x)} I(x;\mathbf{y}_d) = \frac{L}{L+\tau}\,\mathbb{E}\left[\log_2\left(1+\frac{E^s}{G}\left(\mathbb{E}_h\left[\frac{|H_v|^2}{|N_v|^2}+\frac{|H_h|^2}{|N_h|^2}\right]\right)\right)\right] \\ &\cong \mathbb{E}\left[\log_2\left(1+\frac{E^s}{GL}.\mathbb{E}_h\left[\sum_{k=1}^{L}\frac{\left|\tilde{h}_{sd}^v+\sqrt{k_r}\tilde{h}_{rd}^v\tilde{h}_{sr}e^{-j2\pi\frac{k}{L}\tau}\right|^2}{\left|\sqrt{k_r}\tilde{h}_{rd}^v N_r(k)e^{-j2\pi\frac{k}{L}\tau}+N_{d_v}(k)\right|^2}\right.\right.\right. \\ &\qquad \left.\left.\left.+\sum_{k=1}^{L}\frac{\left|\tilde{h}_{sd}^h+\sqrt{k_r}\tilde{h}_{rd}^h\tilde{h}_{sr}e^{-j2\pi\frac{k}{L}\tau}\right|^2}{\left|\sqrt{k_r}\tilde{h}_{rd}^h N_r(k)e^{-j2\pi\frac{k}{L}\tau}+N_{d_h}(k)\right|^2}\right]\right)\right] \\ &= \mathbb{E}\left[\log_2\left(1+\frac{E^s}{GLN_0}\left(\frac{\sum_{k=1}^{L}\left|\tilde{h}_{sd}^v+\sqrt{k_r}\tilde{h}_{rd}^v\tilde{h}_{sr}e^{-j2\pi\frac{k}{L}\tau}\right|^2}{1+k_r\left|\tilde{h}_{rd}^v\right|^2}\right.\right.\right. \\ &\qquad \left.\left.\left.+\frac{\sum_{k=1}^{L}\left|\tilde{h}_{sd}^h+\sqrt{k_r}\tilde{h}_{rd}^h\tilde{h}_{sr}e^{-j2\pi\frac{k}{L}\tau}\right|^2}{1+k_r\left|\tilde{h}_{rd}^h\right|^2}\right)\right)\right] \end{aligned} \tag{31}$$

where $\mathbb{E}_h[.]$ denotes the expectation conditioned on the channel gains, G is a normalization factor that is used to make sure that the transmission energy of the APC scheme is the same as that of non-cooperative scheme, and is given by

$$G = 1 + \frac{PL_{rd}}{PL_{sd}}. \tag{32}$$

Notice that the pre-log factor $\frac{L}{L+\tau}$ can be approximated to be one as the frame length L is much larger than the delay τ. Hence the APC scheme will have a higher capacity than the conventional scheme, which inevitably has the 1/2 pre-log factor.

3.4.2 Polarized ANF

As conventional ANF does not have the cross polarized channels, a polarized ANF system is presented in this subsection for fairer comparison with the APC scheme. The system model of polarized ANF with vertical polarized source antenna, vertical polarized relay antenna and dual polarized destination antennas is given as

$$\mathbf{y}_{pa} = \mathbf{H}_{pa} \sqrt{E^s}\, x(t-u) + \mathbf{n}_{pa}$$

where $\mathbf{H}_{pa} = \left[\tilde{h}^v_{sd} \;\; \tilde{h}^h_{sd} \;\; \sqrt{k_r}\tilde{h}^v_{rd}\tilde{h}_{sr} \;\; \sqrt{k_r}\tilde{h}^h_{rd}\tilde{h}_{sr}\right]^T$ and

$$\mathbf{n}_{pa} = \underbrace{\begin{bmatrix} 1\,0\,0\,0 & 0 \\ 0\,1\,0\,0 & 0 \\ 0\,0\,1\,0 & \sqrt{k_r}\,\tilde{h}^v_{rd} \\ 0\,0\,0\,1 & \sqrt{k_r}\,\tilde{h}^h_{rd} \end{bmatrix}}_{\mathbf{Q}} \begin{bmatrix} n_{d_v}(t) \\ n_{d_h}(t) \\ n_{d_v}(t+T) \\ n_{d_h}(t+T) \\ n_r(t) \end{bmatrix}$$

where $\tilde{h}^v_{sd}$ and $\tilde{h}^v_{rd}$ are the co-polarized channels and $\tilde{h}^h_{sd}$ and $\tilde{h}^h_{rd}$ are the cross-polarized channels. The ergodic capacity of the polarized ANF is thus given by

$$C_{pa} = \frac{1}{2}\mathbb{E}\left[\log_2 \det\left(\mathbf{I} + \frac{E^s}{GN_0}\mathbf{H}_{pa}\mathbf{H}^H_{pa}\left(\mathbf{QQ^H}\right)^{-1}\right)\right] \tag{33}$$

where the normalization factor G is identical to (32). It can be noted that the 1/2 pre-log factor in (33) shows that polarized ANF also requires two timeslots for one complete transmission.

3.5 Energy analysis of asynchronous polarized cooperative system

Cooperative communication achieves diversity through spatially separated cooperating nodes. In most potential applications, these nodes are battery powered. Therefore energy consumption must be minimized without compromising the transmission quality. As more RF front ends are used by polarized antennas in the APC scheme, the total energy requirement to achieve a required quality must be compared to the conventional ANF. In this section we formulate the transmission energy consumption and total energy consumption of the APC scheme.

In the following analysis, the energy consumption model developed by Cui *et. al.* is used (Cui et al., 2004). The total energy consumption model that includes both the transmission energy and the circuit energy consumption per bit is given by

$$E_{bt} = \frac{(P_{PA} + P_C)}{B\, R_b} \tag{34}$$

where P_C is the power consumption of all circuit blocks, B is the bandwidth, R_b is the bit rate, and P_{PA} is the power consumption of all power amplifiers, which depends on the transmit power P_{out},

$$P_{out} = E_T\, R_b\, B \tag{35}$$

where E_T is the sum of transmission energy from both the source and relay nodes. For the APC scheme E_T can be written as

$$E_T = E^s + k_r\, E^r = E^s \left(1 + \frac{\mathbb{E}\left[\left|\tilde{h}^v_{sd}\right|^2\right] + \mathbb{E}\left[\left|\tilde{h}^h_{sd}\right|^2\right]}{\mathbb{E}\left[\left|\tilde{h}^v_{rd}\right|^2\right] + \mathbb{E}\left[\left|\tilde{h}^h_{rd}\right|^2\right]}\right)$$

$$= E^s \left(1 + \frac{1/PL_{sd} + \chi/PL_{sd}}{\chi/PL_{rd} + 1/PL_{rd}}\right) = E^s \left(1 + \frac{PL_{rd}}{PL_{sd}}\right). \quad (36)$$

The power consumption of the power amplifies can be approximated as

$$P_{PA} = (1 + \psi)\, P_{out} \quad (37)$$

where $\psi = (\xi/\eta) - 1$, with η the drain efficiency of the RF power amplifier and ξ the peak to average ratio, which depends on the modulation scheme and the associated constellation size M Cui et al. (2004)

$$\xi = 3\frac{M - 2\sqrt{M} + 1}{M - 1}. \quad (38)$$

The power consumption of all circuit blocks along the signal path is given by

$$P_C \approx M_t\,(P_{DAC} + P_{MIX} + P_{FILT}) + 2P_{SYN} + M_r\,(P_{LNA} + P_{MIX} + P_{IFA} + P_{FILR} + P_{ADC}) \quad (39)$$

where P_{DAC}, P_{MIX}, P_{FILT}, P_{SYN}, P_{LNA}, P_{IFA}, P_{FILR}, P_{ADC} are the power consumption values of the digital-to-analog converter (DAC), the mixer, the active filter at transmitter side, the frequency synthesizer, the low-noise amplifier, the intermediate frequency amplifier, the active filter at receiver side, and the analog-to-digital converter (ADC) respectively. M_t and M_r is the number of RF chains involved in one complete transmission at transmitter and receiver side respectively. Although the APC scheme has two extra physical antennas installed as compared to conventional ANF, both schemes effectively use the same number of RF chains for one complete transmission. It is because conventional ANF takes two timeslots for one complete transmission, which uses the RF chains again at the relay and the destination. Simulation results for energy analysis are shown in the next section under the same throughput and BER requirement.

4. Simulation results of asynchronous polarized cooperative system

Computer based Monte-Carlo simulations are carried out to illustrate the BER performance, capacity and energy consumption of the APC system. In order to provide a fair comparison among different schemes, spectral efficiency is kept constant for all protocols and is set to be 2bps/Hz. The SISO and the APC scheme uses QPSK, whereas the ANF protocol uses 16QAM for one relay network. This is because the SISO and the APC scheme takes approximately one time-slot for complete transmission of one data frame, whereas conventional ANF protocol takes two time-slots. For both the polarized ANF and the APC scheme, the cross-polarized channel power (χ) and receiver correlation coefficient (ρ_r) are set to be 0.4 and 0.5 respectively. The time delay τ is assumed to be one symbol period. To obtain reasonable values of received SNR, the transmitted signal from the source node is amplified by $\sqrt{PL_{sd}}$ to compensate the path loss. The direct link SNR after this normalization is defined as γ_{sd}. For the ANF and APC scheme, normalization factor G in (32) is used to ensure the same total transmission power as the SISO. Hence the normalization SNR γ_{sd} can be used as a reference for all schemes in capacity, and BER analysis. Table 1 summarizes the system parameters for all simulations, which are mostly based on (Cui et al., 2004), and (Cui et al., 2003). The parameter f_c is the carrier frequency, $\bar{P}_b$ is the average probability of error for energy consumption analysis, and M_L is the link margin compensating the hardware process variations and other background interference and noise. The number of transmit antennas M_t and receive antennas M_r involved in one complete transmission are respectively 2 and 3 for conventional and polarized ANF as well as the APC schemes, whereas they are both one for SISO scheme. Table 2 shows the parameters for the tapped delay line channel model derived by Matolak *et. al* for vehicle to vehicle communication (Matolak et al., 2006).

$P_{DAC} = 15.4$mW	$G_t G_r = 5$dBi
$P_{MIX} = 30.3$W	$\alpha = 3$
$P_{FILT} = 2.5$mW	$f_c = 5.12$GHz
$P_{FILR} = 2.5$mW	$\eta = 0.35$
$P_{SYN} = 50$mW	$\bar{P}_b = 10^{-4}$
$P_{LNA} = 20$mW	$M_L = 40$dB
$P_{IFA} = 3$mW	$B = 10$MHz
$P_{ADC} = 6.7$mW	

Table 1. System Parameters.

For capacity and BER analysis, the source to destination node distance d_{sd} is set to be $200m$. The relay node is set at the midpoint between the source and destination node, i.e, $d_{sr} = d_{rd} = 100m$. For energy analysis, various positions of the relay node are considered.

Tap Index	Fractional Tap Energy	Weibull Shape Factor (b)	Weibull Scale Factor (a)
1	0.7018	2.49	0.8676
2	0.1158	1.75	0.3291
3	0.0543	1.68	0.2226
4	0.0391	1.72	0.1903
5	0.0259	1.65	0.1528
6	0.0198	1.60	0.1322
7	0.0118	1.69	0.1040

Table 2. Vehicle to vehicle channel model (Matolak et al., 2006).

The increase in capacity of the APC scheme as compared to the conventional ANF scheme is demonstrated in Fig. 8. The capacity of the APC scheme significantly outperforms the conventional ANF protocols due to the relay's full duplex capability. For polarized ANF, the use of dual polarized antenna at the destination node provides a marginal increase in capacity. Therefore, even if polarized antennas are also used, the APC scheme has a significant capacity advantage over the polarized ANF scheme. The APC scheme without cross-polarization has slightly less capacity than the APC system with cross-polarization but still it is higher than the ANF systems.

The BER performance comparison among the SISO, ANF protocol, and the APC system is presented in Fig. 9. The APC system without cross-polarization has a gain of about 4.5dB over the conventional ANF protocol at BER 10^{-3}. Thus the cost of using dual polarized antennas and separate RF chains at the relay node is justified by the significantly lowered BER. With the presence of cross-polarization, the performance further improves because

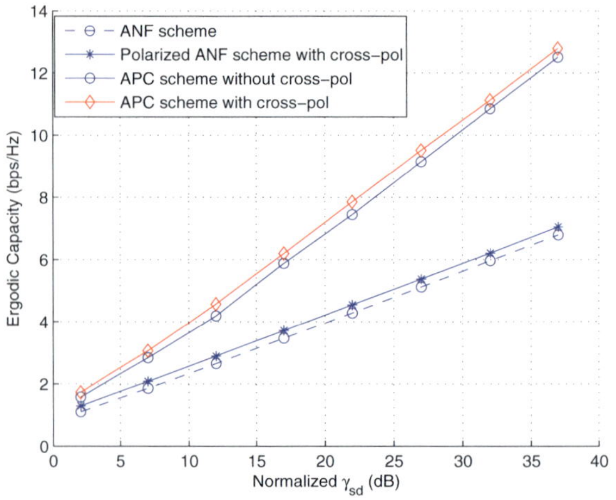

Fig. 8. Capacity comparison of one relay APC scheme.

polarization diversity can be achieved. The polarized ANF also has a marked improvement, but is approximately 3dB worse than the APC scheme. Another observation is the differences in the asymptotic slope of SISO to the APC scheme. It verifies that diversity is achieved for cooperative schemes with and without cross-polarization.

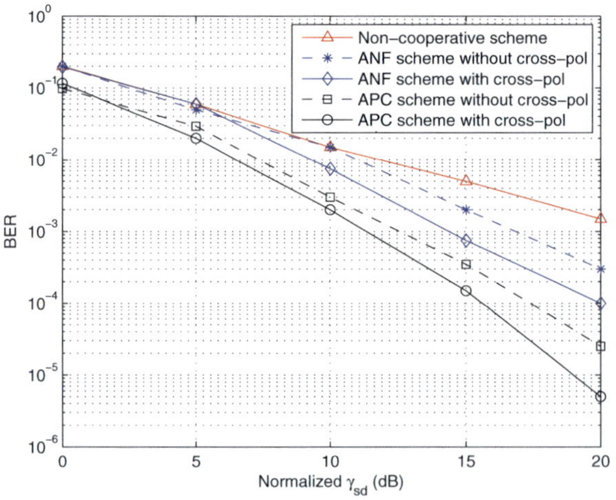

Fig. 9. BER performance of APC system.

As more RF front ends are installed in the APC scheme, the total energy required to achieve a particular quality is compared with the conventional ANF and SISO schemes in Fig. 10. The total energy consumption is calculated using (36), where E^s is obtained using direct link SNR γ_{sd} observed at BER=10^{-4}, where $\gamma_{sd} = E^s / N_0$. The direct link SNR is obtained by evaluating the BER over 10,000 randomly generated channel samples at each transmission distance. It can be observed that the APC scheme becomes more energy-efficient than both the ANF and SISO protocols when $d_{sd} \geq 23m$. The crossover point indicates the distance where the transmission energy saving exceeds the extra circuit energy consumption in the APC scheme comparing to the SISO and ANF scheme. In addition, for practical applications, the source to destination node speration will be mostly larger than 20m. Hence, the APC scheme will consume less energy in realistic scenario.

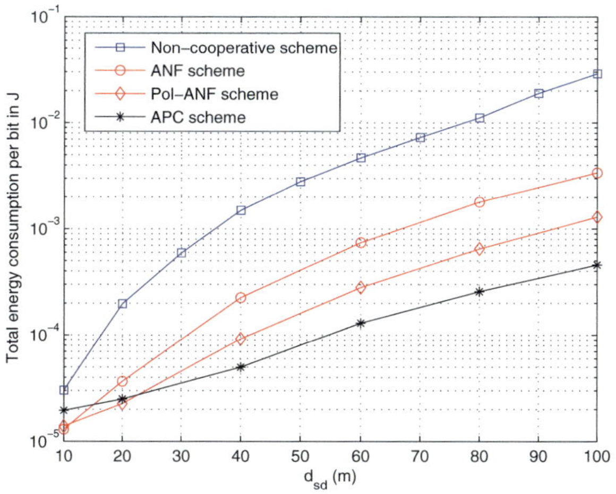

Fig. 10. Total energy consumption per bit over d_{sd} when the relay node is located midway between source and destination nodes.

5. Conclusion

In this chapter, we discuss some of the major asynchronous cooperative communication protocols that can be used in vehicle-to-vehicle cooperative communications. The APC scheme with full duplex relay that completes the data transmission between the source and the destination in approximately same time duration as non-cooperative scheme is discussed in detail. The performance improvement of APC scheme is demonstrated by the BER and the capacity simulation results, which show its superiority over non-cooperative, conventional and polarized ANF protocol. Even with the use of more RF front ends, the APC scheme has less total energy consumption than ANF and non-cooperative schemes over more practical distances between the nodes. Thus, the APC scheme is both spectral and energy efficient, and is suitable for inter-vehicle cooperative communication.

6. References

Brennan, D. (Feb 2003). Linear diversity combining techniques, *Proceedings of the IEEE* 91(2): 331–356.

Cui, S., Goldsmith, A. & Bahai, A. (2003). Energy-constrained modulation optimization for coded systems, pp. 372–376.

Cui, S., Goldsmith, A. & Bahai, A. (2004). Energy-efficiency of mimo and cooperative mimo techniques in sensor networks, *IEEE Journal on Selected Areas in Communications* 22(6): 1089–1098.

Fitzek, F. & Katz, M. (2006). *Cooperation in Wireless Networks: Principles and Applications*, Springers.

Haykin, S. & Moher, M. (2004). *Modern Wireless Communications*, Prentice Hall.

Laneman, J. N., Tse, D. & Wornell, G. (2004). Cooperative diversity in wireless networks: Efficient protocols and outage behavior, *IEEE Transactions on Information Theory* 50(12): 3062–3080.

Matolak, D. W., Sen, I. & Xiong, W. (2006). Channel modeling for V2V communications, *Proc. Third Annual International Conference on Mobile and Ubiquitous Systems: Networking and Services*.

Peters, S. & Heath, R. W. (2008). Nonregenerative MIMO relaying with optimal transmit antenna selection, *IEEE Signal Processing Letters* 15: 421 –424.

Sohaib, S. & So, D. K. C. (2009). Asynchronous polarized cooperative MIMO communication, *Proc. IEEE 69th Vehicular Technology Conference* pp. 1–5.

Sohaib, S. & So, D. K. C. (2010). Energy analysis of asynchronous polarized cooperative MIMO protocol, *Proc. IEEE 21st Personal, Indoor and Mobile Radio Communications Symposium*.

Sohaib, S., So, D. K. C. & Ahmed, J. (2009). Power allocation for efficient cooperative communication, *Proc. IEEE 20th Personal, Indoor and Mobile Radio Communications Symposium*.

Tang, X. & Hua, Y. (2007). Optimal design of non-regenerative MIMO wireless relays, *IEEE Transactions on Wireless Communications* 6(4): 1398 –1407.

Wang, D. & Fu, S. (2007). Asynchronous cooperative communications with STBC coded single carrier block transmission, *Proc. IEEE Global Telecommunications Conference* pp. 2987 –2991.

Wei, S., Goeckel, D. L. & Valenti, M. (2006). Asynchronous cooperative diversity, *IEEE Transactions on Wireless Communications* 5(6): 1547 – 1557.

3

Efficient Information Dissemination in VANETs

Boto Bako and Michael Weber
Institute of Media Informatics, Ulm University
Germany

1. Introduction

Vehicular ad-hoc networks (VANETs) enable promising new possibilities to enhance traffic safety and efficiency. The vision of VANETs is that vehicles communicate spontaneously, in an ad-hoc manner over a wireless medium. Based on this inter-vehicle communication (IVC), vehicles exchange important information, e.g., about road conditions and hazardous situations. Moreover, such information can be propagated via multiple hops, thus making the dissemination of important information possible over longer distances.

This is the key advantage of this kind of safety applications compared to conventional safety systems. Whereas conventional safety systems only rely on information sensed in the direct neighborhood by onboard sensors of a vehicle, active safety applications based on IVC can utilize information generated by nodes multiple hops away. Moreover, such information can be enriched on the way with information sensed by relaying cars. This greatly enhances the potential of VANET applications. The advantage is twofold:

- Having information about distant hazardous situations like an accident ahead or icy road, the driver can be warned in-time, thus being able to completely avoid the dangerous situation.
- Aggregating information from multiple cars enables retaining information on a higher semantic level. This way, applications like cooperative traffic jam warning and cooperative parking place detection can be realized.

The enabling technology for such applications is the wireless ad-hoc communication between vehicles. Especially the dissemination of messages in a specific geographic region represents a fundamental service in VANETs to which we refer to as geographic broadcast (GeoCast). This communication paradigm is used by many applications to enhance traffic safety and efficiency but it can also serve as a basic mechanism for other routing protocols. Because of its relevance in the domain of vehicular networks, it is of key importance that the communication protocol enables efficient message dissemination.

The realization of a robust and efficient broadcast mechanism is a challenging task due to the wide range of applications envisioned to build upon this communication technology, the rigorous requirements of safety applications, and the special network characteristics of vehicular networks. Therefore, the main focus of this chapter is the efficient broadcast of information for VANET applications. We want to give a broad and in-depth review of recent research in this topic and present simulation results of efficient dissemination protocols designed for such applications.

This chapter is organized as follows: In Section 2 we discuss briefly different types of VANET applications, followed by an overview of different communication mechanisms used by these applications. After that, the special network characteristics of VANETs are discussed and the requirements of broadcast protocols are summarized. We conclude that a geographically limited broadcast is one of the most important communication paradigms for VANET applications. Therefore, this communication mechanism is surveyed and classified in Section 3. This is followed by an evaluation of selected protocols by simulations in Section 4. Finally, Section 5 concludes the results and presents possible future works on this topic.

2. Inter-vehicle communication

The main objective of this section is to define the key requirements for IVC (with the focus on broadcast mechanisms) in VANETs. Therefore we first give an overview over different application types with their characteristics, followed by a short description of communication paradigms used to realize such applications. The identified properties of these applications together with the special network characteristics of VANETs allow us to perform a requirements analysis for the dissemination protocols.

2.1 VANET applications

There are many applications envisioned for VAENTs, in (Vehicle Safety Communications Project [VSCP], 2005) e.g., more than 75 application scenarios were identified. A successful deployment of such applications would result in a high benefit. According to this benefit the VANET applications can be classified as follows (Bai et al., 2006):

Safety applications

Active safety applications represent the most important group of VANET applications. The goal of these applications is to reduce the number of injuries and fatalities of road accidents. In the European Union (EU27) e.g., more than 1.2 million traffic accidents involved injury of passengers in 2007 and more than 42,000 accidents ended fatal (EURF, 2009). Hence, there is a high potential benefit in the implementation of such applications (VSCP, 2005). To achieve this goal, safety applications disseminate information about hazardous situations (e.g. about abnormal road conditions or post-crash warning) to vehicles which can benefit from such information to avoid an accident. Thus, this kind of applications rely mostly on the dissemination of information into a specific geographic region, therefore, a geographically limited broadcast – or so called GeoCast – is used (Bai et al., 2006; Bako et al., 2008; Slavik & Mahgoub, 2010). It is also important to note, that safety applications are delay critical, i.e. it is essential that the information is disseminated immediately without any delay. A special case of this broadcast is the one-hop MAC-layer broadcast – also called beaconing – which is periodically sent to neighbors in communication range to exchange information (e.g. position, velocity) sensed by own sensors. This information can be used for safety applications like cooperative collision warning.

Convenience (traffic management) applications

This kind of applications intends to improve the driving efficiency and comfort on roads by means of communications. Driving efficiency applications intend to optimize the traffic flow on roads, i.e. minimizing the travel time by disseminating information about traffic flow conditions on roads. Therefore, cars periodically exchange information, combine information received from neighboring vehicles with information sensed by own sensors, aggregate them, and disseminate the newly gathered information for other vehicles. This way, applications like a cooperative traffic jam detection or travel time estimation for road segments can be realized.

A driver having e.g. information about a traffic jam in advance, can choose an alternative route (with the assistance of the car navigation system), thus reducing greatly the travel time. Therefore, a successful deployment of such applications could greatly reduce the fuel consumption of vehicles, which would have a great impact on cost reduction as well as on the reduction of CO_2 emission. Comfort applications assist the driver in many situations, e.g. for finding free parking spots or merging into the flow traffic and so on. These applications are delay-tolerant, i.e. they don't impose such tight time constraints as safety applications, but mostly need to exchange information periodically into the direct neighborhood.

Commercial applications

Commercial applications provide communication services like entertainment, web access and advertisement. Examples are remote vehicle diagnostics, video streaming, and map download for the navigation system. In contrast to the previously discussed types of applications, commercial applications mostly rely on unicast communication and require a much higher bandwidth than the two other application groups.

A more detailed classification with application examples can be found in (Schoch et al., 2008). In the next subsection we briefly overview the different communication mechanisms used to implement the presented application types.

2.2 Communication paradigms

Considering the three type of application classes from a network perspective, it can be stated that the broadcast/GeoCast, beaconing and unicast communication paradigms are the building blocks for these applications. According to (Bai et al., 2006), broadcast can be further divided into the event-driven, scheduled, and on-demand sub-classes.

Event-driven broadcast is used by delay-critical applications like road hazard condition warning, and the information is disseminated over multiple hops into a specific geographic region. Scheduled broadcast is used for applications like cooperative collision warning and most of the traffic management applications. Information needed by such applications is exchanged periodically and sent as MAC-layer broadcast only to neighbors in communication range. Applications which require a multi-hop dissemination need to apply some efficient aggregation techniques to overcome the limited bandwidth problem (c.f. (Dietzel et al., 2009a; Dietzel et al., 2009b)). Unicast is important especially for commercial and entertainment applications.

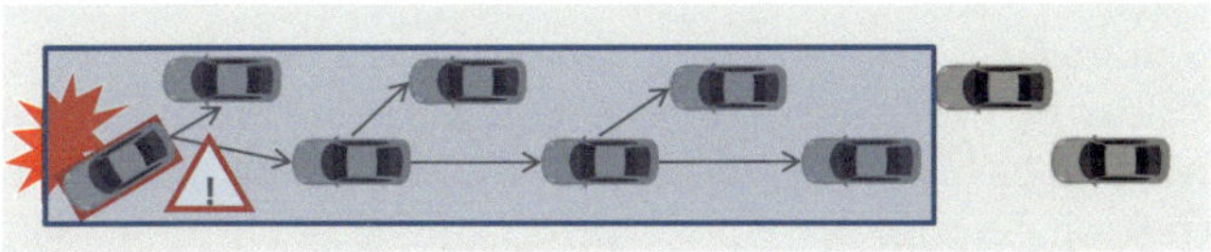

Fig. 1. Example for GeoCast communication.

Figure 1 shows the GeoCast communication paradigm. A broken down vehicle (marked red on the left side of the image) initiates a broadcast message about the hazardous situation. This message is disseminated via multiple hops to inform all vehicles in the specified destination region. Beaconing is shown in Figure 2. The vehicle marked red in the figure sends a MAC-layer broadcast message with data about the own vehicle, like position, heading, and

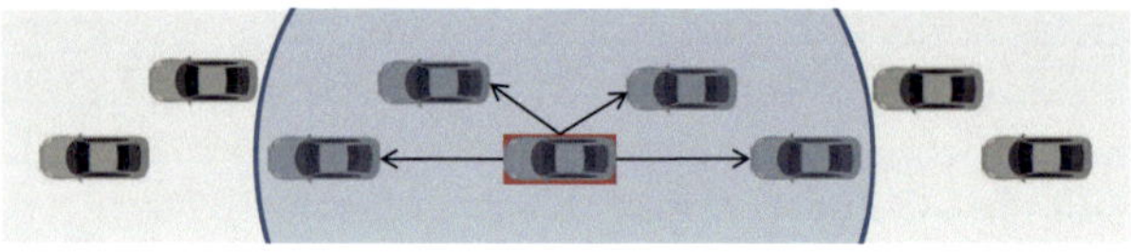

Fig. 2. Example for Beaconing.

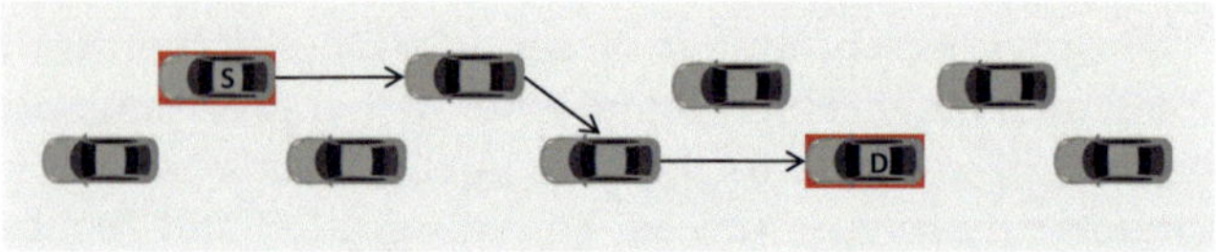

Fig. 3. Example for unicast communication.

velocity. This message is received by all vehicles in communication range and is not further forwarded. The last communication paradigm, unicast communication, is shown in Figure 3. A message is routed hop by hop from a sender (S) to a destination (D).

As already stated, active safety applications are the most important class of envisioned applications and they offer a high potential benefit. Because the majority of these applications rely on broadcast, we can conclude that the broadcast/GeoCast communication paradigm is of eminent importance for a successful application of VANETs. Moreover, broadcast is a basic service used also for route discovery in many reactive unicast protocols like DSR (Johnson & Maltz, 1996), AODV (Perkins & Belding-Royer, 1999), and LAR (Ko & Vaidya, 2000). Therefore, the objective of this chapter is the evaluation of this type of vehicle-to-vehicle communication.

2.3 VANET characteristics

Before going into the details of broadcast protocols, we briefly summarize the VANET characteristics. These characteristics, together with the discussed application classes, reveal some important requirements onto the communication protocols.

High mobility

Vehicles on highways potentially travel at very high speeds. Thus, the communication period between these vehicles can be very short. Moreover, high node velocities cause more frequent topology changes which result in outdated neighbor tables. Thus, when a protocol relies on such a table, the forwarding decision may be incorrect due to old or nonexistent entries in this table.

Dynamic topology

Characteristic for VANETs is a very high dynamic network topology. The reason therefore is twofold: First, the node density ranges from very sparse and partitioned networks e.g. on rural freeways or late night hours to very dense networks at rush hours and traffic jams. Thus, the number of neighbor vehicles in transmission range can vary from zero, up to hundreds of nodes. Second, node mobility can range from static nodes in traffic jams up to very high velocities on free highways. This implies that a routing protocol has to overcome with sparse and partitioned as well as dense scenarios, which are subject to rapid changes over time due to node mobility.

Wireless communication

The dissemination of information in a VANET is based on a wireless medium which represents an error-prone and scarce resource in the network. Especially in dense scenarios where many cars compete for the wireless medium, the limited bandwidth constitutes a severe problem for the routing protocol. Therefore, an efficient broadcast protocol is of eminent importance for a successful deployment of VANET applications.

Delay constraints

As outlined in the previous subsection, most of the safety applications are delay-critical. This means, they rely on broadcast mechanisms which allow the dissemination of information with minimal delay. Thus, a broadcast mechanism designed for such applications has to forward safety critical information immediately, without introducing any delay e.g. for routing purposes.

Geographical addressing

In VANETs a geographical addressing method is used and it fits for most envisioned applications. This means that a broadcast is not performed network wide (which is simply not possible due to the potential size of several million nodes in the network), but is limited by a geographic region (GeoCast). Similarly, unicast protocols make use of position information available via GPS receiver for route decisions.

Mobility patterns

Another important aspect of VANETs is that vehicle movements are constrained by the road topology. This means, node movements obey mobility patterns imposed by the road network. Thus, node mobility is predictable and can therefore be utilized by routing protocols to enhance the dissemination performance. Roughly three main classes of movement patterns can be distinguished, which directly influence the degree of predictability of node movements: inner city roads, rural roads, and highways.

Beaconing

The presence of up-to-date neighborhood information is a prerequisite for many VANET applications (e.g. for cooperative collision warning). This information is exchanged by periodical one-hop MAC-layer broadcast messages, so called beacons. This information can be used by a broadcast protocol to enhance the rebroadcast decision, without introducing additional communication costs.

Pseudonym change

By communicating information, vehicles reveal personal information which results in a sever privacy problem. To solve this problem, vehicles are supposed to communicate using pseudonyms which they change at a given frequency. By changing its pseudonym, a node may be inserted into the neighbor table multiple times under different pseudonyms. Having such incorrect neighbor entries, routing decision cannot be met correctly anymore. Thus, pseudonym changes may heavily affect the underlying protocol if it uses neighborhood information.

Characteristics	Implications
High Mobility	Outdated neighborhood information and short communication periods.
Dynamic Topology	High variance in network density and node velocity: partitioned networks vs. traffic jams.
Wireless Communication	Limited bandwidth and error-prone wireless communication.
Delay Constraints	Messages need to be broadcast immediately, without introducing any delay.
Geographical Addressing	Position information of vehicles are needed and GeoCast is an important communication mechanism for safety applications.
Mobility Patterns	Protocols can benefit from predictable mobility patterns to enhance their routing decisions.
Beaconing	More network load, but protocols can benefit from information exchanged by this basic service.
Pseudonym Change	Pseudonym changes result in incorrect neighbor tables. Thus, pseudonym changes introduce a new challenge to VANET protocols which rely on neighborhood information.

Table 1. VANET characteristics and their implications.

Table 1 summarizes the discussed VANET characteristics together with their implications. Based on these implications, a more exhaustive requirements analysis can be done in the next subsection.

2.4 Requirements analysis

The diversity of VANET applications and the special network characteristics impose several requirements to the broadcast protocols. These requirements are deduced from the previous subsections and summarized in the following.

Scalability

The broadcast protocol has to cope with very dense networks like traffic jams in order to enable correct operation of safety applications in such scenarios.

Effectiveness

The broadcast protocol has to assure that all nodes (or a percentage of nodes, defined by the application) in the destination region receive the disseminated information.

Efficiency

Due to the limited available bandwidth, the broadcast protocol needs to eliminate message redundancy. This is achieved by minimizing the forwarding rate, but still achieving a reception of a message by all nodes in a specific geographic region. This helps to avoid the broadcast storm problem (Ni et al., 1999) and enables the coexistence of multiple VANET applications.

Dissemination delay
Safety applications require the immediate relaying of information, without the introduction of any delay.

Delay-tolerant dissemination
Because vehicular networks are subject to frequent partitioning, it is desirable to cache information in such scenarios and propagate them later when new vehicles are available in the vicinity. Otherwise important information can be lost when the network in the destination region is not fully connected.

Robustness
The communication over the wireless medium is error-prone, nevertheless, the broadcast has to cope with packet losses in order to assure the correct function of vital safety applications.

It has to be noted that not all requirements can be met to a full extent because some requirements are contrary. So, for example, when minimizing the forwarding ratio to achieve a high efficiency, the requirement robustness cannot be fulfilled anymore because relaying nodes represent a single point of failure in this case. Thus, when a relaying node fails to forward a message (which is probable due to the wireless nature of the communication channel) the overall reception rate can also drop significantly. Therefore, in most cases an elaborate tradeoff between such requirements is needed.

3. Review of broadcast protocols for VANETs

As we have seen, due to the diversity of VANET applications and the special network characteristics, the design of an efficient broadcast protocol is a challenging task. The simplest way to implement a broadcast mechanism is the use of naïve flooding. In flooding every node rebroadcasts a message exactly once (given that it is located inside the destination region), thus the message is flooded into the whole region. The downside of simple flooding is that this mechanism is very inefficient. Given the limited bandwidth of the wireless medium, an inefficient information dissemination scheme like naïve flooding leads to redundancy, contention, and collision, to which is referred to as the broadcast storm problem (Ni et al., 1999). To overcome these problems (and the ones identified in the previous sections), many improved broadcast protocols were proposed by the research community.
In this section we first introduce a classification of different broadcast approaches and discuss them. After that, we review novel broadcast mechanisms which were designed to perform well in highly dynamic environments like VANETs. We not only consider VANET protocols here, but also cover protocols for mobile ad-hoc networks. Thus, we present an up-to-date and broad discussion of broadcast protocols from multiple domains.

3.1 Classification of broadcast protocols
One of the first in-depth classification of broadcast approaches was done by Williams and Camp in (Williams & Camp, 2002). They identified four main classes: Simple Flooding, Probability Based Methods, Area Based Methods, and Neighbor Knowledge Methods. More recent works on this topic overtake this thorough analysis and refine it with new properties (cf. e.g. (Heissenbüttel et al., 2006; Khelil, 2007; Yi et al., 2003)). Although such a categorization is useful to evaluate and discuss the properties of broadcast protocols belonging to different classes, it has a significant drawback. Such an exclusive differentiation into rigid classes is not

practicable for many broadcast protocols. We argue that many protocols are not belonging to one fixed class, but combine the properties from different classes. This was also stated by (Slavik & Mahgoub, 2010) and we call them therefore Hybrid Broadcast Protocols.
In the following we give an overview over basic attributes of protocols, which define key characteristics. Knowing such attributes together with their implications, it allows a more thorough analysis of the properties of a protocol. For example, we don't consider area based methods as an attribute class (in contrary to many other classifications in the literature) because it only tells how the rebroadcast decision is calculated (based on the additional coverage), but gives no information about the protocols' properties. To compute the additional coverage, atomic information like position and distance are needed, but it can be also deduced from topology information. If a protocol uses such information, then exact properties can be determined like complexity, weaknesses and strengths. Therefore we consider such information as key attributes which are used in our classification.

Probabilistic
In this scheme, a node rebroadcasts a message with a certain probability. This probability can be fixed a priori (Static Gossip) or adapted dynamically (Adaptive Gossip). In their pure form, probabilistic schemes are very simple and stateless (no need for neighborhood information). They have moderate efficiency but are robust to packet losses due to their probabilistic nature.

Topology based
Topology based protocols use neighborhood information (e.g. 1-hop or 2-hop) to calculate the rebroadcast decision. Such information needs to be exchanged periodically (by so called beacon messages) at a frequency depending on nodes' velocity. This results in higher communication overhead due to the periodical exchange of beacon messages but allows on the other hand very efficient rebroadcast decisions. In dynamic networks this kind of protocols may degrade in performance with increasing node velocity due to outdated neighbor information.

Position/distance based
By using position information, the rebroadcast decision can be calculated more accurately in some cases. E.g. the rebroadcast probability could be adjusted based on the distance to the sender or relays can be selected in a VANET based on their positions.

Local decision
In local decision protocols, a node decides itself on reception to rebroadcast the message or not. This is the contrary of imposed decision and is a desired property of protocols especially in highly dynamic environments like VANETs, because this way rebroadcasts can be decided locally, thus decoupling sender from receiver, which results in a more robust protocol.

Delayed rebroadcast based
This class of protocols introduces a delay before rebroadcasting a message defined by a delay function (randomly or according to some property of the node like distance to the sender). The delayed rebroadcast is useful when nodes overhear the communication channel and gather information about rebroadcasts from other nodes, upon that a more efficient rebroadcast decision can be taken. An example for this type of protocols is the Dynamic Delayed Broadcasting (DDB), introduced by (Heissenbüttel et al., 2006). We consider this

mechanism to improve the broadcast performance as orthogonal to other techniques. Thus, it can be combined with other mechanisms, and therefore, we don't consider them separately in this work.

Clustering, in contrary to other classifications is not considered as a basic attribute of protocols, but is more an aggregation of other properties. A standard clustering scheme utilizes normally topology information to build the clusters and clusterheads utilize the imposed decision scheme to designate the relays. This holds also for more advanced clustering schemes, thus they utilize a combination of the key protocol classes defined above.

3.2 Deterministic broadcast approaches

A subclass of topology based broadcast protocols are the imposed decision protocols, where a sender specifies in the broadcast message which neighbors have to perform a rebroadcast. We refer to this type protocols as deterministic broadcast approaches. Deterministic approaches explicitly select a small subset of neighbors as forwarding nodes which are sufficient to reach the same destinations as all nodes together. Therefore, a relaying node has to know at least its 1-hop neighbors. As finding an optimal subset (i.e. with minimal size) is NP-hard, heuristics are used to find not necessarily optimal but still sufficient relaying nodes.

These type of protocols were one of the first ones suggested by the research community to minimize the broadcast overhead, thus to overcome the broadcast storm problem. Characteristically these protocols achieve a very high efficiency, because based mostly on 2-hop neighborhood information, very accurate rebroadcast decisions can be calculated. Therefore, many variants of deterministic broadcast protocols can be found in the literature. Examples of deterministic approaches are dominant pruning (Lim & Kim, 2000), multipoint relaying (MPR) (Qayyum et al., 2002), total dominant pruning (Lou & Wu, 2002), and many cluster based approaches (see e.g. (Wu & Lou, 2003) and (Mitton & Fleury, 2005)).

Despite the high efficiency they offer, deterministic broadcast has a significant disadvantage: relaying nodes represent a single point of failure. If a relay fails to forward a message (e.g. due to wireless losses, node failure, or not being in transmission range due to mobility) then the overall reception rate of the message may drop significantly. Thus, these kind of protocols lack robustness and perform poorly in dynamic environments like VANETs. Therefore, they can't be used for safety critical applications in VANETs and more robust – but at the same time also efficient – broadcast schemes are needed.

3.3 Probabilistic broadcast approaches

One of the early probabilistic approaches to improve flooding is static gossiping, which uses a globally defined probability to forward messages (Chandra et al., 2001; Haas et al., 2006; Miller et al., 2005). All these variants work best if the network characteristics are static, homogeneous, and known in advance. Otherwise they result in a low delivery ratio or a high number of redundant messages. To overcome these problems, adaptive gossiping schemes have been developed.

Haas et al. (Haas et al., 2006) introduced the so called two-threshold scheme, an improvement for static gossiping based on neighbor count. A node forwards a message with probability $p1$ if it has more than n neighbors. If the number of neighbors of a node drops below this threshold n then messages are forwarded with a higher probability $p2$. The obvious advantage of this improvement is that in regions of the network with sparse connectivity messages are prevented to die out because the forwarding probability is higher than in dense regions.

(Haas et al., 2006) also describes a second improvement which tries to determine if a message is "dying out". Assuming a node has n neighbors and the gossiping probability is p then this node should receive every message about $p \cdot n$ times from its neighbors. If this node receives a message significantly fewer, the node will forward the message unless it has not already done so.

In (Ni et al., 1999), Ni et al. introduced the Counter-Based Scheme. Whenever a node receives a new message, it sets a randomly chosen timeout. During the timeout period a counter is incremented for every duplicate message received. After the timeout has expired, the message is only forwarded if the counter is still below a certain threshold value.

Although all these adaptations improve the broadcast performance, they still face problems in random network topologies. For example, if a node has a very large number of neighbors, this results in a small forwarding probability in all of these schemes. Despite this, there could e.g. still be an isolated neighbor which can only receive the message from this node. An example of such a situation is shown in Figure 4 (example taken from (Kyasanur et al., 2006)).

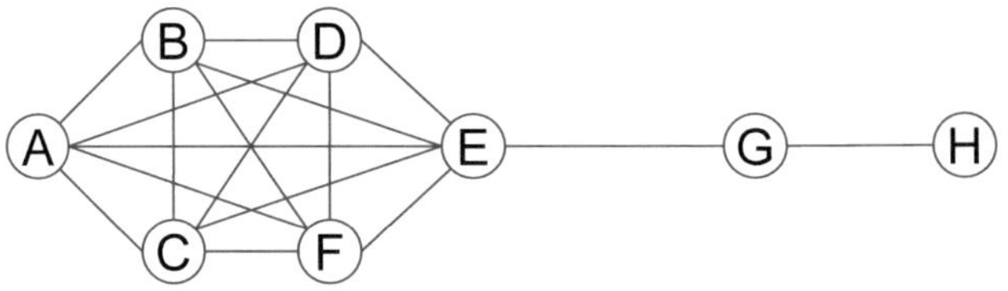

Fig. 4. Sample topology where static gossiping fails

When node *A* sends a message, all nodes in its neighborhood receive it. In this example scenario only node *E* should forward it with the probability of 1 since *E* is the only node that can propagate the message to node *G*. If the gossiping probability is only based on the neighbors count, node *E* will be assigned a low probability since it has many neighbors. So the broadcast message will "die out" with a high probability and never reach *G* and all later nodes. If the part of the network connected only via *G* is very large, the overall delivery ratio will drop dramatically. Such situations can occur quite regularly in dynamic networks of a certain density.

3.4 Hybrid broadcast approaches

As we have seen, deterministic broadcast approaches achieve a very high efficiency but they lack robustness. On the other hand, probabilistic approaches behave much better in the presence of wireless losses and node failures, but have also other limiting disadvantages. E.g. the adaptation of the forwarding probability to actual network condition is a challenging task and is not solved adequately with simple heuristics. Therefore, recently novel probabilistic broadcast approaches were proposed, which combine the strength of both protocol types, becoming this way highly adaptive to the present network conditions. We call this type of protocols hybrid broadcast approaches.

One of the first hybrid broadcast approaches is the so called Smart Gossip protocol, introduced by (Kyasanur et al., 2006). In smart gossip every node in the network uses neighborhood information from overheard messages to build a dependency graph. Based on this dependency graph, efficient forwarding probabilities are calculated at every node. To ensure building up a stable directed graph, the authors make the assumption that there is only one message originator in the whole network. This assumption may be sufficient in a few scenarios, but especially in the case of VANETs this is not applicable, and therefore, as shown in (Bako et al., 2008a; Bako et al., 2007) the performance of the protocol degrades massively in such environments.

To overcome these problems, a novel hybrid probabilistic broadcast was introduced by (Bako et al., 2007). In this so called Position based Gossip (PbG) 1-hop neighborhood information are used together with position information of neighboring vehicles to build a local, directed dependency graph. Based on this dependency graph efficient forwarding probabilities can be calculated which adapts to current network conditions. PbG was designed for message dissemination only into one direction, e.g. for a highway traffic jam scenario, where approaching vehicles have to be informed about the traffic jam. Thus, messages are propagated only against the driving direction. This way only one dependency graph has to be built, and therefore this protocol is denoted as the 1-Table version of PbG.

It is obvious that most VANET applications need to disseminate information in both directions of a road and cannot be restricted only to one direction. For example at an intersection, we face four road segments and therefore a message can be distributed in four directions. Therefore, in (Bako et al., 2008b) a 2-Table version of the protocol was introduced, which fits much better for general highway and intersections scenarios.

Furthermore, in (Bako et al., 2008) two more extension of the PbG protocol was introduced: a network density based probability reduction and a fallback mechanism. The first mechanism reduces the forwarding probability in dense networks, thus reducing the broadcast overhead, at the same time achieving similar reception rates as the original protocol. The second extension aims to prevent message losses: A common problem in wireless networks represents the so called hidden station problem. Because MAC layer broadcast frames are used, techniques like RTS/CTS cannot be used to avoid this problem. Especially in very dense networks the hidden station problem has a significant impact on the performance of the protocol. In such cases, the packet loss rate increases and application level requirements for the delivery ratio cannot be fulfilled any more. To overcome this problem, the second enhancement tries to determine if a message is "dying out". The enhancement works as follows. Each node receiving a new message initializes a counter which is incremented every time it overhears the same message being forwarded by some other node. If the counter is below a certain threshold after a fixed period, the message is rebroadcast with the same probability as if it was received for the first time.

A more general gossip protocol similar to PbG was introduced in (Bako et al., 2008a). In this so called Advanced Adaptive Gossip (AAG) protocol two-hop neighborhood information are used to calculate forwarding probabilities similar to PbG. Thus, no position information are needed, which may be imprecise or even not available in some cases. Moreover, this protocol is not limited to any road topology. Furthermore, this protocol was enhanced by a message loss avoidance mechanism in (Schoch et al., 2010), which is similar to the fallback mechanism from (Bako et al., 2008). With this extension the protocol becomes much more robust and is therefore called robust AAG, or short RAAG. In the mentioned work also beneficial properties of RAAG considering security are discussed and evaluated.

4. Evaluation

In this section we evaluate the performance of selected protocols in different scenarios. Because the simulation of all protocols is very time consuming, we selected one representative protocol for each protocol type discussed in Section 3 and evaluate the impact of mobility, node density, and high broadcast traffic on these schemes. Therefore, we first introduce the simulation parameters and describe the two evaluated scenarios: city and highway. After that, we show that deterministic broadcast schemes are heavily affected by node mobility, thus they are inapplicable for VANETs. The remaining subsections present the results of the selected hybrid broadcast schemes in a highway and city scenario. For comparison we include also the results of naïve flooding and static gossiping. Results of the following protocols are presented:

- Multipoint Relaying (Qayyum et al., 2002)
- Flooding
- Static Gossiping (Chandra et al., 2001; Haas et al., 2006)
- Advanced Adaptive Gossiping (AAG) (Bako et al., 2008a)
- Robust Advanced Adaptive Gossiping (RAAG) (Schoch et al., 2010)

4.1 Simulation setup

For the evaluation of the broadcast protocols we use the JiST/SWANS (Barr et al., 2005) network simulator, including own extensions. JiST/SWANS provides a radio and MAC-layer according to IEEE 802.11b. This is close to the IEEE 802.11p variant, which is planned for vehicular communication. On the physical layer the two-ray ground model is used together with the additive noise model, thus, the effect of packet collisions can be investigated. The radio transmission power is set to achieve a wireless transmission range of 280 meters. For the city scenario a field size of 1000m x 1000m is used, whereas the simulations for the highway scenario are run on a 25m x 3000m field. Node density is varied from 10 up to 300 nodes, thus comparing sparse as well as dense scenarios.

Parameter	Value
Field	City: 1000m x 1000m, Highway: 3000m x 25m
Simulation Duration	120s
Broadcast Start	5s
Pathloss	Tworay
Noise Model	Additive
Transmission Range	280m
Beaconing Interval	1s
Number Messages	3 Messages per node, max 150
MIA Acknowledges	1
MIA Replay Delay	2.5s
MIA Last Replay Offset	100s
Placement	Random
Static	Node Speed: 0
Random Waypoint	Node Speed City: 3 – 20 m/s, Highway: 22 – 41 m/s
Highway Mobility	Node Speed Highway: 0 – 30 m/s

Table 2. Simulation setup parameters.

The number of broadcast messages depends on the node density: Every node generates one broadcast message per second (with a minimal payload), limited by a maximum count of three messages per node. The absolute number of broadcast messages is limited by 150. Thus, in a scenario with 10 nodes 30 messages are initiated, whereas in scenarios with 50 or more nodes 150 messages are created (if not otherwise specified). This way we evaluate the protocols under low as well as under heavy network load. To hold the neighbor tables up-to-date beacons are used which are exchanged with a rate of 1 beacon per second. The beacon size depends on the information required by the broadcast protocol. Thus with AAG and MPR the entire neighbor list is sent in a beacon, whereas in Flooding only a message with minimal size is sent (we assume this is required by the VANET applications).

A setup is simulated over 120s, where the broadcast of messages starts at 5 seconds. For the RAAG protocol, the message loss avoidance (MlA) mechanism is configured to await at least one acknowledge for a sent message, otherwise the message is rebroadcast again once (if new nodes are present in the neighborhood), with a delay of 2.5 seconds. Messages have a timeout of 100s and if a message was not yet acknowledged at least once, the message is rebroadcast one more time.

To evaluate the impact of node mobility on the performance of the broadcast protocols we use three different mobility models:

- Static
- Random Waypoint (RW)
- Highway Mobility (HM)

The static model is used to measure the protocols' performance in a best-case scenario, i.e., nodes didn't move at all, thus all neighborhood information are up-to-date. With the Random Waypoint mobility model a worst-case scenario is investigated where nodes move in arbitrary directions. A more realistic scenario is provided by the Highway Mobility model, which is an own extension inside the JiST/SWANS framework. With this mobility model cars move in the same direction on a 4-lanes highway with random speeds. They hold a safety distance to other cars, change lanes and pass slower cars if necessary. At the end of the simulated highway the lanes are blocked by 4 cars, thus traffic congestion is simulated here. The exact parameters used for our simulations can be found in Table 2.

According to (Bani Yassein & Papanastasiou, 2005), the optimal fixed probability for static gossip is 0.7. Therefore, we use this value for the static gossip protocol in our evaluations. For each simulation setup 20 simulation runs are done and the results averaged.

4.2 Effect of node mobility on deterministic broadcast

Multipoint Relaying (MPR) was selected as a representative for deterministic protocols to evaluate the impact of node mobility onto this protocol type. Therefore, a highway scenario with three different mobility models is used: static, random waypoint, and highway mobility. Because MPR lacks robustness, and therefore the number of broadcast messages heavily influences the performance of the protocol, we also simulated a scenario where only one broadcast message is initiated (Static 1). The other three simulation configurations (Static 2, RW, and HM) use the normal parameters described in 4.1.

Figure 5 shows the results of this evaluation. As we can see, in sparse networks (10 and 25 nodes) the reception rates in all four simulation setups are very low. These results are as expected, because the network is partitioned and therefore not all nodes can be reached by a broadcast without additional mechanisms. With higher node densities and only one broadcast message per simulation (Static 1), MPR achieves quite good reception rates. With 100 and 150 the reception rate is almost 100% and drops slightly with increasing nodes, but stays over 90% which is an acceptable ratio. This slightly decline is due to the higher overhead introduced by the beacon messages.

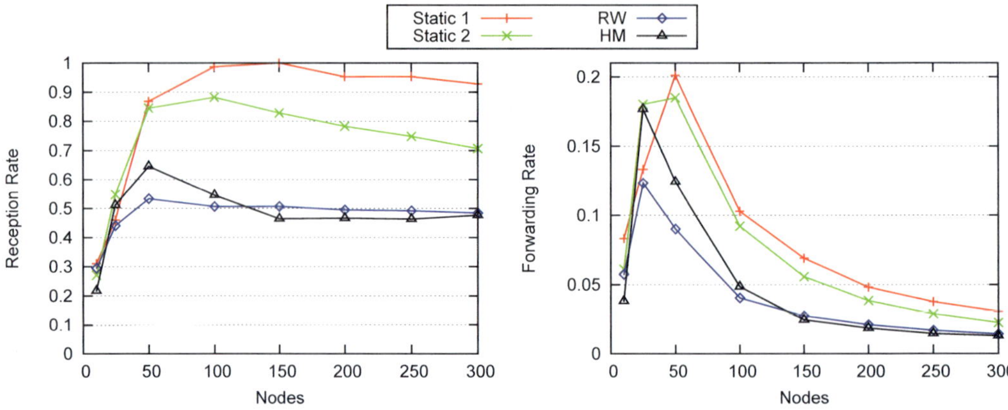

Fig. 5. Performance of MPR in a highway scenario with different mobility models and number of messages.

On the other hand, with a high number of broadcast messages (Static 2), the reception rate drops significantly in higher node densities. With 300 nodes MPR achieves only around 70% reception rate with is clearly unacceptable for safety critical VANET applications. Thus, these results show that heavy network load has a significant influence onto deterministic protocols. Now considering mobility, we can see that with the random waypoint and highway mobility model the reception rate drops even more drastically. With both mobility models in almost all node densities the reception rates are around 50%. Thus, deterministic approaches are inapplicable for dynamic environments like VANETs.

Regarding the forwarding rates, we can see that MPR is highly efficient, needing only around 3% or less rebroadcasts with 300 nodes. Thus, we can conclude that deterministic broadcast approaches are highly efficient but can't meet VANET requirements in the presence of mobility and high network load.

4.3 Hybrid broadcast approaches in a highway scenario

In this subsection we evaluate two hybrid broadcast protocols (AAG and RAAG) in a highway scenario and compare the results with flooding and static gossip (SG). Figure 6 shows the results for this scenario with static nodes. As we can see, in a partitioned network like with 10 nodes in these results, the reception rates of all four protocols are almost identical. Whereas with 25 nodes (here the network is also not completely connected), static gossip already has a significant lower reception rate of around 10%. This gap is even bigger with 50 nodes, where static gossip has a reception rate of around 57% compared with 83% of RAAG. This is because the static gossip probability of 70%, which is too low for sparse networks.

With higher densities, AAG significantly drops regarding the reception rate, reaching not even 70% of other vehicles for the 300 node setup. Here static gossip and flooding achieve better reception rates, both protocols are slightly under 90%. However, RAAG clearly outperforms the other protocols, reaching almost 100% reception rates.

Regarding the forwarding rates, we can see that flooding has the highest forwarding rates except for the scenario with 10 nodes. Here the message loss avoidance mechanism of RAAG generates more overhead, but has not much impact onto the reception rate because the nodes are static. The rebroadcast rate of flooding is way too high in higher densities, and that is a serious problem causing the so called broadcast storm. We will discuss this effect later in a

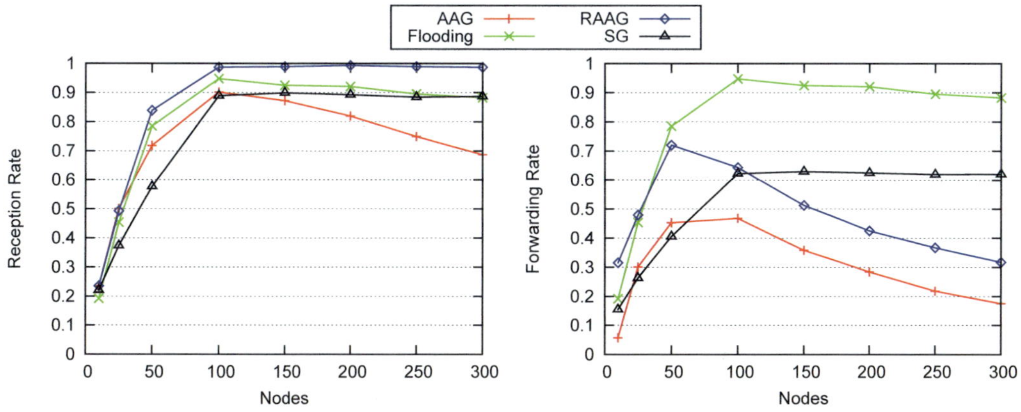

Fig. 6. Performance of hybrid broadcast approaches in a static highway scenario.

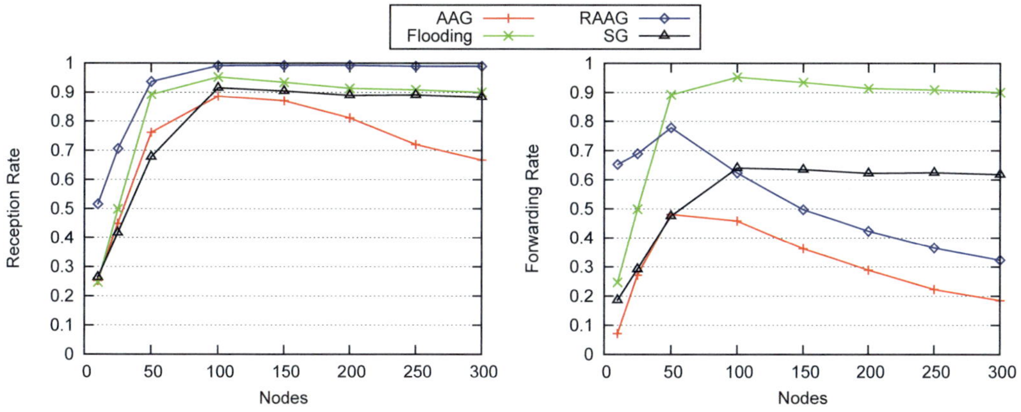

Fig. 7. Performance of hybrid broadcast approaches in a highway scenario using the random waypoint mobility model.

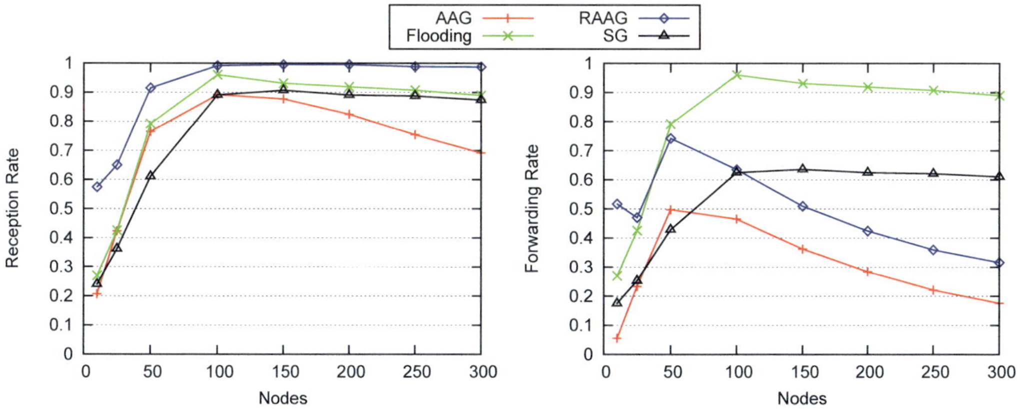

Fig. 8. Performance of hybrid broadcast approaches in a highway scenario using the highway mobility model.

scenario with higher network load. AAG achieves the best forwarding rate, but as we saw, the performance is insufficient for this scenario. Static gossip has a lower forwarding rate as RAAG with few nodes, but remains constant slightly about 60% with higher node densities. Thus, static gossip doesn't scale well with increasing node density. On the other hand, the forwarding rate of RAAG decreases constantly with increasing density and is constantly around 10% higher as AAG due to the message loss avoidance mechanism.
Figure 7 and 8 show the same scenario with random waypoint and highway mobility models. As we can see, there is almost no difference in the reception and forwarding rates compared with the static scenario. This means, that all these protocols are not affected at all by node mobility. This is a very important property which makes these protocols well suited for VANETs. The only difference compared with the static scenario is the reception and forwarding rates of the RAAG protocol in low densities. Due to node mobility, the cached messages are here physically transported and rebroadcast later. Thus, RAAG manages to overcome network partitions and achieves a much higher (at a cost of more rebroadcasts) reception rate.

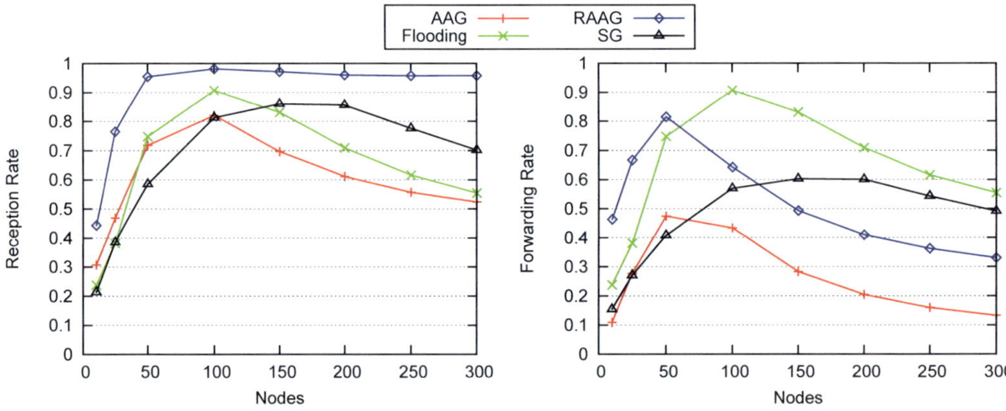

Fig. 9. Performance of hybrid broadcast approaches in a highway scenario under high message load using the highway mobility model.

In the next simulation setup we evaluate the performance of these protocols under high network load. Therefore, we increased the payload of broadcast messages to 512 bytes and raised the limit of the absolute number of messages to 300. This means, every node creates exactly 3 messages, with a rate of one message per second. The results for this simulation setup are shown in Figure 9. As we can see, AAG and flooding can't cope with increasing network load, thus the reception rate is dropping significantly, reaching almost only 50% of the nodes in the 300 node setup. The reception ratio of static gossip also declines constantly with increasing node densities. Thus, these protocols are not scalable and can't be used for VANET applications in such scenarios. Only RAAG manages to reach good reception ratios in the tested setup, and as can be seen, it clearly outperforms the other protocols. Thus we can conclude, that RAAG allows an efficient and effective dissemination also in scenarios with extreme high network load. The forwarding rates can be compared with the other results. AAG, flooding, and static gossip have lower forwarding ratios due to the packet losses.

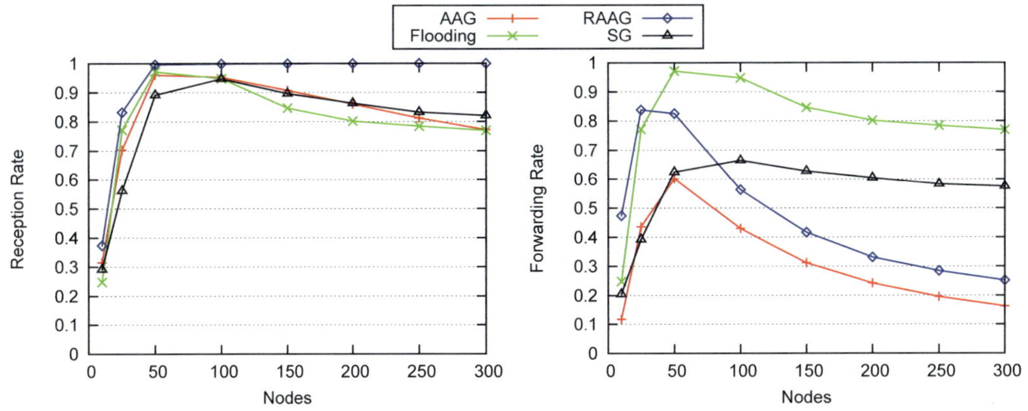

Fig. 10. Performance of hybrid broadcast approaches in a static city scenario.

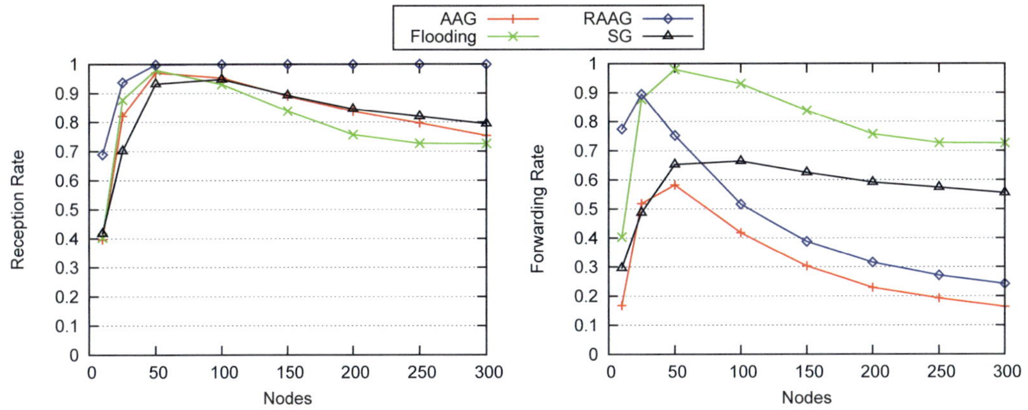

Fig. 11. Performance of hybrid broadcast approaches in a city scenario using the random waypoint mobility model.

4.4 Hybrid broadcast approaches in a city scenario

For the city scenario we simulate a field of 1000m x 1000m with static and random waypoint mobility. Figure 10 shows the results for the static scenario. As we can see, the results are similar to the static highway scenario. RAAG achieves the best reception rates for all node densities, reaching almost 100% with 50 and more nodes. The reception rates of the other protocols drop constantly with increasing nodes, and reach only around 80% with 300 nodes. This is clearly not sufficient for critical safety applications in VANETs. The forwarding rates are also similar to the previous scenario: flooding and static gossip have very high forwarding rates and these rates don't scale well in contrary to RAAG and AAG.

Considering the mobile city scenario shown in Figure 11, we can here also conclude that mobility has almost no effect on these protocols. Except for the RAAG protocol, where the message loss avoidance mechanism positively benefits from nodes' movements. In highly partitioned networks, like with 10 nodes in this figure, RAAG manages to achieve a reception rate of around 30% higher than the other protocols, or RAAG itself in a static scenario. This is a significant gain and these results underline the need of a message loss avoidance mechanism for partitioned networks.

5. Summary and outlook

In this chapter we gave an overview over possible VANET applications and showed different communication paradigms used for such applications. We also pointed out the importance of broadcast mechanisms for active safety applications. This was followed by an overview of the special network characteristics of VANETs. From that, we deduced a set of requirements for broadcast protocols which have to be fulfilled for a successful deployment of VANET applications.

Also a classification of broadcast protocols was introduced which enables a more systematic analysis of broadcast mechanisms. Based on this, we have reviewed state-of-the-art broadcast protocols designed for inter vehicle communication. The main focus here was on hybrid protocols, which combine positive properties of more protocol classes and offer thereby promising characteristics for broadcast applications in vehicular networks.

The theoretical evaluations were confirmed by extensive simulations. We have shown that deterministic protocols are heavily affected by node mobility and network load, and they are therefore not suitable for VANET applications. Furthermore, we have shown that pure flooding, as well as static gossip, is not scalable, i.e. they cause the so called broadcast storm problem. Thus, with increasing node density and network load their performance drop significantly and they are therefore unfeasible for VANETs.

On the other hand, the RAAG achieves very promising results in sparse as well as in dense networks. We have shown that the message loss avoidance mechanism yields a significant performance gain in sparse scenarios and increases the robustness of the protocol also in dense networks. Moreover, RAAG is not affected by node mobility which is a very desirable property of VANET protocols. Thus we can conclude that RAAG is predestinated for dynamic networks like VANETs and satisfies the requirements in such networks also in the presence of critical safety applications.

Although the presented results are very promising, there are some issues we want to address in future work. First of all, RAAG requires 2-hop neighborhood information which generates more overhead. We aim to reduce this required knowledge to 1-hop neighbors, similar to the PbG protocol but in a more general way. Moreover, we have to evaluate the performance of RAAG in the presence of pseudonym changes, which may have a significant effect on broadcast protocols. Also a detailed evaluation of the message loss avoidance mechanism in partitioned networks and its optimization could result in a significant gain in delay-tolerant networking.

6. References

Bai, F., Krishnan, H., Sadekar, V., Holl, G. & Elbatt, T. (2006). Towards characterizing and classifying communication-based automotive applications from a wireless networking perspective, *In Proceedings of IEEE Workshop on Automotive Networking and Applications (AutoNet)*, San Francisco, USA.

Bako, B., Kargl, F., Schoch, E. & Weber, M. (2008a). Advanced Adaptive Gossiping Using 2-Hop Neighborhood Information, *IEEE Globecom 2008 Wireless Networking Symposium (GC'08 WN)*, New Orleans, USA.

Bako, B., Kargl, F., Schoch, E. & Weber, M. (2008b). Evaluation of Position Based Gossiping for VANETs in an Intersection Scenario, *4th International Conference on Networked Computing and Advanced Information*, Gyeongju, Korea.

Bako, B., Rikanovic, I., Kargl, F. & Schoch, E. (2007). Adaptive Topology Based Gossiping in VANETs Using Position Information, *3rd International Conference on Mobile Ad-hoc and Sensor Networks (MSN 2007)*, Beijing, China.

Bako, B., Schoch, E., Kargl, F. & Weber, M. (2008). Optimized Position Based Gossiping in VANETs, *2nd IEEE International Symposium on Wireless Vehicular Communications*, Calgary, Canada.

Bani Yassein, Ould Khaoua, M. M. & Papanastasiou, S. (2005). Improving the performance of probabilistic flooding in manets, *Proceedings of International Workshop on Wireless Ad-hoc Networks (IWWAN-2005)*, London UK.

Barr, R., Haas, Z. J. & van Renesse, R. (2005). JiST: an efficient approach to simulation using virtual machines: Research Articles, *Softw. Pract. Exper.* 35(6): 539–576.

Chandra, R., Ramasubramanian, V. & Birman, K. (2001). Anonymous Gossip: Improving Multicast Reliability in Mobile Ad-Hoc Networks, *Technical report*, Ithaca, NY, USA.

Dietzel, S., Bako, B., Schoch, E. & Kargl, F. (2009). A fuzzy logic based approach for structure-free aggregation in vehicular ad-hoc networks, *VANET '09: Proceedings of the sixth ACM international workshop on VehiculAr InterNETworking*, ACM, New York, NY, USA, pp. 79–88.

Dietzel, S., Schoch, E., Bako, B. & Kargl, F. (2009). A structure-free aggregation framework for vehicular ad hoc networks, *6th International Workshop on Intelligent Transportation (WIT 2009)*, Hamburg, Germany.
URL: *http://www.kargl.net/docs/mypapers/2009-03-wit.pdf*

EURF (2009). European road statistics 2009, *Technical report*, The European Union Road Federation.
URL: *http://www.irfnet.eu*

Haas, Z. J., Halpern, J. Y. & Li, L. (2006). Gossip-based ad hoc routing, *IEEE/ACM Trans. Netw.* 14(3): 479–491.

Heissenbüttel, M., Braun, T., Wälchli, M. & Bernoulli, T. (2006). Optimized Stateless Broadcasting in Wireless Multi-hop Networks, *Proceedings of the 25th Conference on Computer Communications (IEEE Infocom 2006)*, Barcelona, Spain.

Johnson, D. B. & Maltz, D. A. (1996). Dynamic source routing in ad hoc wireless networks, *Mobile Computing*, Kluwer Academic Publishers, pp. 153–181.

Khelil, A. (2007). *A Generalized Broadcasting Technique for Mobile Ad Hoc Networks*, PhD thesis, Universität Stuttgart.

Ko, Y.-B. & Vaidya, N. H. (2000). Location-aided routing (lar) in mobile ad hoc networks, *Wirel. Netw.* 6(4): 307–321.

Kyasanur, P., Choudhury, R. R. & Gupta, I. (2006). Smart Gossip: An Adaptive Gossip-based Broadcasting Service for Sensor Networks, pp. 91–100.

Lim, H. & Kim, C. (2000). Multicast tree construction and flooding in wireless ad hoc networks, *MSWIM '00: Proceedings of the 3rd ACM international workshop on Modeling, analysis and simulation of wireless and mobile systems*, ACM, New York, NY, USA, pp. 61–68.

Lou, W. & Wu, J. (2002). On reducing broadcast redundancy in ad hoc wireless networks, *IEEE Transactions on Mobile Computing* 1(2): 111–123.

Miller, M. J., Sengul, C. & Gupta, I. (2005). Exploring the Energy-Latency Trade-Off for Broadcasts in Energy-Saving Sensor Networks, *icdcs* pp. 17–26.

Mitton, N. & Fleury, E. (2005). Efficient broadcasting in self-organizing multi-hop wireless networks, *ADHOC-NOW*, pp. 192–206.

Ni, S.-Y., Tseng, Y.-C., Chen, Y.-S. & Sheu, J.-P. (1999). The Broadcast Storm Problem in a Mobile ad hoc Network., *Proceedings of the 5th annual ACM/IEEE international conference on Mobile computing and networking, MobiCom '99*, ACM, New York, USA, pp. 151–162.

Perkins, C. E. & Belding-Royer, E. M. (1999). Ad-hoc On-Demand Distance Vector Routing, *2nd Workshop on Mobile Computing Systems and Applications (WMCSA '99)*, New Orleans, USA, IEEE Computer Society, pp. 90–100.

Qayyum, A., Viennot, L. & Laouiti, A. (2002). Multipoint Relaying for Flooding Broadcast Messages in Mobile Wireless Networks, p. 298.

Schoch, E., Bako, B., Dietzel, S. & Kargl, F. (2010). Dependable and secure geocast in vehicular networks, *VANET '10: Proceedings of the seventh ACM international workshop on VehiculAr InterNETworking*, ACM, New York, NY, USA, pp. 61–68.

Schoch, E., Kargl, F., Leinmüller, T. & Weber, M. (2008). Communication Patterns in VANETs, *IEEE Communications Magazine* 46(11): 2–8.

Slavik, M. & Mahgoub, I. (2010). Stochastic broadcast for vanet, *CCNC'10: Proceedings of the 7th IEEE conference on Consumer communications and networking conference*, IEEE Press, Piscataway, NJ, USA, pp. 205–209.

VSCP (2005). Task 3 final report: Identify intelligent vehicle safety applications enabled by dsrc, *Technical report*, Vehicle Safety Communications Project, U.S. Department of Transportation.

Williams, B. & Camp, T. (2002). Comparison of Broadcasting Techniques for Mobile Ad Hoc Networks, MobiHoc '02: Proceedings of the 3rd ACM international symposium on Mobile ad hoc networking & computing, ACM Press, New York, USA, pp. 194–205.

Wu, J. & Lou, W. (2003). Forward-node-set-based broadcast in clustered mobile ad hoc networks, *Wireless Communication and Mobile Computing* 3: 155–173.

Yi, Y., Gerla, M. & Kwon, T. J. (2003). Efficient flooding in ad hoc networks: a comparative performance study, *Proc. IEEE International Conference on Communications ICC '03*, Vol. 2, pp. 1059–1063.

4

Reference Measurement Platforms for Localisation in Ground Transportation

Uwe Becker
University Braunschweig, Institute for Traffic Safety and Automation Engineering
Germany

1. Introduction

Although the importance of satellite based localisation has been rising in the last years to a significant level in safety related applications like transportation systems, it seems obvious that saturation of sales volume for those systems has not been reached yet.
Due to the development and set up of the European satellite based localisation system GALILEO, new services and accuracy in localization open new market sectors. GALILEO will provide five different services with different performances and characteristics that will be suitable for different ranges of applications. These services will be: Open Services, Safety of Life Services, Commercial Services, Public Regulated Service and Rescue Service. The Safety of Life Service is the key service for most safety-related applications due to its guaranteed characteristics integrity, availability and accuracy.
A new important aspect for the usage of these services is their formal approval by a safety authority. Regarding this approval some intensive analysis has to be undertaken regarded from the angle of safety and reliability.
Considering this analysis, following leading points are of eminent importance

- possibilities for the (practical) analysis of the reliability and availability of the localisation information provided by a receiver
- set up of a regulatory framework in combination with specific procedures for the evaluation according to the requirements of the safety authorities.

For the field of transportation it is important to know, that there exist multiple safety authorities: (nearly) each mode of transportation (road, rail, water, air) has at least one of them. In the past it was not necessary to set up a domain spanning approval for components because it was not of economic interest. In the future this interest will emerge from new technological possibilities. According to this, the present situation may change with the upcoming GALILEO system because of its new functionalities and accuracy: at least the key component of all applications will have to be used in all modes of transportation - this will be a receiver for the satellite signals.
The two aspects mentioned above will be discussed. In the section "Reference Localization Platforms", the systems for the analysis for the availability and accuracy of the satellite signals by means of a special reference localization platform will be discussed. Figure 1 shows the generally concept. The first results on the topic of the fusion/structural comparison of regulatory frameworks will be subject of the second section "Regulatory Frameworks".

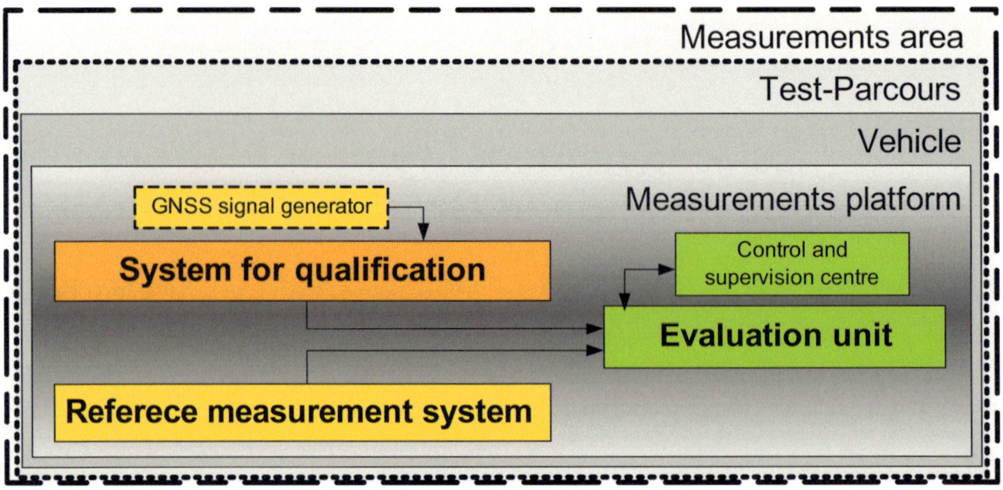

Fig. 1. Generally concept for the measurements area

2. Reference localisation platforms

Satellite-based positioning systems are used in many applications for air, maritime and land transport. In either case, at least four distinct satellites are used to obtain a four dimensional position, consisting of three coordinates in space and one in time (fourth coordinate). This position can be (almost) anywhere on the surface of the earth or in the airspace above it (Grewal et.al., 2001).

When used on the surface of the earth, the reception of the necessary amount of satellites (namely four) can be difficult due to environmental barriers (e.g. buildings, trees, etc.) in close range of the object to be localised. This problem arises, because quasi-optical wave propagation occurs in the frequency range used for satellite positioning. If an object obstructs the necessary direct line of sight to the satellite, no signal can be received (Hänsel et al., 2005). This fact reduces the availability of satellite based positioning in places shadowed by other objects, which cannot be avoided in railway environment.

For the usage in railway systems, a high reliability is necessary for the use in safety related functions (e.g. the train control system). On the other hand, the implementation of the reference positioning system is simplified by the special domain inherent constraint that the vehicle cannot leave the track.

For the experimental evaluation of the availability and accuracy, two generic reference localisation platforms have been set up (Becker et.al., 2008a). Because of the focus to ground based transportation (i.e. road and rail transportation) this section is confined to two generic platforms "CarLa" and "CarRail" which were developed at the Institute for Traffic Safety and Automation Engineering (iVA)

2.1 Road based traffic

CarLa (Car Laboratory) is a test bed for several sensors and control algorithms. The basic hardware setup of CarLa is depicted in Figure 2.

The sensors used in the platform can be classified into the ones for the GNSS localization system and the ones for environmental perception which are used in the control algorithms.

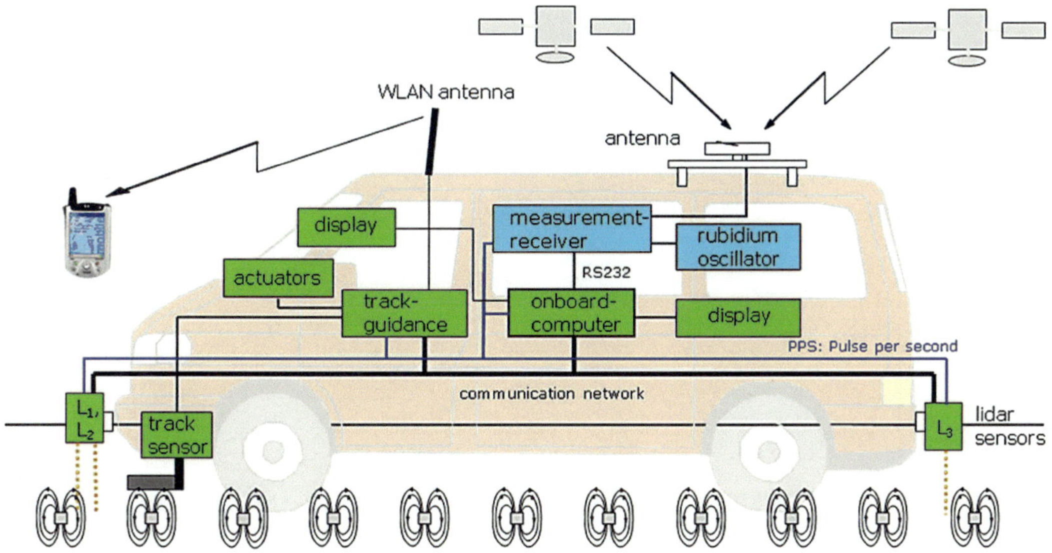

Fig. 2. Measurement Platform for Road Application

The GNSS localization system consists of a GNSS receiver which is coupled with a rubidium oscillator as an extremely precise clock.

The environmental perception is represented by a special tracking concept and additional sensors for the measurement of the vehicle's orientation.

The central component of the system is a small but powerful computer, running the Linux operating system. All the sensors are connected to this by appropriate hardware interfaces (e.g. CAN-bus, Ethernet, serial interface).

The computer offers two displays for different purposes. One is placed near the driver's seat (where usually the navigation system is located) and the other one is placed in such a way that an operator, sitting on the back seat, can see it. The first one is a touch screen, so that input can easily be made by the driver (if it is necessary for the application); the second display is equipped with a standard keyboard and mouse.

Because of the use of a classical operation system and hardware that is near to a standard desktop PC, the development of the software is quite easy and fast.

The tracking concept is based on a reference track which is installed into the ground via equally spaced magnets. These sensor signals are used to control the vehicle's dynamics in lateral and longitudinal direction.

The intention of the CarLa reference platform is to guarantee a position and velocity control of high accuracy to the reference track. The attained accuracy of the lateral control is +/- 1 cm whereas the one of the longitudinal control is +/- 0.1 m/s inside the range from 0 up to 100 kph.

Therefore, ensuring high accurate positions of the track magnets, the GNSS measured position can be compared to the reference position, which is estimated via the recognition of the track magnets and the vehicle's complete state vector.

Figure 3 shows an example of a measurement run. Shown is the error of measurement of the GNSS antenna position from the calculated reference antenna position as lateral deviation e_y in metre.

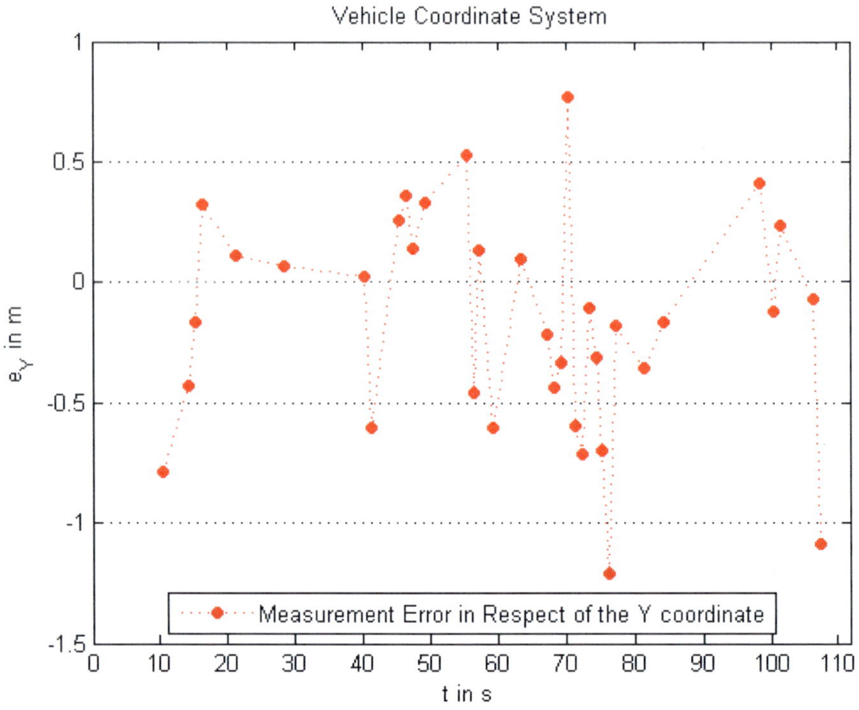

Fig. 3. Error measurement of lateral deviation in driving direction

2.2 Guided traffic

For the evaluation of satellite based applications in rail transport a generic reference measurement platform has been set up as well, CarRail (Figure 4).

The platform uses two different sensor systems together with an precise map of the track.

The first sensor used is a Doppler radar sensor for a continuous measuring of the relative position along the track. This sensor has the disadvantage of a relatively high drift (0.2%) that has to be stabilised.

The RFID-based absolute positioning is used as second sensor to stabilise the drift of the radar sensor. For the RFID system, transponders are located along the track, which represent absolute marks. The positions of the transponders (or at least some) are known with high accuracy (about 0,5 cm) and they are unambiguously identifiable. Their positions are stored in an XML based electronic track map to obtain information about the matching between the topology of the track and the (geographic) positions of the transponders.

The first test track has been equipped with this reference measurement system. It is situated near the Braunschweig main railway station. This track is 3 km long and includes 8 switches, straight sections as well as curved sections. This topographical constellation is suitable for tests of track selectivity. The topographically constellation with a bridge and a big manufacturer building near the track provides areas with critical reception of satellite signals. Hence, it is well suited to test the accuracy and availability for safety related applications in rail transportation.

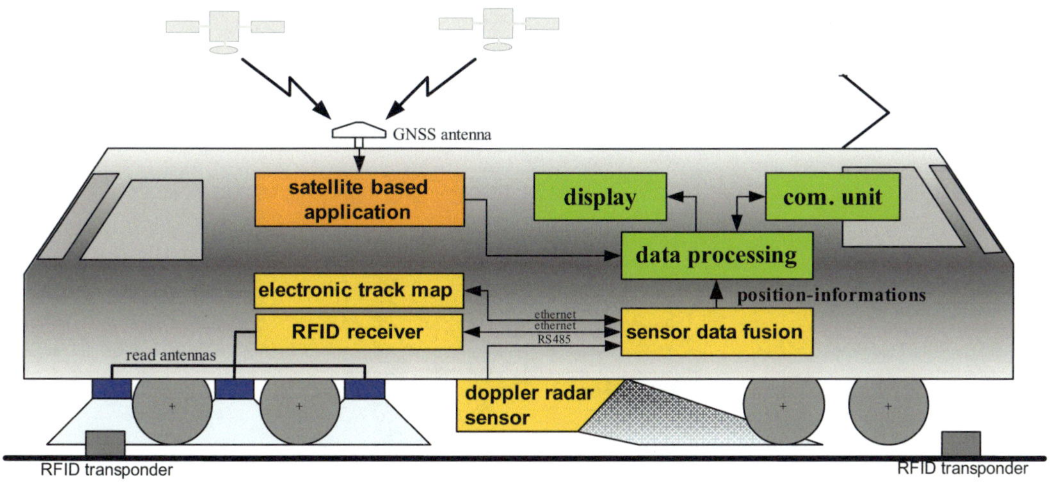

Fig. 4. Measurement Platform for Rail Application

Currently, the accuracy of the reference measurements platform is 50 cm. With a modified arrangement of the transponders, i.e. a modified distance between two transponders, it will be possible to achieve an accuracy of about 15 cm at maximum speeds of up to 200 kph. By the combination of these two sensors (together with the map), a continuous positioning with a very accurate absolute position information is set up.

3. Regulatory framework

The spectrum of possible applications for the vehicle and the rail reference platform is quite wide. We are focusing on questions on the accuracy and the availability of positioning systems for safety related applications in road traffic. In road transport these will become important in the field of advanced driver or traffic assistance systems that have the possibility to intervene in traffic maneuvers or even to outvote the driver. Hence a certification will be obligatory for the systems (Becker et.al., 2008b). The reference platforms can be used for the investigation of the positioning systems.

For the transportation domain(s), the field of systems and applications regarding safety responsibility will use the GALILEO based localization. A complete set of regulations exist for the certification of safety related applications due to the European focus for safety improvement in transportation. For the upcoming satellite based systems, adequate regulations have to be set up as well. The existing ones have to be kept or adopted to keep them interoperable with existing systems and for the purpose of migration.

Instead of a simple approach by setting up GALILEO concerning requirements for each domain of transportation, the authors propose an intermodal (or domain spanning) approach, so that those requirements occurring in all domains may be separated and taken as a "generic kernel" of requirements to be the basis of a first step of a certification system, containing two stages. The first stage will concern the domain independent certification as stated above. The second one is domain dependant and contains those special requirements which are specific for each domain.

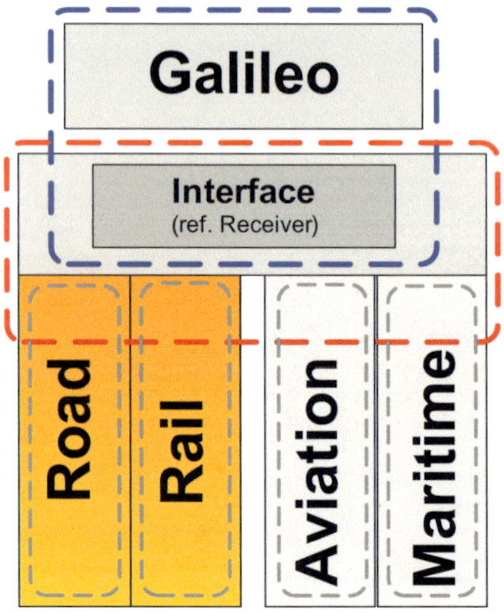

Fig. 5. Galileo and its Application Domains

For the transportation domain(s), the field of systems and applications regarding safety responsibility will use the GALILEO based localization. A complete set of regulations exist for the certification of safety related applications due to the European focus for safety improvement in transportation. For the upcoming satellite based systems, adequate regulations have to be set up as well. The existing ones have to be kept or adopted to keep them interoperable with existing systems and for the purpose of migration.

Instead of a simple approach by setting up GALILEO concerning requirements for each domain of transportation, the authors propose an intermodal (or domain spanning) approach, so that those requirements occurring in all domains may be separated and taken as a "generic kernel" of requirements to be the basis of a first step of a certification system, containing two stages. The first stage will concern the domain independent certification as stated above. The second one is domain dependant and contains those special requirements which are specific for each domain.

This splitting into two stages avoids the tests of the first stage to be done multiple times when certifying equipment for more than one domain. This opens the opportunity to suppliers to broaden their market by having the stage one done and being able to undergo domain specific tests for lower costs.

Within the approach, formal techniques from the field of internet technology (semantic web resp. ontology) are used to develop models of the different domain languages and the processes required for the certification. This allows the building of relationships between the (same) processes and different terminologies from the different domains. By using this relationship, experts from different domains can view the processes under the "eyeglasses" for their respective domain, which helps them to understand the processes. As a result, the communication between the experts from different domains can be improved and discussions can be freed from terminological discussions (due to misunderstandings) and accelerated to bring results. Another advantage is the increase in confidence in the certification from different domains, because the processes can be seen and compared with those still understood by the experts (DIN 1319-1).

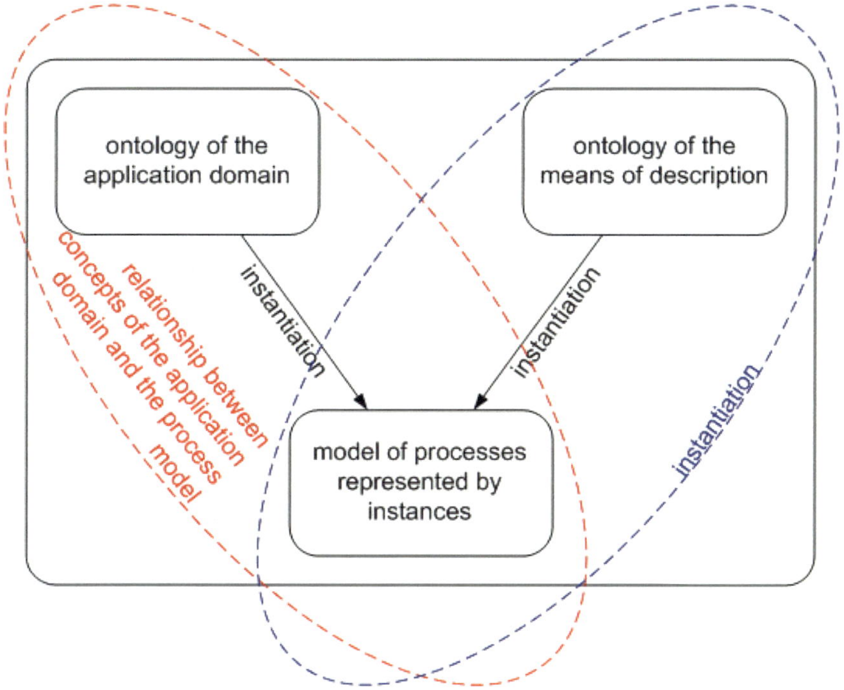

Fig. 6. Methodological approach for the integration of different domain specific terminologies with process descriptions

This methodological approach is implemented in new projects funded by the European Commission, European space agency and German federal ministry of research and education. The promising framework for certification processes shown here is still under development and the paper will show the current status of the project and it will be show on one GNSS based application for low density secondary lines.

Today's standards and regulations are formulated as texts in natural language. This is adequate for documents being used in one domain (having its own domain specific language e.g. terminology) and one country (having its own natural language). Different domains often have languages (terminologies) that are incompatible with each other because they use identical terms for different concepts. This leads to misunderstanding between incorporated persons from different domains and makes a joined work very hard.

To overcome this problem, the terminologies used have to be described in a way known by all incorporated persons to be understood easily. If this common language is a formal one, formal methods and techniques can be applied to check the results for consistency and for correctness. Different possibilities exist for formal descriptions. In Figure 7, a petrinet is shown, that describes several processes defined in the standard IEC 17000 for the conformity assessment (DIN EN ISO/IEC 17000).

For the approach to be used in certification, the terminologies used as well as the processes to be applied have to be described formally. After having identified this, new standards or advancements can be specified, by using "inheritance" like mechanisms (as speaking in terms of object oriented methods). By that formal description, the concepts and structures contained by the documents are formulated in an explicit and unambiguously way.

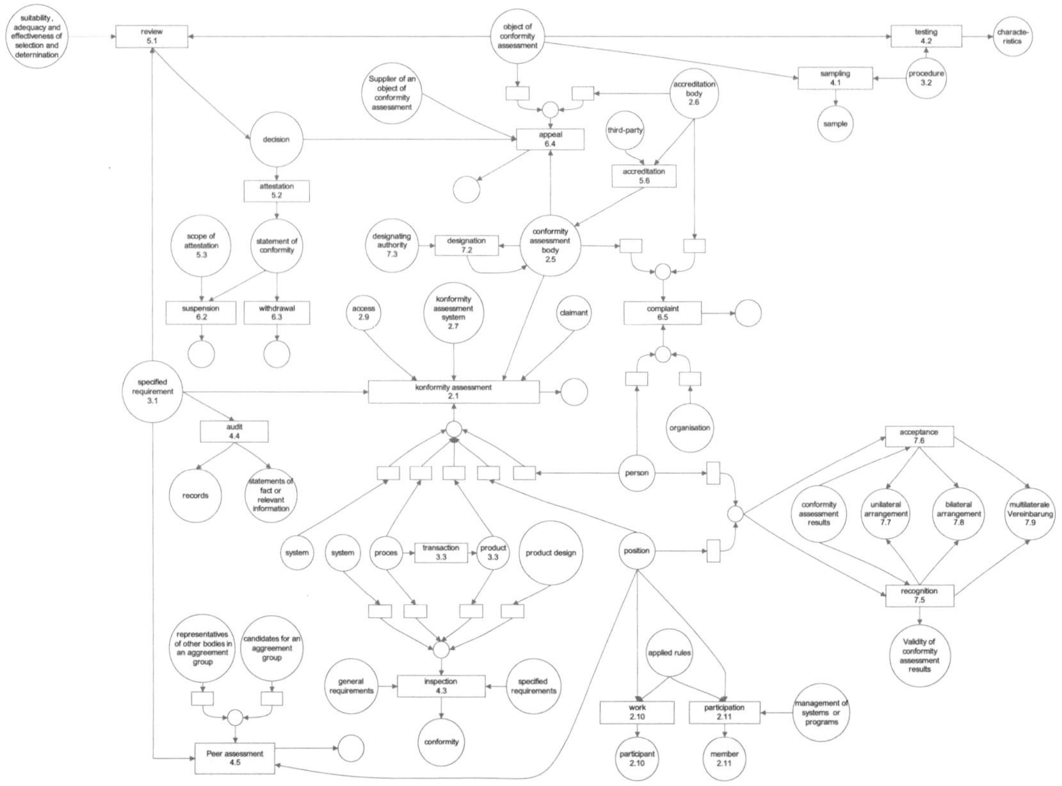

Fig. 7. Formal description of processes defined in the standard IEC 17000 by means of a petrinet

From modelling existing documents (standards, regulations …) from different domains, the common core can be identified more easily to be extracted for the use in the mode independent certification.

An additional advantage of having the formal description is, that the translations to natural languages can be realised in an easy way and avoids the usage of different terms for the same concept which can lead to confusion or misunderstanding.

3.1 Modelling the contend of standards

The contents of normative documents can be modelled according to the approach of ontological modelling described above. Two methods were developed (Hänsel, 2008). One describes the retrospective modelling of existing normative documents to clarify the contents. The starting point for this method is an existing technical standard. A second method describes the modelling of new standards to avoid ambiguities right from the beginning of the development of standards.

Both methods are similar in the setting up of the different parts required for the overall model according to the ontological modelling as described above. The means of description has to be selected in advance. The resulting ontology of the application domain will be modelled as a network of concepts with their terms and relations, e.g. a taxonomy.

As an example for the modelling of a new normative document, Figure 8 shows a part of a taxonomy that was set up for the certification approach of GNSS-receiver for different transportation modes.

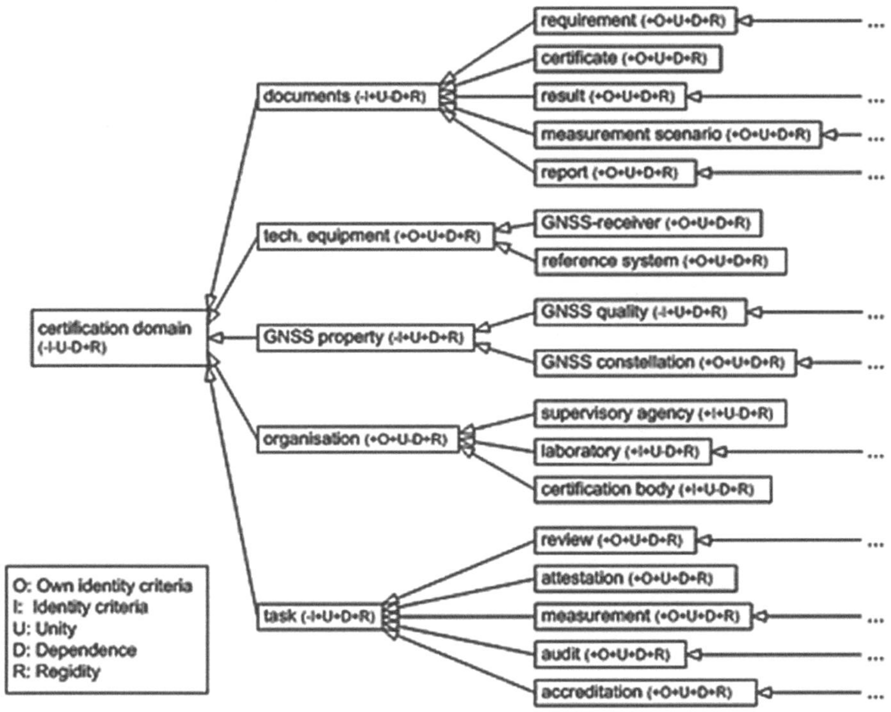

Fig. 8. Part of a taxonomy describing the certification of GNSS-receivers

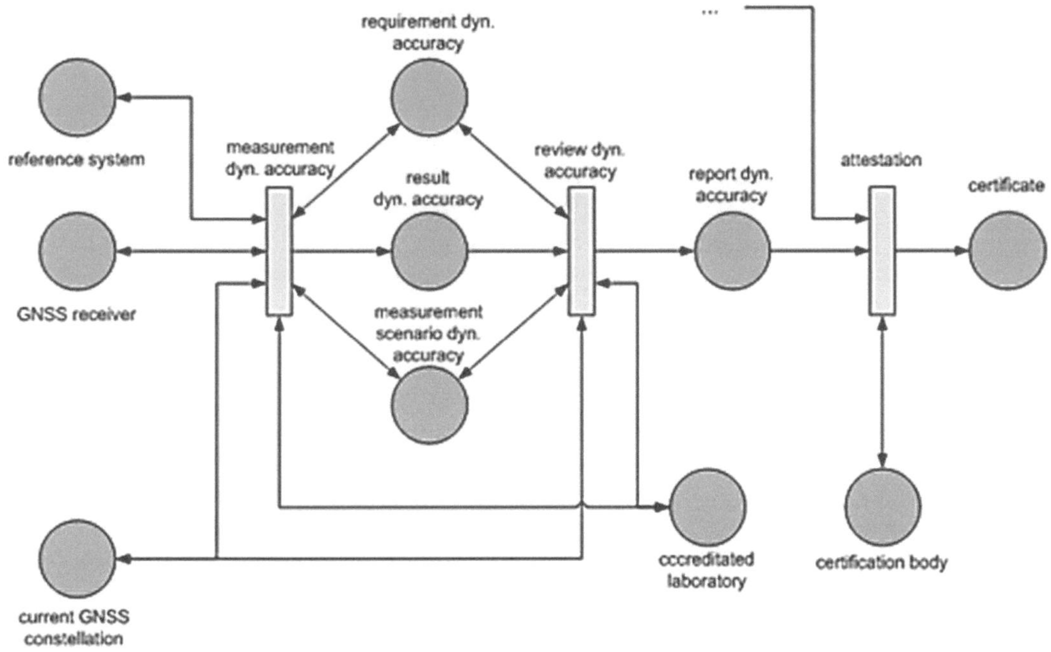

Fig. 9. Part of a process for the certification approach of GNSS-receiver

With the application of the ontological modelling, the clarity of the definitions is improved compared to the "classical" method of just writing natural language text. In Figure 8 the results of the OntoClean analysis are included in the diagram in brackets.
A part of the process using the terms from the application domain is shown in Figure 9.
Here the means of Petrinets is used for the process modelling. This explicit modelling of the processes helps the reader to understand the meaning or the intention of the author and improves clarity.

4. Concepts of metrology

Two aspects are describes that will become important as soon as safety related assistance The GNSS-System is a complex dynamic system. Its satellites and onboard receivers move continuously. Many different sources of influences in terms of metrology can be identified, which result in several GNSS system errors. Especially, a lot of principles of terminology have to be analysed in detail to be compliant with the existing standard documents.
Besides the exact understanding of the standards in metrology, i.e. ISO 5725 and GUM, for an intended certification, the metrology domain has to be linked to the requirements from the application domain, for which the certification is intended. Such requirements are the framework, in which the observed values have to fit in.
To understand the concepts of metrology and to be able to communicate with other persons about this, a formal modelling, as described in the previous chapters, is very helpful. One aspect needing clarification is the existence of multiple concepts termed as "result of measurement" in the standards. In natural language texts it is common, that the same term is associated with different concepts. However, the formal description makes the meaning of the (intended) concepts clearer.
Figure 10 shows a preliminary result of a conceptual and formalised analysis of the standards ISO 5725 and GUM based on UML class diagrams and explains the relationship between the individual terms.
As can be seen, each measurement has a true value. The true value is by nature indeterminate. Uncorrected observations $P_{IND,k}$ are obtained by measurement. An uncorrected arithmetic mean of observations includes uncorrected arithmetic mean of observations P_{IND} and standard uncertainty of the uncorrected mean μ_{PIND}. A corrected result is a result of a measurement after correction for systematic error. The correction ΔP and the uncertainty of the correction $\mu_{\Delta P}$ are determined by a calibration.

5. Conclusion

Two aspects are describes that will become important as soon as safety related assistance systems in rail or road based transportation will integrate satellite based localization (esp. GALILEO). These are, on the one hand, adequate reference localization systems for the realization of measurements of accuracy and availability of the location information provided by satellite systems. On the other hand, the regulatory framework will be discussed with the focus on a domain spanning approach, where the regulations from multiple domains have to be integrated. In addition to the description of standards, the approach is applied to the field of metrology, where, in some cases, it is also of high importance to exactly specify what is meant when doing measurements and evaluating the results especially for the safety cases.

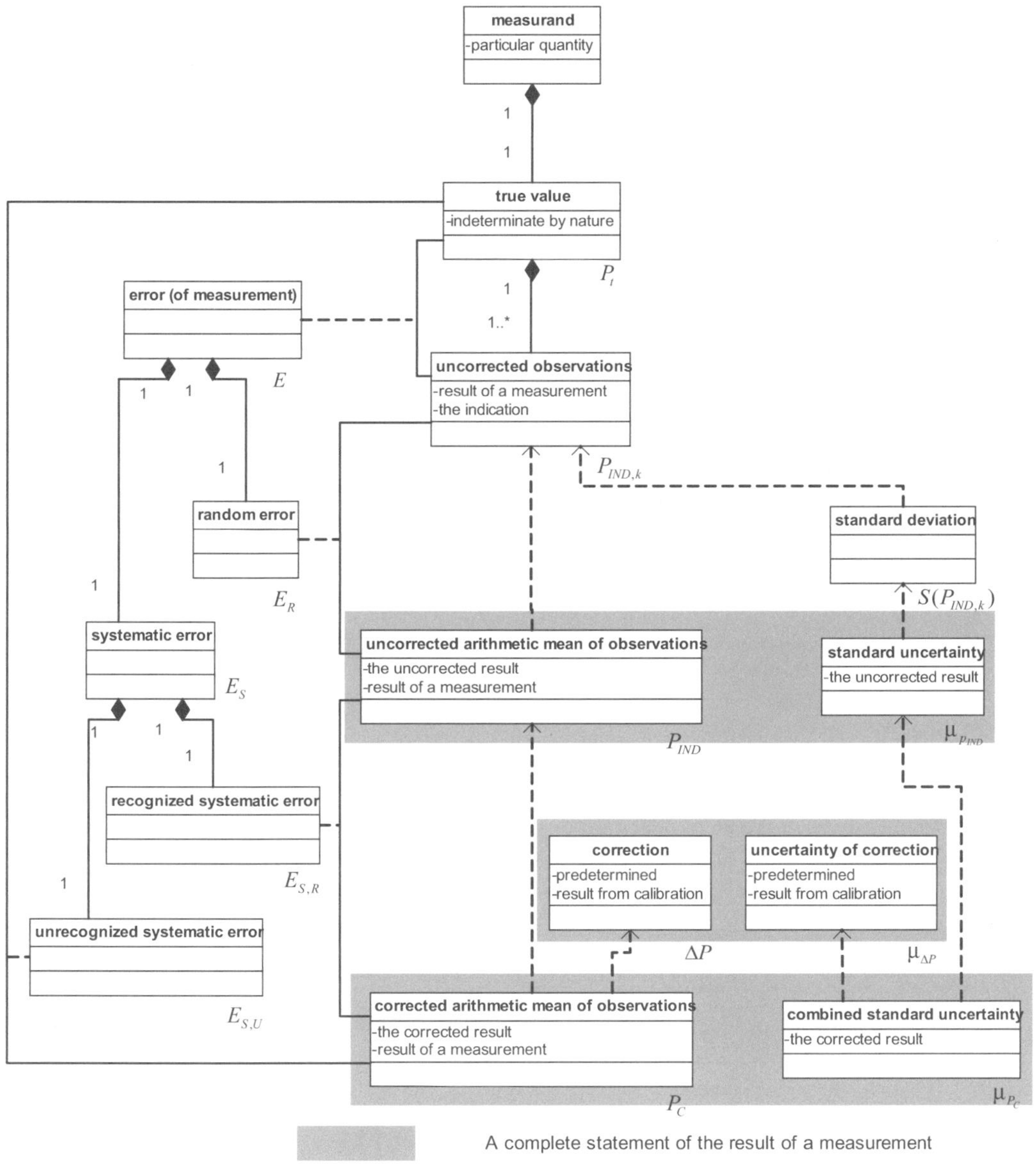

Fig. 10. Formalised taxonomy of the standards ISO 5725 and GUM by UML class diagram representation

6. Acknowledgement

This contribution results from a project, which has been supported by the German Federal Ministry of Economics and Technology (BMWi) under grant no. 50NA0614 and 50NA0615 which is gratefully acknowledged.

7. References

Becker, U.; Hänsel, F.; Hübner, M.; Poliak, J.; Schnieder, E.; Weng, Y.; Zhou, Y.(2008a). Dynamic validation of satellite based positioning systems using reference measurement platforms, *Tagungsband der POSITIONs 2008,* Gesamtzentrum für Verkehr Braunschweig e. V., Dresden

Becker, U.; Beisel, D.; Hänsel, F.; Poliak, J.; Schnieder, E. (2008b). Approach for the Certification of safety related Satellite Based Transport Application across modes *ITS World Congress 2008,* ITS Congress Association, New York

DIN 1319-1. *Grundlagen der Messtechnik – Teil 1: Allgemeine Grundbergriffe.* Beuth-Verlag, Berlin.

DIN EN ISO/IEC 17000: *Conformity assessment – Vocabulary and general principles,* Beuth-Verlag, Berlin

Grewal M. S., Weill L. R., Andrews A. P. (2001). *Global Positioning Systems, Inertial Navigation, and Integration",* John Wiley & Sons, ISBN 0-471-35032-X, New York

Hänsel, F. (2008). *Towards formalization of technical standards* Technical University of Braunschweig, Institute for Traffic Safety and Automation Engineering, PhD-dissertation.

5

Coupling Activity and Performance Management with Mobility in Vehicular Networks

Miguel Almeida[1] and Susana Sargento[2]
[1]*Nokia Siemens Networks, Universidade de Aveiro*
[2]*Instituto de Telecomunicações, Universidade de Aveiro*
Portugal

1. Introduction

We live in a mobile, fast paced world, where users are constantly on the move. Transportation plays a major role in this matter. Users thrive for services being promptly delivered anytime and anywhere. Nevertheless, business models still focus around Content Service Providers (CSP) and Network Service Providers (NSP), who, as trusted entities, provide more than the connection, further focusing, as time evolves, on the service delivery capitalization. As the trends start to position these providers as the relay points for the information to be conveyed into 3rd party cloud services, the delegation of management functions is also outsourced to the 3rd party entities. It is in this view that the remote management of vehicles becomes of the utmost importance, since connectivity allows the delivery of novel services built around the monitoring of the vehicles' conditions, location and user preferences. The immediate benefits would result in presence/location awareness for retrieval of additional information of the surroundings, or even mechanical support, mechanical failure prediction or detection, based on the continuous monitoring of the vehicles hardware sensors, as well as a whole plethora of new advantages, propelled by the collection of performance and behavior information.

Vehicular networks are inherently associated with high mobility scenarios and this fact introduces new requirements. Usually associated with high velocity patterns, the requirements to support these networks are mainly positioned around the enabling of fast mobility management protocols, and hence interfaces, gifted with the extensibility potential for the exchange of additional information. Furthermore, when considering vehicular scenarios, network mobility and efficiency are two crucial features which need to be kept in mind at all times. They have special influence over the choice of the protocol used to gather information from the vehicles towards the network. These requirements lead us to consider a framework that was originally designed for the management of the mobility of the terminals, and which therefore supports mobility with a high efficiency ratio in terms of resource consumption. This framework, the IEEE 802.21 Media Independent Handovers (802.21-2008, 2009), contains functionalities and elements that can be extended with advanced reporting capabilities to provide seamless reporting in heterogeneous technologies and environments. Using IEEE 802.21, it is also possible to integrate the actions of reporting with the actions of network decisions enforcement. We show that this approach provides a significant set of functionalities not achieved with current approaches, while reducing the overhead on cross-layer reporting. The typical approach is to perform such procedures above the IP layer.

Besides reducing the overhead, gathering performance and action reports at lower layers also saves on signaling and simplifies the protocol stack. When bringing mobility into the picture, these concerns become even more crucial.

Knowing that different types of devices have different groups of requirements in terms of network, hardware and applicational capabilities, the primitives with which all of them interface should be the same: a common Application Programming Interface (API) which is mobility driven and that cleanly exposes management functions (already under evaluation in current research) for seamless mobility and reporting. Management frameworks today also introduce, as a requirement, the definition of interfaces to 3rd party entities. We consider the central management entity to be a cloud of functionalities and of centralized intelligence, which allows interfaces for other cloud services, thus empowering the CSPs with new advanced services and new ways of capitalizing the management functions. By considering the several existing approaches, we derive a solution which combines the most commonly used web services in order to provide a Cloud view of the performance of the several vehicles.

In this chapter we present a solution to collect performance and behavior related information from different communication layers of the vehicles, while keeping in mind the major requirements associated with the inherent properties of the technology which will be discussed along with relevant use cases. Performance management represents a topic which is not widely covered when dealing with vehicular networks, and very little information can be found regarding this subject in the literature. Most research is being conducted in topics related with Vehicular networks focus on mobility management and communication techniques. It is our main goal to evaluate the performance penalties introduced by several layers: hardware, network, session and application. Service performance can be evaluated at any layer without depending on a specific technology, while enabling media independent service reports. It is in this context that vehicles, connected via multiple network access technologies, will report user activities (user behavior related), performance metrics of the mechanical hardware, of the network and of the applications running.

The chapter is organized as follows. Section 2 describes the relevant approaches to deal with the collection of performance information and user behavior activities. Section 3 presents the architecture and the main functional entities to support the mobile user vehicle reports integrated with network reconfiguration trigering. Section 4 depicts the performance comparison of the reporting approach against existent mechanisms, both on a qualitative and quantitave basis. Finally, section 5 presents the most important conclusions from the chapter.

2. Background

This section details the relevant approaches to deal with the collection of performance information and user behavior activities. It details several means to gather information and to present it in a cloud oriented solution (Voas & Zhang, 2009). While considering that vehicles are moving and exposed to different environments, different contexts and different conditions, information can be extracted and conveyed into a platform which, Extracts, Transforms and Loads (ETL), processes it according to predefined metrics, or Key Performance Indicators (KPI), and allows the management parties to evaluate the performance and to take actions. The proposals presented bellow represent efforts in trying to bring the devices closer to the cloud in terms of performance management features. The following subsections detail the technologies which are employed (Section 2.1), compare and contextualize the approaches to provide a clear view of their adequacy in terms of offered features and usage drawbacks (Section 2.2), and present the current reporting architecture and KPIs (Section 2.3).

There is already some work related to the interconnection of the devices with cloud services for monitoring and management purposes. These studies are, however, not extensive. Most of the efforts were related to sensors as the main analysis use cases, which are not necessarily mobile. Mobile devices introduce additional concerns, since mobility requires maintaining connectivity upon movement. Even in those scenarios, management related issues are still little explored, as stated in Gurgen & Honiden (2009). In Gurgen & Honiden (2009), the authors provide the major requirements for the definition of a platform to manage such devices. In Jung et al. (2007), the work is more focused on a security aware, technology agnostic framework, using Simple Network Management Protocol (SNMP) (J. Case, 1990) to gather information into a command center. This last work is more connected with our proposal, but our list of requirements goes beyond security, having mobility on the top of the list.

Another major concern is the management of the devices in the Cloud, i.e., in an online distributed tool, which appears to the end users as a centralized Graphical User Interface (GUI). This vision on the management of devices is more related to the concept of the Internet of Things (IoT), in which each vehicle can be seen as a thing. In Mohinisudhan et al. (2006), SNMP is used to incorporate hybrid automobiles with a performance monitoring system. Johansson et al. (2005) underlines the usage of a Controller Area Network (CAN) in the automotive market. CAN is a serial bus communications protocol with the purpose of interconnecting sensors, actuators, controllers and other elements. It defines the physical and data link layers for an efficient and reliable communication between the entities. In Johansson et al. (2005) it is presented an integration example with a passenger car, a truck, a navy boat and a spacecraft. In this work the authors also describe the concept of CAN gateways, which provide a way to integrate CAN-based networks with other networks and protocols. This approach is useful in the context of coupling vehicular devices with a performance management platform, since it allows the integration of industry deployed lower layer mechanisms (very oriented to specific vehicle parts' sensors) with network management solutions such as the one here presented.

Extensible Messaging and Presence Protocol (XMPP) (Peter Saint-Andre, 2009) was created for user communication purposes and has already been used for device integration with the cloud, even if only as a protocol capable of interconnecting sensors in an asynchronous wireless environment (Hornsby et al., 2009) (without yet being used for the IoT potential it carries). More recent work, (namely Miguel Almeida (2010a)), takes into account requirements for remote management and its procedures. This approach will be further evaluated in the upcoming sections, since it employs a mechanism to easily integrate devices into the cloud. Although Miguel Almeida (2010b) work does not focus on mobility, since it is merely the definition of a framework and of the required extensions to support enhanced reporting capabilities, it defines extensions and allows us to use them to couple reporting and vehicular device management along with mobility. It takes a more lower layer approach to deal with the problem we are solving and, because of that, it will also be detailed in the sections bellow. Next we detail the technological solutions that are used in the aforementioned proposals.

2.1 Description of the involved technologies

In this subsection we describe the technologies involved in the process of collection information from devices, first detailing an array of data collection mechanisms which are considered the most relevant. Then, we present the trend in the protocols used in the web environments.

2.1.1 Data collection mechanisms

Simple Network Management Protocol (SNMP) (J. Case, 1990) is one of the most relevant data collection mechanisms with wider acceptance, and which is represented by a large scale adoption in a multitude of scenarios. In SNMP the information is collected from the agents in the managed devices according to the meta-data detailed in the Management Information Bases (MIBs), from where the information can be polled. MIBs group parameters that are accessible via SNMP. The SNMP agents and stations use a request/reply protocol to communicate which supports standard messages (Get-Request, Get-Response, Get-Next-Request, Set-Request and Trap). The SNMP station uses Get-Request to solicit information from the SNMP agent, which answers with a Get-Response message. SNMP has been evolving over time with increased security and efficiency. Also, one important aspect is the addition of an unsolicited mechanism via Traps. SNMP-Trap is an unsolicited message sent by SNMP agents to the manager. These messages inform about the occurrence of a specific event, and can be used to inform that a link is down or that the agent is reinitializing itself. Traps allow for reactiveness and simplify scenarios where polling is not the best option.
Remote Monitoring (RMON) (Waldbusser, 1995) extends this concept by introducing probes and, instead of measuring Network Elements (NE), it focuses more on the analysis of traffic flows. This approach is particularly useful for the identification of third party services or servers, troubleshooting the network problems, security breaches or simply keeping logs of user activities for accounting or profiling.
The Common Management Information Protocol (CMIP) (J. Case, 1990) provides a complete network management framework over many, diverse network machines and computer architectures. CMIP's mode of operation differs from SNMP's, in the sense that the latest was designed for simplicity and ease of implementation. Besides the same functions provided by SNMP, CMIP contains more functionalities, thus allowing a wider range of operation sets. In this framework, any relevant information can be requested from the managed object and can be interpreted according to the managing system. A main drawback of CMIP is its complexity, and therefore, its adoption did not fall in the networking enviornment.
Call Detailed Records (CDR) (Breda & Mendes, 2006) include information of the call duration and failure causes, and are generally used with some lightweight data mining processes to withdraw immediate conclusions. They are largely used by cellular operators to perform some minimal profiling computation in their business intelligence solutions. CDRs are typically generated on a per-call basis: each call can originate a CDR. Although originally the CDR was designed to describe call details for billing purposes, it can be used to trace the call at the business level and retrieve service assurance relevant information. This information complements the Performance Management information by extending the network behavior analysis to the service/subscriber scope, providing the ability to propose new analysis scenarios (e.g. to assess if network is accurate in the service delivery, or which services are more suitable for that network considering the traffic model and user behavior). In order to support our requirements of supporting seamless reporting through inter-technology

environments, CDRs would need to be extensively expanded leading to a high increase in the overhead. The three most common procedures to collect CDR comprise: (1) the real-time transfer of a single CDR each time a call occurs; (2) the near-to-real-time transfer which takes place after several CDRs have been grouped into a single Blocked Generation Log and subsequently sent via Event Forwarding Discriminator (EFD); (3) and the collection of CDR records being stored in File Generating Log.

Other approaches, such as Mobile QoS Agents (Soldani, 2006), installed in the mobile devices, are also being implemented to gather end-user related information. The agents are executed remotely and gather a limited set of performance metrics that provide information to derive the Quality of Experience (QoE), as they are physically near the users. Other proprietary protocols over IP are also being implemented, where the data structure is XML, and the meta-data is defined by the hardware manufactures and interpreted by the performance monitoring solutions with knowledge on the specifications. The agents can gather very specific metrics depending on the device or on the analysis use case, and thus these solutions should be seen as very implementation dependent and customizable. Their usage can be applied to functions such as measure the user feedback from an application (via a pop up questionnaire), substitute road-tests by measuring signal power and SNR to evaluate coverage, analyze network metrics (throughput, delay, bit rates over time), estimate location and deduce trends.

The IEEE 802.21 Media Independent Handovers (MIH) framework deals with the exchange of information for mobility management in heterogeneous environments. This information includes: events, which typically deal with the changes on the link layer level and which may prompt for handover; commands, which serve the management purpose by indicating control information about handovers; and information messages, which provide details on the status of the extended services of the network and information on the available networks.

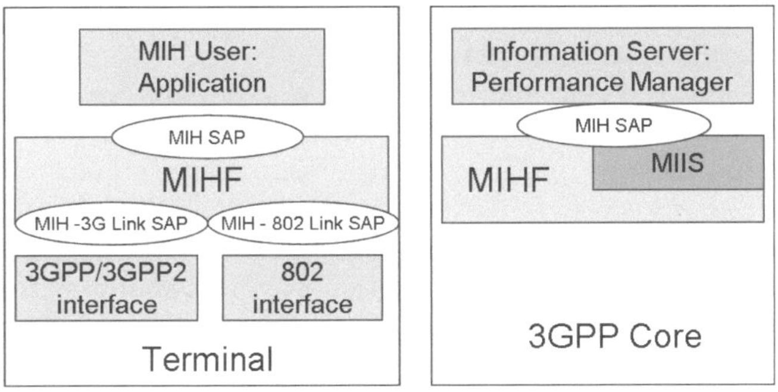

Fig. 1. IS Server Location

Media Independent Information Services (MIIS) were defined to support various Information Elements (IEs), which can be used to provide further information for handover decisions. Fig. 1 shows how the MIH Users can interact with the MIH Function (MIHF), and how the Service Access Points (SAPs) are implemented in order to allow communication with the lower layers.

2.1.2 Integrating with the webcloud

Up to now, the presented approaches are focused on ways to exchange information using existent technologies that were primarily created for that particular purpose. The requirements now invoke wider concerns, namely those related with the gathering of information into a cloud of services. In this sense, from lower layer related protocols we now move into solutions which were defined to integrate computers connected to the Internet, which is the de-facto environment for the cloud based applications, and the substratum for all presentation layers.

The typical model for communicating with the cloud, or web service guideline, is based on Representational State Transfer (REST) interfaces over HTTP (Fielding, 2000). REST provides a clear interaction model that enables a powerful and flexible solution through simple interfaces in a scalable environment. REST and Simple Object Access Protocol (SOAP) (Don Box, 2000) are associated to HTTP as means to convey information understandable by the cloud. SOAP provides the support of objects over HTTP; however, SOAP faces a scalability issue because it usually requires a large amount of technology to establish bidirectional invocation: it usually requires an HTTP web server, coupled with an application server to enable the web service environment. Moreover, current trends resort to REST over HTTP due to its simplicity and ease of usage given the mapping with of the HTTP methods (GET, PUT, DELETE and POST). It also provides a long-lasting interface that is not coupled with the business logic behind the interface. When deploying these mechanisms, manufacturers are concerned about assuring a future proof solution, and hence look at the choices which grant them a more secure bet on the long term.

XMPP (P. Saint-Andre, 2004) offers good conditions as a transport protocol for applications within the web services' (WS) scope since it offers reliability, synchronous and asynchronous delivery of messages, and does not require a complex set of features such as WS-Routing and WS-Referral to ensure identity trace back (Fabio Forno, 2005) within private domains, since addressing is not only IP based. XMPP was conceived as an alternative Instant Messaging protocol, but has been evolving to a broader concept. Given the fact that it is open and XML based, it became easy extensible and became an IETF standard.

2.2 Taking advantage of the existing approaches

In this section we evaluate the several approaches to deal with the problem of collecting performance management and behavior related data, and presenting it in the cloud - a place which is spatially and software distributed, but which creates an abstraction to present a centralized logic to the users accessing it. Users access a GUI which hides all the hardware and software complexity behind it. Bellow are two major contributions that provide an answer to this problem and which will be evaluated and used. The first handles performance collection on higher layers and takes advantage of existing solutions at the applicational layer, namely using web oriented protocols, that are user oriented. The second provides a MAC layer solution for the gathering of information using a protocol which was originally conceived for the management of the devices' mobility. We will take advantage of both solutions as inputs for the definition of the architecture presented in this chapter. The way both are integrated is explained in Chapter 3.

2.2.1 Using XMPP and REST

As stated, one good approach for collection of performance related information is to perform it in a web oriented environment. Using XMPP and REST (Miguel Almeida, 2010a) brings advantages in terms of collecting user information. According to (P. Saint-Andre, 2004), there are three Core Stanza types defined by XMPP: The <message/>, <presence/> and <iq/>. The first works as a push mechanism to immediately send messages if the destination is online.

Presence relies on publish-subscribe mechanisms through which nodes inform the server of their availability (e.g.: online, away, do not disturb), and is usually distributed among the other nodes in the roster. The last one is a stanza responsible for entities making requests and receiving responses (hence Info/Query) from each other for management, feature negotiation and remote procedure call invocation.

One of the biggest advantages of XMPP is the fact that the addresses can be associated with people or devices such as computers, mobile phones, sensors, routers or cellular network elements (3GPP RAN and Core Network Elements). This is achieved by the use of a Jabber ID (JID), a uniquely addressable ID, which is a valid Uniform Resource Indicator (URI) (Berners-Lee & Masinter, 1998), created according to the following format: person@domain/resource, where person usually represents the user entity; domain represents the network gateway or "primary" server to which other entities connect for XML routing and data management capabilities; and resource, which is of special interest since it allows to identify a specific device associated with the person. Security can be achieved by using Transport Layer Security (TLS) for channel encryption, while authentication is achieved through Simple Authentication and Security Layer (SASL). Regarding the portability and interoperability requirements, XMPP uses the "over-IP" approach and allows the binding of resources to streams for network-addressing purposes. This feature also allows to perform Identity Management via the relationships of the user and of the resource.

One of the requirement of our vehicle scenario is the communication across multiple domains (e.g. across two operators). XMPP (P. Saint-Andre, 2008) allows multi domain management that can be achieved while making use of server-to-server communication. It also allows the capabilities' exchange and location awareness features via the presence stanzas. Regarding the efficiency of the protocol, several activities are being conducted to improve XMPP performance, namely new lightweight version such as (Hornsby & Bail, 2009); however, the major performance issues drive from the presence signaling which can be optimized. This concern can be overcome with proposals like SOAP over XMPP (Fabio Forno, 2005), that would even enrich the performance concerns, since XMPP and SOAP are two XML based protocols, running one on top of the other.

2.2.2 Using media independent handovers

(Miguel Almeida, 2010b) focus on the possibility to merge reporting with mobility intrinsic protocols. Since IEEE 802.21 was developed and is used to provide a lower layer communication framework to deal with the exchange of information in heterogeneous environments, our aim is to further extend it to enable the exchange of end user reports independent from the underlying technologies. Moreover, this extension will allow the seamless integration and activation of network reconfiguration procedures.

By extending the IEEE 802.21 MIIS, end user reporting can be performed at lower layers, using one single protocol to carry all user, vehicle and network related information, which will increase the efficiency of resource consumption. The MIIS is expected to provide mainly static information but, for the envisioned approach, real-time and dynamic information is required. The IEEE 802.21 standard also mentions that dynamic information such as available resource levels, state parameters and dynamic statistics, can be obtained directly from the respective access networks. However, this information usually does not provide a clear view on the end-to-end service performance. Also, the gathering of user behavior related information from the network requires a means to access this information: this can be supported through the IEEE 802.21, by loosening the concept of the MIIS and supporting new features and functionalities.

To determine the QoS and QoE, it is required to assess the impact of the lower layer information on higher layers at the core side. This information can be related via the cross relation of PoA (Points of Access) with the terminal identification via the SAP (Service Access Points). Therefore, it allows the collection of most of the information locally (either from lower or higher layers), pre-evaluate it and then send it to the network. This view is aligned with IEEE 802.21 which, through the MIIS, can provide an indication of higher layer services supported by different access networks and other relevant information that may aid in making handover decisions. Such information may not be available (or could not be made available) directly from MAC/PHY layers of specific access.

Finally, the support of user and device (vehicular) reports through IEEE 802.21 allows the seamless activation of network reconfiguration procedures, such as session and terminals handover to networks that better match the user/device requirements, to jointly optimize network resources and user experience. In sections 3.1 - 3.3, we further detail and propose extensions to the MIIS to support a more detailed communication of application level parameters, and introduce more intelligence upon handover decision.

2.3 Performance and behavior management

As defined by ITU-T, the Telecommunications Management Network model (TMN) (ITU, 1996), used for managing open systems in a communications network, establishes four management layers comprising: (1) the element management, which entities are hierarchically above and gather the information which is collected by each Network Element; (2) the Network management system, which evaluates these metrics (after a Transform and Load process); (3) the Service Management, which is in charge of taking into account the previous layer and extrapolate conclusions that can lead to active changes in the network; (4) and the business management layer, which introduces the agreement levels that need to be accomplished. The Network Elements (NE) are typically the network nodes which interact with the delivery systems. Operations, Administration and Maintenance (OAM) describes a set of management levels and their interactions. The concept has more recently evolved to include Provisioning and Troubleshooting. It ideally would imply the cross view of the TMN model with the Fault, Configuration, Accounting, and Performance and Security (FCAPS) functionalities.

To provide a clear view on the performance of the vehicles, indicators need be defined according to the relevant metrics in the vehicle network. Key Performance Indicators (KPIs) are a set of selected indicators used for measuring the current performance and trends. KPIs highlight the key factors of the current performance and warn of potential problems. Considering a counter as the most elementary value which is collected from a vehicle, a KPI can simply be equal to a counter or to an arithmetic abstraction of counters that can be applied to monitor a certain part of the network, functionality or protocol. KPIs play a major role in creating immediate and relevant feedback on the performance of a certain element (may it be network, hardware, or behavior).

2.3.1 Generic reporting tool architecture

Since we are proposing a remote management platform, the whole system would not be complete without the inclusion of an architecture to evaluate the performance of the vehicles in the cloud. This Reporting Tool receives the information from the devices and allows an online verification of their performance by the end users. Fig. 2 shows the main components which are typically included in a generic architecture for a reporting tool. Bellow we explain the major functionalities of each component and their relevance to the architecture. The Reporting Engine is the mind behind the Reporting Tool (Fig. 2). It is responsible for the database queries, it processes the results and displays them in a defined format. It provides all the data visualization capabilities, offering different pre-defined models and allowing the user to create their own. These pre-defined visualization models are important, because they allow the manipulation of data in different dimensions, providing different reports for different types of end users, and even for different type of analysis, starting from a unique data set. Related to these models, there is an important reporting component, which is the KPI set. KPIs are defined in configuration components and can be either calculated on the fly by reporting engine, or pre-calculated and stored in the Reporter Database. The Automated Knowledge Discovery model is another important part of this reporting engine, and provides the very important feature of automatic data monitoring, searching for patterns in the network behavior for the purpose of forecasting upcoming events, such as the Operation, Maintenance and Optimization needs.

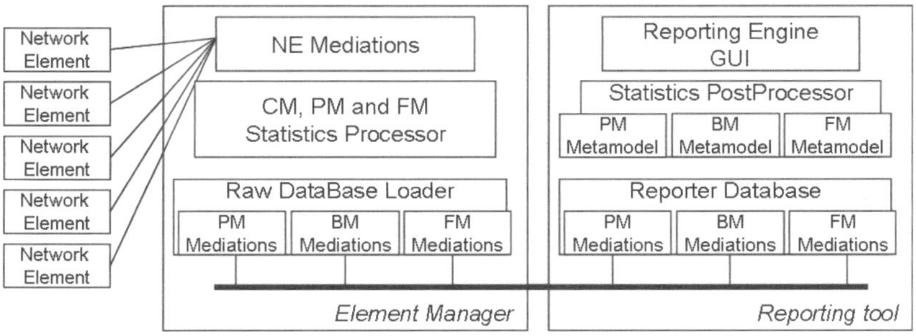

Fig. 2. Generic Reporting Tool Architecture

The Reporter Database is a data warehouse designed for the coherent integration of diverse data sources, dimensioned to optimize the data discovery and reporting. This data repository modulates all the network topology into a hierarchal object structure, which provides the capability to analyze the entire network. This analysis can focus on the correlation of different parameters that can be Configuration, Performance or Fault Management related. A possible use case would be to assess what kind of configuration optimizes better the performance of the network, by improving network capacity and reducing its faults. This analysis can be extended in time and from different network perspectives, as historical and object data aggregation is possible. Moreover, the database primitives allow for the storage management and provide all the data access information to other layers. This database is modulated based on NE specific metadata, which defines the Object Class (OC) structure, and for each OC all the PM Measurements and related list of PM Counters and PM counters aggregation rules. The FM metadata is generic for all the OC, defining a list of failures that can occur.

The Statistics Post-Processor is a software component that plays a decisive part on the reporting process. It is responsible for the entire object and time aggregations, which enhances the analysis capabilities, allowing the time trend analysis and drilling through the network objects, enabling a great diversity of network analysis. The aggregation rules are all defined through metadata specific for each NE, and provide information on how PM counters must be

aggregated. The statistics Processor is responsible for converting all the diverse data gathered from the NEs according to a structured and generic meta-model. This particular module processes Configuration, Performance and Failure Management Information.
The Raw Database loader is responsible for providing interfaces for the access relating to data storage management features. It is another function of an ETL procedure which uploads the gathered information into the raw databases. This module includes interfaces for mediation of the interactions between the EM and the analysis tools that evaluate the collected data. These interfaces answer to the Reporting Tool for requests related to Performance Management (PM), Configuration Management (CM) and Fault Management (FM) data.
The NE Mediations manage the interactions between the Element Manager Module and the several Network Elements in the network. They are responsible for the collection of the Performance and Fault Management functions existing in each of the elements of the network. The NE Mediation Module implements the Extraction part of an ETL procedure. Each network element monitors its performance through the Performance Mediation. A subset of that module is responsible for the communication with the Element Manager. That interface is divided into three types of primitives relating to the type of data which is to be transported: PM, CM and FM. The first presents metrics related with the continuous operation of the equipment, while the second indicates the configuration setup, including information such as topology and capabilities. FM is a more urgent type of data as it indicates critical issues to be evaluated.

2.3.2 Types of metadata

As stated, the main objective of this chapter is to focus on the end-to-end reporting capabilities between the devices and the cloud, while providing mechanisms and information so that decisions can be made and measures can be taken if problems occur. However, the decision making and acting components are not discussed. When considering reporting functionalities, the typical supported metadata types are: CM, PM and FM. The work presented in this chapter also considers Behavior Management (BM), in the sense that it allows the gathering of metrics associated with the behavior of inhabitants of the vehicles and their interactions with the vehicles.
Configuration Management metadata is responsible for the mapping between the different NEs present in the network and their components into a coherent and structured Object Class model. This way, CM metadata is used to identify objects with the same properties and to maintain possible occurrences of an object in the object class hierarchy. Two types of objects can de defined for this model, Managed Objects (MO) and Reference Objects (RO). Managed Objects refers to objects that are directly related elements present in the network that can be managed, configured, manipulated, which are obviously the NE elements and their components e.g. a Node B and its Cells. Reference Objects refers to virtual reporting dimensions, i.e. elements that are virtually created to ease the network analysis by dividing and grouping the network into smaller segments thus reducing the analysis complexity. This Reference Objects are created and stored in the Reporter Database by the Statistics PostProcessor module, using the CM metadata information.
Performance Measurement metadata is responsible for defining, for each network element, all the PM measurements and Counters and relating them to the CM data, i.e. to the OC structure. As network element represents a specific role in the network, there will be a different set of measurements/counters for each NE. The number of measurements and counters needed

to monitor a specific NE is dependent on the NE complexity, ranging with the number of functionalities. A PM measurement is a logic representation of a NE functionality that defines a set of counters that monitors the network performance behaviour. A PM counter is the fundamental element of the performance monitoring process, as it provides detailed information ranging from specific procedures up to group functions. As counters are the basis of PM, they are used to develop different kinds of aggregations such as KPIs and Reports. This way, different kind of users and analysis can be satisfied with only single tool.
Fault Management metadata defines the mapping between all the NE components and the fault events that describe system failures, either hardware or software driven. FM metadata thus relates OC with incoming network failure notifications. These failures are categorized and ranked by severity, which can range from debug to emergency state. The special characteristic of this type of data is the fact that it typically has an unsolicited behavior and requires near real time functionalities.

3. Architecture description

The following section depicts the architecture and the main functional entities that need to be included to sustain the previously defined requirements, namely, the inter-technology scenarios and the support of end user terminal reports integrated with network reconfiguration triggering. Fig. 3 presents the vision explained in this chapter. Vehicles are moving freely and through a wireless connection, which can be WiFi, WiMAX or 3GPP based (UMTS, iHSPA or LTE), and are connected to a CSP. By using a mobility management protocol like the Media Independent Handover with extended reporting capabilities, the performance measurements of the vehicles and the behavior of the users can be gathered, stored and evaluated within a cloud. This allows early problem detection, location and context awareness, remote assistance, and a plethora of services from which fleet management functionalities should be underlined.

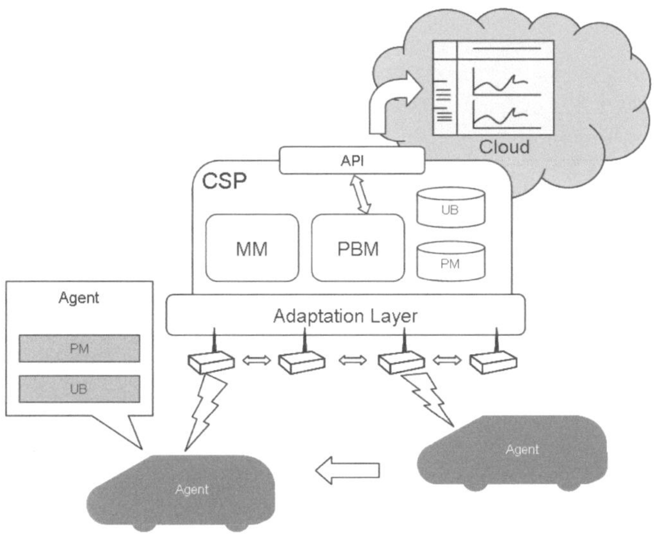

Fig. 3. Interactions between the vehicles and the cloud

Each vehicle has an agent installed which sends information to the cloud, and is handled by entities connected to a logic bus, the Media Independent Handover Function (MIHF). On the other side, the Performance and Behavior Management module (PBM), collects that information and stores it according to the type of data (User Behavior (UB) or Performance Management (PM), which will be detailed in the next subsection). 3rd Party Cloud services access this information and apply analysis algorithms (data mining procedures can be applied but are not within the scope of this work), to present graphics explaining occurrences of problems in certain vehicle models or in certain zones.

3.1 Architecture specification

Fig. 4 shows the main required entities for the proposed reporting architecture. End user Behavior reports are communicated by the Multimedia Application to the Behavior Manager at an agent (BM@Agent) installed in the vehicle. The Performance Manager (PM@Agent) is a MIH User as well, and collects information from both the lower layers (QoS information) and upper layers (QoE related information).The PM also interacts with the running applications and the vehicle's mechanical parts as well as the software/firmware for monitoring purposes. The agent is a MIH entity that is responsible for gathering information and communicating performance and behavior metrics. Lower layers report metrics that are extracted from the technology drivers, including link and network layer values, such as throughput, bit rate and SNR. From the upper layers, the PM will receive the information regarding service performance, mainly related with end-to-end performance and QoE feedback. To achieve this, these modules have open interfaces, which can be used through specified primitives, as will be explained in the next subsection. Lower layer information can also be retrieved directly by the Media Independent Information Service (MIIS). The MIIS (see Fig. 1) collects the data on the network side and feeds the relevant user behavior and performance values to the Performance and Behavior Manager (PBM@network).

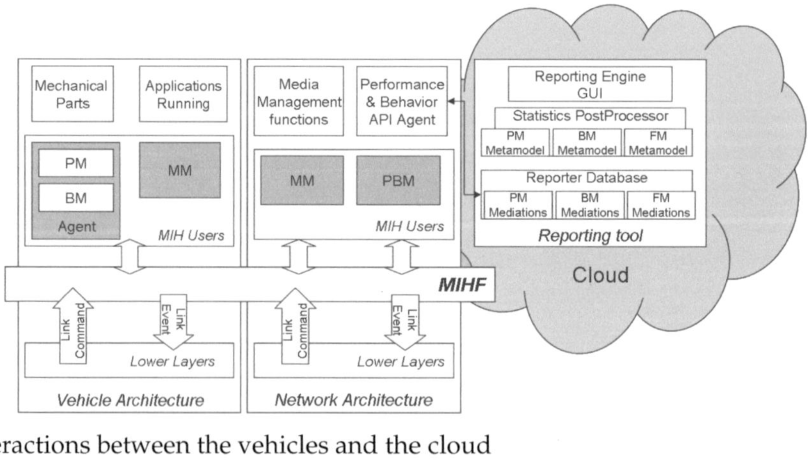

Fig. 4. Interactions between the vehicles and the cloud

The PBM@network is responsible for the interaction with the multiple terminals to perform profiling analysis for individual and group behaviors. It comprises a database and interacts with the API Management Agent, which is responsible for allowing access to the reporting tool and 3rd party services. The typical approach to evaluate a user's opinion on a service, and hence depict his profile, is to collect end user reports after a service has been delivered, and

converge the opinion with the provided service's characteristics. However, other methods can provide an equally efficient evaluation of the user's acceptance to performance trade offs. The QoE and expected experience form a couple of properties that cannot be considered separately. A user may be willing to accept lower performances if the contract fee is lighter. This conclusion and consequent profiling can be drawn from the user behavior. After applying the profiling analysis algorithm, the PBM formats the information to feed the Mobility Manager (MM) and stores the results.
The MM receives the inputs from terminals and decides if an action is required. The MM can use this input to take decisions, activating events in the terminal or events in the network for optimization purposes. This process will make use and extend the IEEE 802.21 signaling. Other proposals (Chung et al., 2008), (Jesus et al., 2007) deal with the mechanisms involving mobility decisions and mobility signaling more deeply. To better understand this process, the core network should be seen as a mediator of information. The vehicles will send information to the CSP infrastructure to be handled by the PBM, and this information will be made available to the in-cloud applications through the Performance and Behavior API Agent (see subsection 3.4). It is also through that agent that the in-cloud applications can interact with the vehicles. Fig. 4 shows an application in the cloud performing the remote management of the performance of the devices. This scenario shows how the mediated data collected from the devices can be outsourced to other services.

3.2 Signaling

To better understand the message flow, we will consider the scenario where a user contains a multi-homed terminal connected to two wireless networks (e.g. WiFi and UMTS), but is using the WiFi one (Fig. 5). Periodically, the multimedia applications report activity updates and performance metrics. These messages are not IEEE 802.21 messages (Action messages in Fig. 5), but are internal primitives. The same application will periodically issue another message (Performance Report) informing about the relevant performance metrics for that particular service. As shown in Fig. 5, an Action message is sent from the user application to the BM@agent indicating that the user wants to receive a video and has issued for a VoD request. The message should be according to the type: Action (Application ID, Type of Request, Timestamp), thus specifying the application type which is being used (Video, Audio, Gaming, Browsing, etc.) and the type of request (VoD, Streaming, Conference, Starting Game, etc.). Following that procedure, different Performance messages will also be sent regularly. The application issues a message containing the following structure: Performance (SourceID, MetricType, Value, Timestamp), thus depicting the type of metric being reported and the value for that metric. These messages are issued locally and then mapped to IEEE 802.21 by the MIH Users (both BM and PM) for transmission to the network side (in the form of the messages and procedures depicted in Section 3.3. Moreover, the PM@Agent receives Performance updates from the hardware adaptation as explained in Section 3.3.2.
Ideally, the agent@terminal has well defined interfaces with the applications, and each application can be responsible for reporting the user's activities and performance feedback. However, we introduce these entities as functional blocks for a better comprehension and easier compliance with legacy applications. Hence, both message types Performance and Action do not reflect any type of signaling protocol but are instead internal primitives. The BM and the PM are now ready to report this information to the network using the MIH Function. When the Information Service requests for a UE update (MIH_Get_Info.request), it gets a

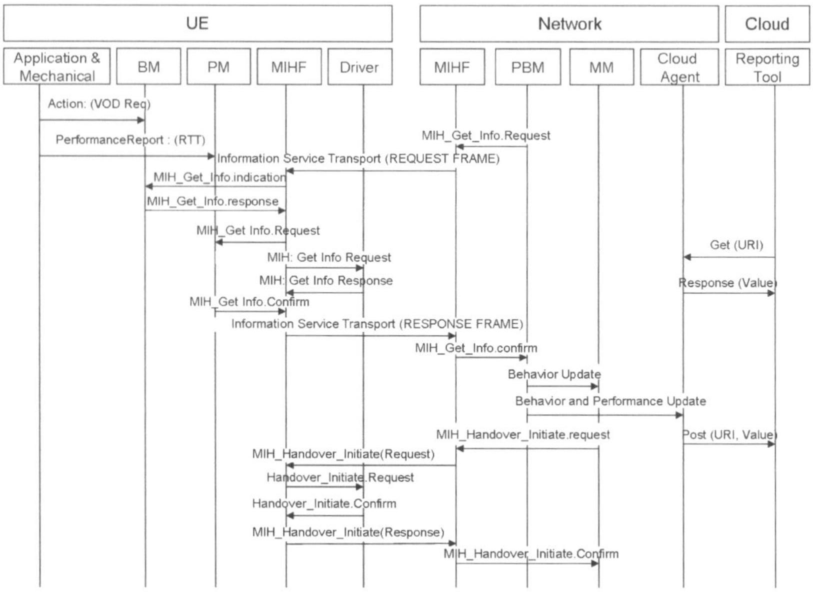

Fig. 5. Signaling diagram. URI is constructed from through the vehicle ID as explained in Fig. 7. (*Mechanical and Application Adaptations which are the Agent interfaces)

response containing the QoS performance from the application, from the lower layers, and also the reported user action (via the Information Service Transport message). The IS receives an update on the user status, and forwards this information to the PBM which evaluates the QoS parameters, increments the user profile and communicates the changes to the MM. The MM will evaluate the feasibility of this network for the desired service. Since it already knows that the terminal has another interface with different properties (via standard IEEE 802.21 signaling: link_up message), it decides that a link with more bit rate would be better for the services in use. It then issues a handover request to the terminal, so that it performs a handover to LTE.

Whenever an action is taken by a user, the system needs to identify if the user is satisfied with the current quality of the service (according to QoE parameters): the desired characteristics and the used application's requirements should be taken into account to assess if they are being met. If not, a change is required, by using another available interface (performing a session handover) or switching to another PoA in order to enhance the terminal reception conditions. This makes it possible to optimize the network or re-allocate users on different PoAs. When Behavior and Performance Updates are collected by the PBM@Network, the Cloud Agent is notified. It should then lookup the 3rd party entities which have interest in receiving this update and use the POST method to transfer that information into the destination web applications. This information can also be requested from the 3rd Party services (which we designate as cloud in Fig. 5), using the Get method. Both GET and POST methods use the URI, which is formulated via the identification of a user and the element of a vehicle belonging to that user as well as the metric which is to be retrieved (this mapping is done via the SAP ID and the metric type as explained n Section 3.4). To support this view, it is required that both the Cloud Agent and the 3rd party entities (in Fig 5 the web service

is exemplified by a performance reporting tool) are running a web server, since the REST methods are used via HTTP.
The way the mapping of web resources is done is detailed in subsection 3.4. In this section, we demonstrate the signaling flow to better explain how the information is conveyed to and from a web cloud based application. In general terms, the main concern is to cache the information locally and send an update to the web applications with which the CSP has an SLA, and which has expressed intention in receiving a particular device's instruction. It is assumed that this process is preceded by a pre-enrollment phase, by means of which the owner of the vehicle agrees on the availability of his data, as well as remote management functionalities being sent to the vehicle. The defined signaling will allow both synchronous and asynchronous reporting messages that are of relevance in order to support two types of information: a) periodic performance or behavior data that has no crucial impact on the performance of the vehicle; b) relevant and urgent information that might indicate a malfunction or a possible problem and thus should be sent in an unsolicited way. Although the use case of sending a command from the cloud is not expressed in Fig. 4, it is supported as explained in subsection 3.4, via the usage of a Rest command which is received at the API on a particular resource which unequivocally identifies the vehicle and the command to be remotely executed. The API agent then conveys that command to the destination vehicle through the UBM.

3.3 Information elements

The IEEE 802.21 already defines general IEs, access network specific IEs, PoA specific IEs and other IEs. Information Service elements are grouped into three categories: a) General Information and Access Network Specific Information, which give an overview of the different networks; b) PoA Specific Information that provides information about different PoAs for each available access network; c) Other information that is access network specific, service specific, or vendor/network specific. Next, we propose to include the Service Performance IEs and the User Behavior IEs to be used in the previously presented architecture, thus extending the ones defined in the standard, while taking (Miguel Almeida, 2010b). The agent at the vehicles handles the following 3 types of messages:

- Action (SourceID, Type of Request, Timestamp)
- Performance (SourceID, MetricType, Value, Timestamp)
- Alarm (SourceID, MetricType, Value, Timestamp)

The first two can be easily handled by normal MIIS procedures, while the last one, given its nature, would benefit from a more unsolicited behavior. One way to solve this problem is to use the MIH_Get_Information.request message carrying data that would indicate the occurrence of an alarm situation. The MIH_Get_Information.request is sent to the MIHF in the termina,l and then in the network side, a MIH_Get_Information.indication notifies the PBM@network (corresponding MIH User).
The MIH_Get_Information.response brings the confirmation that the alarm has been received. This provides a good workaround for the lack of unsolicited messages between MIH Users. The Alarm reflects the over crossing of a threshold value; the format of the message is the same as the one of the Performance, but only indicates the urgency of the problem to the network. For the purpose of supporting interfaces with the Controller Area Network (CAN) existing in the vehicles (Johansson et al., 2005), the agent should also be able to act as a CAN gateway. Since CAN and IEEE 802.21 are both lower layer based approaches, the overhead is minimal,

thus presenting a resource efficient solution. The way our framework handles this integration is explained bellow in Section 3.3.2.

3.3.1 User behavior information elements

The type of active applications is usually communicated in the bootstrap of an application. We define it to be in the form TYPE_IE_UB_ACTIVE_APP_ID. It contains an index of the application regarding the content a user is requesting. After reporting the active application, several requests can be made by the user. The message type that should be used for this type of report will be: TYPE_IE_UB_ACTION (ApplicationID, UserAction, Timestamp). The User Action field is defined in Table 1, which contains information on the actions that are required to be supported for the previously mentioned services. The Timestamp is a time reference.

ApplicationOn	ApplicationOff	RequestChannel	IdleMode
EndConversation	InitiateConversation	RequestURL	ActionMode
SendMessage	ReceiveMessage	MovementMode	RetrieveNeighboursList
JoinServer	LeaveServer	LeaveGame	JoinGame

Table 1. User/Service Interactionsn

3.3.2 Performance information elements

Performance metrics from lower layers are already supported by the IEEE 802.21; the ones being proposed relate to the application aware QoS and QoE values, which are directly reported from the application layer to the MIH User. The agent collects information from the several sensors in the vehicles as well as from the applications running. Vehicles become multimedia oriented as time evolves, and now include displays for video and gaming purposes, network connectivity for the retrieval of additional network based services, or even simple internet access by itself. The QoE values will require additional metrics that are relevant to characterize the service delivery quality.

QoS Performance Information Elements

As stated, it is required to have the complete view of the vehicles' performance in the management platform. Usually the metrics associated with the vehicles performance are related with the components of each type of vehicle. Different sensors are applied to the several parts and monitor temperature, pressure, speed, etc. Table 2 shows an example of Distributed Control Architecture using CAN (Johansson et al., 2005).

Powertrain and Chassis	Body electronics
Transmission control module	Driver information module
Engine control module	Steering wheel module
Brake control module	Rear, Frontal and Central modules
Door module	Climate control module
Steering angle sensor	Steering wheel module
Suspension module	Auxiliary electronic
Audio module	Infotainment control

Table 2. example from (Johansson et al., 2005) for a particular automotive integration solution

The defined information elements do not need to be gathered via the CAN solution. They can be collected via any customized solution as long as the agent receives them in the pre-defined format and is able to send them to the network-side modules of management. For the purpose of integrating the evaluation of the vehicle's analysis in the cloud, we consider Information Elements to be of the type: TYPE_IE_VP - Type Information Element Vehicle Performance.

QoE Performance Information Elements

The QoE Performance Information Elements are related with the services being accessed by the users on the vehicle. The services can be bundled by the CSPs as part of the overall package and can be evaluated individually. These parameters were first described in Miguel Almeida (2009), where a more extensive study is performed, employing a 3GPP view while underlining the relevant parameters at each layer of a 3GPP network. In fact, these parameters are a first glance at the performance view of a service, and could be narrowed down, for problem identification purposes, until the lower layers. It defines the way in which the KPIs should be constructed and how they can be evaluated in a Cross Layer View, while in Igor Pais (2009), a more QoE centric analysis is performed. For each service we consider metrics such as the total setup waiting time for a service to be received (TYPE_IE_SP_WT), the Mean Opinion Score (TYPE_IE_SP_MOSQ), the Service Availability (TYPE_IE_SP_AVAILABILITY), the Lost Packets (TYPE_IE_SP_LOSS), the Time Resolution (TYPE_IE_SP_TIMERES for voice and TYPE_IE_SP_FPS for video), and, of course, the Bit Rate (TYPE_IE_SP_BR). All primitives include the following fields: SourceID (or application ID), Application Type ID, Time and Value.

3.4 Integrating with the cloud

In the previous sections we have been debating the collection mechanisms which allow to gather and convey the information from the vehicles into the network. This would allow the Mobile Network Operators (MNO) which own the gathering technology to access the data and evaluate it. In order to make it publicly available and, in this way, further capitalize this solution, the MNO would greatly benefit from a seamless way to make it available to 3rd party entities (e.g. 3rd Party CSPs), which could hire the access to the data. For instance, fleet management functions could be employed and then the vehicles' performance information might be outsourced. Subsection 2.2.1 shows a way to gather information using XMPP and then relaying it into the cloud using REST. By extending that proposal in order to also convey the information carried in the MIH messages, we create a compliant system. To do so, we need to create an adaptation in the Cloud Bridge Server (Miguel Almeida, 2010a), which we denote as Performance and Behavior API Agent. That agent interfaces with the MIH agents which collect the messages from the vehicles generating REST alike messages which update the 3rd party webservices accordingly. In this way, instead of using XMPP as the transport protocol, we use 802.21 to convey the performance messages, which is a more efficient solution for medium to high mobility scenarios (see Fig. 6). The web services expose an interface that allows information to be asynchronously supplied, and commands requested, when necessary, due to network management operations.

We gather the SAP ID and cross match it with the vehicle's information within the operator. There are two solutions to cope with the device identification. We couple the information of the IDs of the Service Access Points and of the MIH Users that coexist in the same vehicle and then translate them into a resource. By employing the translation shown in Fig. 7, we can guarantee the required consistency between the adaptation and the CBS, enabling the exposure of MIH resources to the REST interface, which can now be fully controlled on a

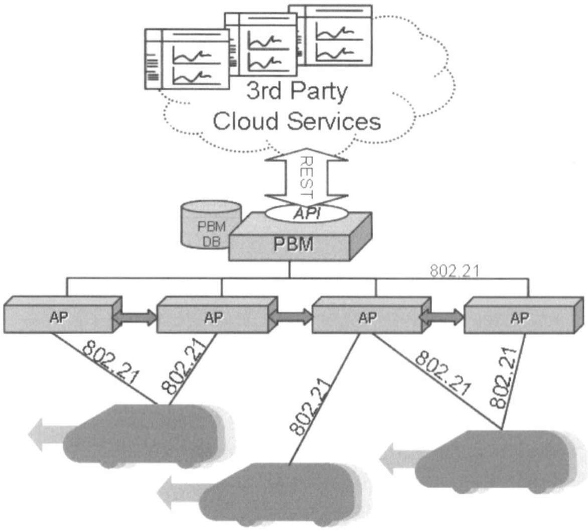

Fig. 6. Protocols Used for the interaction between the Vehicles and the Cloud

per-vehicle policy determined by the bridge. All messages are sent to the Cloud seamlessly, given that any web like environment can support RESTful primitives.

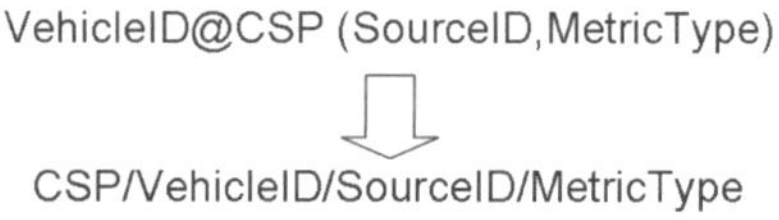

Fig. 7. Creating the ID for the REST resource

As the entry point towards the Cloud, the Performance and Behavior API Agent enables the communication between the devices and the Software as a Service, taking on a vital role for authentication and authorization. Each service must register, define an SLA, and authenticate to gain access to information pertaining to the vehicles. Given the control over the information, the Performance and Behavior API Agent is able to define a granular access control to the information exposed. The Performance and Behavior API Agent supports HTTP compliant messages (GET, PUT, UPDATE, DELETE), which are also the core of REST functionality, thus assuring the integration with minimal effort on the Cloud Bridge Server. This secure setup even allows customizing the devices towards a specific web service. If the devices are configured accordingly, they can opt to send reports to the custom Web Service URLs. The 802.21 signaling will transport the report, and then the Performance and Behavior API Agent will perform a HTTP POST on the destination Web Service. The Web Services need to be aware of the data model being communicated.

4. Performance comparison

The main advantage of the described approach is to save on signaling and overhead while simultaneously allowing end-to-end heterogeneous reports, and seamlessly integrating both reporting and network enforcement processes. The proposed architecture aims to provide an accurate profiling of users for an enhanced network optimization and resource consumption prediction. In this section, it is presented a quantitative evaluation in terms of traffic generated by different approaches, and a qualitative evaluation in terms of supported functionalities.

4.1 Qualitative evaluation

In this section we include a qualitative evaluation of the functionalities provided by current approaches and the IEEE 802.21 based approach (denoted as MIHR). A summary of the supported features is presented in Table 3. We first start by comparing the different approaches in terms of features support. We can see that for the purposes of integrating the devices with a web environment, SNMP is the most inadequate, since HTTP based solutions are already web based, and XMPP is XML based which enables a direct transformation of the objects. The 802.21 approach requires a special adaptation on the network side. Since it only defines a way for the devices to intercommunicate with the network on a lower level, a MIH user is required at the network side to behave as a proxy towards the web cloud.

Regarding the security features, XMPP creates a secure unique channel using TLS. Earlier versions of SNMP present serious security constraints, and this has been addressed in SNMP v3; still, the common applications use IPSec bellow the SNMP communication. Although this feature does not reside within the main focus of the 802.21, it can use 802.1x for the IEEE based networks and use the channel security procedures of the 3GPP networks. SNMP also allows authentication to verify that the message is from a valid source, while XMPP with REST supports authentication of an Identity and its merging with accounting information. With respect to Identity Management, XMPP is the best proposal to link several devices to someone's identification. Considering that the platform is to be deployed at a Network Operator's site, then it would be good to allow the cross matching of accountings with the operator's database, only feasible using the possibilities offered by XMPP. Since we are focusing on the device management more than on the Identity management functions, it is simpler to use a lower layer device management system, which takes into account the SAP ID or simply an IMEI.

Feature:	MIHR	XMPP + REST	HTTP Based	SNMP
Security	Yes	TLS	SSL	IPSec
Reliability	Yes	Yes	No	Yes
Authentication	No	SASL	No	No
Web Cloud Integration	High	Easy	Easy	Medium
CSP Integration	Medium	Easy	N/A	Complex
Identity Management	No	Yes	No	No
Bi-Directionality	Yes	Yes	No	Yes*

Table 3. Management Features Comparison; *not when using traps

Being able to support asynchronous communication allows the deployment of a very relevant feature for fault management functions, which is sending alarms in a near real-time way. There are several applications that take advantage from this possibility, and XMPP is the best approach to deal which this type of events. Moreover, it allows bidirectional communication without the need to run a web server in the devices, which is the only way to support bidirectional communication using HTTP with REST or with SOAP. The way 802.21 handles this problem is through commands, which allow sending information to the devices. In this sense, by gifting the network with the appropriate intelligence at the MIH user, it is possible to enable bi-directionality of the management applications, in a very resource effective way.

IEEE 802.21 considers reliability to be a requirement from the underneath media, which needs to have a reliable message delivery procedure in order to allow the MIH Protocol Acknowledgment Service (802.21-2008, 2009). XMPP and HTTP run on top of TCP, which ensures the reliability mechanisms. SNMP runs on top of UDP, and therefore it employs the mechanism of request/response: if the response is not obtained, the request is again sent (and can be customized). Obviously, this a problem for the trap-enabled version of the protocol.
CDRs are call oriented and transport the information of the sessions and of the users involved. Their major advantage is the fact that they were created for an easy usage within the operator's domain. The MIHR requires an effort to integrate the solution with the Mobile Network Operator's (MNO) Home Location Register (HLR), while employing a mapping between the devices and the owners. For the integration of the HTTP based solutions, this would be difficult and would require a great deal of customizations, namely assuring that the agents collect such information on the client side, and then, convey it to operator. The degree of customization would be so high that we opt to define it as non applicable, since it is a concept too broad and too subjective. SNMP also would require an adaptation mechanism that is aware of the identification of the device with which the network is communicating. The network would then need to map this information with something previously known, i.e., it would need to previously be aware of the agent ID, and map it into a specific user.
A summary of the performance metrics is presented in Table 4. Using the IEEE 802.21 method, it is not required to exchange the full performance details, but only those which are required. Since we are dealing with a media independent proposal, our reports can be applied in any type of technology even in a switched domain. SNMP is technology oriented and MIBs are defined for specific types of hardware. Typically, CDR analysis is very technology-oriented and includes details relevant only for the source of the type of access. When using an approach on top of HTTP, it also becomes depends on the network information, and the typical approach is to perform QoS measurements and establish a TCP connection towards a collecting server. In IEEE 802.21, the MIIS already handles that interface seamlessly. XMPP uses TCP and runs on top of IP, so although it could be considered technology agnostic, it brings large amount of of signaling, from the presence stanzas. For cellular networks, this would require optimization procedures, and additionally, login/logout functionalities.

Metric	MIHR	XMPP + REST	CDR	HTTP Based	SNMP
Overhead	Good	Bad	Bad	Bad	Medium
Signaling	Good	Bad	Bad	Bad	Medium
Heterogeneity	Yes	Yes	Yes	Yes	Yes
Synchronous	Yes	Yes	Yes	Yes	via Traps
Asynchronous	Yes	Yes	Yes	Yes	Yes
Multilayer	Yes	Yes	No	N/A	No

Table 4. Qualitative Performance Comparison

Multilayer analysis relates to the ability of the different procedures to evaluate the performance at different layers, and in parallel, to deal with the end user behavior issues. Using the IEEE 802.21-based approach, information can be collected from lower layers and from upper layers, simultaneously or separately depending on which information is required. Information can be collected locally at the NEs and then be reported to a central server: an approach with reporting over IP will still work, but will be less seamless for the intermediate

NEs. The same concept applies to SNMP; however, usually this procedure will not be applied to upper layer analysis or behavior related parameters. On the other hand, CDRs will not focus on the network information. XMPP is dependent on what the agent is collecting, but it introduces no constraints at this level, relying only on the capabilities of the agent@terminal. IEEE 802.21-based approach relies on the IEEE 802.21 mechanisms, which do not support events from upper layers. Having that in mind, it is only possible to support reactive events for some of the typical performance parameters. However, as explained before, we overcome this problem for the support of alarms via the issuing of customized messages from the terminals. Asynchronous events could be supported by implementing event triggering from upper layers for end user behavior analysis. Since event triggering from lower layers is still valid, this proposal partially supports synchronous reporting. XMPP was created for instant messaging purposes, so it excels at both approaches, while SNMP can only be considered to be asynchronous when using traps. CDRs are implemented on a per-call basis, so it can be considered an asynchronous approach. The HTTP based approach can also be considered proactive, since it requires inputs, but it supports both asynchronous and synchronous methods (e.g. Get method will request for information, while the Post will add a new resource). In order to employ a synchronous procedure with requests from the network to the client, this would require the devices to have a webserver installed, which is clearly a downside. When considering the heterogeneity support, one issue arises: assuming that most of the approaches can run on top of the IP/TCP stack, we would need to state that XMPP, HTTP and SNMP approaches that use TCP and UDP are heterogeneous as long as the technologies use IP. This is a fairly accepted assumption today. However, the only technology that was created having in mind heterogeneity support was the IEEE 802.21.

In terms of signaling, overhead and performance comparison, we evaluated how the access links that connect the devices to the cloud would behave. As can be seen in Section 4.2, SNMP outperforms the XMPP approaches. This happens mainly because in XMPP we defined an Object Class and used the objects within an XML to make the transactions. The results presented for the XMPP approach were obtained from experimental evaluation, while in the SNMP related results, we computed the overhead induced by raw data transportation, when conveying binary information. This information would require post processing at the network management system. The exchange of performance records in CDRs is typically fixed size oriented and depends on the record detail. Moreover, the overhead is significantly high, since it includes information related to the user device for identification (e.g. IMEI), which is not required with IEEE 802.21, since the SAP ID already matches the device ID. The major inconvenience of using CDRs with additional inputs from the terminals is that it requires additional overhead and signaling, since in 3GPP networks it is required to consider the tunneling effects and the establishment of IP or GPRS connections for data transmission. In the future 3GPP releases this problem may be mitigated; however, reporting at this level will always introduce an overhead larger than the IP+UDP one. The best approach is to use the IEEE 802.21 solution, which handles the problem on a lower layer basis, thus reducing overhead and signaling.

4.2 Quantitative evaluation

Figure 8 presents the traffic generated by the several approaches with the size of the object of information being reported. We compare the performance of different protocols on the wireless link, i.e., the link between the vehicles and the network's infrastructure nodes, which is the most problematic part of the network, when considering high mobility scenarios. This is in fact the link which introduces more concerns, in terms of resource consumption efficiency. The evaluated protocols include: SNMP Polling or Trap-based methods, HTTP using SOAP (Simple Object Access Protocol), IEEE 802.21 and XMPP transporting REST methods. Regarding XMPP and REST, we used the results detailed in Miguel Almeida (2010a), which were experimentally obtained. The other metrics were obtained as detailed bellow. Typically, Mobile Web Services introduce high overheads in general. (M. Tian, 2003), (Pras et al., 2004) include details on the overhead impact under various conditions. As Figure 8 shows, when using an object based transport like the one used in MQA using SOAP over HTTP, one must consider the overhead introduced by the signaling and also the headers of the IP, TCP (with timestamps), SOAP and SOAP envelope. The calculated overhead is presented in Equation 1. As the object size increases, packet segmentation occurs at the TCP layer, thus significantly increasing the size of generated traffic.

$$length_{Soap} = header_{IP} + header_{TCP} + header_{HTTP} + +header_{Soap} + envelope + object_{size} \quad (1)$$

When using SNMP, the overhead per object is decreased for both scenarios: we consider both the best case scenario with unsolicited messages' exchange (via the usage of traps), and also in the case of polling-based approach. According to (de Lima et al., 2006), the header size of SNMPv2c is approximately 25 octets. The overhead of SNMPv3 is given by Equation 2, where the Header Data of the SNMP is given by Equation 3 as 17 octets. This means that SNMPv3 adds a minimum of 17 octets to SNMPv2c. Considering the signaling generated by both versions, the total generated traffic is given by Equation 4 and Equation 5, respectively for SNMPv2c and SNMPv3. In Figure 8, SNMPv3 traffic was generated using ASN.1 (Abstract Syntax Notation One), and Get processes (Request + Response messages) were taken into account for overhead measurement purposes. For the transmission of more objects, the performance of SNMP is decreased, as shown in (de Lima et al., 2006) and in Figure 9. Using a bulk procedure would greatly reduce the overhead in this case, but this would be true for all other proposals.

$$Traffic_{SNMPv3} = HeaderData + SecurityParameters + ScoopedPDUdata \quad (2)$$

$$Header_{SNMPv3} = Version + MSGID + MaxSize + Flags + SecurityModel = 17Octets \quad (3)$$

$$TrafficSNMPv2c = 54 + n(10 + 2.Objectlength + Value) \quad (4)$$

$$Traffic_{SNMPv3} = 88 + n(10 + 2.Object_{length} + Value) \quad (5)$$

When considering the trap-based approach, we determine the traffic of SNMP in Equation 6 for SNMPv2. Then, we calculate the minimum traffic generated by SNMPv2 traps in Equation 7 and add the additional minimum overhead (considering Communitysize=6 octets and Traplength=1 octet) to get Equation 8 for the traffic generated by SNMPv3 traps (OID refers to Object Identifier). As can be seen in Figure 8 and Figure 9, Traps reduce the generated traffic, especially when a large number of events are being generated.

$$Traffic_{TrapSNMPv2c} = 63 + Community_{size} + TrapOID_{size} + n.(3 + OID + Value) \quad (6)$$

$$Traffic_{SNMPv2Trap} = 70 + (3 + OID + Object_{size}).numObjects \quad (7)$$

$$Traffic_{SNMPv3Trap} = 87 + (3 + OID + Object_{size}).numObjects \quad (8)$$

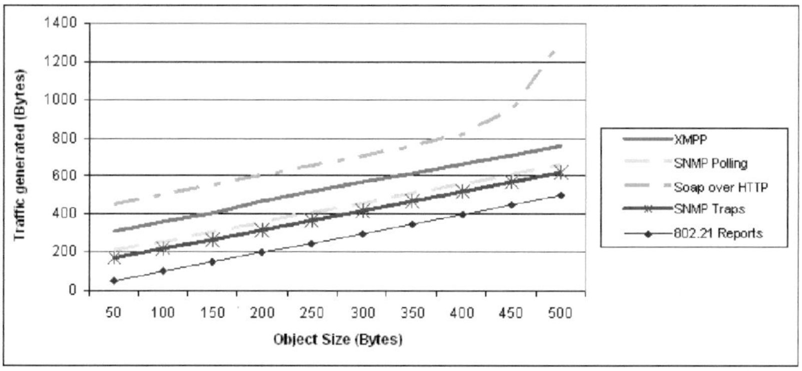

Fig. 8. Traffic Generated vs number of Objects

We also evaluated the minimum traffic generated by an approach which only uses UDP over IP without any additional signaling. Besides the application port, the OID and the value of the object, we only considered the values for an unsolicited procedure with the overhead of the IP and UDP headers and a variable field using one octet for the OID. The traffic generated consequently decreases when compared to SNMP. Considering our IEEE 802.21 approach, we can observe that it generates similar or lower traffic as an IP unicast transmission of the raw information over UDP. This becomes more clear in Figure 6, since MIH reports do not require the specification of a way to request for specific objects above the IP and UDP layer; the major overhead saving comes from the lack of requirement for the usage of the IP header and signaling, which would be required in order to support the request of an object and consequent transport. (Melia et al., 2007) further details this aspect with emphasis on the low overhead introduced by the protocol during mobility procedures.

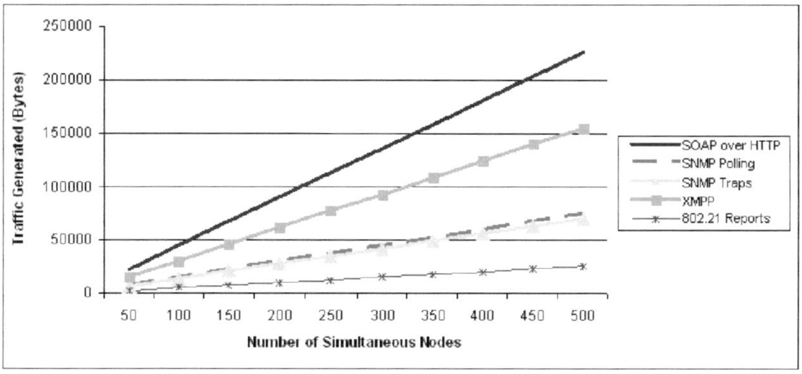

Fig. 9. Traffic generated by number of Nodes

5. Conclusions

In this chapter, we presented an architecture based on IEEE 802.21 that is able to gather information from vehicles and allows access via a web cloud. It provides seamless support at different levels: reporting of cross-layer information, support of inter-technology environment, and integration of the actions of reporting with those of network reconfiguration. This approach combines the basic mechanisms of the IEEE 802.21 to gather information from the devices on the wireless link and the REST primitives to convey that information into the Cloud. Using the IEEE 802.21 Information Service, we underlined the required basic signaling to integrate both upper and lower layers information, regarding both the QoE the user is experiencing and the QoS parameters of the services. Moreover, using IEEE 802.21, the network can then act on the vehicles, basing its decision on profiling algorithms which estimate the future actions of a group of users, seamlessly supporting both mechanisms of reporting and reaction. The results presented show that this approach significantly decreases the reporting overhead, while it introduces a set of functionalities not present in current approaches.

Through the analysis of several transport technology (and as the core driver for interactions within the core domains), we consider a XMPP based solution to be the best approach when envisioning the integration of end users interacting with terminals, gaming consoles, cell phones, IP enabled sensors, etc, with web environments and also with the operator's infrastructure. However, for more mobile environments, where wireless resources are the major concern, and where fast connectivity maintenance procedures play a major role, its performance decreases. The number of features supported allows the authentication using the account identification of the devices' owners within the operator's Charging Gateways, allowing the easy deployment of charging per usage.

The integration with the cloud environment provided through REST interfaces allows the interaction with 3rd party web services, increasing the possibilities of applicability and revenue. By providing common and consistent interfaces to act and report on devices, we enable a new array of business relationships and opportunities that put the telecommunication and infrastructure operator back in the driver seat of the network, while enabling a clear interaction with the Cloud world, a feature which has been profoundly lacking from the operators portfolio. We believe that these paradigms will be a key revenue system where both operators and service providers can capitalize by using the adequate tools to unite the common approaches. This view greatly facilitates the interactions with vehicles on the move, via several technologies seamlessly reporting behavior and performance metrics using a lightweight reporting mechanism coupled with mobility

6. References

802.21-2008, I. S. (2009). Ieee standard for local and metropolitan area networks- part 21: Media independent handover, *IEEE Std 802.21-2008* pp. c1 –301.

Berners-Lee, T., F. R. & Masinter, L. (1998). *Uniform Resource Identifiers (URI): Generic Syntax*, IETF RFC 2396.
URL: *http://www.ietf.org/rfc/rfc2396.txt*

Breda, G. & Mendes, L. S. (2006). Qos monitoring and failure detection, *Telecommunications Symposium, 2006 International*, pp. 243 –248.

Chung, T.-Y., Yuan, F.-C., Chen, Y.-M., Liu, B.-J. & Hsu, C.-C. (2008). Extending always best connected paradigm for voice communications in next generation wireless network, *Advanced Information Networking and Applications, 2008. AINA 2008. 22nd International Conference on*, pp. 803 –810.

de Lima, W., Alves, R., Vianna, R., Almeida, M., Tarouco, L. & Granville, L. (2006). Evaluating the performance of snmp and web services notifications, *Network Operations and Management Symposium, 2006. NOMS 2006. 10th IEEE/IFIP*, pp. 546 –556.

Don Box, David Ehnebuske, G. K. A. L. N. M. H. F. N. S. T. D. W. (2000). *Simple Object Access Protocol (SOAP) 1.1*, W3C Note.

Fabio Forno, P. S.-A. (2005). *XEP-0072: SOAP Over XMPP*, 1999 - 2010 XMPP Standards Foundation.
URL: *http://xmpp.org/extensions/xep-0072.html*

Fielding, R. T. (2000). *Architectural Styles and the Design of Network-based Software Architectures*, Doctor of Philosophy Dissertation, University of California, Irvine.

Gurgen, L. & Honiden, S. (2009). Management of networked sensing devices, *Mobile Data Management: Systems, Services and Middleware, 2009. MDM '09. Tenth International Conference on*, pp. 502 –507.

Hornsby, A. & Bail, E. (2009). x03bc;xmpp: Lightweight implementation for low power operating system contiki, *Ultra Modern Telecommunications Workshops, 2009. ICUMT '09. International Conference on*, pp. 1 –5.

Hornsby, A., Belimpasakis, P. & Defee, I. (2009). Xmpp-based wireless sensor network and its integration into the extended home environment, *Consumer Electronics, 2009. ISCE '09. IEEE 13th International Symposium on*, pp. 794 –797.

Igor Pais, M. A. (2009). End user behavior and performance feedback for service analysis, *Intelligence in Next Generation Networks - ICIN*.

ITU (1996). *Telecommunications Management Network Reference Model: CCITT Recommendation M.3010*, International Telecommunication Union.

J. Case, E. A. (1990). *A Simple Network management Protocol (SNMP)*, RFC 1157.

Jesus, V., Sargento, S., Corujo, D., Senica, N., Almeida, M. & Aguiar, R. (2007). Mobility with qos support for multi-interface terminals: Combined user and network approach, *Computers and Communications, 2007. ISCC 2007. 12th IEEE Symposium on*, pp. 325 –332.

Johansson, K. H., TÃűrngren, M. & Nielsen, L. (2005). *Vehicle Applications of Controller Area Network*, BirkhÃd'user.

Jung, S.-J., Lee, J.-H., Han, Y.-J., Kim, J.-H., Na, J.-C. & Chung, T.-M. (2007). Snmp-based integrated wire/wireless device management system, *Advanced Communication Technology, The 9th International Conference on*, Vol. 2, pp. 995 –998.

M. Tian, e. A. (2003). Performance considerations for mobile web services, *IEEE Communication Society Workshop on Applications and Services in Wireless Networks*, Vol. 2, pp. 741 –746.

Melia, T., de la Oliva, A., Vidal, A., Soto, I., Corujo, D. & Aguiar, R. (2007). Toward ip converged heterogeneous mobility: A network controlled approach, *Comput. Netw.* 51(17): 4849–4866.

Miguel Almeida, A. M. (2010a). Bridging the devices with the web cloud: A restful management architecture over xmpp, *6th International Mobile Multimedia Communications Conference*.

Miguel Almeida, Rui Inacio, S. S. (2009). Cross layer design approach for performance evaluation of multimedia contents, *International Workshop on Cross Layer Design*.

Miguel Almeida, S. S. (2010b). Media independent end user behavior and performance reports, *IEEE GLOBAL COMMUNICATIONS CONFERENCE*.

Mohinisudhan, G., Bhosale, S. & Chaudhari, B. (2006). Reliable on-board and remote vehicular network management for hybrid automobiles, *Electric and Hybrid Vehicles, 2006. ICEHV '06. IEEE Conference on*, pp. 1 –4.

P. Saint-Andre, E. (2004). *Extensible Messaging and Presence Protocol (XMPP): Core*, IETF RFC 3920.
URL: *http://www.ietf.org/rfc/rfc3920.txt*

P. Saint-Andre, E. (2008). *XEP-0238: XMPP Protocol Flows for Inter-Domain Federation*, 1999 - 2010 XMPP Standards Foundation.
URL: *http://xmpp.org/extensions/xep-0238.html*

Peter Saint-Andre, Kevin Smith, R. T. (2009). *XMPP: The Definitive Guide Building Real-Time Applications with Jabber Technologies*, O'Reilly Media.

Pras, A., Drevers, T., van de Meent, R. & Quartel, D. (2004). Comparing the performance of snmp and web services-based management, *Network and Service Management, IEEE Transactions on* 1(2): 72 –82.

Soldani, D. (2006). Means and methods for collecting and analyzing qoe measurements in wireless networks, *World of Wireless, Mobile and Multimedia Networks, 2006. WoWMoM 2006. International Symposium on a*, pp. 5 pp. –535.

Voas, J. & Zhang, J. (2009). Cloud computing: New wine or just a new bottle?, *IT Professional* 11: 15–17.

Waldbusser, S. (1995). *Remote network Monitoring management Information Base*, RFC 1757.

6

Ultra-Wideband Automotive Radar

Akihiro Kajiwara
The University of Kitakyushu
Japan

1. Introduction

A lot of progress has been made for automotive radar during the last years. There are two types of automotive radar; "long-range radar at 77GHz with a range capability up to 200m" for automatic cruise control (ACC) and "short-range radar at 24/26 and 79GHz up to 30m" for anti-collision. Long radar with narrow radiation beam enables a automobile to maintain a cruising distance, while short-range radar has recently attracted attention because of many applications such as pre-crush warning, stop-and-go operation and lane change assist. The short-range radar with a very broad lateral coverage has a few significant problems to be overcome such as target detection and clutter suppression. This is because the widely radiated radar echo contains not only automobile echo, but also unwanted echoes called clutter. It is actually not easy to detect a target echo in increased clutter. Ultra-wideband impulse-radio (UWB-IR) radar with high range-resolution has recently attracted much attention for automotive use, because it offers many applications such as pre-crush warning and lane change assist.

The followings provide an overview of this chapter;

1. Section 2 introduces various radar systems for automotive use. It begins with a discussion of radar technologies such as Pulse Doppler, FM-CW and UWB-IR.
2. UWB-IR radar requires high speed A/D devices which can directly process the received nanosecond pulse. For example, A/D devices of several GS/s or more should be required for the UWB-IR radar with a bandwidth of 1GHz, which have not been available yet. The use of wideband may also cause unacceptable interference on existing narrowband systems. Therefore, some interference mitigation scheme may be required for the radar emission in the future. To solve these problems, a stepped-FM radar scheme is introduced in Section 3, which does not require high speed A/D device and provides the co-existence with exisiting narrowband systems.
3. Short range radar is expected to provide a ***wide*** coverage in azimuth ***angle.*** Therefore, increased clutter makes it difficult to detect automobile target accurately. The clutter can be classified from automobile by the Doppler, but it will not be applicable to the UWB-IR. In Section 4, a scheme is introduced which estimates the Doppler by using the time-trajectory of radar echo and the measurement results are presented.
4. Automotive radar is required to detect automobile accurately, but not to detect clutters falsely, even in complicated traffic conditions. In order to satisfy the requirement, a target discrimination scheme with range profile matching is introduced in Section 5 and the measurement results are presented. The results show that the automobile type can be discriminated.

2. Fundamentals of radar technologies

2.1 Radar detection

The basic principle of UWB-IR automotive radar detection is illustrated in Fig.1 where the received signal includes many echoes scattered from desired and undesired objects (Skolnik, 2001) (Taylor, 1995). The one-dimensional signal, which is referred to as range profile, is generally presented by multiple impulses with gains $\{\beta_k\}$ and propagation delays $\{\tau_k\}$, where k is the impulse index. Suppose a nanosecond pulse of $s(t)$, the range profile, $y(\tau, t)$, is the time convolution of $s(t)$ and the impulse echo response $\Sigma\beta_k\delta(t-\tau_k)$ as follows;

$$y(\tau,t)=\sum_k \beta_k s(t-\tau_k) \tag{1}$$

Fig.1 (b) shows an example of received power range profile for a bandwidth of 1GHz (corresponding to 1 nanosecond pulse) on a roadway.

When a target echo exceeds a given threshold, it can be recognized. However if a clutter echo exceeds the threshold, it is mistaken as a target. This is called a missed detection. Consider the target detection in increased clutter as shown in Fig.1, it is not easy to recognize the automobile target since the echoes over a threshold can't be classified usually as target or clutter.

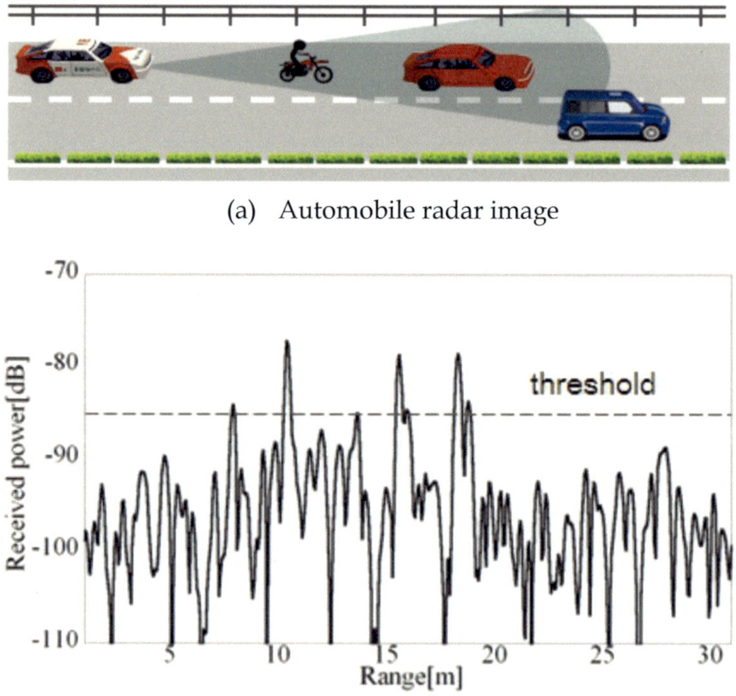

(a) Automobile radar image

(b) Power range profile

Fig. 1. Principle of radar detection

The typical radar utilizes a pulse waveform. The transmitted pulse is intercepted by some objects and re-radiated in many directions. The re-radiation directed back towards the radar is collected by the directional antenna. The received signal is then processed to detect the presence of target on the threshold basis. The range to a target is estimated by measuring the travelling time from the transmitter to the receiver. The range is given by $d = c\tau / 2$ where τ is the time and c is the speed of light. The radar equation generally relates the detectable range. It is useful not only for determining the maximum range at which the radar can detect a target, but it can serve as a means for understanding the factors affecting radar performance. For the transmit power P_t in a particular direction, the maximum gain G of antenna and the received signal P_r at a range d is given by (Skolnik,2001)

$$P_r = \frac{P_t \cdot G^2 \cdot \sigma \cdot \lambda^2}{(4\pi)^3 d^4}, \tag{2}$$

where λ is the wavelength and σ denotes the reflectivity of the target which is also called radar cross section (RCS).

The RCS depends on the target shape, λ and radar bandwidth. An example of the measured RCS of a sedan type of automobile is shown in Fig.2, where we considered a CW of 25.5GHz and UWB of 22-29GHz (Matsunami et al., 2008). Change in the RCS of the CW is found to be significant to the azimuth angle, while the RCS forof the UWB is much less sensitive to the azimuth angle relative to the CW.

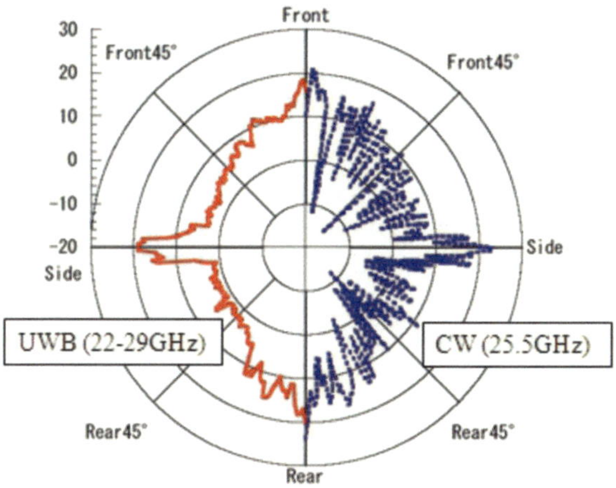

Fig. 2. Azimuth variation of the RCS of a sedan typed automobile

2.2 Automotive radar system

There are several types of radar systems and a large number of different applications. In this section, three types of radar system are discussed: Pulse Doppler, FM-CW and UWB-IR radar systems.

Pulse Doppler radar

Pulse Doppler radar sends out a pulse train. When the transmit pulse is long enough and the target's Doppler is large enough, it may be possible to detect the Doppler shift on the basis of the frequency change within a single pulse. Fig.2-3 (b) shows that there is a recognizable Doppler shift in frequency domain. To detect the Doppler on the basis of a single pulse of width T_p generally requires that there be at least one cycle of the Doppler frequency f_d within the pulse. Also, the transmit frequency source must have very good phase stability and the system is required to be coherent. Fig. 3 illustrated the principle of Pulse Doppler where the pulse repetition period is T. For a moving target, the received echo experiences a Doppler f_d as shown in Fig.2-3. Therefore the Doppler can be estimated by processing the received echo in frequency domain.

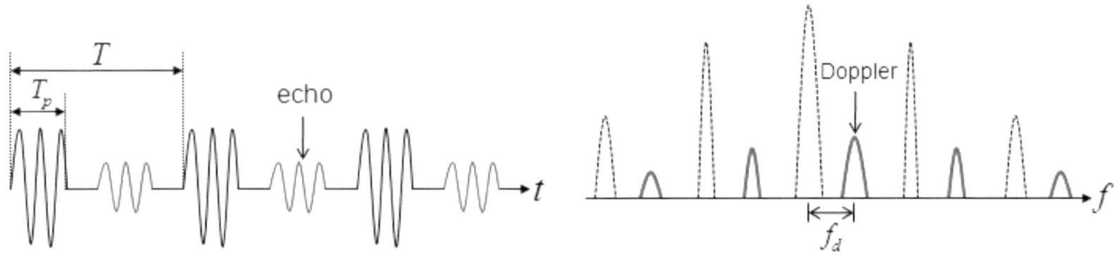

Fig. 3. Principle of Pulse Doppler radar

FM-CW radar

FM-CW (Frequency-modulated continuous-wave) radar system varies the frequency of the transmit signal and measures the range based on the frequency difference between the instantaneous transmit signal and received echo. The typical modulations includes triangular and saw-tooth. Current applications of FM-CW include high-resolution systems. Fig. 4 (a) shows the block diagram of a typical triangular FM-CW radar where the super-heterodyne-receiver is assumed. The frequency-analyzed beat signals for up-frequency and down-frequency are passed through an A/D device, and digital processing of FFT is then conducted. Please note that the beat signal exhibits a peak at which the intensity becomes large corresponding to the target. And the peak frequency corresponding to this peak carries information concerning the distance, and the peak frequency differs between the up portion and down portion of the triangular FM-CW wave due to the Doppler associated with the relative velocity of a target.
The up-beat α is given by the followings;

$$\alpha = f_1 - f_2 = \frac{\Delta f}{T_m} \cdot \frac{2d}{c} - f_d \tag{3}$$

$$f_1 = \frac{\Delta f}{T_m} t \tag{4}$$

$$f_2 = \frac{\Delta f}{T_m}\left(t - \frac{2d}{c}\right) + f_d, \tag{5}$$

where Δf is the frequency excursion.
The down-beat β is also given by

$$\beta = f_2 - f_1 = \frac{\Delta f}{T_m} \cdot \frac{2d}{c} + f_d \tag{6}$$

$$f_1 = -\frac{\Delta f}{T_m}(t - T_m) + \Delta f \quad (7)$$

$$f_2 = -\frac{\Delta f}{T_m}\left(t - T_m - \frac{2d}{c}\right) + \Delta f + f_d \quad (8)$$

From the above α and β, the Doppler f_d and the distance d to a target are derived as follows;

$$\begin{aligned} f_d &= \frac{1}{2}(\beta - \alpha) \\ d &= \frac{cT_m}{4\Delta f}(\beta + \alpha) \end{aligned} \quad (9)$$

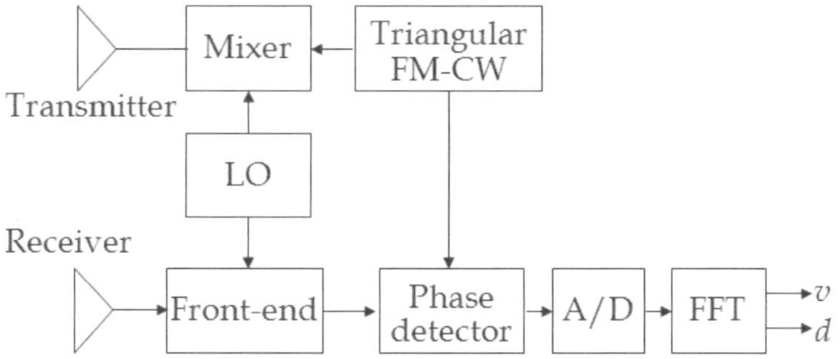

(a) Block diagram of triangular FM-CW radar

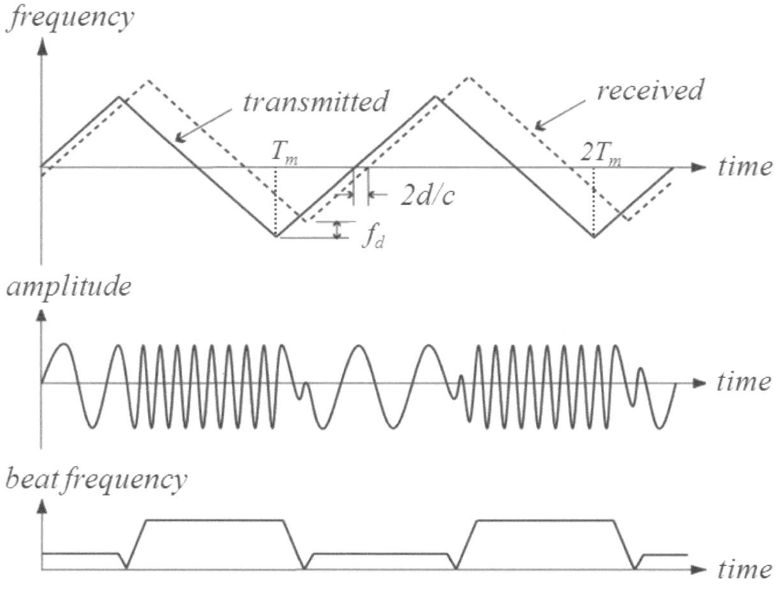

(b) Frequency-time relation for triangular FM-CW

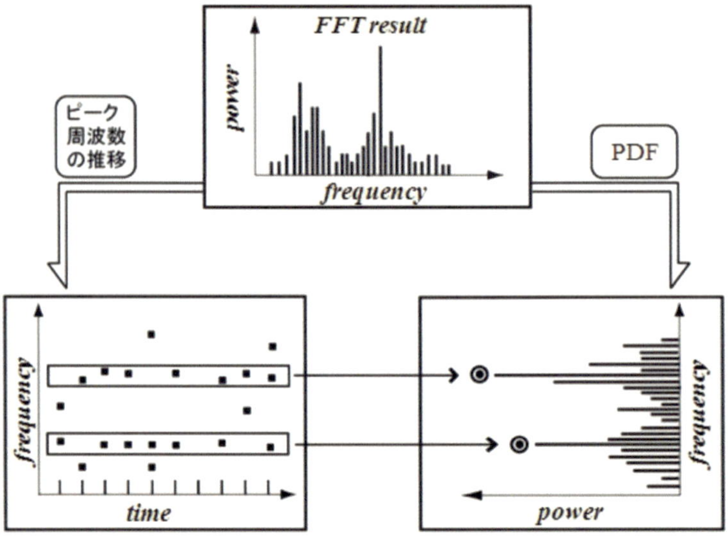

(c) Frequency-time relation

Fig. 4. FM-CW radar

If there is more than one autobile traveling in front, one pair of peak frequencies, one in the up portion and the other in the down portion, occurs for each automobile. The pairing of the peak frequencies between the up and down portions is done on an automobile-by-automobile basis. The pairing is performed based on the peak frequencies as shown in Fig. 4 (c), and the range and relative velocity of the corresponding automobile are determined. However the selection requires complicated processing.

UWB-IR radar

UWB-IR radar system transmits signals across a much wider frequency than conventional signals. The transmit signal is significant for its very light power spectrum which is typically lower than -41.3dBm. The most common technique for generating a UWB-IR signal is to

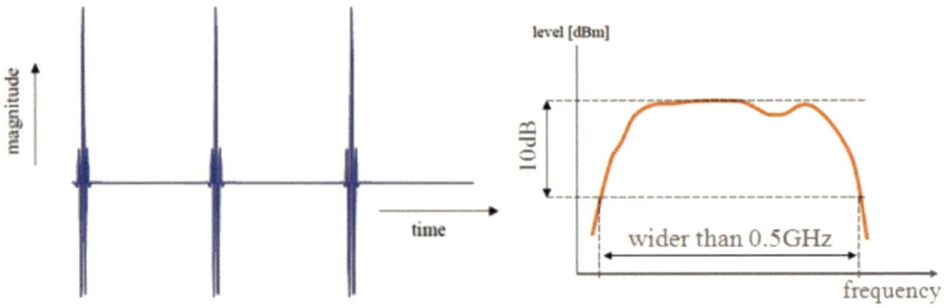

Ultra-short pulse (b) Power spectrum

Fig. 5. Train of ultra-short pulses and its spectrum

transmit a very short pulse train (less than 1nanosecond). The spectrum of a very narrow-width pulse has a very large frequency spectrum approaching that of white noise as the pulse becomes narrower and narrower. These very short pulses need a wide bandwidth as shown in Fig.5. The amount of spectrum is at least 25% of the center frequency. For the high range-resolution radar with wider than 1GHz, each scattered echo should be separated. Fig.6 shows the range profiles of a roadway where an automobile target was located at 10m. The measurement was conducted for various bandwidths of 100MHz to 5GHz (Matsunami et al., 2008). It is seen that the echoes for various objects can be separated using a shorter pulse.

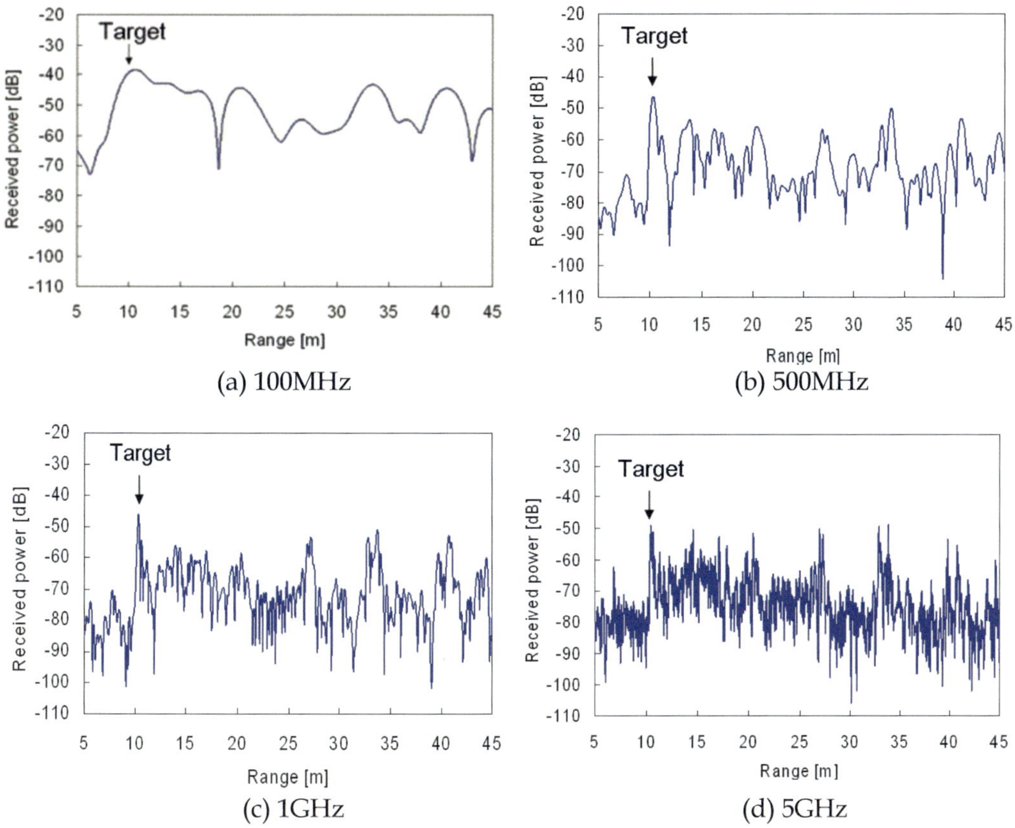

Fig. 6. Power delay profiles for automobile

3. Stepped-FM radar

UWB-IR radar provides multipath tolerability and high range-resolution, but it requires high speed A/D devices which can directly process the received nanosecond pulse. For example, A/D devices of several GS/s or more should be required for the UWB-IR radar with a bandwidth of 1GHz, which have not been available yet. Please note that it is difficult to realize analogue devices which can process such wideband pulse. The use of wideband would also cause unacceptable interference on existing narrowband systems. Therefore, some interference mitigation scheme may be required for the wideband radar emission in the future. To solve these problems, stepped-FM radar is introduced which does not require high speed A/D device and provides the co-existence with exisiting narrowband systems.

3.1 Stepped-FM scheme

The block diagram and each signal waveforms are shown in Fig.7. The stepped-FM radar transmits a series of bursts of narrowband pulses, where each burst is a sequence consisting of N pulses shifted in frequency from pulse to pulse with a fixed frequency step Δf. The received echo from a target is phase-detected into a train of narrowband base-band pulses and is then I-Q sampled by a relaxed speed of A/D. Each complex sample is applied to the inverse discrete Fourier transformation (IDFT) device in order to obtain an N-element synthetic range profile, which is called range spectrum, where the range resolution becomes approximately $1/N\Delta f$ (Nakamura et al., 2010) (Wehner, 1995). For example, suppose Δf =34.5MHz and N=30, the resolution is approximately 30 cm which is equivalent to a very short pulse with 1GHz. Therefore it does not require high speed A/D devices at the receiver.

The received stepped pulses are phase-detected and the resulting video pulses are then I/Q sampled. The n-th complex sample R_n is given by

$$R_n = A_n \exp(-j\phi_n) \tag{11}$$

$$\phi_n = 2\pi\left(f_c + (n-1)\Delta f\right)\cdot\frac{2d}{c}, \tag{12}$$

where A_n is amplitude of n-th pulse (For a stationary target, A_n should be approximated by A), f_c is fundamental frequency and c is velocity of light. Therefore received pulses (n = 1, 2,…,N) are denoted by $N\Delta f$ at frequency domain and the range-resolution ΔR is given by

$$\Delta R = \frac{c}{2N\Delta f} \tag{13}$$

Then, N complex sample is applied to the following IDFT device in order to obtain an N-element range spectrum.

$$\begin{aligned} R(\varphi) &= \left|\sum_{n=1}^{N} R_n \cdot \exp\left(j\frac{2\pi}{N}(n-1)\cdot\phi\right)\right| \\ &= N\cdot A\cdot\left|\frac{\operatorname{sinc}\left[\pi\left(\phi - N\Delta f\frac{2d}{c}\right)\right]}{\operatorname{sinc}\left[\frac{\pi}{N}\left(\phi - N\Delta f\frac{2d}{c}\right)\right]}\right| \end{aligned} \tag{14}$$

It is clear from Eq. 14 that the peak of range spectrum appears at $\Phi = 2dN\Delta f/c$, the estimated distance d is given by

$$d = \frac{c\phi}{2N\Delta f}, \tag{15}$$

where the maximum detectable range d_{max} is given by $c/2\Delta f$.

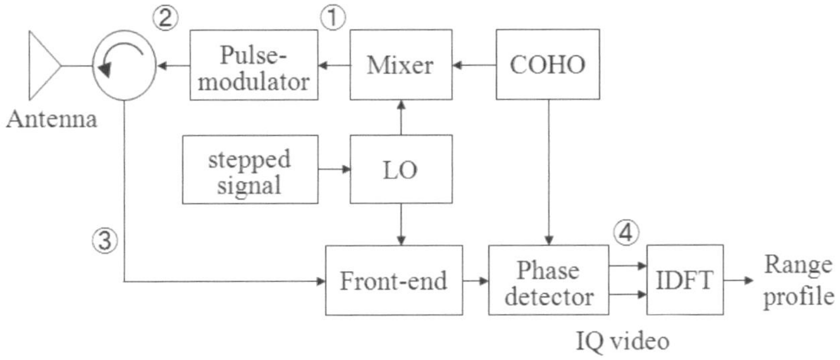

(a) Block diagram

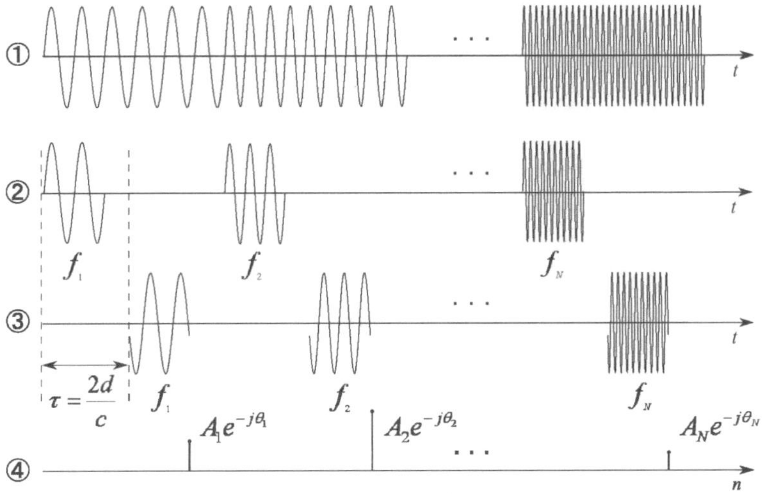

(b) Signal waveforms

Fig. 7. Block diagram and signals

3.2 Spectrum hole

The stepped-FM scheme also offers spectrum hole (non-activated within a portion of the radio spectrum) since it consists of independent pulses with different frequency. Thus it is expected to coexist with the existing systems (Nakamura et al., 2011).

Fig.8 shows the power spectrum for $\Delta f = 34.5\text{MHz}$ and N=30 where the stepped-FM pulses with the spectrum-hole of 69MHz from 3.655 to 3.724GHz 6.6% (corresponding to two consecutive stepped pulses) are not transmitted. The holed band is seen to be by approximately 13dB suppressed relative to the signal spectrum. Consider the FCC regulation of -41.3dBm/MHz, the normalized spectrum of the holed band is less than -55dBm. Therefore it should not interfere with the existing systems unlike a UWB-IR. Fig. 9 shows the range spectrum for a point target at 2.2m where we assumed a spectrum-hole of 6.6% (69MHz band as 3.655~3.724GHz) and a zero-padding for the IDFT processing (1,024points). Remarkable degradation of range-resolution is not seen, although the range side-lobe is degraded.

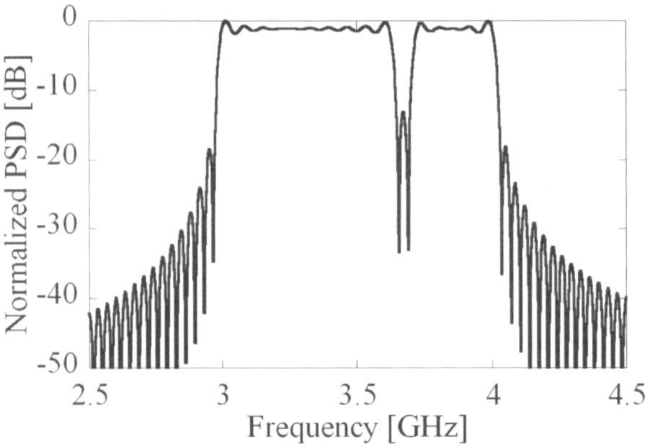

Fig. 8. Power spectrum of transmit signal with spectrum-hole of 6.6%

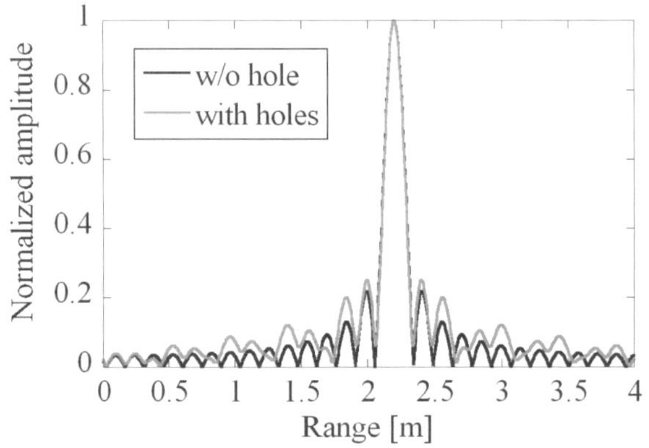

Fig. 9. Range spectra with and without hole. A point target is placed at 2.2m

3.3 Effect of spectrum-hole

The effect of spectrum-hole on the range spectrum is presented where the measurement specification is shown in Table 1. Two sphere targets with -9dBsm and -15dBsm are measured in an RF anechoic chamber (Skolnik, 2001) (Nakamura et al., 2011). The measurement was conducted in an RF absorber where theses targets on turn table were placed at 2.2m and 3m from the antenna. Fig.10 shows the range spectrum with spectrum-hole of 6.6%s, which is compared with that without hole. Please note that the other echoes at 0.8m and 1.6m are from the turn table. Fig.11 shows the range spectrum as a function of rotation angle where the distane from thsese targets to the antenna are almost equal at the rotation angle of 90 degree. These targets are found to be discriminated because of the range resolution of approximately 15cm. The measurements were conducted for $\Delta f = 34.5$MHz and N=30. Consider $\Delta f = 7.5$MHz and N=133, however, the maximum detectable range d_{max} is 20m and the range-resolution is approximately 15cm which is applicable to the short-range automotive radar.

From the measurement results, it can be concluded that the stepped-FM radar without high speed A/D devices can be coexistent with other narrowband wireless applications.

Frequency	3~4GHz
Stepped width Δf	34.5MHz
Number of step N	30
Stepped cycle	10msec
A/D	10kS/sec
IDFT point	1024

Table 1. Measurement specifications

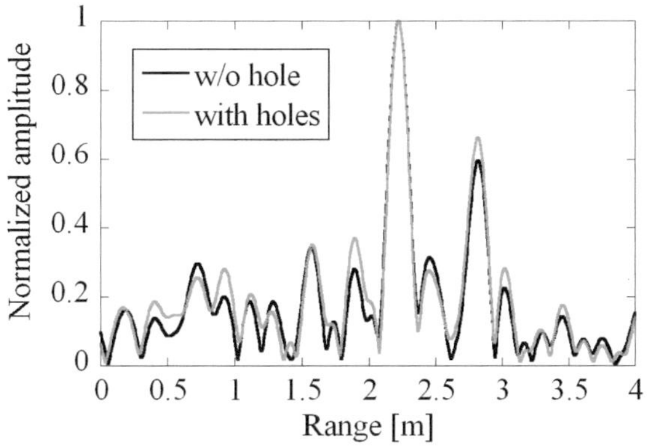

Fig. 10. Range spectra for two targets when the spectrum-holes is 6.6%

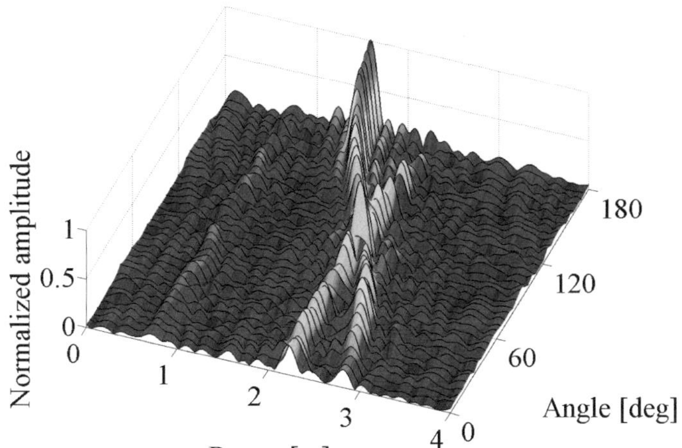

Fig. 11. Range spectrum as a function of for two sphere targets

4. Detection using trajectory estimation

Short-range automotive radar with high range-resolution should suffer from clutter because of its very broad lateral coverage. It is therefore an important issue to detect moving automobile in heavy clutter conditions. The clutter may be generally classified from automobile by the Doppler, but it will be difficult for a very short-pulse of UWB-IR radar. This is because a shorter pulse will have better range-resolution, but poorer Doppler resolution. Observing the range profile during several micro-seconds, however, each object echo's trajectory is estimated using Hough transformation and the Doppler is then calculated (Okamoto et al., 2011). When the speed of object is almost constant during the time, for example, the trajectory is regarded as linear on the time-range coordinate (Hough space). As a result, moving automobiles are separated from stationary clutter in the Hough space and detected/tracked with high range. The field measurement results at 24GHz are presented.

4.1 Time-range profile

Fig.12 shows an example of received range profile on a roadway for a bandwidth of 1GHz. The profile includes many echoes distinguishable with different delay. Detection, recognition and tracking of automobile in clutter are very important issues in automotive radar. Traditionally, the received range profile for each transmit pulse is compared against a given threshold and a detection decision is made. And once the decision is successfully done, the range profile is discarded and the next one is considered. This is called threshold detection. However it is not easy to detect some automobiles simutaneously in heavy clutter because the automobiles can't be distinguished from clutter in frequency domain. A time-range profile based detection is useful for the UWB-IR radar where moving automobiles are classified from clutter by observing the range profile. Fig.13 shows the range profiles as a function of transmit pulse number, which is called time-range profile. It is seen from Fig.13 that each echo's trajectory may be estimated and the Doppler is then calculated.

4.2 Hough transform

Hough transform (HT) has been widely applied for detecting motions in the fields of image processing and computer vision. Consider the time-range profile as shown in Fig.13, the time trajectory of each object echo can be estimated by the HT, which is a computationally efficient algorithm in order to detect the automobile on time-space data map. For example, the trajectory would be linear for a short duration of 0.1 second or less, thereby the Doppler can be calculated from the inclination of line.

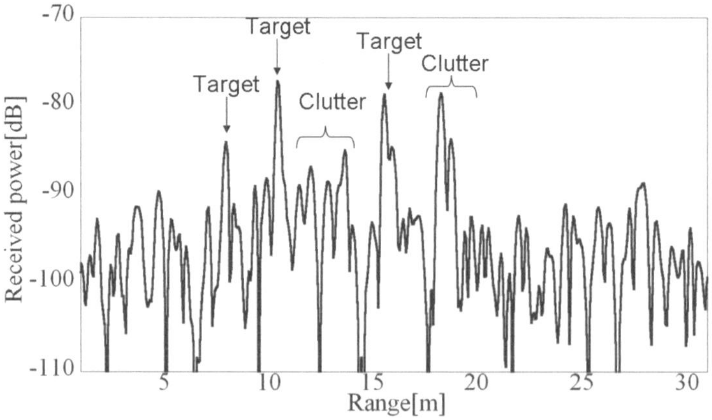

Fig. 12. Power range profile for a roadway

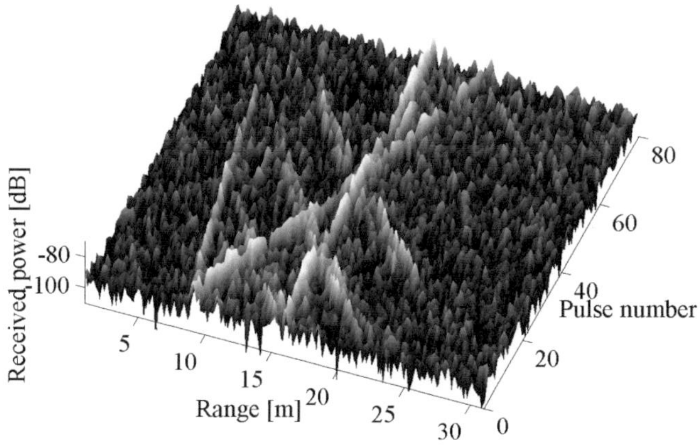

Fig. 13. Time-range profiles for 50 nanosecond pulses

4.3 Automobile classification

A. Measurement set-up and procedure

The measurements were conducted on a roadway as shown in Fig.14. The detail specification is shown in Table 2. The four automobiles were driven along the roadway and the received signals were processed on board. A pulse repetition interval (PRI) of 15ms is considered for the scenario of Fig.14. The antennas with a beam-width of 70° in horizontal direction were placed 60cm above the ground. Please note the anti-collision radar is designed for short-range/wide-angle object detection.

(a) Measurement scene

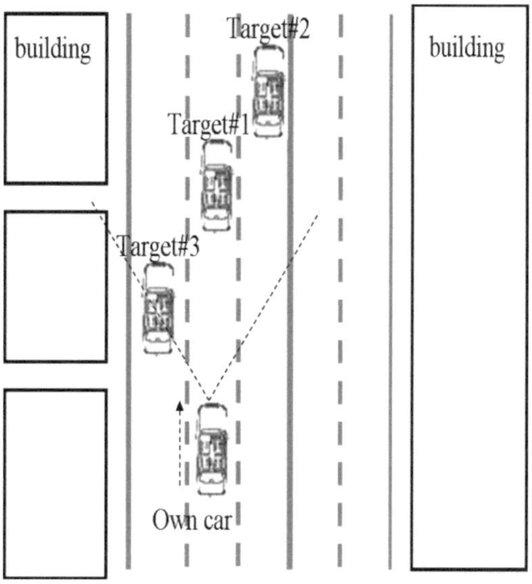

(b) Measurement scenario

Fig. 14. Measurement scenario

Bandwidth	5GHz, 1GHz (centered at 24GHz)	
Antenna	Polarization	H-H. plane
	Type	Double-ridged Horn
	Gain	12.5dBi (24GHz)
	Height	60cm
Target	Sedan: #1	4.64m×1.72m×1.34m
	SUV: #2	4.42m×1.81m×1.69m
	Mini-van: #3	4.58m×1.69m×1.85m

Table 2. Measurement parameters

B. Measurement results

Fig.15 shows the flow of HT algorithm from time-range profile to trajectory line. The quasi-images (8bits time-range image) for BW=300MHz and 500MHz are shown in Fig.15(a) and (b) respectively. Many trajectories are plotted by the Hough space translation. The number of trajectory lines depends on the signal-to-clutter ratio (SCR) and the window size to observe the time-range profile. Some trajectory lines of a time-range profile would be connected to the lines of the following profile. Therefore the trajectory of object echo can be selected using the continuity between the consecutive time-range profiles, while the quasi trajectory should be discarded. Fig.16 (a) shows the estimated trajectory lines for a BW of 500MHz. It is seen that many lines are depicted because of significant clutter. Fig.17(b) shows the survived lines by the algorithm of Fig.15 where three time-range profiles for 20 pulses are used. It is seen that clutter can be estimated from the Doppler. Fig.18 also shows the estimated lines for a BW of 300MHz. The results of Figs. 17and 18 are found to agree with the scenarios. The measurements were also conducted for different scenarios of side-looking and back-looking radar and the trajectory estimation scheme is found to be useful in order to classify the automobile from heavy clutter.

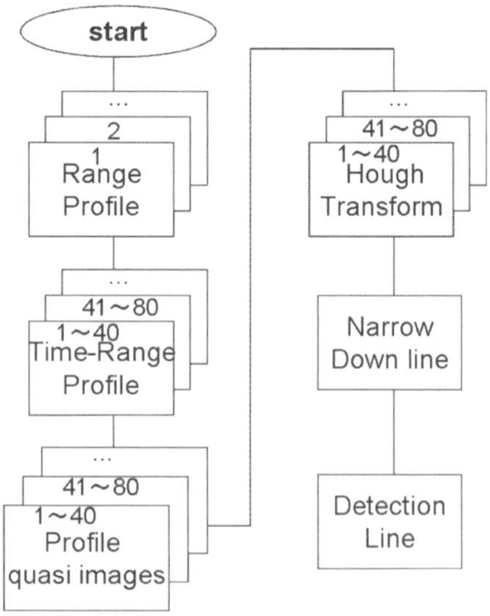

Fig. 15. Signal flow for HT algorithm

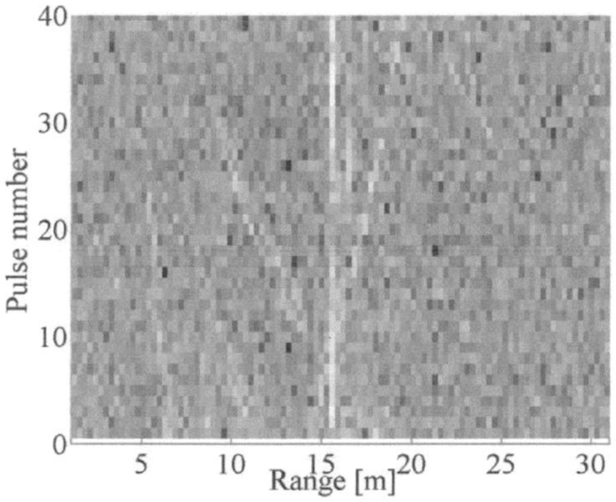

(a) BW=500MHz

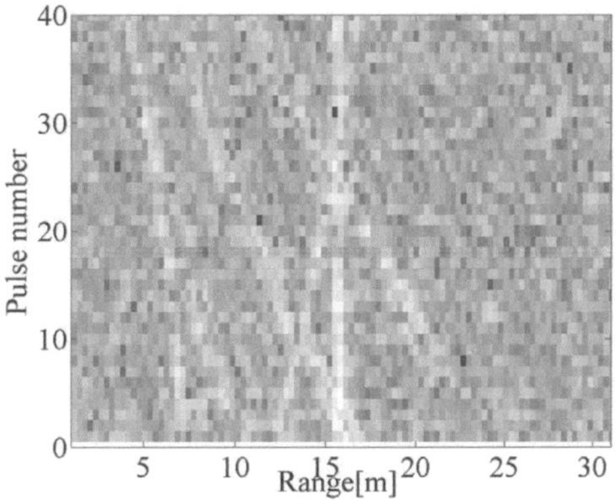

(b) BW=300MHz

Fig. 16. Quasi-images of time-range profile

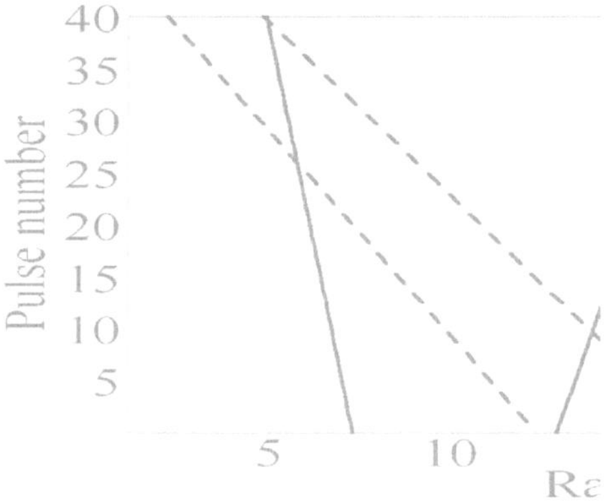

(a) Estimated trajectory lines by HT

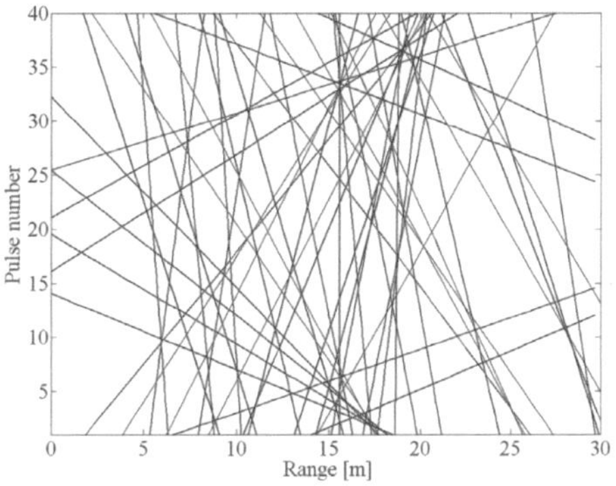

(b) Survived trajectory lines

Fig. 17. Estimated trajectory line (BW=500MHz)

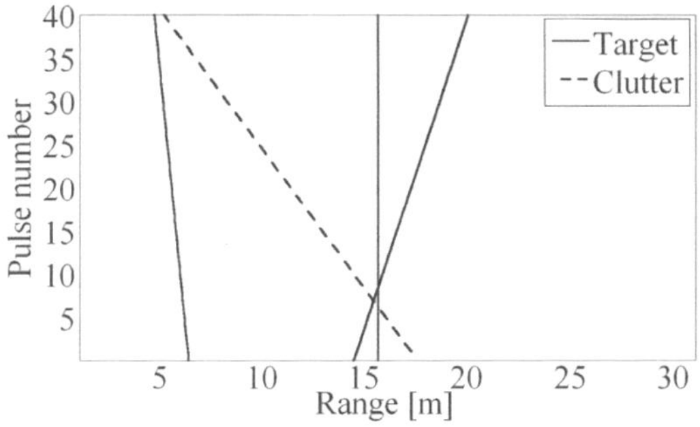

Fig. 18. Estimated trajectory line (BW = 300MHz)

5. Target discrimination

Automotive radar is required to detect automobile accurately, but not to detect clutters falsely, even in complicated traffic conditions. One-dimensional range profile of an automobile target has dependence on the shape because it has some remarkable scattered centers. Therefore the different types of automobile has different range profile feature which can be used as a unique template for automobile target discrimination/identification purpose in tracking mode. That is, the target is detected accurately by the correlation of received signal with template. The scheme also offers real-time operations unlike two-dimensional image processing (Overiez et al., 2003) (Sato et al., 2006). The measurement results are presented for various types of automobile (Matsunami et al., 2009) (Matsunami et al., 2010).

5.1 Target discrimaination and identification

Figs.19(a)-(c) show the measured range profile for various bandwidths where a sedan typed automobile was placed at approximately 10m. Please note that the profiles are expressed as a function of range-bin corresponding to the range-resolution (=1/BW). Echoes from various objects are found to be distinguished for wider bandwidth. It is seen that there exist some remarkable scattered centers. However the feature is not so clear because of scintillation and noise. Figs.20-22 show range profiles for various bandwidth where the non-coherent integration of 50 pulses was conducted in order to reduce the scintillation and noise. For the dedan, some strong echoes are seen from the side mirror and interior, and the SUV shows a unique feature.

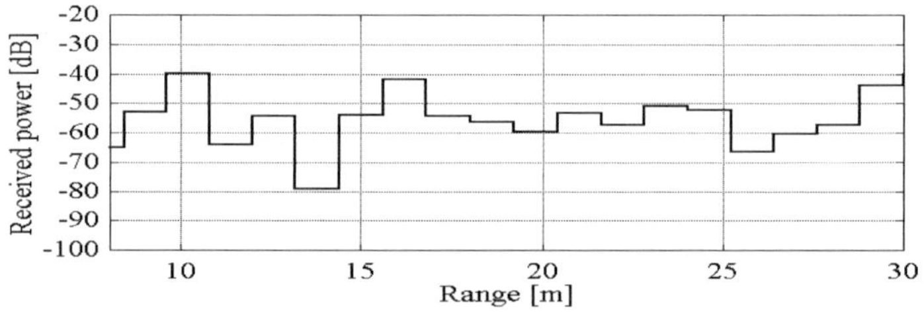

(a) BW=500MHz

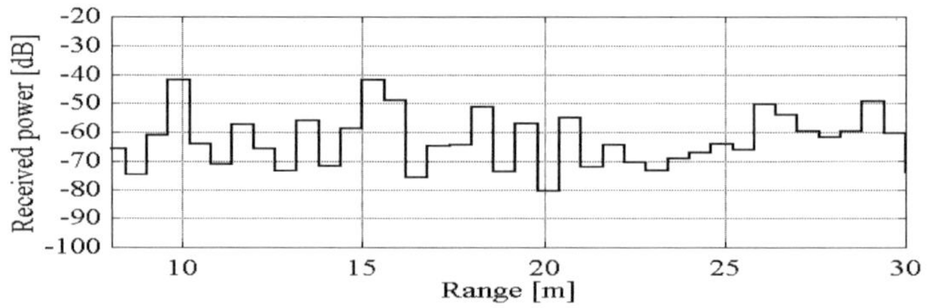

(b) BW=1GHz

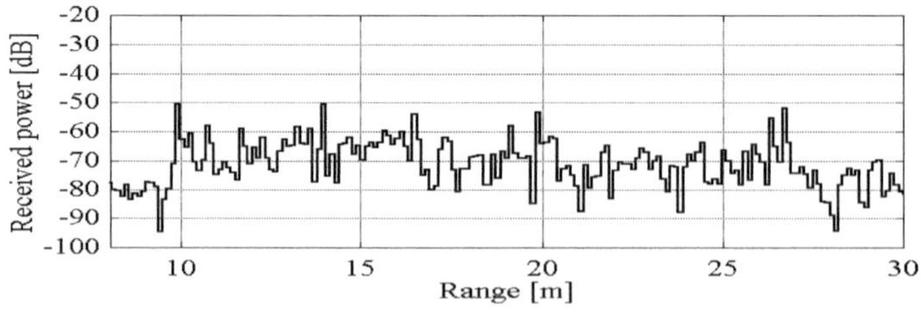

(c) BW=5GHz

Fig. 19. Power range profiles for various values of BW. A sedan was placed forward the radar antenna where the antenna to target separation was approximately 10m

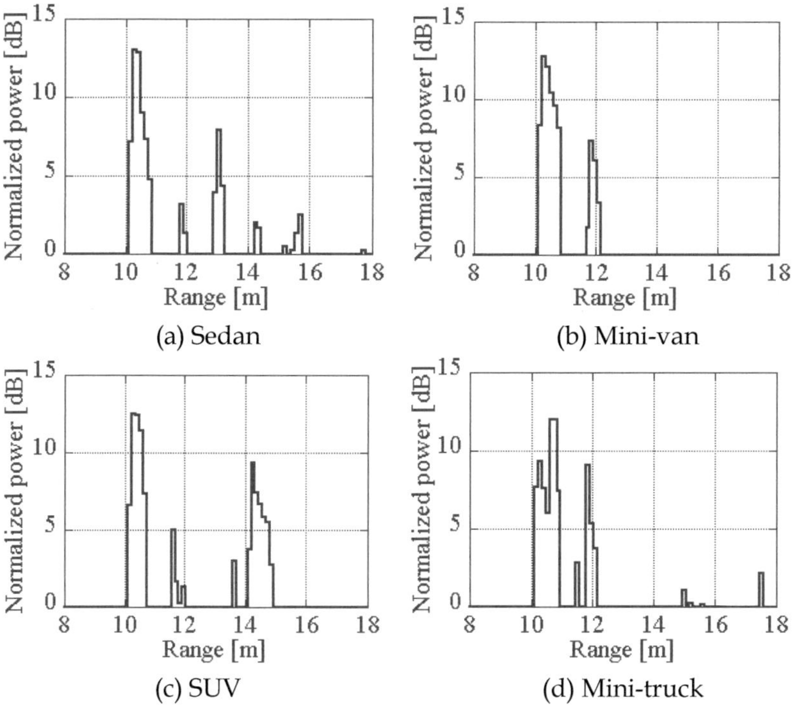

Fig. 20. Unique profiles of automobile (BW=5GHz)

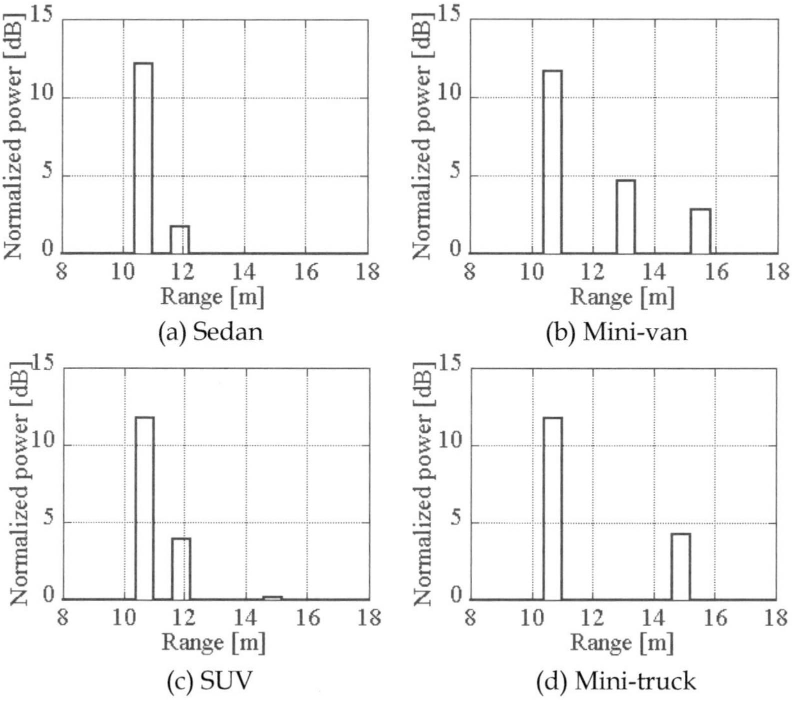

Fig. 21. Unique profiles of automobile (BW=1GHz)

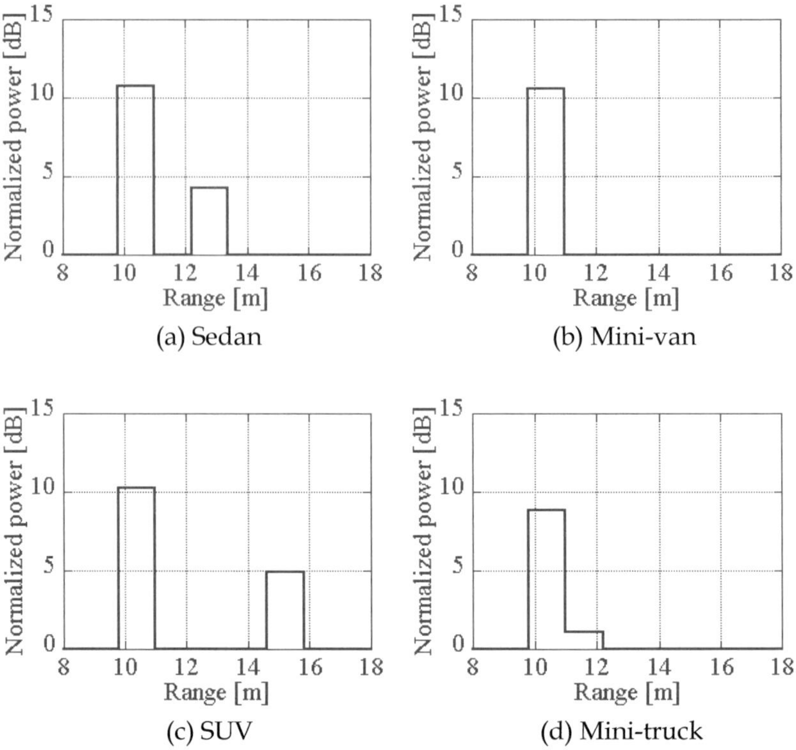

Fig. 22. Unique profiles of automobile (BW=500MHz)

5.2 Profile matching

Range profiles have been measured for four automobile #1~#4 (sedan, mini-van, SUV and mini-truck) which have been processed as the template. And the profile matching rate is calculated for various unknown automobiles. The matching rate is shown in Table.3-5. For BW=500MHz or more, it is higher than 96% when the automobile is the same as the template and each automobile can be detected. Assuming a correlation value of 0.6 for the discrimination, each profile can be identified in clutter since it has unique feature with god cross-correlation.

Template	Subject vehicle			
	Sedan	Mini-van	SUV	Mini-truck
Sedan	**99.1**	22.5	26.9	19.9
Mini-van	13.1	**96.9**	14.9	15.5
SUV	8.6	4.6	**98.6**	21.6
Mini-truck	16.8	20.8	15.3	**98.2**

Table 3. Matching rate [%]（BW=5GHz）

Template	Subject vehicle			
	Sedan	Mini-van	SUV	Mini-truck
Sedan	**99.4**	30.4	53.7	6.2
Mini-van	19.5	**96.0**	24.8	28.3
SUV	45.1	19.3	**99.3**	18.4
Mini-truck	14.7	24.8	31.7	**98.6**

Table 4. Matching rate [%] (BW=1GHz)

Template	Subject vehicle			
	Sedan	Mini-van	SUV	Mini-truck
Sedan	**99.9**	38.0	76.6	33.5
Mini-van	38.0	**98.9**	19.2	31.4
SUV	55.3	25.3	**98.2**	31.7
Mini-truck	31.2	20.0	33.0	**99.3**

Table 5. Matching rate [%] (BW=500MHz)

6. Conclusion

UWB-IR short-range radar a5 24/26GHz will be used for various applications such as pre-crash detection and blind spot surveillance. The short-range radar has a few significant problems to be overcome such as multiple targets detection and clutter suppression.
This chapter has presented how to detect multiple automobile targets in clutter. The presented results are as follows;

- UWB-IR radar requires high speed A/D devices to synchronize and detect the received nanosecond echo, thereby the system becomes very complicated and expensive. In section 3, the use of stepped-FM scheme which does not require high speed A/D has introduced for UWB-IR radar. In addition it offers spectrum hole to coexist with existing wireless systems.
- UWB-IR short-range radar is expected to provide a ***wide*** coverage in azimuth ***angle.*** Therefore, increased clutter makes it difficult to detect multiple automobile targets. Section 4 has introduced a multiple target detection scheme in heavy clutter using the trajectory of radar echoes.
- Section 5 has introduced a target identification scheme in order to improve the detection performance where a power delay profile matching is employed and the usefulness has been demonstrated by the measurement at 24GHz. The results have showed that automobile targets can be recognized and identified.

7. References

Skolnik, M. (2001). *Introduction to Radar systems, 3rd ed.*, McGraw-Hill, ISBN0-07-288138-0, New York

Taylor, J. D. (1995). Introduction to Ultra-wideband Radar Systems, CRC Press LLC, ISBN0-8493-4440-9, Wsshington, D.C..

Matsunami,I.; Nakahata, Y.; Ono, k. & Kajiwara,A. (2008). Empirical Study on Ultra-wideband Vehicle Radar, *Proc. of IEEE Vehicular Technology Conference,* ISBN 978-1-4244-1722-3, 8G-5, Calgary, Sept. 2008.

Nakamura,R.; Yokoyama,R. & Kajiwara,A. (2010), Short-Range Vehicular Radar Using Stepped-FM Based UWB-IR, *Proc. of IEEE Radio and Wireless Symposium*, ISBN 978-1-4244-4726-8, New Orleans, Jan. 2010.

Wehner, D. R. (1995). *High-Resolution Radar,* Artech House, ISBN978-0-89006-727-7, pp.197-255, 1995.

Nakamura,R. & Kajiwara,A.(2011), Empirical Study on Spectrum-Hole Characteristics of Stepped-FM UWB Microwave Sensor, to be appeared in *Proc. of* IEEE *Radio and Wireless Symposium*, Jan. 2011.

Okamoto,Y.; Matsunami,I. & Kajiwara,A.(2011), Moving vehicle discrimination using Hough, transformation, to be appeared in *Proc. of IEEE Radio and Wireless Symposium*, Jan. 2011.

Ovariez,J.P.; Vignaud,L.; Castelli,J.C.; Tria, M., & Benidir,M.(2003). Analysis of SAR image by multidimensional wavelet transform. *IEE Proc. Radar Sonar Navig.*, pp.234-241, Aug.2003.

Sato,T. & Sakamoto,T(2006). Reconstruction Algorithms for UWB Pulse Radar Systems, *IEICE Trans. Comm.*, ISBN1344-4697, vol.J88-B, pp.2311-2325, Dec.2006.

Matsunami,I. & Kajiwara,A.(2009). Power Delay Profile Matching for Vehicular Radar, *Proc. of IEEE Vehicular Technology Conference,* ISBN 978-1-4244-2514-3, 5E-1, Anchorage, Sept. 2009.

7

An Ultra-Wideband (UWB) Ad Hoc Sensor Network for Real-time Indoor Localization of Emergency Responders

Anthony Lo[1], Alexander Yarovoy[1], Timothy Bauge[2], Mark Russell[2], Dave Harmer[2] and Birgit Kull[3]

[1]Delft University of Technology,
[2]Thales Research & Technology Limited,
[3]IMST GmbH,
[1]The Netherlands
[2]UK
[3]Germany

1. Introduction

A localization system is a network of nodes, which is used by an unknown-location node to determine its physical location. The Global Navigation Satellite System, GNSS (Hofmann-Wellenhof, 2008) is an example of a widely used outdoor localization system. However, outdoor localization systems perform poorly in indoor environments due to strong signal attenuation and reflection by building materials, and no line-of-sight propagation. Thus, Indoor Localization Systems (ILSs) are needed to provide similar localization inside buildings. ILSs have many potential applications in the commercial, military and public safety sectors. This chapter focuses on the public safety application. The considered ILS is used to track emergency responders, e.g. fire-fighters and policemen, who carry out search and rescue missions in the disaster zone such as building fires and collapsed tunnels. Such an ILS was first crystallized in the EUROPCOM (Emergency Ultra wideband RadiO for Positioning and COMmunications) project (Harmer, 2008; Harmer et al., 2008). The EUROPCOM system is an ad hoc sensor network which comprises a small number of base or reference nodes deployed outside surrounding a building, and the rest of the nodes are unknown-location nodes which are worn and deployed by emergency responders entering the hostile building. The unknown-location node is self-localized by collectively determining its position relative to base nodes. Additionally, the unknown-location node is also allowed to determine its position relative to neighboring unknown-location nodes. This greatly enhances the accuracy and robustness of the ILS. It is fully autonomous and can be rapidly deployed with little human intervention.

Ultra-WideBand (UWB) is the radio transmission technology used by the EUROPCOM system. A UWB signal is defined to be one that possesses an absolute bandwidth of at least 500 MHz or a fractional bandwidth larger than 20% of the center frequency. Currently, several UWB technologies exist, namely direct sequence UWB, impulse radio UWB, Multiband Orthogonal Frequency Division Multiplexing (MB-OFDM) UWB, Chaotic UWB, and Frequency Hopping (FH-UWB). The EUROPCOM system selected FH-UWB because it offers significantly better range and position accuracy than other technologies such as pulse UWB (Frazer, 2004).

A great deal of effort has been expended on localization algorithms, but the Medium Access Control (MAC) and routing protocols for ILS have received very little attention yet. Unlike other ad hoc sensor networks, the considered ILS exhibits unique characteristics. Therefore, it poses new technical challenges in the MAC and multi-hop routing protocol design. Firstly, the ILS is heterogeneous in the sense it is composed of different types of nodes with varying capability, processing power and battery energy. Secondly, the ILS operates in a highly dynamic and hostile environment. Lastly, emergency applications require fast localization in the order of seconds. In order to address these challenges, we propose a novel Self-Organizing Composite MAC (SOC-MAC) protocol and a Lightweight and robust Anycast-based Routing (LAR) protocol. Cross-layer approach is present in the design to attain highly optimized, bandwidth- and energy-efficient protocols.

2. Network architecture of an Indoor Localization System (ILS)

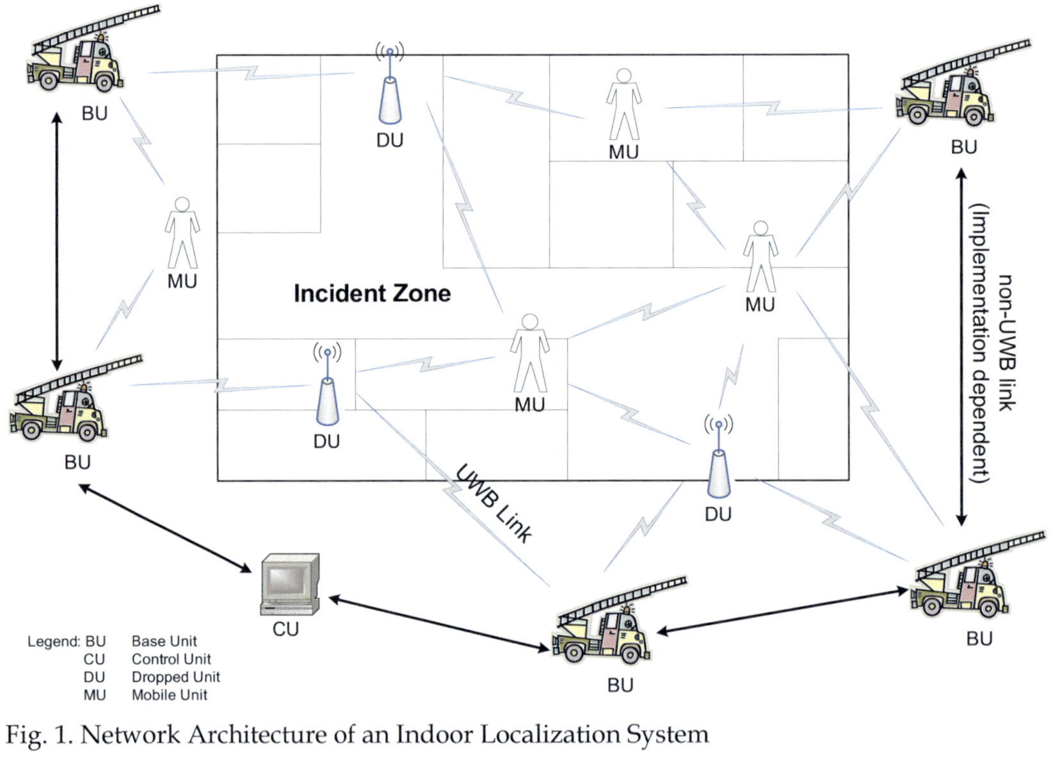

Fig. 1. Network Architecture of an Indoor Localization System

The assumed ILS, which is an ad hoc sensor network, consists of four types of nodes: a Control Unit (CU), Base Units (BUs), Dropped Units (DUs) and Mobile Units (MUs), as shown in Fig. 1. The MU is a sensor that is worn by every emergency responder. The MU has the capability to calculate its position which is in turn delivered to the CU. The BUs are located outside and around the incident area, while maintaining wireless connectivity with the emergency responders inside the building. Unlike other units, the position of BUs is known and most likely to be acquired through GNSS. Furthermore, the BUs will remain stationary throughout the entire mission. The DUs are strategically placed in the incident area by emergency responders to serve as relay nodes once the MUs lose wireless connectivity with the BUs. Similar to MUs, the DUs can determine their positions and relay them to the CU. The CU provides the main visual display to the rescue coordinators, showing the current position and direction of movement of individual emergency responders with respect to the incident area topology, e.g. a building. As shown in Fig. 1, the ILS is composed of a UWB subnetwork and a non-UWB subnetwork. The reason for two separate subnetworks is that the CU is not involved in the localization process. Thus, more radio resources are available for the UWB subnetwork, in particular, when the number of MUs increases.

2.1 System assumptions

In this subsection, we state several assumptions made in the design of the MAC and routing protocols. The MAC and routing protocol design assumes the FH-UWB technology is employed by the Physical layer of the BU, the DU and the MU. The operating bandwidth of the FH-UWB units is 1.25 GHz which consists of 125 carrier frequencies. This means, the carrier spacing is 10 MHz. The center frequency is located at 5.1 GHz. Each unit follows a fixed hop pattern. The pair CU-BU communicates over a non-UWB link. Similarly, the BU-BU transmission is also over non-UWB links. The rationale for using a non-FH-UWB technology is that more radio resources are available to the FH-UWB subnetwork. Since the non-FH-UWB technology is implementation-dependent, we will not further deal with the specifics of the non-UWB technology in the rest of the chapter. The design of the MAC and routing protocols is described in subsections 2.2 and 2.3, respectively.

2.2 A Self-organizing Composite Medium Access Control (SOC-MAC) protocol

As each MU is mobile, it will determine and transmit its position information to the CU periodically. For instance, in order to cope with user mobility in the order of 0.5 m/s (walking speed), an MU needs to measure and transmit position information to CU at a rate of one position packet per second. As a result, SOC-MAC is based on the Time Division Multiple Access (TDMA) because such a MAC is particularly suited to the periodic nature of localization process. Unlike traditional TDMA, SOC-MAC is designed for ad hoc networks with no requirement for a central controller for allocating time slots as it is self-organizing.

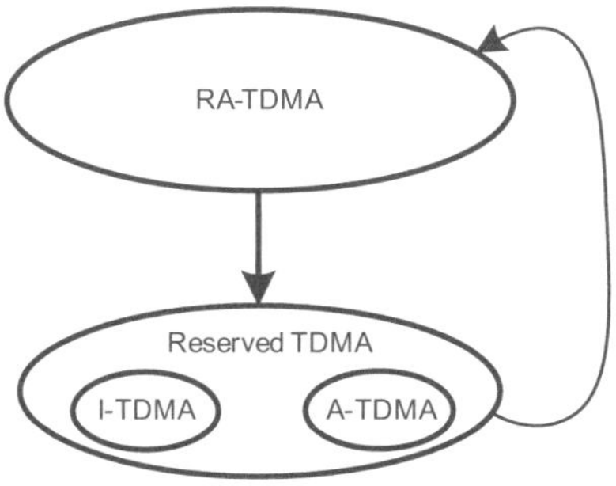

Fig. 2. SOC-MAC

Using SOC-MAC, each unit can autonomously select and reserve time slots based only on local network knowledge without the need of dedicated signaling messages.
As shown in Fig. 2, the SOC-MAC protocol operates in two phases: a Random Access TDMA (RA-TDMA) phase and a Reserved TDMA phase. The former phase is invoked by a unit prior to joining the network or when the unit has not used any time slots in the previous superframe; thus, there are no reservations in the current frame. In this phase, a unit acquires a time slot through a random access mechanism. Once a time slot has been acquired, SOC-MAC will enter the Reserved TDMA phase. The RA-TDMA phase and the Reserved TDMA phase are described in the following subsections.

2.2.1 Random Access TDMA (RA-TDMA) phase

The UWB medium is segmented into SOC-MAC superframes, each of which has a constant period of *T* seconds. Each superframe is in turn partitioned into *N* orthogonal time slots of duration *T/N* seconds. The start of the superframe is provided by one of the BUs, known as the Master BU (MBU). Naturally, MBU will occupy the first time slot. Units unable to hear the transmissions of MBU will synchronize to the TDMA frame by monitoring the transmissions of other neighboring units, which will identify the time slot in which they are transmitting. From this time slot number information, the start of the frame can be inferred.
Fig. 3 depicts the structure of the superframe and time slot. Each time slot can be used for either data transmission (referred to as "data slot") or ranging (referred to as "ranging slot"). The latter time slot format is specifically used for determining the range between two units. Thus, the ranging slot can accommodate a very limited payload. The MAC header includes identifiers of up to five ranging units, denoted as "R1 ID" to "R5 ID", which have been selected to respond to ranging requests. The final part of the ranging slot is reserved for the corresponding "pong" responses from "R1 ID" up to "R5 ID". Unlike ranging-slot, data-slot is purely utilized to transmit user data and can accommodate larger payload. The MAC header of the two slot structures contains similar fields except the ranging-slot includes the ranging unit identifiers and the position data of the transmitting unit (i.e., TX ID POS). Hence, the MAC header of the ranging-slot is longer and has to be split into two parts separated by two pilot tones as shown in Fig. 3.
In general a unit enters the RA-TDMA phase prior to joining the network. Fig. 4 contains the flow chart of the RA-TDMA phase. Before the unit can transmit in a time slot, it must listen to the physical channel for at least one complete TDMA superframe period. During this period, the unit constructs a list of one-hop neighbors and a map of their time-slot usage. Based on the time-slot usage map, the unit derives a list of vacant time slots in the forthcoming superframe. The number of vacant slots in the list is denoted as candidate slot counter (*csc*) in Fig. 4. When a first vacant time slot in the next superframe arrives, the *p*-persistent algorithm is applied to determine if this vacant time slot can be used for transmission. The *p*-persistent algorithm defines two parameters, namely *P1* and *P2. P2* is inversely proportional to *csc,* and *P1* is randomly selected from an interval [0 .. 1]. If *P1* is equal to, or less than *P2,* then the vacant time slot is reserved and transmission should occur in the reserved time slot. If not, the number of vacant time slots *csc* in the list is decremented by one and the same procedure is repeated for the next vacant time slot. The *p*-persistent algorithm minimizes the chance that two or more units in the RA-TDMA phase are contending for the same time slot. A low *csc* increases the probability of selecting the next vacant time slot. The number of unsuccessful attempts in reserving a time slot is recorded in attempt count (*ac*). Once a vacant time slot has been successfully reserved, the RA-TDMA phase ends and the reserved TDMA phase sets in to complete the channel access procedure.

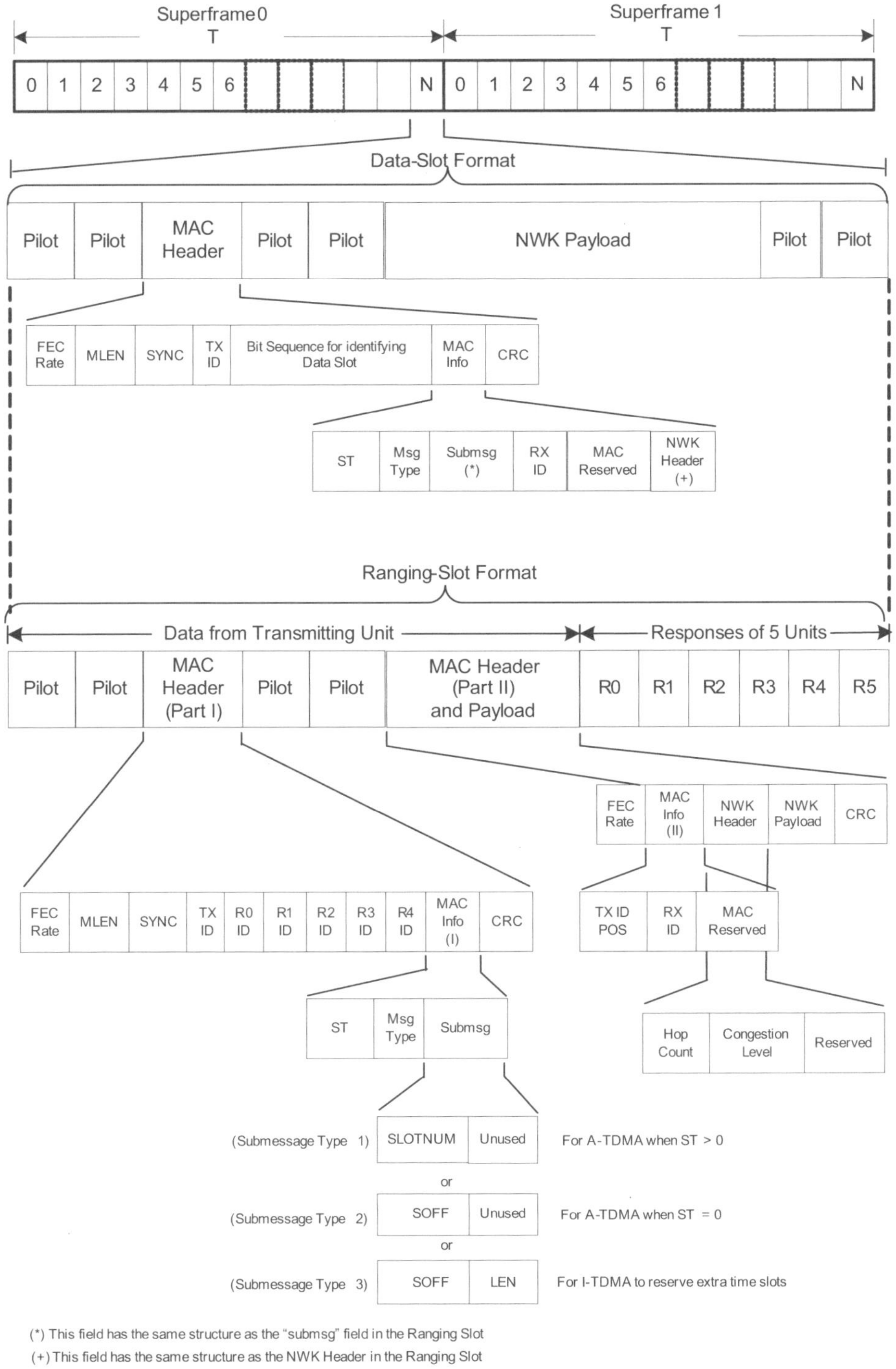

Fig. 3. SOC-MAC Superframe

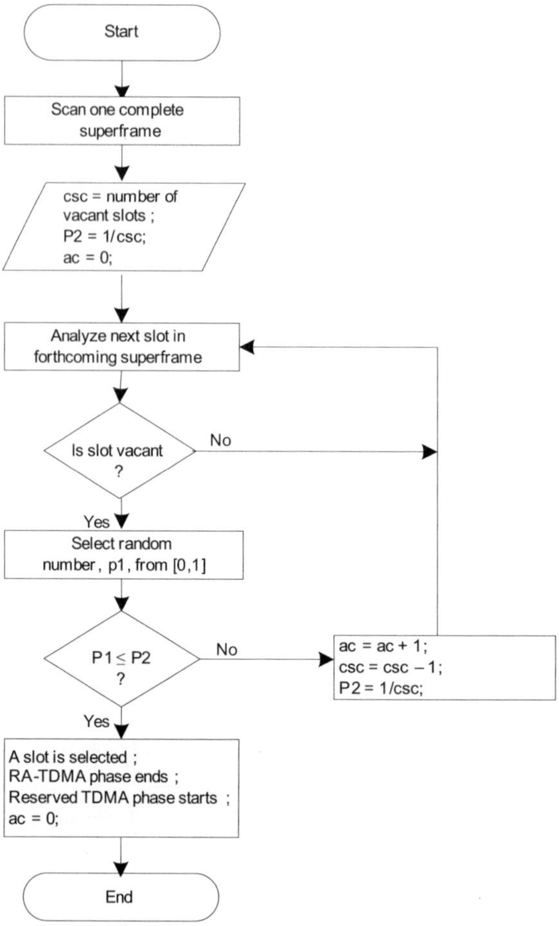

Fig. 4. Flow Chart of RA-TDMA

2.2.2 Reserved TDMA phase

The Reserved TDMA mode comprises two operations, namely the Autonomous TDMA (A-TDMA) and the Incremental TDMA (I-TDMA). The latter is used to acquire additional slots in the same superframe in addition to the one acquired in the RA-TDMA phase. A-TDMA is responsible for managing the acquired time slots.

A-TDMA

Once a slot has been acquired through RA-TDMA and/or I-TDMA, the same time slot is automatically reserved for the next *Slot_Timeout* (ST) superframes, where ST is randomly picked from an interval [*1* .. *MAX_TIMEOUT*]; *MAX_TIMEOUT* is a MAC design parameter. A-TDMA is responsible for keeping ST up-to-date. That is, ST is decremented by one in each new superframe. ST is included in the MAC header so that other units can determine when the time slot will be free. When ST > 0, the *Submessage Type 1* is used, which contains the time-slot number (SLOTNUM) of the currently reserved time slot as illustrated in Fig. 3. When a time slot expires (i.e., ST = 0), A-TDMA randomly chooses a vacant time slot in the next superframe from a list of vacant time slots in the time-slot usage map, and pre-announces to the other units the offset between the present time slot and the newly selected time slot (SOFF expressed in number of time slots) using the format *Submessage Type 2* in the current superframe as shown in Fig. 3. This allows other units to find this unit in the next superframe without searching and to update the time-slot usage map. The new time slot will only be used in the next superframe. The continuous change of time-slot positions ensures that if two or more units had chosen the same time slot in the RA-TDMA phase, the collision can only persist for a maximum of *MAX_Timeout* superframes before one or all involved units must choose a different time slot. Thus, the collision is resolved through a probabilistic means. Hence, *MAX_TIMEOUT* must be small in order to reduce the number of collisions that are energy-wasting. On the other hand, if *MAX_TIMEOUT* is too small then neighbor units need to perform frequent updates on the time-slot usage map, which in turn increases power consumption. The new time slot is assigned a new *ST* value which is obtained using the same process as described above. The A-TDMA algorithm is depicted in Fig. 5.

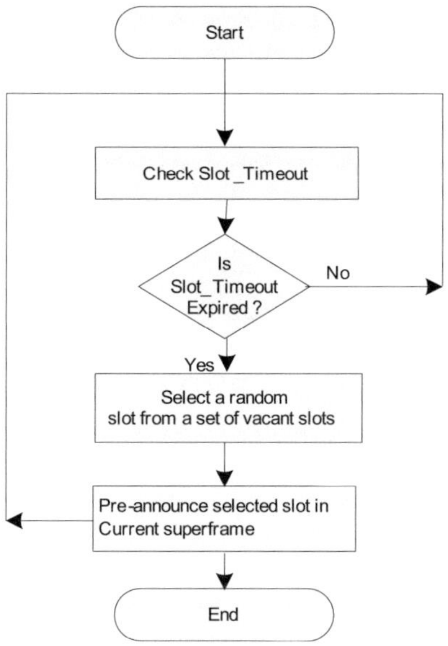

Fig. 5. Reserved TDMA Phase: ATDMA Operation

I-TDMA

The time slot acquired during the RA-TDMA phase is the first and only one for each unit. If a unit needs extra time slots, then I-TDMA is employed to reserve the extra time slots in the same superframe to increase the data rates. I-TDMA calculates the number of required time slots N_I based on the actual queue length provided by the Network layer. It searches for a block of N_I successive vacant time slots in the time-slot usage map. If not available, N_I is reduced until the search is successful. The number of reserved time slots (LEN) and the offset (SOFF) between the current and the first new time slot are advertised using the format *Submessage Type 3*, refer to Fig. 3, so that all other units are informed about the new reservations. In principle each new time slot can be used for another I-TDMA operation so that the number of time-slot reservations can grow more rapidly. Hence, the usage of I-TDMA needs to be restricted if the channel is busy and the number of vacant time slots is small. Note that in almost all cases the A-TDMA operation is required, while the I-TDMA operation is only sporadically needed to increase the data rates by reserving additional time slots. In order to free reserved time slots, the time slots are simply not renewed by A-TDMA after *Slot_Timeout* superframes.

2.3 A Lightweight and Robust Anycast-based Routing (LAR) protocol

The Lightweight and Robust Anycast-based Routing (LAR) protocol routes data packets from MUs or DUs to the nearest BU. There is no exact destination BU for a data packet. Thus, routing decisions must rely on routing parameters and packet types. LAR defines two routing parameters, namely *hop count* and *congestion level*. Hop count indicates the distance of a unit (in terms of the number of hops) to a reference BU. It increases monotonically at each hop. Congestion level is used to indicate the buffer occupancy of a unit. These routing parameters are not disseminated using dedicated routing packets but carried and propagated in the Network (NWK) header of data packets. Thus, LAR does not incur routing packet overheads. The format of the NWK header is depicted in Fig. 3. This means that irrespective of the data type, the NWK header always contains the mandatory routing parameters. The NWK header occupies 12 bits in a total of 1831 bits in one time slot of the SOC-MAC superframe. Therefore, the overheads of the NWK header for routing are less than 1%, which conserves bandwidth and energy.
Route establishment is initiated by BUs to form spanning trees rooted at each BU. This is a natural choice because each BU periodically broadcasts its position which is known beforehand, while DUs and MUs just listen to the BU broadcasts since they need to determine their position. The BU sets the initial value for the hop count and congestion level. From the BU broadcasts, the DUs/MUs create a new entry in the routing table if it does not exist. The routing table entry contains the following fields: *neighbor unit id, hop count, congestion level, FEC level* and the *expiration time of the entry*. The first field identifies the address of the unit that broadcasts the data packet, which represent the next-hop unit for the route towards a destination BU. The neighbor unit id is contained in the MAC header. Note that the unit maintains only the next-hop routing state, which provides the routing protocol with a high degree of scalability. The hop count in the routing table is incremented by one with respect to the received hop count. For instance, if the incremented hop count is $n+1$ then the unit is $n+1$ hops away from the destination BU. The congestion level field is extracted from the NWK header. FEC (Forward Error Correction) level determines the channel bit rate for communicating with the next hop of the neighbor unit id. Four FEC levels, viz., FEC-1 to FEC-4, are defined. FEC-4 provides the highest bit rate but no or the lowest level of error protection. The FEC level is also contained in the MAC header. Once a DU/MU has determined its position, it can broadcast its position. The hop count in the NWK header is obtained from the selected route in the routing table entry, while the congestion is set to the maximum of its congestion level and that in the routing table entry. In the case of multiple entries in the routing table, a route selection algorithm with load balancing is used to choose the next hop. The algorithm will be described in subsection 2.3.1. So far, we have focused on route construction from an MU/DU to a destination BU, which is referred to as forward route. A reverse route (from a BU to an MU or DU) can be constructed using data packets sent on the forward route. One such data packet is position reporting which is used to transport position data to the BU. Position reporting packets are periodically sent by an MU and DU. The position reporting packets are transmitted using a forward route selected by the unit to a destination BU. All units along the forward route store the source and forwarding unit identifiers in their routing table. The latter identifier is

the address of the intermediate unit that forwards the data packet while the source identifier is the address of the unit which generates the position reporting packets. No other routing parameters are needed for the reverse route. Since the position broadcasting and position reporting are periodic, the forward and reverse routes are always up-to-date. Therefore, no specific route recovery or maintenance functions are required.

2.3.1 Load balancing

Load balancing is achieved using the congestion level parameter, which is based on the occupancy of queues in a unit. The queues allocated by a unit are assumed to be fixed size. The congestion level is then deduced from the queue occupancy as shown in Table 1.

Congestion Level	*Queue Occupancy*	*Definition*
0 - 2	20% - 40% full	Not congested
3 - 4	50% - 60% full	Slightly congested
5 - 6	70% - 80% full	Congested
7	90% full	Heavily congested

Table 1. Congestion Levels for Load Balancing

2.3.2 Route selection

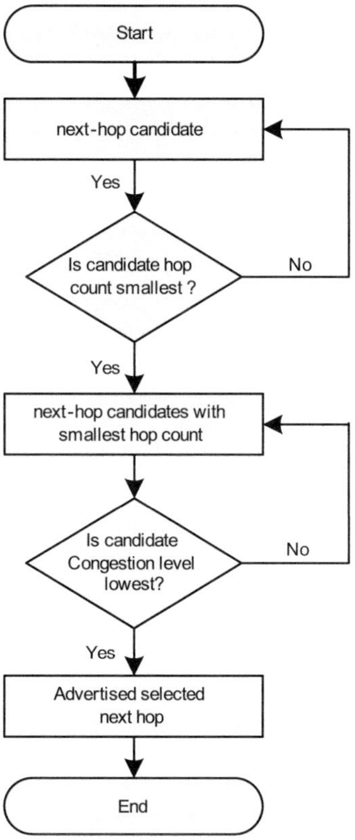

Fig. 6. Next-hop Selection Algorithm with Load Balancing

Hop count is the primary routing metric, while *congestion* is the secondary metric due to the delay at the MAC layer, which cannot be tolerated by real-time data packets. In the case of multiple entries in the routing table, LAR must select the candidate route with the smallest hop count. If there are several candidate routes with the same hop count then the candidate with the lowest congestion level is picked. By selecting the candidate route with the smallest hop count the selection algorithm can guarantee loop-free delivery as a data packet is always forwarded from a unit with a higher hop count to a unit with lower hop count. The selection algorithm is shown in Fig. 6.

3. Simulation set-up

The feasibility and performance of SOC-MAC and LAR are evaluated by means of simulation. To this end, we extended the Mobility Framework (MF) (Mobility Framework) module by incorporating a model for a UWB Physical layer, the SOC-MAC protocol, the LAR protocol and the Application layer, and the ILS network entities. MF is an add-on package for simulating mobile and wireless networks on the OMNeT++ platform (OMNeT++) which is a powerful generic, object-oriented and discrete-event simulation tool. Naturally, MF can be easily extended for simulating the ILS network. Thus, three new simulation nodes, namely BUhost, DUhost, and MUhost, were defined. These nodes correspond to the units BU, DU and MU, respectively. CU was not modeled because it is in the non-UWB subnetwork which is implementation-specific. Fig. 7 depicts a sample of the simulation network, which consists of four BUhosts, two DUhosts and four MUhosts. In each of the simulation nodes, three protocol models, viz., the application, the network and the Network Interface Card (NIC) were defined as extensions to the corresponding models in MF. The internal structure of the node is shown in Fig. 8(a). The Blackboard and Mobility models were used without extensions. Note that BUhost, DUhost and MUhost have the same internal node structure. The application model, EuropAppLayer, the network model, EuropNetwLayer, the MAC model, EuropMacLayer, and the Physical model are described in the next subsections.

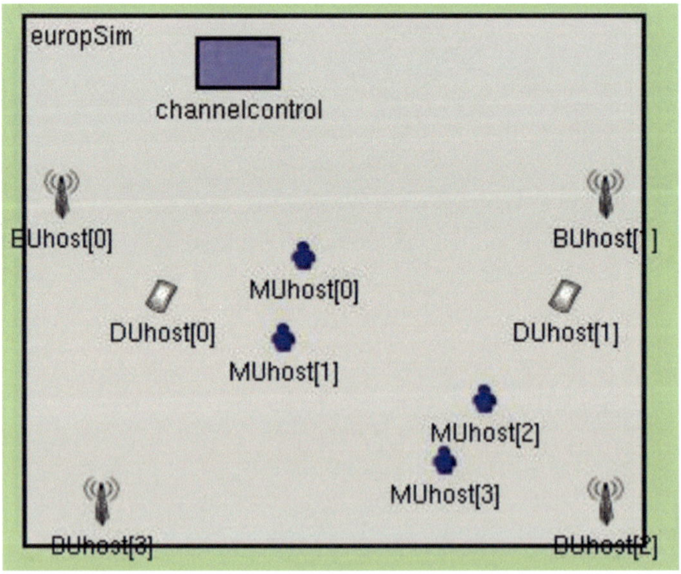

Fig. 7. Simulation Network

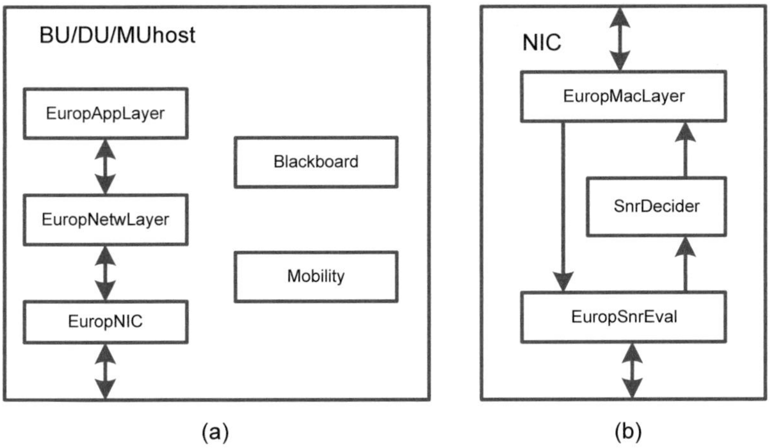

Fig. 8. Node and NIC Structure

3.1 Physical layer model

The Physical model is divided into EuropSnrEval and SnrDecider as shown in Fig. 8(b). The former was extended from SnrEval in MF while the latter was used as it is. EuropSnrEval is used to calculate the Signal-to-Interference-plus-Noise (SINR) of a received MAC frame. The SINR is defined as

$$SINR = 10\log\frac{P_{r,\max}}{P_n + I} \tag{1}$$

where $P_{r,max}$ is the strongest received signal power among the received signals, based on the capture effect (Rappaport, 2001). P_n is the Additive White Gaussian Noise (AWGN). I is the interference power which is defined as the sum of all received signal power excluding $P_{r,\,max}$. The interference power I is expressed as

$$I = \sum_{\neq P_{r,\max}} P_r \tag{2}$$

In case of collision-free transmission, the term I is null. Hence, Equation (1) is reduced to

$$SINR = 10\log\frac{P_{r,\max}}{P_n} \tag{3}$$

The computed SINR is passed to SnrDecider which determines whether the MAC frame is correctly received or not. A MAC frame is considered to be correctly received, if SINR ≥ $SINR_{th}$, where $SINR_{th}$ is the SINR threshold. A correctly received frame is delivered to the EuropMacLayer, otherwise it is discarded. $SINR_{th}$ was obtained through physical layer simulation, which produces Bit Error Rate (BER) plots as a function of SINR. Given a target BER, $SINR_{th}$ is deduced. The physical layer simulation was carried out separately using another tool since OMNeT++ and MF lack the support for simulating physical layer functions such as frequency hopping, channel coding, modulation, and signal processing.

The received power P_r in Equation (1) is characterized by large-scale fading and small-scale fading. Large-scale fading represents the average signal power attenuation when transmitted through the medium. The attenuation or commonly known as Path Loss (PL) as a function of distance is expressed as (Rappaport, 2001).

$$PL(d) = PL(d_0)(\frac{d}{d_0})^{\gamma}\, 10^{x_\sigma / 10}, \qquad d \geq d_0 \tag{4}$$

where d_0 is the reference distance, γ is referred to the path loss exponent, and X_σ denotes the log-normal shadowing effect with a zero-mean normal distribution (in dB) and standard deviation σ (also in dB). $PL(d_0)$ is evaluated using the free-space path loss equation or by conducting measurements. In our work, $PL(d_0)$ was determined using the free-space path loss equation which is given by (Rappaport, 2001)

$$PL(d_0) = \left(\frac{4\pi f_c d_0}{c}\right)^2 \tag{5}$$

where $f_c = \frac{f_{\max} + f_{\min}}{2}$. f_{min} and f_{max} are the lower and upper boundary of UWB transmission frequency band, respectively. Substituting Equation (5) into Equation (4), and let d_0 = 1 m in our case, we can rewrite Equation (4) as

$$PL(d) = \left(\frac{4\pi f_c}{c}\right)^2 d^{\gamma}\, 10^{x_\sigma/10}, \qquad d \geq d_0 \tag{6}$$

Small-scale fading represents the wide variations in received signal strength caused by interference between two or more versions of the transmitted signal arriving at the receiver at slightly different times. It is typically modeled by the Ricean distribution or the Rayleigh distribution when there is a line-of-sight or non-line-of-sight, respectively. In UWB systems, the signal power variations due to small-scale fading are not severe due to the ultra-large bandwidth of UWB signals and diversity techniques used in the physical layer. Thus, in our physical channel model, we are only concerned with the large-scale fading. Hence, the received power $P_{r,max}$ in Equation (1) and P_r in Equation (2) can be calculated using

$$P_r = \frac{P_t}{PL(d)} \tag{7}$$

where $PL(d)$ is given in Equation (6), and P_t is the transmit power.

3.2 MAC layer model

The MAC model, EuropMacLayer, captures the complete functionality of SOC-MAC described in Section 2.2. It was derived from the BasicMacLayer model of MF. The model definition consists of three parts, referred to as a EuropMacLayer module definition, a EuropMacLayer protocol data unit definition, and a EuropMacLayer module implementation. The EuropMacLayer module definition, which is specified using the OMNeT++ NED language. The EuropMacLayer protocol data unit definition, called EuropMacPkt, was derived from the MacPkt definition of MF. The derived module contains the fields of the EuropMacLayer protocol data unit only. The EuropMacLayer module implementation contains the algorithms of the composite MAC. Unlike the EuropMacLayer module definition and EuropMacPkt definition, this module was directly written in the C++ programming language. The EuropMacLayer module definition and EuropMacPkt are translated into C++ code when an executable of the simulation program is built.

3.3 Network layer model

The Network model, EuropNetwLayer, implements the LAR protocol described in Section 2.3. It was derived from the SimpleNetwLayer model of MF. Similar to the MAC model, it consists of three parts: a EuropNetwLayer module definition, a EuropNetwLayer protocol data unit definition, and a EuropNetwLayer module implementation. The EuropNetwLayer protocol data unit definition, called EuropNetwPkt, was derived from the NetwPkt definition of MF.

3.4 Application layer model

The application traffic model generates dummy position packets of fixed size at regular intervals. The dummy position packets carry no real position information and the simulated nodes do not perform position estimation. This does not affect the performance of SOC-MAC and LAR as long as the application model can mimic the traffic behavior of the real system. The application traffic model, called EuropApplLayer, which was derived from BasicApplLayer of MF. The application traffic model also consists of three parts: a EuropApplLayer module definition, a EuropApplLayer protocol data unit definition, and a EuropApplLayer module definition.

4. Simulation results

4.1 SOC-MAC performance

We analyze the performance of SOC-MAC. The performance measures for SOC-MAC are the successful SOC-MAC packet reception rate and the network throughput. An SOC-MAC packet consists of a header and payload for both the data- and ranging-slot as illustrated in Fig. 3. Thus, in one time slot, only one SOC-MAC packet is transmitted. The successful packet reception rate P in the network is defined as the total number of SOC-MAC packets received by all units divided by the total number of SOC-MAC packets transmitted by all units in the network. Hence, P is expressed as

$$P = \frac{\sum_{i=1}^{M} r_i}{(M-1)(\sum_{j=1}^{M} b_j)} \tag{8}$$

where r_i is the number of MAC packets received by the ith unit, and b_j is the total number of MAC packets transmitted by the jth unit. M is the total number of units in the network. The scale factor in the denominator of Equation (8) is due to the fact that a packet transmitted by jth unit is received by all the other $M - 1$ units in the single hop case. Therefore, P is unity in an ideal case.

Network throughput is defined as the total throughputs of all units, where the throughput of a unit is the amount of successfully received MAC frames in bits per second. The network throughput is normalized to the channel capacity. Thus, the normalized network throughput S is defined as

$$S = \frac{\sum_{i=1}^{M} x_i}{C} \tag{9}$$

where x_i is the throughput (in bits per second) of the ith unit, C is the channel capacity in bits per second, and M is the total number of units in the network.

Parameter		Value
Number of units M = (MU + DU + BU)		From 10 to 360
Area $X \times Y \times Z$		40m × 40m × 3m
MAC	Superframe duration T	4s
	Number of time slots N	160
	Time slot duration T/N	25ms
	Number of time slots allocated per Superframe	1 and 4 time slots
	MAX_TIMEOUT	2, 4, 6, 8, 10
Propagation model	Transmit Power P_t	0.11 mW
	AWGN P_n	-115.1 dBm
	Receiver Sensitivity	-120 dBm
	Path loss exponent γ	3.5
	Shadowing standard deviation σ	0dB, 2dB, 4dB, 8dB
	$SINR_{th}$	-5 dB
	Bandwidth	1.25 GHz
	f_{min}, f_{max}	6 GHz, 7.25 GHz
	Center frequency f_c	6.625 GHz
Application Traffic model		0.25 packet/s, 1 packet/s
Simulated time		600 s
Number of simulation runs		30

Table 1. Simulation Settings

The simulation parameter settings are given in Table 1. In the simulation, the total number of units was varied from 10 to 360 units. Since we fixed the number of BUs to four units, the number of MUs and DUs was varied but always at an equal quantity. The DUs and MUs were randomly distributed in a square region with area $X \times Y$ m^2. The length of X and Y is calculated such that signals can still be detected by the receiving units which are at the maximum distance from the transmitting units using the transmit power, AWGN, path loss exponent γ, receiver sensitivity, and $SINR_{th}$ given in the table. The BUs were placed around the edge of the area in order to reflect the arrangement of the real system.

4.1.1 Effects of shadowing

Figs. 9 and 10 depict the successful MAC packet reception rate and network throughput as a function of the number of units without the shadowing effect, i.e., $\sigma = 0$ dB. In an area of 40m × 40m, all units are within the radio transmission range of each other. This means that the transmission of a unit is heard by all the other $M - 1$ units. In Figs. 9 and 10, two sets of similar simulation runs were carried out. In the first set, each unit reserved only one time slot in each SOC-MAC superframe, while in the second set, four time slots were reserved by each unit per SOC-MAC superframe. In the second set of the simulation run, the RA-TDMA phase is followed by I-TDMA to reserve the extra three time slots. The I-TDMA phase is not triggered in the first run since only one time slot is required, which is already reserved in the RA-TDMA phase. The plots for one time slot and four time slots, which are denoted as 1-timeslot (solid line) and 4-timeslot (dashed line), respectively in Figs. 9 and 10, exhibit similar behavior except the roll-off of the frame reception rate and the maximum network throughput occurs at different number of units. When the time slot occupancy is less than the number of time slots in a superframe, i.e., $N = 160$, the network throughput increases linearly and the successful frame reception rate is 100%. For $N = 160$, the network can accommodate a maximum number of $M = 160$ or 40 units for 1-timeslot and 4-timeslot, respectively. When $M = 160$ or 40, the successful reception rate drops to 90% for 1-timeslot and 4-timeslot, respectively, and the network throughput reaches the peak, which is 90%, as observed in Fig. 10 and then gradually falls as M increases. When $M < 40$ or 160, the effect of time slot collisions during the RA-TDMA phase is negligible. Thus collisions are resolved when the involved units enter the A-TDMA phase, which randomly picks a free time slot in the next superframe.

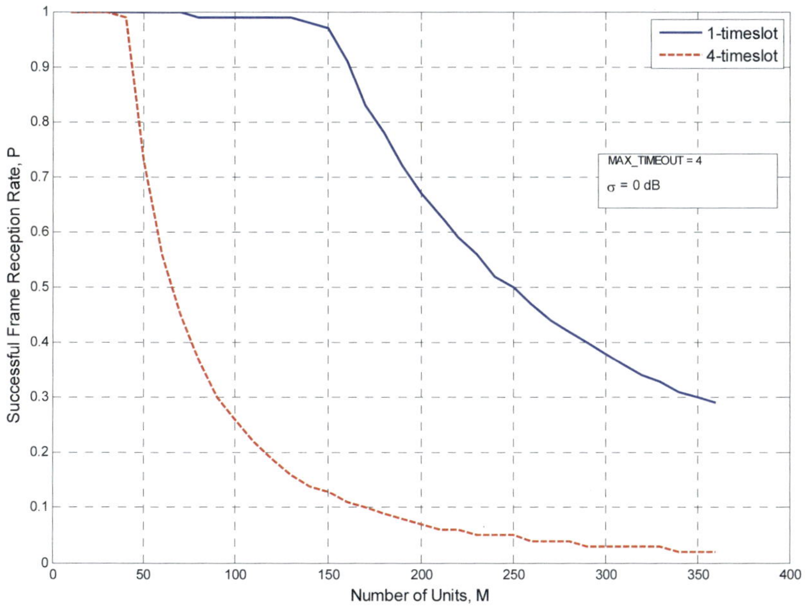

Fig. 9. Successful SOC-MAC Packet Reception Rate versus Number of Units for σ = 0 dB, and MAX_TIMEOUT = 4, and σ = 0 dB

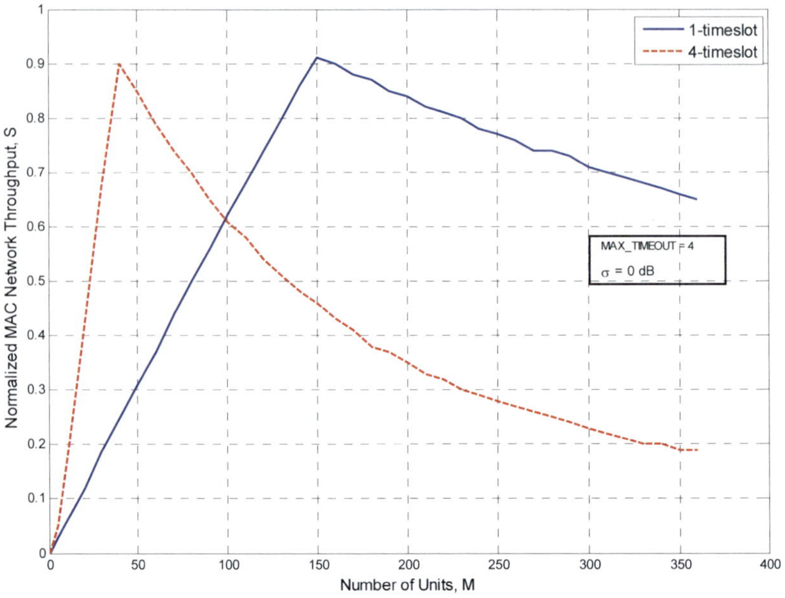

Fig. 10. Normalized Average SOC-MAC Network Throughput versus Number of Units for MAX_TIMEOUT = 4, and $\sigma = 0$ dB

When $M > 40$ or 160, both the successful frame reception rate and the network throughput rapidly deteriorate due to a larger number of collisions occurred in the RA-TDMA phase. Unlike in the case of $M \leq 40$ or 160, the A-TDMA is unable to resolve all the collisions because the number of time slots in the superframe is insufficient to accommodate the number of time slots needed by all the units. Simulation traces reveal that the collisions persisted through the entire simulation duration. In the case of collisions, the SINR of a received frame by all the receiving units is given by Equation (1). If the SINR of a received MAC frame is less than $SINR_{th}$, then the information on newly reserved time slot is lost in the collided time slot and it is considered to be free by other units. Hence, the set of free time slots built by each unit will consist of spurious free time slots since all the actual free slots are occupied. Furthermore, when the unit density is above the number of time slots in the superframe, the probability that two or more units select the false time slot is reasonably high.

Next, the same sets of simulation as above were repeated with the shadowing effect of $\sigma = 2$ dB, 4 dB and 8 dB. These values were chosen based on UWB channel measurements for indoor environment (Irahhauten et al., 2006). Figs. 11 and 12 show the successful SOC-MAC packet reception rate and the normalized network throughput as a function of the number of units. As shown in the figures, shadowing has detrimental effect on the performance. The performance degrades as σ increases. For $\sigma = 8$ dB, the maximum achievable network throughput is around 75% as compared to 90% for the case without shadowing. As observed in Fig. 12, the shadowing effect is more pronounced when $M = [35 .. 50]$ and [140 .. 200] for 4-timeslot and 1-timeslot, respectively. In this region of M, the channel capacity is saturated. Therefore, SOC-MAC packet losses are due to both collisions and shadowing. When M > 50 and 200 for 4-timeslot and 1-timeslot, respectively, the majority of the SOC-MAC packet losses are due to collisions rather than shadowing. This is evidenced by the convergence of the three plots.

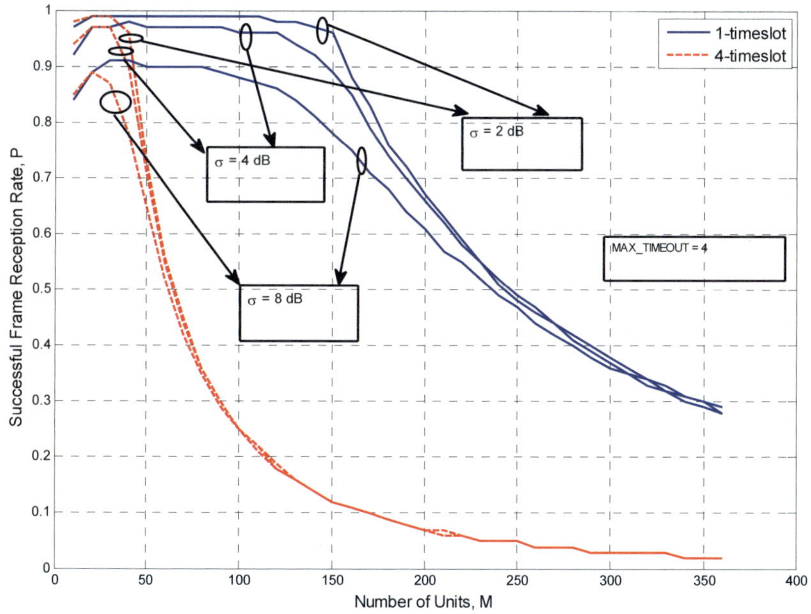

Fig. 11. Successful Packet Reception Rate versus Number of Units for MAX_TIMEOUT = 4, and σ = 2 dB, 4 dB and 8 dB

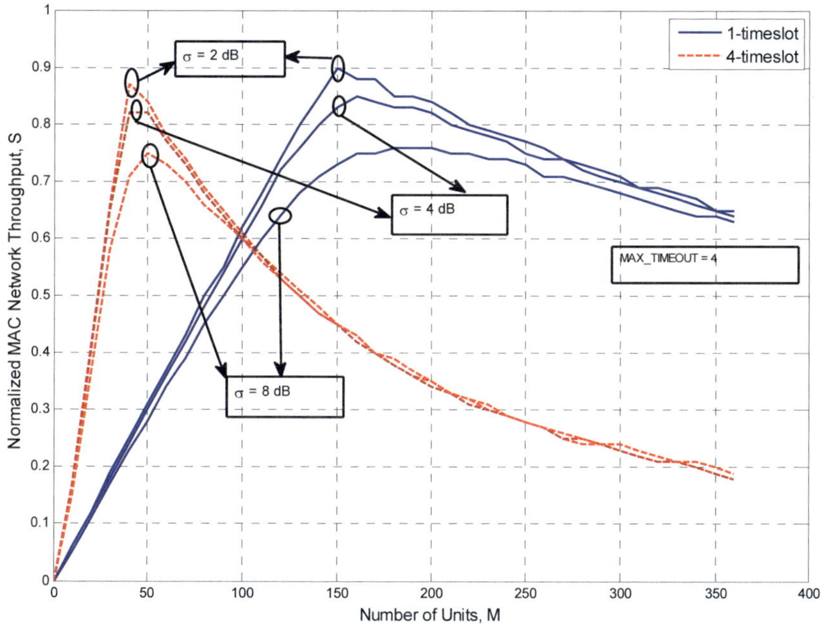

Fig. 12. Normalized Average SOC-MAC Throughput versus Number of Units for MAX_TIMEOUT = 4, and σ = 2 dB, 4 dB and 8 dB

4.1.2 Effects of timeout

In this subsection, we investigate the influence of the *MAX_TIMEOUT* parameter on the performance. As mentioned in subsection 2.2, *MAX_TIMEOUT* is the maximum number of superframes a unit may occupy a particular time slot. The same sets of simulation, as in subsection 4.1.1, were repeated and the *MAX_TIMEOUT* value was varied from 2 to 10 superframes in a step of 2. Figs. 13 and 14 depict the plots for the successful MAC packet reception rate and the network throughput. We can make two observations. Firstly, each of the *MAX_TIMEOUT* values delivers the same performance at low unit density, which is for $M < 40$ and 150 for 4-timeslot and 1-timeslot, respectively. Secondly, at high unit density, larger *MAX_TIMEOUT* values give a slight performance advantage than smaller ones. The performance of *MAX_TIMEOUT* = 2 is the worst, and *MAX_TIMEOUT* = 8 and 10 achieve the best performance. For *MAX_TIMEOUT* = 2, it means that the frequency of time slot renewal is the highest. Thus, at high unit density, a frequent renewal is not preferred because all of the time slots are fully occupied. If there is an available time slot, two or more units would select the same time slot, which results in collision as evidenced by the lowest successful MAC frame reception rate in Fig. 13. Additionally, information on the newly reserved time slot is lost in the collided time slot at high unit density. Other units would have mistaken these time slots for vacant. At high unit density, the time-slot occupancy map constructed by any unit would mainly consist of spurious time slots. As a result, selecting any of these spurious time slots would prolong collisions.

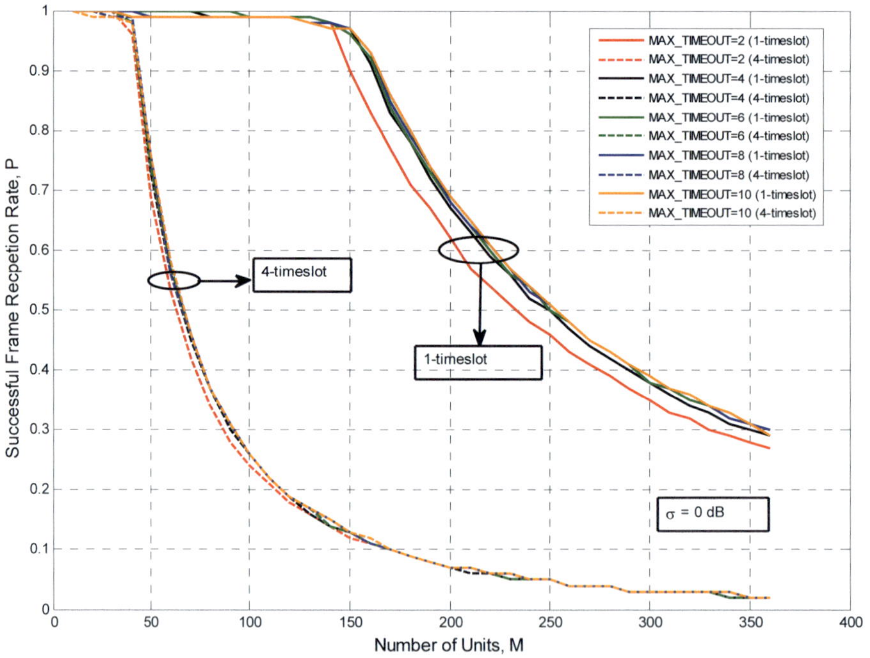

Fig. 13. Successful SOC-MAC Packet Reception Rate for σ = 0 dB, and MAX_TIMEOUT = [2, 4, 6, 8, 10]

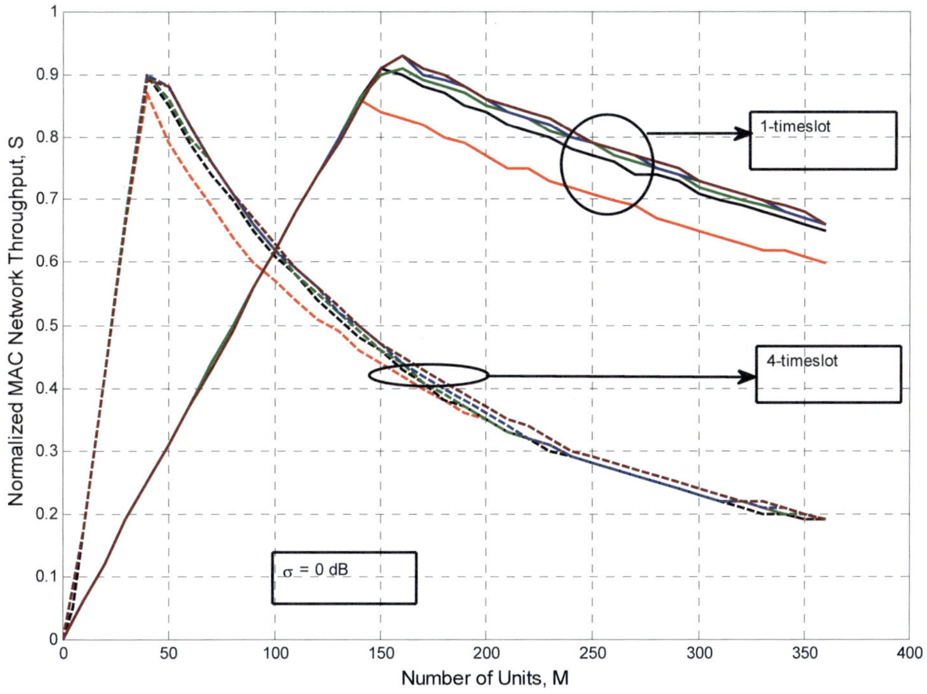

Fig. 14. Normalized Average SOC-MAC Network Througput for σ = 0 dB, and MAX_TIMEOUT = [2, 4, 6, 8, 10]

4.2 LAR performance

In this subsection, we investigate the performance of LAR. First, we analyze the performance of LAR route establishment algorithm without the influence of route selection. Thus, a chain topology was used. With the chain topology, a unit usually has two neighbors except the unit at either end of the chain, which has only one. Each unit is stationary and spaced at an equidistant of 50m, which is just below the maximum transmission range that was determined using Equation (7) and the parameter values of Table 1. A single BU is placed at one end of the chain. A chain topology, which comprises (h+1) units, consists of h hops. In the chain topology, the BU is always assumed to be the *1*st unit and the (h+1)th unit is the last unit. The BU is responsible for initiating the route construction by broadcasting its position packets. The rest of the units in the chain are either MUs or DUs. The composition of MUs and DUs is irrelevant since they are functionally equivalent from the routing protocol point of view. The *route discovery* and *end-to-end packet delays* were examined. Route discovery delay is defined as the time in seconds when a unit (except the BU) found a route (i.e., next-hop neighbor) on the forward path. End-to-end delay is defined as the time in seconds taken by a data packet to traverse from an MU or a DU to the BU. Fig. 15 shows that both the route discovery and end-to-end packet delays are linearly proportional to the number of hops. The results prove that LAR became highly scalable. For a network diameter of 50 hops, it took the last unit at the end of the chain less than 30 SOC-MAC superframes to discover a route since the BU started broadcasting the routing information. End-to-end delays are determined using position reporting packets which are sent by the last unit, i.e. the (h+1)th unit, to the BU for an h-hop network where h varies from 1 to 50. Note that the (h+1)th unit only starts transmitting the position reporting packets once a route is found to remove the queuing effect due to route discovery.

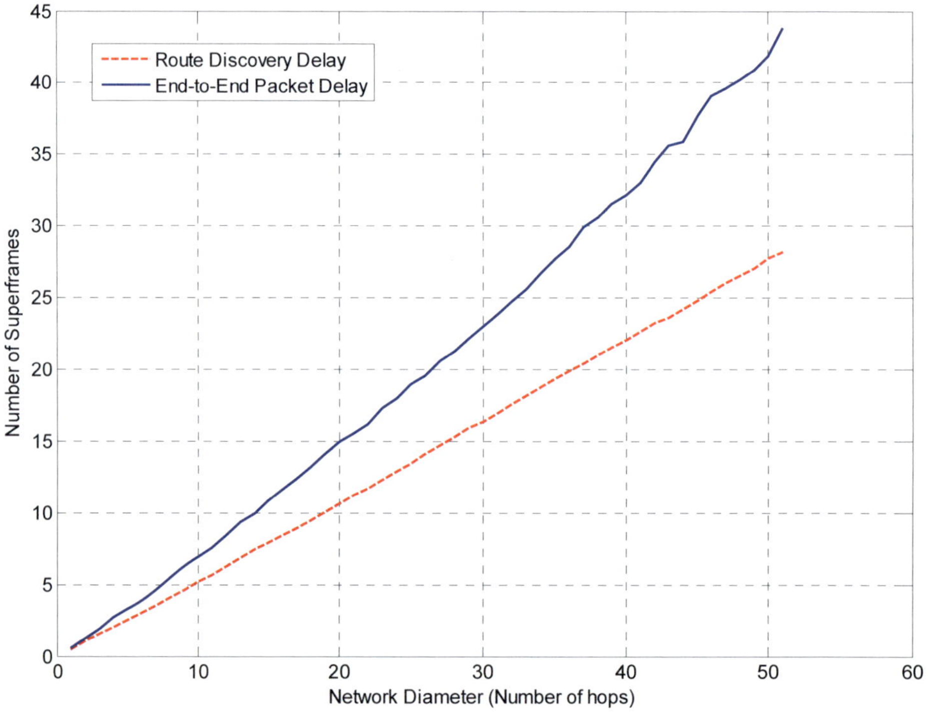

Fig. 15. Route discovery and End-to-End Packet Delay

Next, we will analyze the performance of the LAR route selection algorithm. For the analysis, we used a star topology as shown in Fig. 16. Each unit was stationary and spaced at an equidistant of 50m from its adjacent neighbors. BU0, BU1 and BU2 initiated the route construction simultaneously by broadcasting their position packets, and triggered neighboring units to transmit their positioning packets. In the star configuration, the unit at the center, MU0, received position packets from three different neighboring units, namely DU1, DU2 and MU1, see Fig. 16. Consequently, MU0 created three forward routes in its routing table. These routes are referred to as Route 1, Route 2 and Route 3, respectively, as shown in Fig. 16. The hop count of Route 1 and Route 2 is three hops while Route 3 is four. Since hop count is the primary routing metric, the routes with the least hop count would be selected by MU0. In this case, Route 1 and Route 2 were picked by the route selection algorithm of MU0. In the simulations, each unit broadcast position packets at a fixed interval of 4s. Hence, the traffic load was uniformly distributed across the network. In other words, none of the MUs or DUs were more congested than others. Therefore, the route selection algorithm would arbitrarily choose between Route 1 and Route 2. The MU0 was set to transmit position reporting packet at time t = 50s after the BUs started the route construction. Simulation traces show that Route 2 was selected by MU0 for transporting its position reporting packets to BU1. And the end-to-end packet delay is approximately 2 superframes, which conforms to the 3-hop delay in Fig. 15. At t=100s, MU2 was set to send position reporting packets, which introduced extra traffic on to the network. MU2 used Route 2 for transporting its position reporting packets since Route 2 was the shortest. Fig. 17 shows the congestion level seen by MU0.

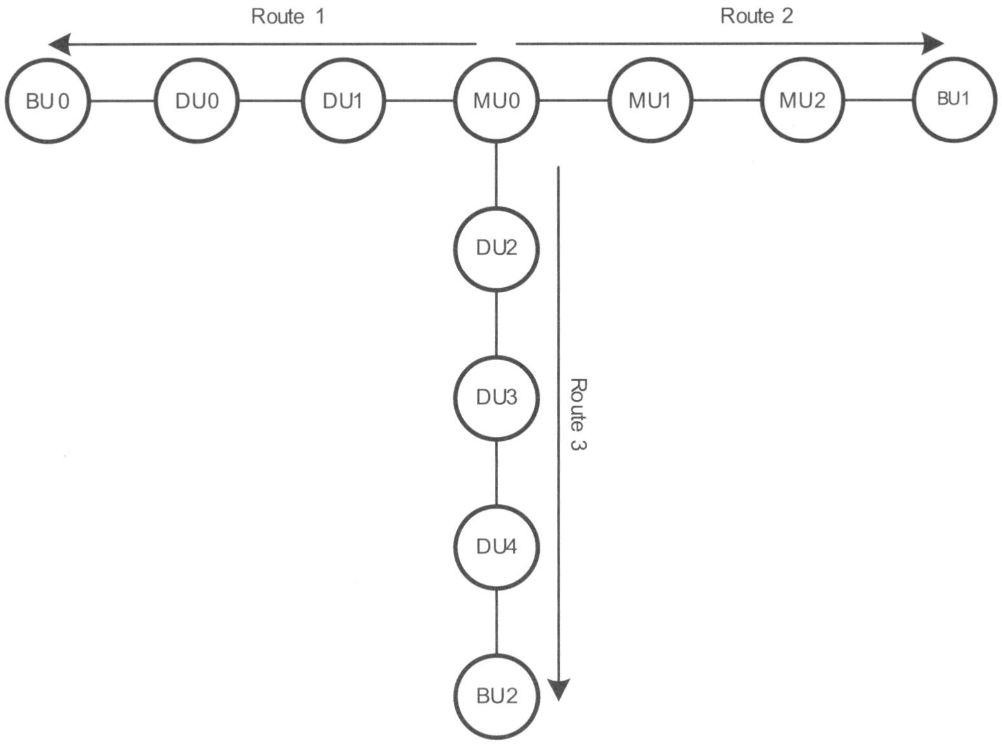

Fig. 16. Star Network Topology

Fig. 17 depicts the position reporting packets received by BU0 and BU1. As shown in Fig. 17, initially MU0 selected Route 2 for transporting its position reporting packets until the time was approximately 110s, where it switched to Route 1. The switching occurred when MU0 detected the congestion level on Route 2 was increased to 3. The increase in congestion was caused by MU2 when it started transmitting its position reporting packets at t=100s. Due to congestion, some in-flight packets on Route 2 were experiencing excessive delays and arrived at BU1 later than packets sent on Route 1. The congestion level of both Route 1 and Route 2 continued to rise, and on Route 2, the congestion level reached the maximum at about 150s. When both MU0 and MU2 stopped transmitting position reporting packets at 250s, the congestion level did not drop until t = 350s for Route 2 and t = 410s for Route 1 because of a large number of packets already in the queue. At t = 350s, the congestion level of Route 2 dropped to 5, which was the same as Route 1. At this point a route change occurred since MU0 selected Route 2 again. All the remaining packets in its queue were sent on Route 2. After time t = 450s, the congestion level of both MU0 and MU2 dropped sharply.

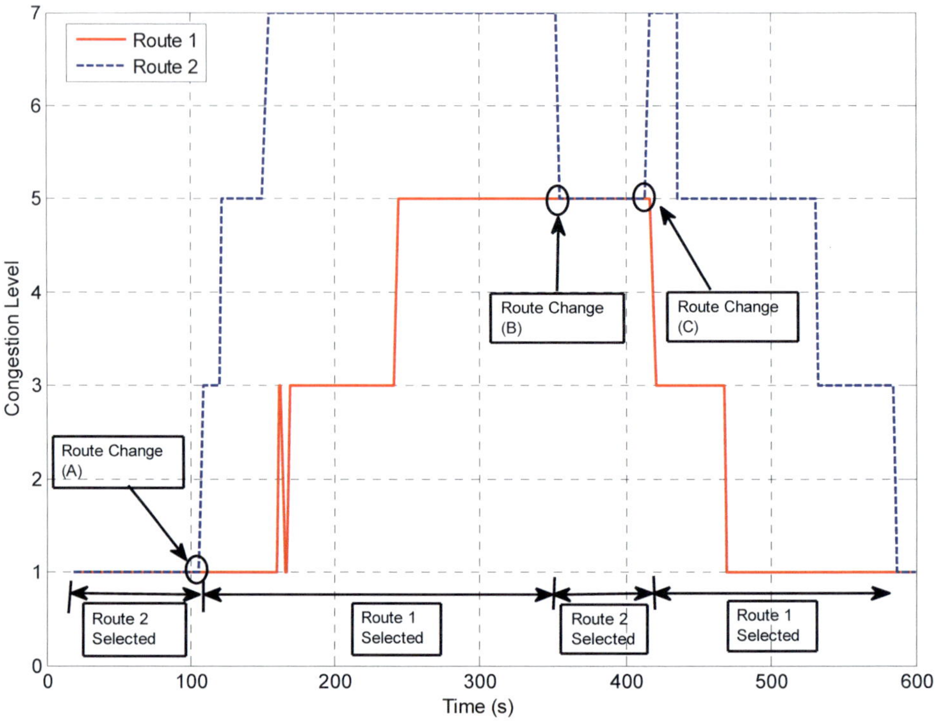

Fig. 17. Congestion Level

5. Related work

This section reviews the MAC and routing protocols developed for UWB-based ad hoc sensor networks.

5.1 UWB-based MAC protocols for ad hoc sensor networks

In the past few years, a number of MAC protocols have been proposed for UWB-based systems. (Legrand et al., 2003) and (Zhu & Fapojuwo, 2005) proposed a modified version of the IEEE 802.15.3 Wireless Personal Area Network (WPAN) MAC protocol, which rely on a centralized controller. These MAC protocols can provide guaranteed Quality of Service (QoS) but are difficult to scale. The WHYLESS.COM project (Cuomo et al., 2002) proposed a distributed UWB MAC, which supports QoS and is scalable but has high complexity. (Chu & Ganz, 2004) described a hybrid MAC for WPAN, which combines the advantages of both centralized and distributed protocols. The MAC protocol assumes that every node in a WPAN is one hop away from every other node. Consequently, the MAC is foreseen to face scalability issues when operating in multi-hop scenarios. Furthermore, a separate control channel is used for signaling purposes, which increases the complexity and is not lightweight for low bit-rate channels. Ultra-Wideband MAC (U-MAC) (Jurdak et al., 2005) is a proactive and adaptive protocol. Similar to (Chu & Ganz, 2004), a separate signaling channel is needed for exchanging a node's state information with its direct neighbors. (Broutis et al., 2007) and (Benedetto et al., 2005) outlined a multi-channel MAC in which communication between two nodes takes place on orthogonal channels. The complexity and overheads incurred by such a MAC protocol are higher than single-channel MAC protocols. (Merz et al., 2005) proposed a combined Physical and MAC layer for very low power UWB system. No separate control channel is needed. However, the signaling overheads incurred by the MAC can be significant for short data packets and low bit-rate channels. In summary, all of the above-mentioned MAC protocols were not designed for localization application in mind. The IEEE 802.15.4a standard (Karapistoli et al., 2010; IEEE 802.15.4a, 2007) specifies a Physical layer and a MAC layer which support localization. The IEEE 802.15.4a MAC supports two different modes of channel access: beacon-enabled and nonbeacon-enabled. The latter is suited for localization application. Unlike SOC-MAC, the nonbeacon-enabled mode of the IEEE 802.15.4a MAC is based on the classical Aloha scheme or the CSMA/CA scheme.

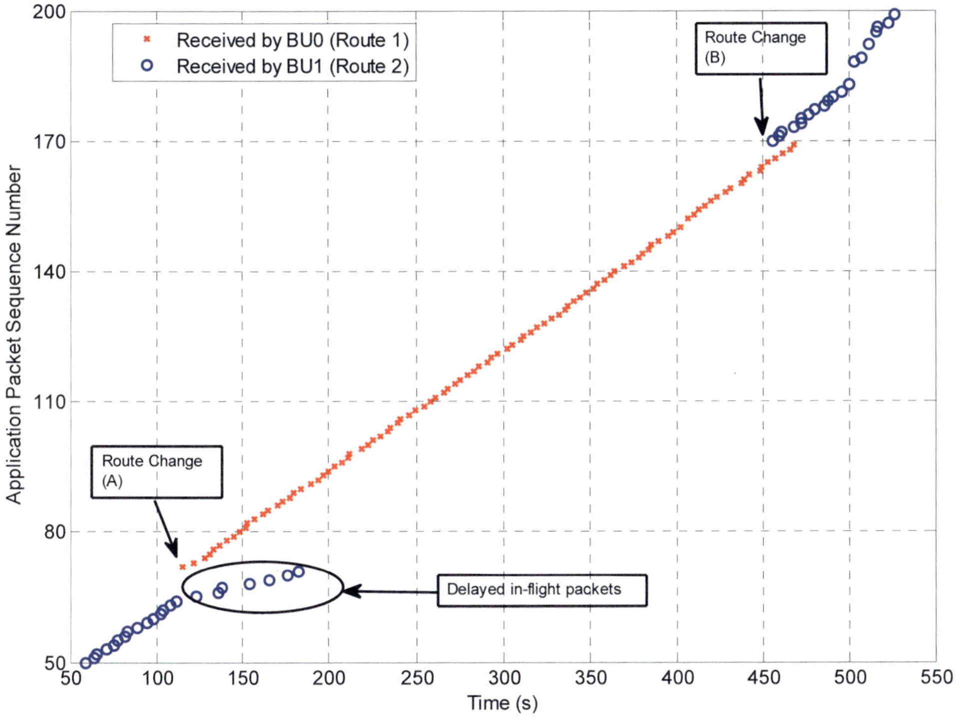

Fig. 18. Position Reporting Packets

5.2 Routing protocols for ad hoc sensor networks

A large number of routing protocols, e.g. (Kulik et al., 2002; Intanagonwiwat et al., 2000; Schurgers & Srivastava, 2001; Shah & Rabaey, 2002; Lindsey & Raghavendra, 2002; Manjeshwar & Agarwal, 2001), have been developed for ad hoc sensor networks. Although the considered ILS is an ad hoc sensor network, it has some profound distinctions which mean existing ad hoc sensor routing protocols are not directly applicable. Firstly, sensor nodes are generally assumed to have very low mobility after deployment (Al-Karaki & Kamal, 2004) in comparison with ILS. Lastly, the relative size of ad hoc sensor networks is huge in the order from thousands to millions of nodes (Al-Karaki & Kamal, 2004) as compared to ILS.

6. Summary

In this chapter, we described the SOC-MAC and LAR protocols that are tailored for indoor localization systems used to track emergency responders. The cross-layer approach is present in the protocol design in order to optimize bandwidth and battery-energy consumption. As a result, SOC-MAC is simple and self-organizing, which is composed of two phases, namely RA-TDMA and reserved TDMA. The former is for initial acquisition of time slots while the latter is for management and maintenance of time slots. In addition to simplicity, LAR is extremely lightweight. No dedicated routing packets are needed. Instead, routing information is carried in the network header of localization packets, which constitutes less than 1% of the total channel capacity. We validated and studied the performance of SOC-MAC and LAR by simulations under varying SOC-MAC and LAR parameters.

7. Acknowledgement

The work was partially funded by the IST-004154 EUROPCOM project.

8. References

Al-Karaki, J. N. & Kamal, A. E. (2004). Routing Techniques in Wireless Sensor Networks: A Survey, *IEEE Wireless Communications Magazine*, Vol. 11, No. 6

Benedetto, M.-G.; De Nardis, L.; Junk, M. & Giancola, G. (2005). (UWB)^2: Uncoordinated, Wireless, Baseborn Medium Access for UWB Communication Networks, *Mobile Networks and Applications (MONET)*, Vol. 10, No. 5

Broutis, I.; Krishnamurthy, S. V.; Faloutsos, M.; Molle, M. & Forester, J. R. (2007). Multiband Media Access Control in Impulse-based UWB Ad Hoc Networks, *IEEE Transactions on Mobile Computing*, Vol. 6, No. 4

Chu, Y. & Ganz, A. (2004). MAC Protocols for Multimedia Supporting UWB-based Wireless Networks, *Proceedings of 1st Int'l Conference on Broadband Networks (BROADNETS)*

Cuomo, F.; Martello, C.; Baiocchi, A. & Fabrizio, C. (2002). Radio Resource for Ad Hoc Networking with UWB, *IEEE Journal on Selected Areas in Communications*, Vol. 20, No. 9

Frazer, E. L. & Tee, D. (2004). A Comparison of UWB Technologies for Indoor Positioning as an Augmentation to GNSS, *Proceedings of 2nd European Space Agency (ESA) Workshop on Satellite Navigation User Equipment Technologies (NAVITEC)*, Noordwijk, The Netherlands, 2004

Harmer, D. (2008). EUROPCOM: Ultra-WideBand Radio for Rescue Services, *Proceedings of 2nd Int'l Workshop on Robotics for Risky Interventions and Surveillance of the Environment (RISE)*, Benicassim, Spain, 2008

Harmer, D., et al. (2008). EUROPCOM: Emergency Ultra-WideBand (UWB) Radio for Positioning and Communications, *Proceedings of IEEE International Conference on Ultra-WideBand (ICUWB)*, 2008

Hofmann-Wellenhof, B.; Lichtenegger, H. & Wasle, E. (2008). *GNSS – Global Navigation Satellite Systems: GPS, GLONASS, and more*, Springer, Vienna

IEEE 802.15.4a (2007). Wireless Medium Access Control (MAC) and Physical Layer (PHY) Specifications for Low-Rate Wireless Personal Area Networks (WPANs)

Intanagonwiwat, C.; Govindan, R. & Estrin, D. (2000). Directed Diffusion: a Scalable and Robust Communication Paradigm for Sensor Networks, *Proceedings of ACM MobiCom*, Boston, MA, 2000

Irahhauten, Z.; Janssen, G. J. M., Nikookar, H., Yaravoy, A. & Lighart, L. P. (2006). UWB Channel Measurements and Results for Office and Industrial Environments, *Proceedings of Int'l Conference on Ultra-WideBand (ICUWB)*, MA, 2006

Jurdak, R.; Baldi, P. & Lopes, C. V. (2005). U-MAC: A Proactive and Adaptive UWB Medium Access Control Protocol, *Wiley Wireless Communications and Mobile Computing*, Vol. 5, No. 5

Karapistoli, E.; Pavlidou, F.; Gragopoulos, I. & Tsetsinas, I. (2010). An Overview of the IEEE 802.15.4a Standard, *IEEE Communications Magazine*, Vol. 48, No. 1

Kulik, J.; Heinzelman, W. R. & Balakrishnan, H. (2002). Negotiation-based Protocols for Disseminating Information in Wireless Sensor Networks, *Wireless Networks*, Vol. 8

Legrand, J.; Bucaille, I.; Hethuin, S.; De Nardis, L.; Giancola, G.; Di Benedetto, M.; Blazevic, L. & Rouzet, P. (2003). U.C.A.N.'s Ultra Wideband Medium Access Control Schemes, *Proceedings of Int'l Workshop on Ultra Wideband Systems (IWUWBS)*, 2001

Lindsey, S. & Raghavendra, C. (2002). PEGASIS: Power-efficient Gathering in Sensor Information Systems, *Proceedings of Aerospace Conference*, 2002

Manjeshwar, A. & Agarwal, D. P. (2001). TEEN: a Routing Protocol for Enhanced Efficiency in Wireless Sensor Networks, *1st Int'l Workshop on Parallel and Distributed Computer Issues in Wireless Networks and Mobile Computing*, 2001

Merz, R.; Widmer, J.; Le Boudec, J. Y. & Radunovic, B. (2005). A Joint PHY/MAC Architecture for Low Radiated Power TH-UWB Wireless Ad Hoc Networks, *Wiley Wireless Communications and Mobile Computing*, Vol. 5, No. 5

Mobility Framework, http://mobility-fw.sourceforge.net

OMNeT++, http://www.omnetpp.org/

Rappaport, T. (2001). *Wireless Communications*, 2nd edition, Prentice Hall

Schurgers, C. & Srivastava, (2001). Energy-efficient Routing in Wireless Sensor Networks, *MILCOM Proceedings on Communications for Network-Centric Operations: Creating the Information Force*, McLean, VA, 2001

Shah, R. C. & Rabaey, J. (2002). Energy Aware Routing for Low Energy Ad Hoc Sensor Networks, *Proceedings of WCNC*, Orlando, FL, 2002

Zhu, J. & Fapojuwo, A. O. (2005). A Complementary Code-CDMA-based MAC Protocol for UWB WPAN System, *EURASIP Journal on Wireless Communications and Networking*, Vol. 2005, No. 2

8

Hybrid Access Techniques for Densely Populated Wireless Local Area Networks

J. Alonso-Zárate[1], C. Crespo[2], Ch. Verikoukis[1] and L. Alonso[2]
[1]*Centre Tecnològic de Telecomunicacions de Catalunya (CTTC),Castelldefels, Barcelona*
[2]*Universitat Politècnica de Catalunya (UPC), Castelldefels, Barcelona*
Spain

1. Introduction

The IEEE 802.11p Task Group has recently released a new standard for wireless access in vehicular environments (WAVE). It constitutes an amendment to the 802.11 for Wireless Local Area Networks (WLANs) to meet the requirements of applications related to road-safety involving inter- and intra-vehicle communications as well as communications from vehicle to the roadside infrastructure. Indeed, the importance of the targeted applications has forced authorities to allocate some dedicated bandwidth (nearby the 5.9GHz) to ensure the security of the communications. However, despite the suitability of this standard for use in high-speed vehicular communications, it is not possible to pass over the unprecedented market penetration of the popular 802.11 networks, the so-called WiFi networks. Before we can see a world where all the cars are equipped with 802.11p devices, current and near-future applications might probably run on the original 802.11. Moreover, interaction between humans and vehicles will probably be carried out by means of the 802.11, which is the standard that is flooding most of personal tech devices, such as laptops, mobile phones, gaming consoles, etc. Therefore, it is important to keep on working in the improvement of the 802.11 Standard for its use in, at least, some vehicular applications.

This is the main motivation for this chapter, where we focus on the Medium Access Control (MAC) protocol of the 802.11 Standard, and we propose a simple mechanism to improve its performance in densely populated applications where it falls short to provide users with good service. Envisioned applications include those were a high number of vehicles and pedestrians coexist in a given area, such as for example, a crossing in a city where all the cars share information to coordinate the drive along the crossing and prevent accidents.

Into more detail, the Distributed Coordination Function (DCF) is the mandatory access method defined in the widely spread IEEE 802.11 Standard for WLANs [1]. This access method is based on Carrier Sensing Multiple Access (CSMA), i.e., listen before transmit, in combination with a Binary Exponential Backoff (BEB) mechanism. An optional Collision Avoidance (CA) mechanism is also defined by which a handshake Request to Send (RTS) - Clear to Send (CTS) can be established between source and destination before the actual transmission of data. This CA mechanism aims at reducing the impact of the collisions of data packets and to combat the hidden terminal problem. The DCF can be executed in either ad hoc or infrastructure-based networks and is the only access method implemented in most commercial hardware. Despite the doubtless commercial success of the DCF, the simplicity of a CSMA-based protocol comes at the cost of a trial-and-error approach where a transmission attempt can result in a collision if several users contend for the access to a common medium as the traffic load of the network increases. Therefore, those networks based on the 802.11 suffer from really low performance when either the number of users or the traffic load is high.

In this chapter, we introduce the idea of combining the DCF with the Point Coordination Function (PCF), also defined in the 802.11 Standard, to overcome its limitations under heavy load conditions. The PCF is defined as an optional polling-based access method for infrastructure-based networks where there is no contention to get access to the channel and the access point (AP) polls the stations of the network to transmit data. Therefore, collisions of data packets can be completely avoided and the performance of the network can be boosted.

The hybrid approach of combining distributed access with reservation or polling-based access has been already used in the context of infrastructure-based networks [2]-[6] combining static Time Division Multiplex Access (TDMA) with dynamic CSMA access. Most of these works propose different alternatives to use the empty slots of TDMA in the case that the user allocated to a given slot has no data to transmit. However, to the best knowledge of the authors, there are very few works in the literature dealing with this approach in a distributed manner, i.e., for ad hoc networks without infrastructure. This is the main motivation for the work presented in this chapter, where we define the Distributed Point Coordination Function (DPCF) as a hybrid combination of the distributed access of the DCF and the poll-based access of the PCF to achieve high performance in highly populated networks with heavy traffic load. Indeed, the work presented in this chapter has been motivated by the successful results presented in [7]. In that paper, a spontaneous, temporary, and dynamic clustering algorithm has been integrated with a high-performance infrastructure-based MAC protocol, the Distributed Queuing Collision Avoidance (DQCA) protocol, in order to extend its near-optimum performance to networks without infrastructure. Upon the conclusion of that work, we realized that the same approach could be applied to the IEEE 802.11 Standard access methods and thus be able to extend the high-performance of the PCF under heavy load conditions to the distributed environments where the DCF runs.

We have observed that there are very few works dealing with the PCF, which can indeed potentially achieve better performance than the DCF under heavy traffic conditions. Some contributions related to the PCF improve the overall network performance through novel scheduling algorithms [8]-[12] or by designing new polling mechanisms that can reduce the overhead associated to the polling process [13]. However, there have been almost no efforts in extending the operation of the PCF to ad hoc networks in order to provide them with some degree of QoS. The only exception can be found in [14] where a virtual infrastructure is created into a MAC protocol called Mobile Point Coordinator MAC (MPC-MAC) in order to achieve QoS delivery and priority access for real time traffic in ad hoc networks maintaining both the PCF and the DCF. In summary, a clustering based mechanism is used to achieve the correct operation of the PCF in a distributed environment. The duration of the PCF and DCF periods and the criterion upon which a terminal is chosen to be the MPC (acting as AP) are fixed and they are determined by the MAC protocol configuration. This approach works well in low dynamic environments where the topology does not vary frequently. In this situation the overhead associated to the "hello" messages required for the clustering mechanism can be kept to a minimum. However, it may not be convenient for spontaneous and highly dynamic environments, such as those present in some vehicular applications, where the clustering overhead could impact negatively on the efficiency of the network. In addition, this protocol does not consider that the responsibility of becoming cluster head should be shared among all the users of the network to ensure certain fairness regarding the extra energy consumption associated to the role of coordinating a cluster.

Taking into account this background and motivated by the success of extending DQCA to become DQMAN [7], we contribute to the field by presenting the DPCF as an extension of the PCF to operate over infrastructure-less networks through smooth integration with the DCF. By combining the DCF and the PCF using a spontaneous and dynamic clustering mechanism at the MAC layer it is possible to extend the higher performance of the PCF to networks without infrastructure. We present a description of the protocol as well as a comprehensive performance evaluation based on computer simulation both for single-hop and multi-hop networks.

The chapter is organized as follows. The DCF and the PCF of the IEEE 802.11 Standard are overviewed in Section II. The DPCF protocol is then described in Section III. In Section IV, we present a comprehensive performance evaluation of the protocol by means of computer simulation. Finally, Section V concludes the chapter and outlines some future lines of research.

2. IEEE 802.11 MAC protocol overview

An overview of the operation of the DCF and the PCF of the IEEE 802.11 Standard is included in this section. A comprehensive description of them can be found in [1]. Following the naming of the standard, we will refer herein to a vehicle or pedestrian equipped with a communications terminal as a mobile station, or simply, a station.

2.1 DCF overview

The DCF is the mandatory coordination function implemented in all standard compliant devices. Two access modes of operation are defined in the DCF:

1. *Basic access (BASIC) mode;* the station which seizes the channel transmits its data packets without establishing any previous handshake with the intended destination.
2. *Collision avoidance access (COLAV) mode;* a handshake RTS/CTS is established between source and destination before initiating the actual transmission of data. These RTS and CTS get the form of special control packets. The COLAV access mode is aimed at reducing the impact of collisions of data packets and at combating the presence of hidden terminals.

Two examples are illustrated in Figure 1 and Figure 2 representing the operation of the BASIC and the COLAV access modes, respectively. In summary, any station with data to transmit listens to the channel for a DCF Inter Frame Space (DIFS). If the channel is sensed idle for this DIFS period, the station seizes the channel and initiates the data transmission (or the RTS transmission in the COLAV mode). Otherwise, if the channel is sensed busy, the station backs off and executes a BEB algorithm by which the size of the contention window is doubled up upon any transmission failure and reset to the minimum value upon success. When a data packet is received without errors, the destination sends back an ACK packet after a Short Inter Frame Space (SIFS). This SIFS is necessary to compensate propagation delays and radio transceivers turn around times to switch from receiving to transmitting mode. It is worth noting that since a SIFS is shorter than a DIFS, acknowledgments have more priority than regular data traffic.

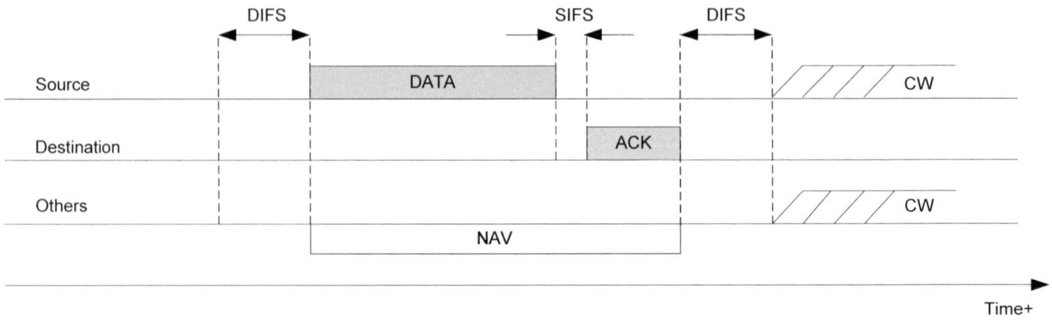

Fig. 1. Example: DCF Operation (Basic Access mode)

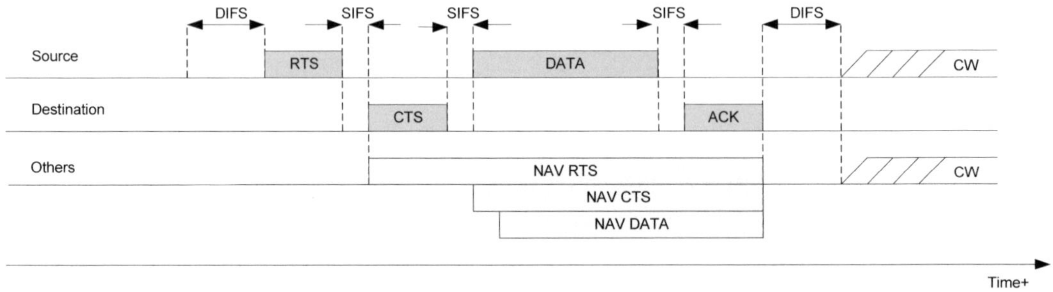

Fig. 2. Example: DCF Operation (Collision Avoidance mode)

A relevant feature of the DCF is the Virtual Carrier Sensing (VCS) mechanism. Stations not involved in an ongoing transmission defer from attempting to transmit for the time that the channel is expected to be used for an effective transmission between any pair of source and destination stations regardless of the actual physical carrier sensing. To do so, stations update the Network Allocation Vector (NAV) which accounts for the time the channel is expected to be occupied. This information is retrieved from the duration field attached to the overheard RTS, CTS, and data packets. This mechanism is mainly aimed at combating the presence of hidden terminals.

2.2 PCF overview

The PCF can only run on infrastructure-based networks wherein an AP sequentially polls stations to transmit data and thus collisions are totally avoided. This mechanism was initially designed for the provision of QoS over WLANs.

When the PCF is executed, time is divided into Contention Free Periods (CFP), wherein the AP sends poll messages to give transmission opportunities to the stations, and Contention Periods (CP), where the DCF is executed. Since the PCF is an optional coordination function and is not implemented in all standard-compliant devices, DCF periods are necessary to ensure access to DCF-only stations. The interleaving of CFPs and CPs is illustrated in Figure 3. As also shown in this figure, a CFP is initiated and maintained by the AP, which periodically transmits a beacon (B). The first beacon after a CP (DCF access) is transmitted after a PCF Inter Frame Space (PIFS). The duration of a PIFS is shorter than a DIFS but longer than a SIFS, providing thus the initialization of a CFP with less priority than the transmission of control packets, but with higher priority than the transmission of data packets. The periodically transmitted beacons contain information regarding the duration of both the CFP and the CP and allow a new arrived station to associate to the AP during a CFP. The CFP is finished whenever the AP transmits a CFP End (CE) control packet.

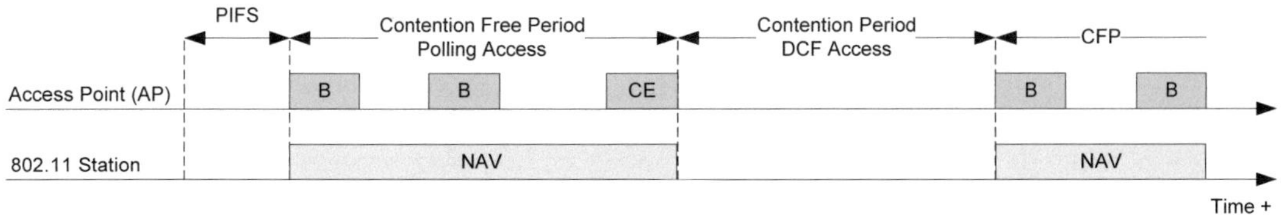

Fig. 3. IEEE 802.11 PCF Interleaves CFPs with CPs

During a CFP, the only station allowed to transmit data is the one being polled by the AP or any destination station which receives a data packet and has to acknowledge (ACK) it, if applicable, and can combine the ACK with data in a same packet. In PCF, some packets can be combined together in order to reduce the number of MAC and PHY headers and thus increase the efficiency of the communications. In any case, the access to the channel is granted one SIFS after the reception of either the poll or the data packet, respectively. A polled user can either transmit a data packet to the AP or to any other station in the network, establishing a peer-to-peer link. If a polled station has no data to transmit, it responds with a special type of control packet, referred to as NULL packet.

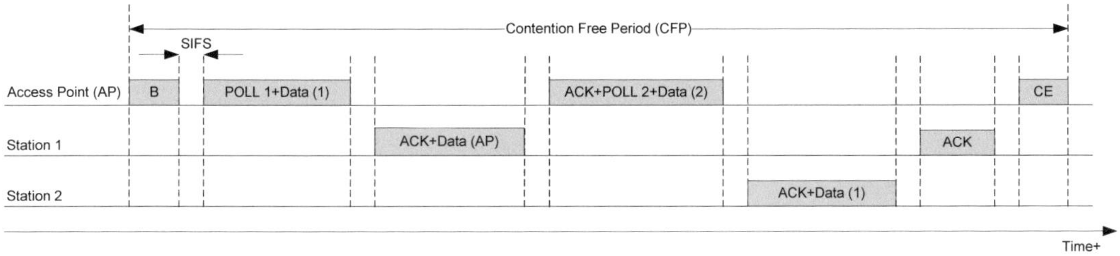

Fig. 4. Example: PCF Operation

An example of PCF operation is illustrated in Figure 4. In this example, the AP initiates a CFP by transmitting a beacon (B). After a SIFS, it combines a poll packet with data to station 1. Upon the reception of this combined packet, station 1 acknowledges the data packet received and responds to the poll by transmitting a data packet to the AP. Note that this is also a combined packet. Then, the AP acknowledges the data packet received from station 1 and combines a poll packet with data to station 2. Upon the reception of the packet, station 2 acknowledges the packet to the AP and transmits data to station 1. Upon the reception of the packet, station 1 acknowledges the received packet. The CFP is finished with the transmission of a CE packet.

3. A new MAC protocol: DPCF

The Distributed PCF (DPCF) protocol is presented in this section as an adaptation and extension of the PCF to operate on distributed infrastructureless wireless ad hoc networks.

As already mentioned before, the main idea is to use the DCF to create spontaneous and temporary clusters wherein the PCF can be executed, having a station acting as the AP for the life time of each cluster.
We consider a set of terminals equipped with WLAN cards forming a spontaneous ad hoc network. Any station must be able to operate in three different modes regarding the clustering mechanism: *idle*, *master*, and *slave*. Initially, all the stations operate in idle mode but they must be able to change the mode of operation when necessary.
Idle stations with data to transmit get access to the channel using the regular DCF. Whenever a station gets access to the channel, it transmits an RTS targeted to the intended destination of the data packet. This packet initiates a clustering process. Upon the reception of the RTS, the intended destination of the packet becomes master and responds to the RTS with a beacon (B) followed by a poll targeted to the station which transmitted the RTS. A cluster is established and a CFP is initiated inside this cluster. All the idle stations which receive the beacon become slaves and get synchronized to the master at the packet level. Cluster membership is spontaneous and soft-binding: there are no explicit association and disassociation processes and a station belongs to a cluster as long as it can receive the beacons broadcast by the master. As in the PCF, a cluster is broken when the master transmits a CE packet. Upon the reception of this CE packet, all the slaves revert to idle mode and execute a backoff in order to avoid a certain collision if more than one station has data to transmit and initiates the DCF access period. Therefore, according to this operation, the clustering algorithm of DPCF is spontaneous in the sense that the first idle station with data to transmit initiates the clustering algorithm.

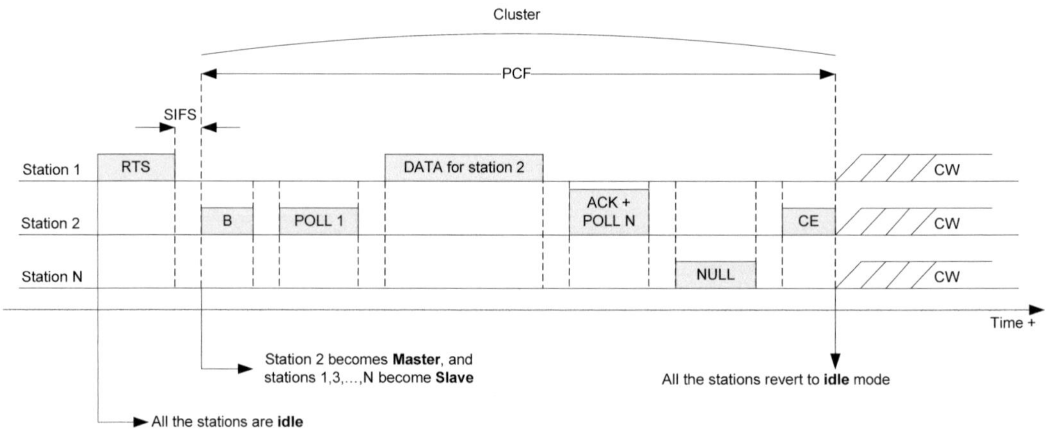

Fig. 5. Example: DPCF Operation

An example of operation is represented in Figure 5. In this example, station 1 has data to transmit to station 2. Once the station 1 successfully seizes the channel executing the rules of the DCF, it transmits an RTS to station 2. Upon the reception of the packet, station 2 becomes master and transmits a beacon. The first poll is then sent to station 1, which has a data packet ready to transmit. Station 1 transmits the data packet to station 2. Then, station 2 acknowledges the reception of the packet and polls station *N* with a combined packet. Since station *N* has no data packets to transmit, it sends a NULL packet. Finally, station 2 transmits the CE packet to indicate the end of the cluster phase. All the slave stations revert to idle mode and execute a backoff to reduce the probability of collision if more than one station has data to transmit.

Within a cluster, the master can poll the slaves following any arbitrary order. Regardless of the specific polling policy, the master has to have some knowledge of the local neighborhood in order to be able to carry out the polling mechanism. To do so, all the stations overhear the ongoing packet transmissions in their vicinity in order to create a neighbor table with an entry for each station in the local neighborhood. This table should be updated along time. The specific scheduling of the polling mechanism is out of the scope of the basic definition of DPCF. Only as an example, a round robin polling scheme can be executed following the entries of the neighbor table. In any case, once a station is polled by the master, it may transmit a data packet to any other slave (peer-to-peer communication model) without routing all the data through the master. Therefore, the master only acts as an indirect coordinator of the communications, but not necessarily as a concentrator of traffic (as the AP does in a regular centralized network).

The duration of a cluster is variable and depends on the traffic load of the network. An *inactivity mechanism* is considered to avoid the transmission of unnecessary polls when there are no more data packets to be transmitted. This mechanism consists of the following: any master maintains a counter that is incremented by one unit upon each NULL packet received from a polled station with no data to transmit. This counter is reset to zero whenever a station responds to a poll with the transmission of a data packet. If the counter gets to a specified value (tunable), the cluster is broken and a CE packet is sent.

On the contrary, it may happen that under heavy traffic conditions once a station becomes master it operates as such for the whole operation of the network due to the absence of idle periods. This would be unfair in terms of sharing the responsibility of being master in the network among all the stations. Therefore, it is necessary to upper-bound the maximum time that a station can operate as master without interruption. This limit is especially important in infrastructureless networks where fair energy consumption is a must. The approach in DPCF is the following: any master has a *Master Time Out* (MTO) counter which determines the maximum duration of a cluster. The value of the MTO corresponds to the maximum number of beacons (MTO=$N_{beacons}$) that a master can transmit without interrupting the operation of its cluster. The MTO counter is decremented by one unit after each beacon is transmitted. Whenever the MTO counter expires, a CE packet is transmitted and the cluster is broken regardless of the traffic load or activity of the stations. Therefore, the maximum time that a station can operate as master is denoted by T_{MAX} and can be computed as

$$T_{MAX} = N_{beacons} \cdot N_{polls} \cdot MIFS = MTO \cdot N_{polls} \cdot MIFS. \tag{1}$$

N_{polls} denotes the number of polls transmitted between beacons, which can also be tuned, and MIFS is the Maximum Inter Frame Space whose duration corresponds to the *maximum* time between two consecutive polls. The duration of a MIFS can be computed as the time elapsed when:

1. The master station combines an ACK of a recently received data packet with a poll and a data packet.
2. The station polled acknowledges the reception of the data packet from the master and combines the ACK with data for a third station.
3. The third station transmits the ACK of the data packet received from the second station.

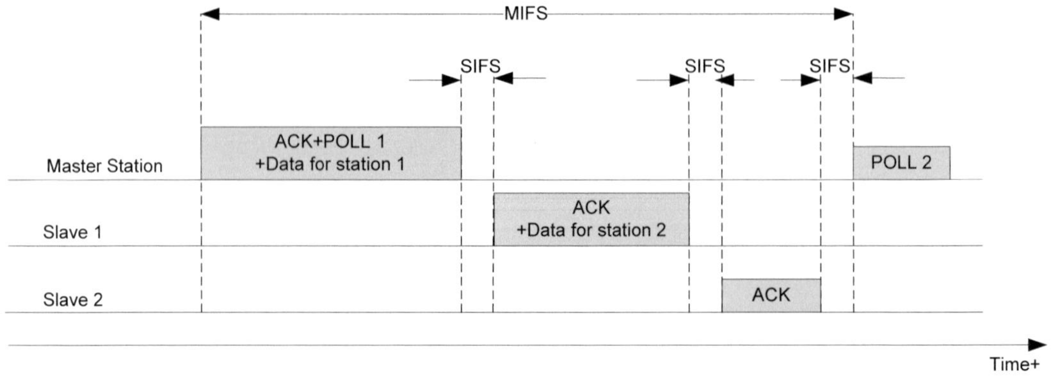

Fig. 6. Definition of MIFS

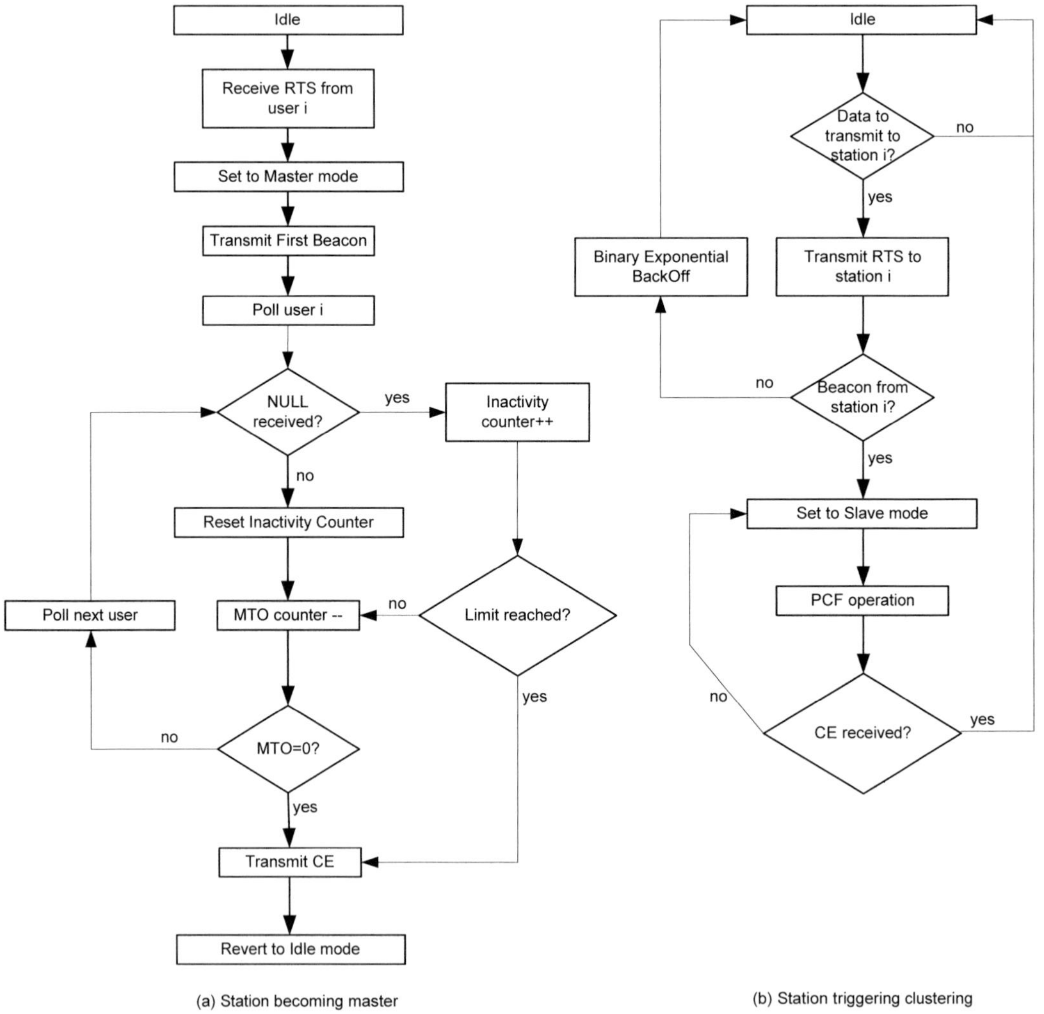

Fig. 7. DPCF Clustering Flowchart

The definition of a MIFS is illustrated in Figure 6. Note that it also corresponds to the minimum period of time that a station has to listen to the channel before establishing a new cluster in order to reduce the probability that another master is present.
In order to summarize the whole operation of DPCF, a general flowchart is shown in Figure 7. The left branch of the chart models the operation of any station becoming master when requested by any other stations and the right branch of the chart represents the operation of the station initiating the clustering algorithm when it has data to transmit.

4. Performance evaluation

In order to evaluate the performance of DPCF, we have implemented the protocol rules in a custom-made C++ link-level simulator. The simulator works in an object-oriented basis and the source code of each station runs in parallel. The implemented code could be directly integrated in a wireless card to execute the protocol rules. The main motivations for implementing the protocol in a custom-made C++ simulator rather than in any other well known system simulation platform (such as ns-2, for example) are:
1. The faster execution of the simulations.
2. The possibility of isolating the MAC protocol performance from the rest of the network.
3. The possibility to implement the protocol in a hardware testbed.

The system parameters have been set according to the PHY layer of the IEEE 802.11g Standard [1] and they are summarized in Table 1.

Parameter	Value	Parameter	Value
Data Packet Length (MPDU)	1500 bytes	Constant Message Length	1500 bytes
Data Tx. Rate	54 Mbps	Control Tx. Rate	6 Mbps
MAC header	34 bytes	PHY preamble	96 μs
SIFS, PIFS, DIFS	10, 30, 50 μs	SlotTime (σ)	10 μs
RTS, BEACON, CF_END and POLL packets	20 bytes	CTS and ACK packets	14 bytes
CW_{min}	16	CW_{max}	256
MTO	3	Polls per beacon	19

Table 1. System Parameters for Evaluation of DPCF

4.1 Single-hop networks

We first consider the case of a single-hop network composed of 20 stations, all of them within the transmission range of each other. All the stations generate data packets of fixed-length following a Poisson arrival distribution and they contribute equally (homogeneously) to the total aggregate data traffic of the network. The destination of each packet is randomly selected among all the stations of the network with equal probability. In order to focus on the MAC layer, all the packets are assumed to be received without errors and thus the results herein presented correspond to an upper-bound of the performance of the protocol.

It is also assumed that an ideal round robin scheduling is performed to poll all the stations once a cluster is established. Three different networks have been studied (they all have been implemented in the simulator):

1. **DCF:** a network wherein all the stations only execute the DCF with the collision avoidance access method.
2. **PCF:** a network wherein an AP manages the access to the channel. However, stations transmit directly to the intended destination without routing traffic through the AP. In this network, we consider that the AP also has data to transmit as any other regular station.
3. **DPCF:** a network wherein all the stations execute the proposed DPCF protocol.

According to the parameters presented in Table 1, the number of polls between beacons has been set to 19 and it indicates that all the slaves within a cluster are polled exactly once by the master between the transmission of two consecutive beacons. In addition, the setting MTO=3 indicates that all the slaves are polled at most three times when a cluster is established unless the inactivity mechanism is triggered by the master.

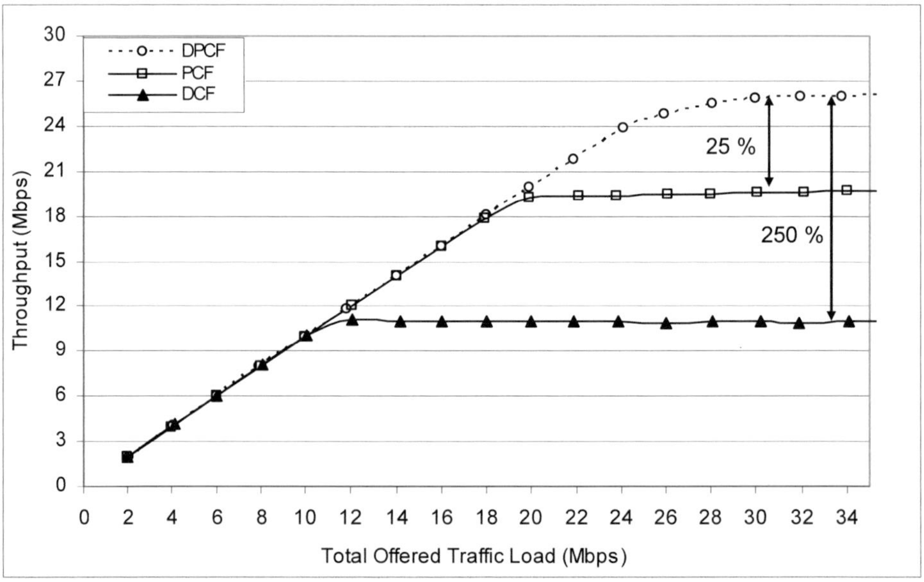

Fig. 8. Throughput Comparison DPCF, PCF, and DCF in a Single-hop Network

The throughput of the three different networks is plotted in Figure 8 as a function of the total aggregate offered load to the network. As expected, the three curves grow linearly until they reach the saturation throughput. The three protocols are stable for heavy traffic conditions without entering in congestion and thus they can operate under sporadic situations of peak high traffic loads without collapsing the network. The saturation throughput of DPCF is remarkably higher than that of DCF, achieving an improvement of approximately 250%. Collisions and backoff periods are reduced in the DPCF network compared to the DCF network, thus yielding higher performance. In addition, the performance of DPCF is even superior to the regular PCF, attaining 25% higher saturation throughput.

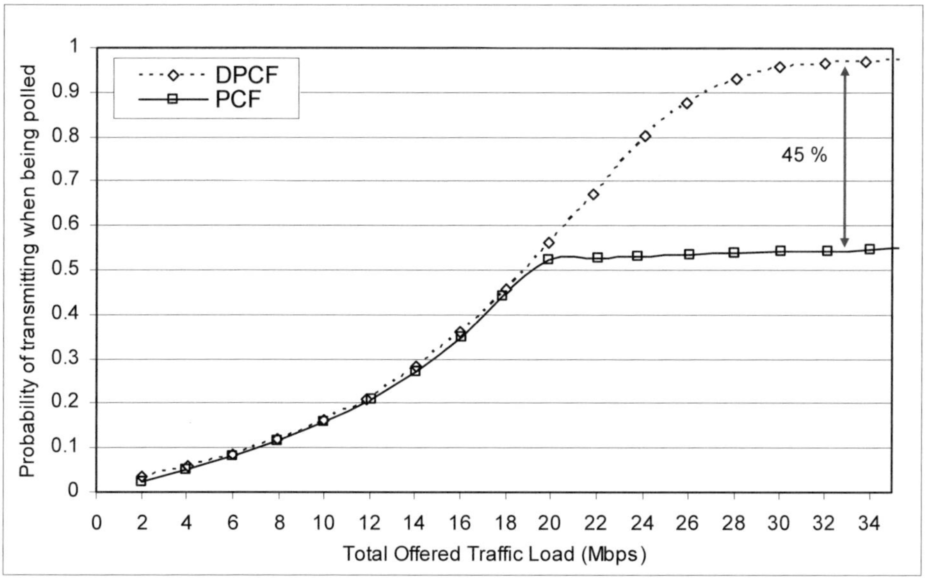

Fig. 9. Probability of Transmitting when Being Polled in a Single-hop Network

In order to further analyze this apparently counter-intuitive result, Figure 9 shows the probability that a station transmits a data packet when it is polled. It has been considered for this calculation that the AP (in the PCF network) and the masters (in the DPCF network) are *virtually* polled every time they poll a station as they have the possibility to combine the polls with data and ACK packets. The probability of transmitting data when being polled is quite similar in the two networks for low traffic loads. However, this probability is much higher in the DPCF network than in the PCF network for high traffic loads. While the efficiency of the polling in DPCF gets close to 98% for high traffic loads, it remains close to 55% in the PCF network. This efficiency translates directly into a higher efficiency of DPCF, since the ratio of data packets transmitted per control overhead is higher. The reason for these figures is that there is a severe unbalance between the channel access opportunities between the AP and the regular stations in the PCF network. This can be seen in Figure 10, where we plot again the probability of actually transmitting when being polled. Now, two different curves for the PCF network are represented corresponding to the average probability among of all the regular stations and to the probability for the AP alone, separately. The AP has a channel access opportunity every time it polls another station, but most of these transmission opportunities are not used for the actual transmission of data (note that the probability of transmitting when being polled is below 10% in all cases for the AP), decreasing the overall efficiency of the polling mechanism.

This unbalance between the AP and the stations is avoided in DPCF by sharing the responsibility of being master among all the stations of the network. It is well known that the DCF is fair in the long-term, and so is the clustering algorithm of DPCF. Since all the stations of the network get the role of master periodically, the unbalanced access of the AP in the PCF network is shared in the DPCF network. Every time a station is set to master it can transmit all its backlogged data packets and thus take advantage of the prioritized access to empty its data buffers while operating as master. Indeed, the fact that a station operating in master mode has more channel access opportunities than a slave station can be seen as an implicit mechanism to provide with some incentive to stations to become master despite the extra actions they must carry out and the corresponding increase in energy consumption.

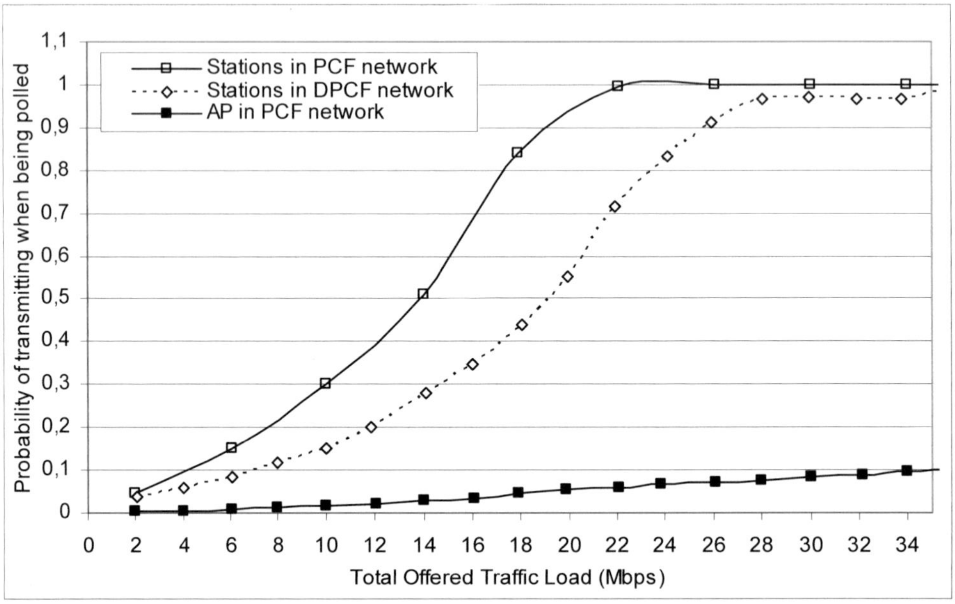

Fig. 10. Probability of Transmitting when Being Polled in a Single-hop Network

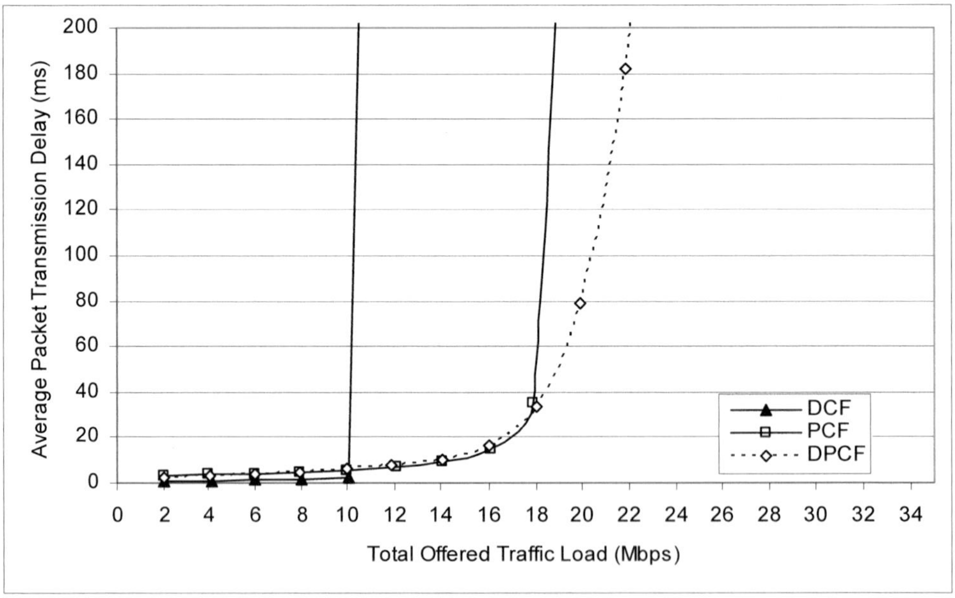

Fig. 11. Average Packet Transmission Delay in a Single-hop Network

The performance in terms of average packet transmission delay is plotted in Figure 11. We define this delay as the average time elapsed since a packet arrives at the MAC layer until it is successfully acknowledged by the intended destination. It is worth seeing that for low offered loads, the best performance is attained in the DCF network. This is an expected result since every time a station has data to transmit it can successfully seize the channel immediately without needing to wait for being polled (the probability of finding the channel busy and the probability of collision are low due to the low offered traffic load). However, as the offered load grows, the average packet transmission delay in the DCF network grows sharply for traffic loads over 10 Mbps. On the other hand, the DPCF attains average delays below 200 ms for traffic loads up to 22 Mbps, increasing the throughput of the standard DCF network and attaining superior performance than the PCF. These results confirm the idea that PCF-like mechanisms are worthy when the traffic load and the number of transmitting stations are relatively high.

4.2 Multi-hop networks

We now consider a multi-hop network. Without loss of generality and as a representative example, we consider a tandem network formed by 5 static stations set in line and equally spaced as the one represented in Figure 12. The distance between the stations, the transmission powers, and the channel propagation parameters have been adjusted so that:

1. Every station can transmit directly to immediate neighbors at one-hop distance.
2. Every station at two hops of a transmitting station can sense the channel busy, but cannot decode the transmitted information.
3. Every station at three hops of a transmitting station is oblivious to the transmission.

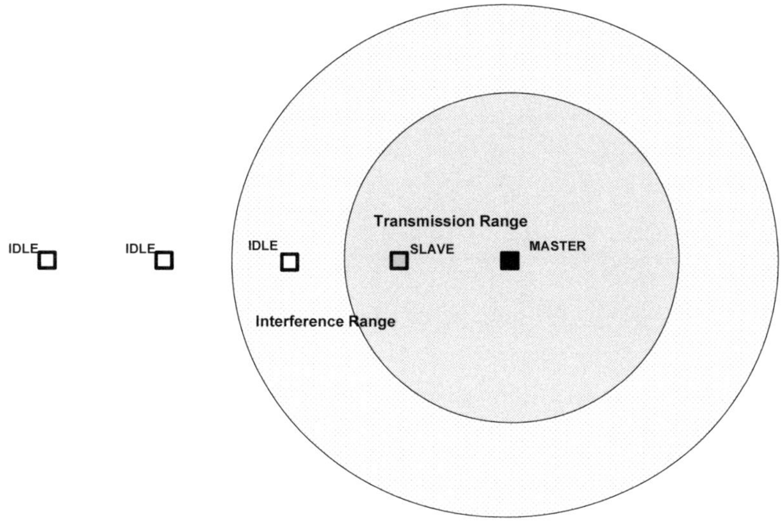

Fig. 12. Tandem Multi-hop Network

A collision occurs if two simultaneous transmissions are received within either the transmission or the interference range of the transmitters. We assume that all the stations have perfect routing information and thus route the packets through the station in its transmission range that is closer to the intended destination. The rest of the parameters have been set as in the previous section for the single-hop evaluation.

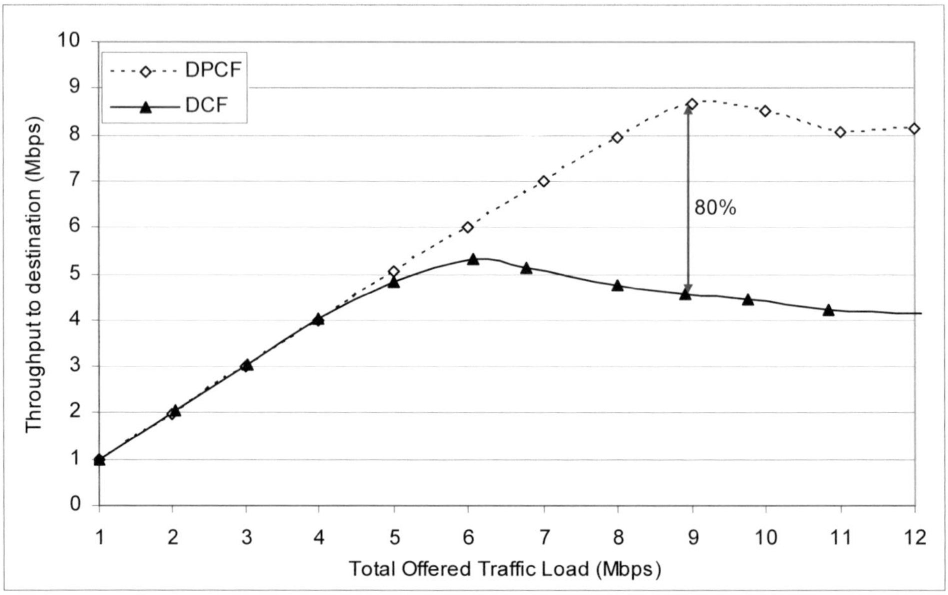

Fig. 13. Throughput to Destination of DPCF in a Multi-hop Network

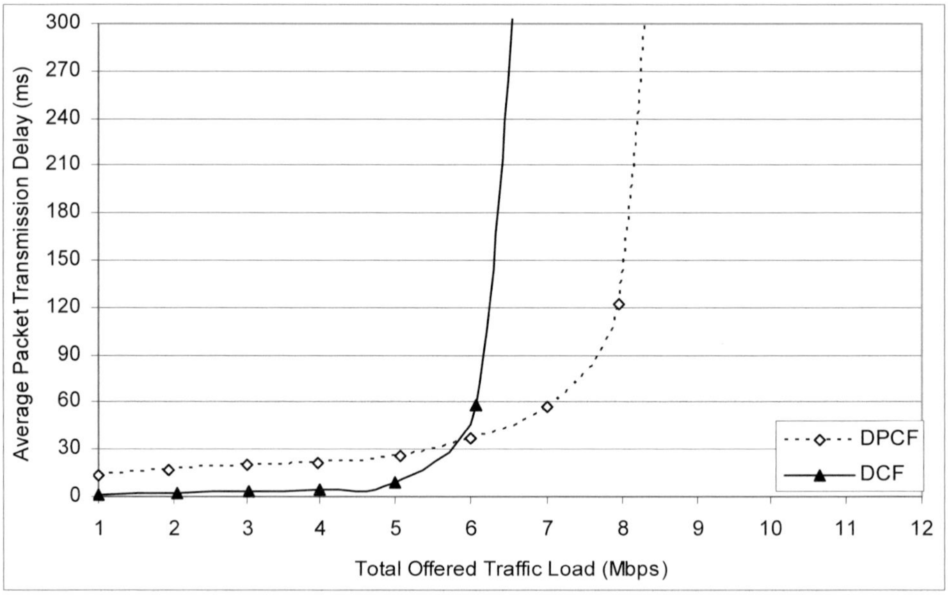

Fig. 14. Average Packet Transmission Delay of DPCF in a Multi-hop Network

The total throughput delivered to destination is plotted in Figure 13 as a function of the total offered load to the network. Note that the traffic delivered to the intermediate stations in a multi-hop route is not accounted for this calculation. The curves show that DPCF outperforms DCF for all traffic loads. Indeed, for low traffic loads both protocols behave almost identically delivering all the data traffic offered to the network. However, while DCF saturates around 5 Mbps, DPCF is capable of delivering up to 9 Mbps (80% higher saturation throughput), almost doubling up the capacity of the legacy DCF. Comparing these results to the ones obtained for the single-hop case, it is possible to see that the total offered load that can be conveyed in the multi-hop network is considerably lower. This is mainly due to the fact that in the multi-hop environment some packets need to travel along several hops to get to the final destination.

The average packet transmission delay is plotted in Figure 14 for both the DPCF and the DCF networks. In this case, this measure is defined as the average time elapsed from the moment a packet arrives at the MAC layer of the source station until it is successfully delivered to the final destination (end-to-end time). The curves show that the DCF attains lower average packet transmission delay for low traffic loads. Two are the main reasons for this lower average delay. First, the longer MIFS of DPCF (compared to the DIFS of DCF) adds latency to all the transmissions, increasing the average packet transmission delay in the DPCF network for low traffic loads. In addition, in the DPCF network, slaves cannot transmit immediately whenever they have data to transmit but they have to wait to be polled by a master, increasing thus the average access delay. However, note that the average delay is lower than 300 ms for loads up to 8 Mbps in the DPCF network and it gets unbounded in the DCF network for traffic loads over 6 Mbps. Therefore, the DPCF protocol attains better performance when the traffic load of the network is higher, attaining up to 25% better performance than the DCF in this multi-hop setting.

5. Conclusions

We have presented in this chapter a simple mechanism to improve the performance of the 802.11 Standard under heavy loaded conditions. These conditions appear in some vehicular scenarios, such as in traffic-light crossings, where vehicles and pedestrians meet together and a number of safety applications may arise.

The key idea consists in combining both distributed and point coordinated access methods to manage the access of the users to the wireless channel. The specific approach has been based on an extension of the PCF of the IEEE 802.11 Standard to operate over distributed wireless ad hoc networks without infrastructure. The main idea of DPCF is that the stations of the network get access to the channel by executing the rules of the DCF. Any station which seizes the channel transmits its data and also establishes a temporary dynamic cluster to manage the pending transmissions of all the neighbors with data ready to be transmitted. The key of this mechanism is that there is no cluster head selection, but clusters are created in a spontaneous manner. This reduces the control overhead to establish a fixed clustering architecture and increases the capability of the network to dynamically adapt to the unpredictable nature of ad hoc networks. Comprehensive performance evaluation of the protocol through link-level computer simulation shows that the new proposal improves the performance of ad hoc networks when compared to current standards.

The results presented in this chapter are rather promising and, in fact, future work will be aimed at theoretically evaluating and optimizing the design of DPCF and at implementing the protocol in a testbed to evaluate its actual performance in a real environment. Ongoing work is being carried out to evaluate the coexistence feasibility of this new approach with legacy implemented networks based on the 802.11.

6. Acknowledgments

The research leading to these results has received funding from the research projects NEWCOM++ (ICT-216715), CO2GREEN (TEC2010-20823), CENTENO (TEC2008-06817-C02-02), and GREENET (PITN-GA-2010-264759).

7. References

[1] IEEE, Part 11: Wireless LAN Medium Access Control (MAC) and Physical Layer (PHY) Specifications, *IEEE Std.* 802.11 - 2007.

[2] D. J. Goodman, R. A. Valenzuela, K. T. Gayliard, and B. Ramamurthi, Packet Reservation Multiple Access for Local Wireless Networks, *IEEE Trans. on Communications*, vol. 37, no. 8, Aug. 1989.

[3] I. Chlamtac, A. Faragó, A. D. Myers, V. R. Syrotiuk, and G. V. Záruba, ADAPT: A dynamically self-adjusting media access control protocol for ad hoc networks, in *Proc. of the GLOBECOM*, pp. 11 - 15, Dec. 1999.

[4] A. Rhee, M. Warrier, and J. Min, ZMAC: A hybrid MAC for wireless sensor networks, in *Proc. of Sensys* 2005, San Diego, California.

[5] M. Shakir, I. Ahmed, P. Mugen, and W. Wang, Cluster Organization based Design of Hybrid MAC Protocol in Wireless Sensor Networks, in *Proc. of the Third International Conference on Networking and Services*, pp. 78 - 83, Jun. 2007.

[6] A. Muir and J. J. Garcia-Luna-Aceves, An efficient packet sensing MAC protocol for wireless networks, *Springer Mobile Networks and Applications*, vol.3, no. 2, pp. 221–234, Aug. 1998.

[7] J. Alonso-Zárate, E. Kartsakli, L. Alonso, and Ch. Verikoukis, Performance Analysis of a Cluster-Based MAC Protocol for Wireless Ad Hoc Networks, *EURASIP Journal on Wireless Communications and Networking*, Special Issue on Theoretical and Algorithmic Foundations of Wireless Ad Hoc and Sensor Networks, vol. 2010, Article ID 625619, 16 pages, March 2010.

[8] A. Kanjanavapastit and B. Landfeldt, A performance investigation of the modified PCF under hidden station problem, in *Proc. of the ICCCAS 2004*, vol. 1, pp.428 - 432, Jun. 2004.

[9] B. Anjum, S. Mushtaq, A. Hussain, Multiple Poll Scheme for Improved QoS Using IEEE 802.11 PCF, in *Proc. of the IEEE INMIC'05*, pp.1 - 6, Dec. 2005.

[10] D. Ping, J. Holliday, A. Celik, Dynamic scheduling of PCF traffic in an unstable wireless LAN, in *proc. of the CCNC. 2005*, pp. 445 - 450, Jan 2005.

[11] K. Byung-Seo, K. Sung Won, W. Yuguang Fang, Two-step multipolling MAC protocol for wireless LANs, *IEEE Journal on Selected Areas in Communications*, vol. 23, no. 6, pp. 1276 - 1286, Jun. 2005.

[12] K. Young-Jae and S. Young-Joo, Adaptive polling MAC schemes for IEEE 802.11 wireless LANs, in *Proc. of the VTC 2003*, vol. 4, pp. 2528 - 2532, Apr. 2003.

[13] A. Kanjanavapastit and B. Landfeldt An analysis of a modified point coordination function in IEEE 802.11, in *Proc. of the IEEE PIMRC'03*, vol. 2, pp. 1732 - 1736, 2003.

[14] Y. Tiantong, H. Hassanein, H. T. Mouftah, Infrastructure-based MAC in wireless mobile ad-hoc networks, in *Proc. of the 27th Annual IEEE Conference on Local Computer Networks*, pp. 821 - 830, 2002.

9

Hybrid Cooperation Techniques

Emilio Calvanese Strinati and Luc Maret
CEA, LETI, MINATEC
France

1. Introduction

A major challenge in the design of next generation wireless communication systems is to achieve both reliable and spectral efficient communication with large coverage range. To tackle this problem, advanced diversity techniques combined with adaptive mechanisms have to be designed in order to combat or even exploit the variability of the radio propagation medium across time, frequency and space. Diversity techniques create signal redundancy, by repeating the information across multiple, independent channel realizations. This is accomplished by allowing the receiver to experience the average channel effect rather than an instantaneous fade. As a consequence diversity techniques improve the link reliability at the expense of the system spectral efficiency. By adjusting the transmission parameters to the momentary link quality, adaptive mechanisms aim at improving both spectral efficiency and link reliability. Nevertheless, in order to guarantee the Quality of Service (QoS) constraints from the upper layers, adaptive mechanisms implement a sub-optimal trade-off between link robustness and bandwidth efficiency (Calvanese Strinati E., 2006). Therefore in this chapter we propose and analyze a novel cooperation protocol, *the hybrid cooperation protocol* and we combine it with link adaptation techniques such as Adaptive Modulation and Coding (AMC) and power allocation. Our task is to minimize the outage probability and maximize the spectral efficiency of transmission, while limiting the cooperation cost in terms of MAC signalling overhead.

The scientific content of this chapter is based on some innovative results presented in three conference papers (E. Calvanese Strinati and S. Yang and J-C. Belfiore, 2007) (E. Calvanese Strinati and Luc Maret, 2008) (M. Baydar and E. Calvanese Strinati and J. C. Belfiore, 2008) presented in 2007 and 2008.

The goals of this chapter are for the reader to have an understanding of cooperative communication issues and challenges and, to be well informed of the state-of-the-art research development. Eventually, the chapter will present what we have done to improve the performance of currently proposed cooperation techniques, comparing performance of our proposed approaches with state-of-the-art one. A critical discussion on advantages and weakness of the proposed approaches, including future research axes, will conclude the chapter.

The innovative contribution in this chapter is threefold.

First, in this chapter we introduce and details challenges and possible solutions for the so-called cooperative diversity (E. Erkip A. Sendonaris and B. Aazhang: Part I, 2003; E. Erkip A. Sendonaris and B. Aazhang: Part II, 2003) techniques where a source terminal cooperates with several relays to exploited the spatial diversity in a distributed manner. From a physical layer viewpoint, cooperation drives to improved transmission diversity and consequent improved outage probability performance. Nevertheless, from a MAC layer viewpoint, fixed cooperation requires to probe the network and acquire channel state information (CSI) about all active relays at least with a channel coherence time frequency. This cooperation probing makes fixed cooperation expensive in terms of signalling overhead, battery consumptions of active relays and protocol delay. Recently, researchers showed that cooperating is not always the best solution in terms of outage probability minimization (D. Gunduz and E. Erkip, 2005). For instance in AF cooperation, when the noise is large, cooperative relays can amplify the noise instead of helping. Alternatively to *fixed* cooperation, we propose to introduce a *cooperation controller* that can decide when and how to cooperate. The basic idea is to cooperate when it is advantageous (*cooperative mode*), and not cooperate otherwise (*non-cooperative mode*). The problem in such approach is how the source-destination pair can decide if it is worth to cooperate for a given channel instance? In fixed topology networks an heuristic approach is to analyze the geometry of the network and determine areas where cooperation can help. In a wireless mobile communication scenario the cooperation protocol should use the momentary channel state information to make its decision. This feedback information introduces a large processing delay and signalling overhead and it is impractical for the destination to acquire full CSI about all active relays. In this chapter we present the innovative approach proposed in (E. Calvanese Strinati and S. Yang and J-C. Belfiore, 2007) of introducing a cooperation controller which makes its decision on non-cooperative/cooperative mode based only on the momentary direct link quality information. This information is directly available at the cooperation controller each time the source sends a request to send (RTS). More precisely, for a selected transmission rate R and direct link channel instance (σ_n^2, f, etc.), the cooperation controller can check if direct non-cooperative transmission will be certainly in outage. If an outage is forecasted, the receiver can switch to cooperative mode trying to avoid transmission outage improving the overall link quality with cooperative diversity. This protocol is called *hybrid cooperation*.

Second, the chapter presents how hybrid cooperation protocol and AMC mechanism can be jointly designed. Eventually, we detail the *hybrid cooperative AMC* mechanism (E. Calvanese Strinati and S. Yang and J-C. Belfiore, 2007) in which adaptation is designed for the hybrid link quality and cooperation is activated only when the instantaneous direct source-destination link quality is not good enough to support the aimed spectral efficiency.

Third, we face the problem of optimal power allocation between source and relay in a cooperative network. We present the interesting results proposed in (M. Baydar and E. Calvanese Strinati and J. C. Belfiore, 2008). The authors succeed in finding a close form power allocation algorithm which can only be applied to OAF cooperative transmission. At a first glance, this solution can be not of interest due to the worse performance of the OAF cooperation strategy. We verify that classical NAF outperforms classical OAF also if an optimal power allocation is done for the OAF cooperation and sub-optimal power allocation is done for NAF. This is due to the suboptimal performance of classical OAF. To solve this problem we further investigated the *hybrid* AF protocol that we propose.

2. Preliminaries on Cooperative transmission techniques

The core topic investigated in this chapter is the improvement of outage probability performance in a cooperation network. In the literature (see, e.g., (H. Bölcskei and R. U. Nabar and F. W. Kneubühler, 2004; H. El Gamal K. Azarian and P. Schniter, 2005; S. Yang and J-C. Belfiore, 2006)), three main cooperative transmission protocols have been proposed:

the amplify-and-forward (AF), the decode-and-forward (DF) ((D. N. Tse J. N. Laneman and G. W. Wornell, 2004) (H. Bölcskei and R. U. Nabar and F. W. Kneubühler, 2004) (H. El Gamal K. Azarian and P. Schniter, 2005)) and the compress-and-forward (CF) for which there has been, recently, a grown interest. Nevertheless, most prior works focused on two principal classes of protocols. The first is the class of AF protocols, where the relay simply amplifies and re-transmits the observed signal. The second is the class of DF protocols, where the relay decodes, re-encodes and re-transmits the message it receives. The DF protocols offer good performance but have clearly a higher complexity compared to the AF protocols which are used in practice, due to their low complexity and low relay power consumption. Actually, for most *ad hoc* wireless networks, it is not realistic for other terminals to decode the signal from a certain user, because the codebook is seldom available and the decoding complexity is unacceptable in most cases.

The second topic treated in this chapter is the design of hybrid cooperation protocol combined with an AMC mechanism. Design of cooperation protocol and AMC algorithms have been extensively investigated separately. Nevertheless, joint design of advanced cooperation protocols with AMC algorithms has not been intensively investigated yet. Aiming at maximizing the physical layer throughput, in (Z. Lin and E. Erkip and M. Ghosh, 2005) the authors study adaptive modulation performance for one relay coded cooperative protocols. In the paper the authors find that coded cooperation combined with adaptive modulation offers better physical layer throughput performance than non-cooperative mode. The authors suggest that cooperative mode MCS selection should be decided based on all direct and relays link quality. However, if the cooperative protocol includes more than one relay, this approach can be complex and catastrophic adaptation can occur as for frequency selective block fading channels (M. Lampe and H. Rohling and W. Zirwas, 2002) since it is hard to obtain a reliable predicted packet error rate (PER_{pred}). In (E. Yazdian and M. R. Pakravan, 2006) the application of adaptive modulation to one relay AF cooperation is investigated. The authors aim at evaluating the energy saving achieved through cooperation due to the improvement in average bit/symbol transmission. Furthermore, the authors study the performance improvement as a function of cooperating user's location to identify areas where cooperation is useful. However, in the paper the possible occurrence of detrimental cooperation is not considered.

The third topic investigated in this chapter is the combination of cooperative diversity techniques with power control algorithms. Optimal power allocation between source and relay in a cooperative network has been studied in (M. Hasna and M-S. Alouini, 2004) (Q. Zhang and C. Shao and Y. Wang and P. Zhang and J. Zhang and Z. Zhang, 2004). A total amount of transmit power over the two slots required for relaying is shared between the source and relay. In (I. Hammerstrom and A. Wittneben, 2006), an iterative joint power allocation method is presented for two-hop communications schemes using OFDM modulation. This method is based on the Karush-Kuhn-Tucker (KKT) conditions. Power allocation optimization for NAF cooperative transmissions is classically done using *waterfilling* techniques. Its effectiveness depends on the *a priori* choice of the power allocated to the relay (P_r). Unfortunately, the optimal selection of P_r can be a challenging task. An iterative search may improve the power allocation algorithm performance at the expense of both search latency and algorithm complexity. Alternatively, the power allocation problem can be faced for OAF schemes for which the a priori knowledge of P_r is not required. In such case, the complexity of the power allocation algorithm is strongly reduced at the expense of performance.

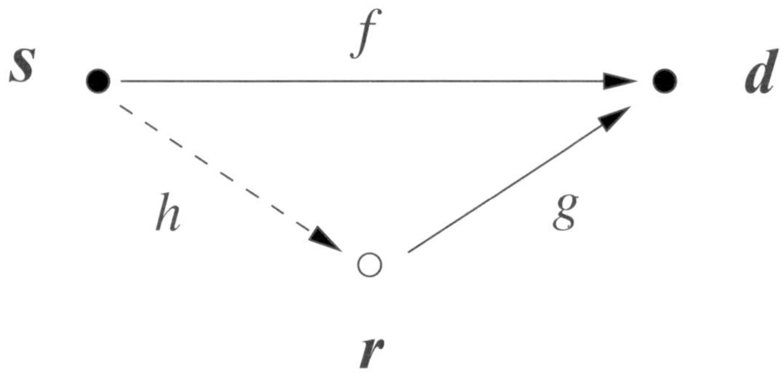

Fig. 1. A relay channel with one source **s**, one destination **d** and one active relay **r**.

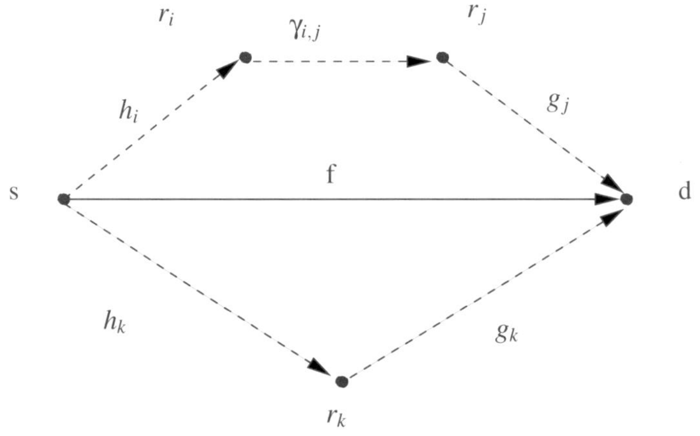

Fig. 2. A relay channel with one source **s**, one destination **d** and N active relays r_i.

3. System model

The considered system model consists of one source s, one destination d and N relays (cooperative terminals) $r_1, \ldots, r_N$. The physical links between terminals are slowly faded and are modelled as independent quasi-static Rayleigh channels, *i.e.*, the channel gains do not change during the transmission of a cooperation frame. This assumption implies that we assume the channel coherence time to be much larger than the maximum delay that can be tolerated by the application. All the terminals (source, relay and destination) are equipped with only one antenna and work in half duplex mode, *i.e.*, they cannot receive and transmit at the same time. Two simple illustrations of the channel model are given in Fig. 1 and Fig. 2 respectively for a cooperative network with only one and N active relays per frame transmission.

The gain of the channel connecting s and d is denoted by f. Similarly, g_i and h_i respectively denote the channel gains between r_i and d and the ones between s and r_i. γ_{ij} is used to denote the channel gain between r_i and r_j. Channel quality between terminals is parameterized by the variance of the channel gains. We assume that the receiver can gain perfect knowledge of the channel gains for the whole network activating the relay probing procedure (S. Yang and J-C. Belfiore, 2006). We consider two cases for the power allocated to source and relays.

First, we impose a total average transmit power constraint and no power control is allowed in our scheme. In this case, in order to simplify the analysis, we consider a suboptimal power allocation scheme where the source transmits at full power in the non-cooperation mode and both the source and the relays transmit at half power in the cooperation mode. Then, we refine this assumption proposing a power allocation algorithm for hybrid OAF cooperative protocols with only one active relay per transmission frame (see section 4.3). Also, we suppose that the terminals are subject to the half-duplex constraint, *i.e.*, they cannot transmit and receive simultaneously. We also assume using the capacity achieving code so that the outage analysis holds. The PER prediction is based on the computation of a link quality metric (LQM) that is linked to the predicted PER by means of a look up table (LUT). Ideally, we consider perfect PER prediction. In our work we consider the Amplify-and-Forward (AF) protocol (orthogonal and non-orthogonal) where the relay simply scales and forwards the received signal. We study half-duplex slotted amplify-and-forward (SAF) cooperative schemes proposed in (S. Yang and J-C. Belfiore, 2006). For an N-relay M-slot scheme, the cooperation frame, composed of M slots of l symbols, is of length Ml. During any slot i, $i = 1, \ldots, M$, the source s transmits a sub-frame of l symbols, denoted by a vector $\mathbf{x}_i \in \mathbb{C}^l$ and the relay r_j, $j = 1, \ldots, N$, can transmit $\mathbf{x}_{r_j,i} \in \mathbb{C}^l$, a linear combination of the vectors it received in previous slots. Under the half-duplex constraint, a relay does not receive while transmitting. For example, the NAF scheme (H. El Gamal K. Azarian and P. Schniter, 2005) is an N-relay $(2N)$-slot scheme and the non-orthogonal relay selection scheme (D. P. Reed and A. Bletsas and A. Khisti and A. Lippman, 2005) is an N-relay two-slot scheme. Obviously, the transmission of a cooperation frame with any SAF scheme is equivalent to l channel uses of the following vector (MIMO) channel

$$\mathbf{y} = \sqrt{\mathsf{SNR}}\,\mathbf{H}\mathbf{x} + \mathbf{z}$$

where $\mathbf{x}$ is the transmitted signal, $\mathbf{z} \sim \mathcal{CN}[\boldsymbol{\Sigma}_{\mathbf{z}}]$ is the equivalent additive coloured noise with covariance matrix $\boldsymbol{\Sigma}_{\mathbf{z}}$ and $\mathbf{H}$ is an $M \times M$ lower-triangular matrix representing the equivalent "space-time" channel between the source and the destination. Moreover, we have $H_{ii} = c_i f$ with c_i being a constant related to the transmission power. Let $\mathbf{H}$ denotes the equivalent channel matrix (S. Yang and J-C. Belfiore, 2006) for a Non-Orthogonal AF scheme[1]

$$\mathbf{H} = \begin{pmatrix} f & 0 \\ \frac{\sqrt{P_r}bgh}{\sqrt{1+P_r\|bg\|^2}} & \frac{f}{\sqrt{1+P_r\|bg\|^2}} \end{pmatrix}.$$

The matrix coefficients, $h_{i,j}$ are functions of f, g, h, P_r the relay transmission power and the normalization factor b which verifies $b^2 = \frac{1}{1+P_s\|h\|^2}$. The input covariance matrix is a diagonal matrix, denoted $\mathbf{Q}$ and whose diagonal elements are P_{s1} and P_{s2}, the source transmission powers in the first and the second slot, respectively.

4. Improving Cooperative transmission protocols effectiveness

This section will be divided in three parts. First we will present and explain the *Hybrid Amplify and Forward Cooperation Protocol* proposed in (E. Calvanese Strinati and S. Yang and J-C. Belfiore, 2007) which has been designed to overcome the suboptimal error rate performance of AF cooperative schemes in the low SNR region. Second, we will describe the Adaptive

[1] An Orthogonal AF scheme is a particular case of the NAF scheme in which the source does not transmit simultaneously with the relay in the second slot (*i.e.*,$h_{2,2} = 0$)

Modulation and Coding Combined with the Hybrid Cooperation protocol proposed in (E. Calvanese Strinati and S. Yang and J-C. Belfiore, 2007; E. Calvanese Strinati and Luc Maret, 2008). In a third part of the section, we will present the Power Allocation Optimization for Hybrid Cooperation Protocols which has been proposed in (M. Baydar and E. Calvanese Strinati and J. C. Belfiore, 2008).

4.1 Improving amplify and forward Cooperation protocol: Hybrid amplify and forward

We present in this section the novel cooperative protocol proposed in (E. Calvanese Strinati and S. Yang and J-C. Belfiore, 2007). The protocol is named *hybrid cooperation* and it has been designed to overcome the suboptimal error rate performance of AF cooperative schemes in the low SNR region. The protocol proposal is based on the observation that cooperating is not always the best solution in terms of outage probability minimization (D. Gunduz and E. Erkip, 2005). For instance in AF cooperation, in the low SNR regime, the relays amplify the noise instead of helping. Alternatively to *fixed* cooperation, a *cooperation controller* can decide when and how to cooperate. The principle of the hybrid cooperation protocol is simple: based on the direct source-destination link quality, a cooperation controller decides if and how to run cooperation. Indeed, cooperation is activated only when the instantaneous direct source-destination link quality is not good enough to support the aimed spectral efficiency. The problem in such approach is how the source-destination pair can decide if it is worth to cooperate for a given channel instance? In fixed topology networks an heuristic approach is to analyze the geometry of the network and determine areas where cooperation can help. In a wireless mobile communication scenario the cooperation protocol should use the momentary channel state information to make its decision. This feedback information introduces a large processing delay and signalling overhead and it is impractical for the destination to acquire full CSI about all active relays. In (E. Calvanese Strinati and S. Yang and J-C. Belfiore, 2007) the authors propose that the cooperation controller makes its decision on non-cooperative/cooperative mode based only on the momentary direct link quality information. This information is directly available at the cooperation controller each time the source sends a request to send (RTS). More precisely, for a selected transmission rate R and direct link channel instance (σ_n^2, f, etc.), the cooperation controller can check if direct non-cooperative transmission will be certainly in outage. If an outage is forecasted, the receiver can switch to cooperative mode trying to avoid transmission outage improving the overall link quality with cooperative diversity.

Classical NAF outperforms classical OAF also if an optimal power allocation is done for the OAF cooperation and sub-optimal power allocation is done for NAF. This is due to the suboptimal performance of classical OAF. To solve this problem we further investigated the *hybrid* AF protocol. Calvanese Strinati *et al.* first propose an OAF hybrid cooperation protocol under the same power constraint adopted above: impose a total average power constraint and no power allocation is considered. If P denotes the total power constraint, in case of NAF cooperation, we impose $P_{s1} = P/2$ for the power allocated to the source in the first slot and $P_{s2} = P_r = P/2$ the power allocated to the source and the relay respectively in the second slot. In the OAF scheme, the authors propose to fix $P_s = P_r = P/2$. The mutual information of the direct channel, the cooperative channel[2] and the OAF channel are respectively:

[2] Factor $\frac{1}{2}$ comes from the fact that two time slots (*i.e.*, two channel uses) are needed to transmit symbols

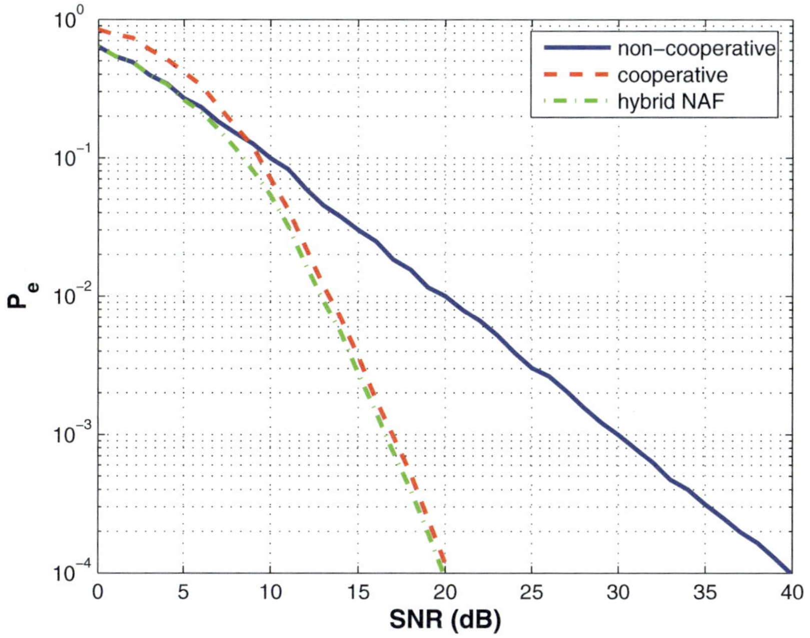

Fig. 3. Outage probability for Cooperative/non-cooperative/hybrid NAF cooperative transmission with $N = 2$, $M = 5$

$$
\begin{aligned}
I_d &= \log_2(1 + \frac{P}{2}\,|f|^2) \\
I_{NAF} &= \frac{1}{2}\log_2\,\det(I + \mathbf{H}\,\mathbf{Q}\,\mathbf{H}^{\dagger}) \\
I_{OAF} &= \frac{1}{2}\,\log_2(1 + P_s|f|^2 + \frac{P_s P_r|bgh|^2}{1 + P_r|bg|^2})
\end{aligned}
$$

Based on these mutual information expressions, in (E. Calvanese Strinati and S. Yang and J-C. Belfiore, 2007) the authors propose to numerically compare non-cooperative, NAF cooperative, hybrid NAF cooperative and hybrid OAF cooperative protocols in terms of outage probability versus average SNR. Let $\mathcal{O}_d$ denotes the direct channel outage event, $\mathcal{O}_d = \{I_d < R\}$, and $\mathcal{O}_c$ denotes the cooperative channel outage event, $\mathcal{O}_c = \{I_c < R\}$. The equivalent channel is in outage if both events, $\mathcal{O}_d$ and $\mathcal{O}_c$, are realized.

4.1.1 Simulation results

We report here some significant simulation results to evaluate effectiveness of the proposed hybrid cooperation protocol when applied to NAF cooperation on Fig 3.
Next, we extend the study of the hybrid cooperation protocol to OAF schemes and we introduce a power allocation algorithm well designed for OAF hybrid cooperative transmission. We find out that transmission outage is slightly smaller adopting hybrid cooperation for OAF scheme than for NAF one. Nevertheless, there are other important

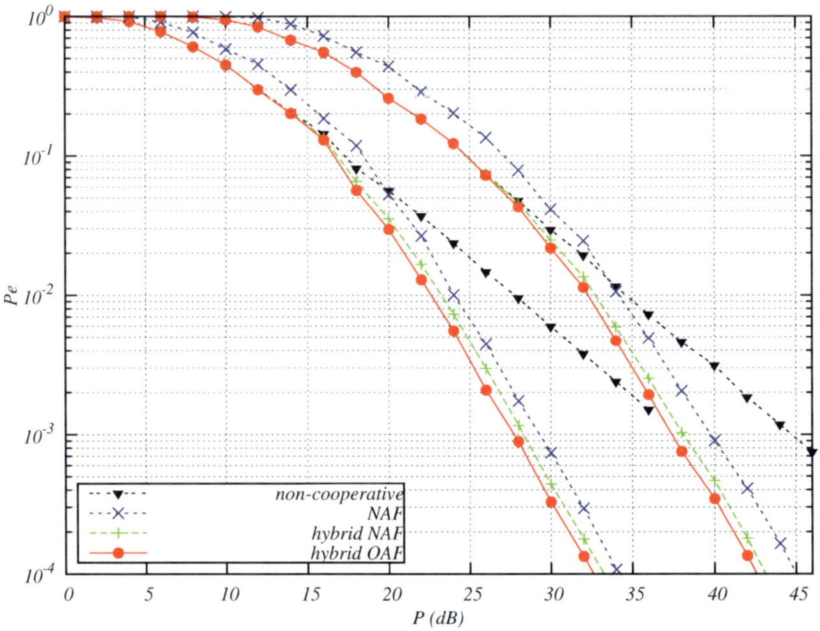

Fig. 4. Outage probabilities for the non-cooperative, NAF, Hybrid-NAF and Hybrid OAF scheme. Considered information rates: 2 and 4 BPCU.

advantages in adopting an OAF hybrid cooperation protocol. First, the cooperation complexity and cost are reduced. Second, the hybrid strategy reduces significantly the complexity of the algorithm implemented to determine the outage probability. This is the key reason for which we succeeded in finding an optimal power allocation algorithm for OAF hybrid cooperation schemes. We now show some simulation results for hybrid cooperative transmission without power allocation. Performance is compared in terms of average outage probability versus average SNR.

Based on these mutual information expressions, we numerically compare non-cooperative, NAF cooperative, hybrid NAF cooperative and hybrid OAF cooperative protocols in terms of outage probability versus average SNR. Let $\mathcal{O}_d$ denotes the direct channel outage event, $\mathcal{O}_d = \{I_d < R\}$, and $\mathcal{O}_c$ denotes the cooperative channel outage event, $\mathcal{O}_c = \{I_c < R\}$. The equivalent channel is in outage if both events, $\mathcal{O}_d$ and $\mathcal{O}_c$, are realized.

Other simulation results are shown in Figure 4 for the case of one active relay and transmission rate of 2 and 4 bits per channel use (BPCU). We find out that, adopting the proposed OAF hybrid cooperation protocol, transmission outage performance is better than for both non-cooperative and NAF hybrid cooperation transmissions. This result confirms our choice of using an orthogonal scheme: since the channel is assumed to be quasi-static, if the direct link is in outage in the first slot, it will remain in outage in the second one. The outage performance improvement is not our major achievement. Combining hybrid cooperation with OAF scheme, we obtain a cooperation protocol with both reduced complexity and cooperation cost. Furthermore, the proposed hybrid strategy permits to reduce the complexity of the outage probability computation. This is the key reason for which we succeeded in finding an optimal power allocation algorithm only for OAF hybrid cooperation schemes.

The Orthogonal AF strategy, sub-optimal in a full time cooperation scheme, is optimal with the hybrid strategy. In fact, since the channels are assumed to be slow fading, if the direct link is in outage in the first slot of the frame, it will be the case in the second. So it is better not to transmit in the second slot, and thus economize power, since we are sure that the reliability of the information is not guarantied. The mutual information is in this case

$$I_{OAF} = \frac{1}{2}\log_2(1 + P_s|f|^2 + \frac{P_s P_r |bgh|^2}{1 + P_r|bg|^2}) \tag{1}$$

4.2 Proposed Adaptive Modulation and Coding Combined with the Hybrid Cooperation Protocol

In this section we present the mechanism proposed in (E. Calvanese Strinati and S. Yang and J-C. Belfiore, 2007) in which the authors propose to combine the *hybrid* cooperation protocol with an AMC mechanism. The protocol is named *hybrid cooperative AMC* mechanism. A flow chart of the proposed algorithm is shown in Fig. 5. $I_{non-coop}$ is the instantaneous mutual information when transmission is done in non-cooperative mode and R is the transmission rate.

The algorithm is summarized as follows:

Step 1: S sends a RTS each time it wants to transmit new data.

Step 2: After receiving a RTS, the AMC mechanism (in D) selects R for next data transmission. R is selected from the set of LUT of PER versus LQM for hybrid cooperation transmission performance, given the LQM computed at previous received packet.

Step 3: D estimates the instantaneous channel conditions of the direct source-destination link (σ^2,f, etc.) and computes $I_{non-coop}(f, \sigma^2)$

Step 4: The *cooperation controller* in D decides if cooperate or not:

- if $I_{non-coop} < R$, non-cooperative transmission is forecasted to be in outage: the cooperation controller starts cooperation (go to step 5)
- otherwise, cooperation mode is not activated (go to step 9)

Step 5: D checks if the *relay probing* is up to date:

- YES (go to step 9)
- NOT (go to step 6)

Step 6: relay probing: D probes the relays available for cooperation and estimates the channel coefficients of the cooperation links.

Step 7 and 8: Each relay calculates the product gain $|g_i h_i|$ and reacts by sending an availability frame after t_i time which is anti-proportional to $|g_i h_i|$. Therefore, the relay with the strongest product gain is identified as relay 1, and so on.

Step 9: D sends a clear to send (CTS) that includes information on transmission rate R, M, relay identifiers, etc.

Step 10: S starts data transmission at rate R

Step 11: After receiving data from S, D derives PER_{pred} from the LUT of hybrid cooperation and selects R for next transmission of S.

Summarizing, based on the direct source-destination link quality, a cooperation controller decides if and how cooperate. We call this cooperation protocol as *hybrid cooperation*. The rate

R is chosen after each received packet by the AMC that aims at maximizing the throughput performance of the hybrid transmission mode meeting the QoS constraints imposed by the upper layers.

Note that the AMC mechanism selects R based on a set of pre-computed AMC switching points that depends on N, M, PER_{target}, transmission scenario, etc. Such switching points are chosen based on the average PER versus average performance of the hybrid cooperation protocol. Given N, M and R, there is a *crossing point* (PER_{cross}) between non-cooperative and cooperative average performance. For $PER \leq PER_{cross}$ cooperation outperforms non-cooperative mode. Hence the gain of hybrid cooperation is high since the direct link results more often in outage that cooperative transmission. When $PER > PER_{cross}$, non-cooperative transmission outperforms cooperation. In such case the gain of hybrid cooperation is reduced and asymptotically (for $PER_{cross} \rightarrow 0$) hybrid cooperation performs as non-cooperative transmission since cooperation is never activated. In order to fully exploit the proposed hybrid cooperative AMC to improve the average system performance, AMC mechanism and hybrid cooperation protocol have to be designed jointly. As an example, given our system model, we computed the minimum values of M (M_{min}) for which hybrid cooperative AMC outperforms both classical non-cooperative and cooperative AMC. A selection of our results are shown on table 1 for maximum transmission rates R_{max} at which the system can operate and typical PER_{target} values imposed to the AMC. Indeed, given

N	M_{min}	PER_{target}	R_{max}
2	9	10^{-1}	10
2	5	10^{-2}	10
2	7	10^{-1}	8
2	3	10^{-2}	8
2	5	10^{-1}	6
2	3	10^{-2}	6
2	5	10^{-1}	4
2	3	10^{-2}	4
2	3	10^{-1}	2
2	3	10^{-2}	2

Table 1. Minimum values of M (M_{min}) for typical PER_{target} values

PER_{target} and R_{max}, we can define an M_{min} from which hybrid cooperation is beneficial. Note that the larger M is the more complex the cooperation protocol is. There is indeed a trade off between cooperation performance and cooperation complexity.

4.2.1 Simulation results

In this section, we show by means of numerical simulations the effectiveness of combining the hybrid cooperation protocol with the AMC mechanism. Results first show how the proposed mechanism drives to improved average system throughput performance. Then, we outline the advantage introduced by the hybrid cooperation protocol in terms of reduction of cooperation signalling overhead, cooperation protocol delay and average power consumed by the active relays. Simulation results are given here for the system model presented in section 3. In the system both AMC and ARQ are implemented. The simulated AMC algorithm selects the MCS which maximizes the throughput while meeting the PER_{target}

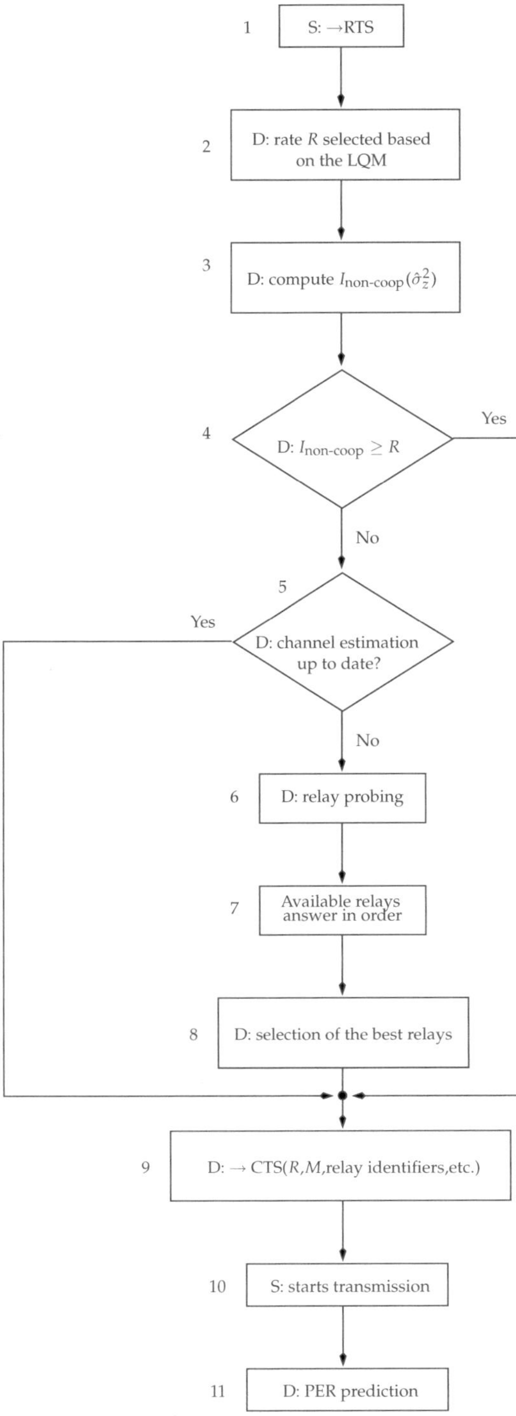

Fig. 5. Flow chart of the proposed hybrid opportunistic cooperation combined with AMC

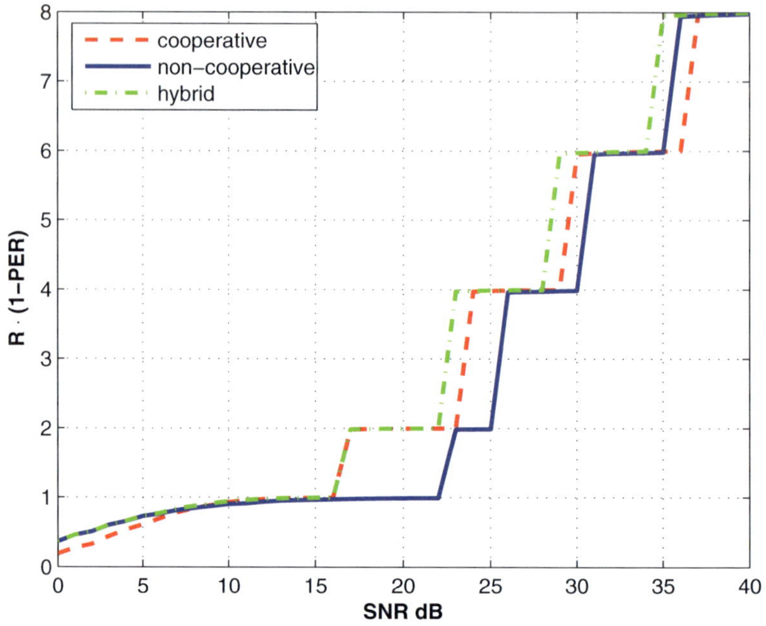

Fig. 6. Cooperative/non-cooperative/hybrid cooperative transmission with $N = 2$, $M = 3$ and $PER_{target} = 10^{-2}$

QoS constraints. The set of MCS corresponds to the transmission rate set $\underline{\mathbf{R}}$=1, 2, 4, 6, 8. We fix the $PER_{target} = 10^{-2}$. Moreover, a total average power constraint is imposed and no power allocation is considered here. We access the average physical layer throughput of a system that can perform data transmission with three different transmission modes: non-cooperative, cooperative and hybrid. Performance is compared in terms of average throughput versus average SNR. The link between source, destination and relays are assumed to be symmetric and with independent fading coefficients.

On Fig. 6 we show the performance of the AMC algorithm combined with cooperation for $N = 2$ and $M = 3$. From these results, we observe three regions for the SNR : the *low*, *medium* and *high* SNR regions. At low SNR, the non-cooperation mode outperforms cooperation mode since the noise power dominates the received power at the relays. In the medium SNR region, the cooperative scheme outperforms the non-cooperative scheme with a gain up to 6 dB. This gain is due to the better diversity-multiplexing trade-off (DMT) of the cooperative scheme. However, this gain decreases for increasing SNR since we fix $PER_{target} = 10^{-2}$ while $R_{max} = 8$ and $M = 3$ (hence $M < M_{min}$, see table 1). Therefore, when $M < M_{min}$, the cooperative scheme is not preferable at high SNR.

On Fig. 7 the performance of the case $N = 2$ and $M = 5$ is shown. As demonstrated in (S. Yang and J-C. Belfiore, 2006), the DMT is improved with the number of slots M. This improvement translates into a better performance in both cases. We observe that the decrease of SNR gain at medium to high SNR is slower than the previous case. Cooperation is always better than the non-cooperation since $M \geq M_{min}$. Best performance is always reached when using *hybrid cooperation*. We remark that the hybrid scheme alleviates the performance loss of cooperation

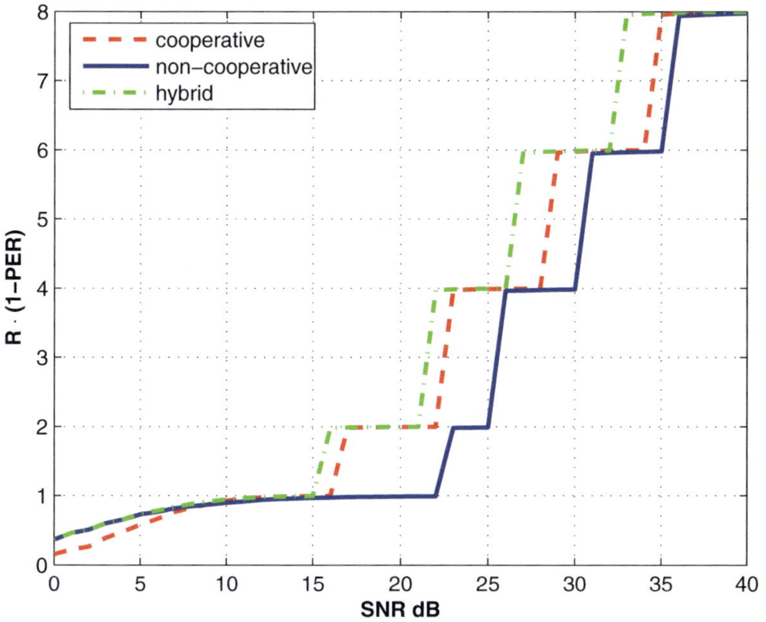

Fig. 7. Cooperative/non-cooperative/hybrid cooperative transmission with $N = 2$, $M = 5$ and $PER_{target} = 10^{-2}$

in both the low SNR and the high SNR regions. In case of $M = 3$ and $M = 5$, we observe respectively up to 5 and 7.5 dB of gap from fixed-cooperation and 1.5 and 2 dB of gap from non-cooperative transmission.

Hereafter we enlarge the investigation on *hybrid* cooperation protocols performance for a realistic communication scenario such as, OFDMA based wireless mobile communication transmission which employs limited modulation alphabets and real FEC codes. We access the effectiveness of *hybrid* cooperation protocol in real communication scenarios in terms of average PER versus average SNR, average system throughput enhancement and average cooperation cost reduction. The set of parameters used in this simulations are chosen according to the IEEE 802.16e standard . The mobile wireless channel is modelled according to (Spatial Channel Model Ad Hoc Group, 2003).

We propose to use an OAF hybrid cooperation protocol under the following power constraint: we impose a total average power constraint and no power allocation is considered. If P denotes the total power constraint, we impose $P_s = P/2$ for the power allocated to the source in the first slot and $P_r = P/2$ the power allocated to the relay in the second slot. Hereafter we adopt the following graphical notation: we represent respectively with the solid blue line, dashed red line and solid green line, non-cooperative, persistent cooperative and hybrid cooperative transmission mode performance.

Simulation results are given here for the system model presented in section 3. We use as Forward Error Correcting (FEC) code the LDPC codes as specified by the standard IEEE 802.16e (IEEE Standards Department, 2005) for the different coding rates.

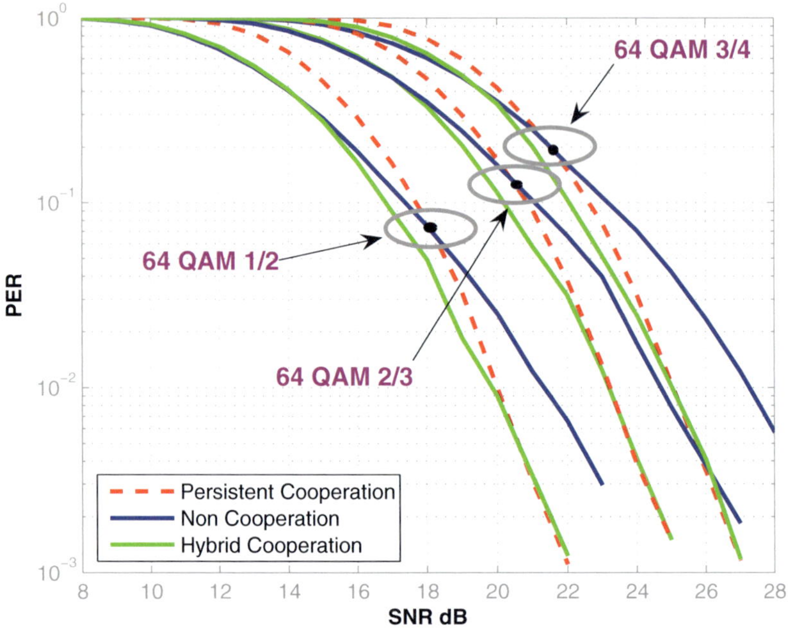

Fig. 8. Cooperative/non-cooperative/hybrid cooperative transmission

On figure 8 we compare the three transmission mode performance in terms of average PER performance versus average SNR. Results are reported here only for 64-QAM modulation with coding rates $R_C = 1/2, 2/3, 3/4$. From these results, we observe that there is a *crossing point* (PER_{cross}) between non-cooperative and cooperative average performance. For $PER \leq PER_{\text{cross}}$ cooperation outperforms non-cooperative mode. Hence the gain of hybrid cooperation is high since the direct link results more often in outage that cooperative transmission. Note that the PER that corresponds to this *crossing point* depends on the code correcting power: stronger codes present the *crossing point* at higher PER. For sake of simplicity we impose same codeword length for each MCS. Therefore, the information block length is larger for higher coding rate which results in a stronger correcting code. This is verified on figure 8. When $PER > PER_{\text{cross}}$, non-cooperative transmission outperforms cooperation. When $PER_{\text{cross}} \rightarrow 0$, hybrid cooperation performs as non-cooperative transmission since cooperation is never activated. Hybrid cooperation notably outperforms both cooperative and non-cooperative transmissions for PER values close to PER_{cross}. Note that in the present simulation we also introduce a feedback delay between $MI_{non-coop}$ estimation and cooperation controller action. Due to this delay, hybrid cooperation performance is slightly decreased comparing to equivalent results presented in (E. Calvanese Strinati and S. Yang and J-C. Belfiore, 2007).

In order to show the effectiveness of hybrid cooperative AMC mechanism, which combines AMC with hybrid cooperation, we compare the three transmission modes in terms of average system throughput versus average SNR. The simulated AMC algorithm selects the MCS which maximizes the throughput while meeting the PER_{target} QoS constraints (Calvanese

Strinati E., 2006). Typical values for the target PER is a few percent. For instance, imposing $PER_{target} \leq 10^{-1}$ results in a residual PER below 10^{-5} after 4 retransmissions.
The set of MCS corresponds to the transmission rate set defined by the IEEE 802.16e standard. In our simulation results we show the per-user performance, having one data region of 24 sub-carriers (in frequency) and 16 data OFDM symbols (in time). Under this assumption, the set of MCS schemes and the related nominal throughputs r_{mcs} and information block lengths N_{Info} are given in table 2.

Modulation	**Code Rate**	**N_{Info}**	**r_{mcs}**
QPSK	1/2	384 (bits)	215 (Kb/s)
QPSK	3/4	576 (bits)	315 (Kb/s)
16-QAM	1/2	768 (bits)	420 (Kb/s)
16-QAM	3/4	1152 (bits)	630 (Kb/s)
64-QAM	1/2	1152 (bits)	630 (Kb/s)
64-QAM	2/3	1536 (bits)	840 (Kb/s)
64-QAM	3/4	1728 (bits)	945 (Kb/s)

Table 2. Modulation and Coding Schemes of IEEE 802.16e

When $PER_{\text{target}} < PER_{\text{cross}}$, then cooperation is always better than the non-cooperation. Otherwise, non-cooperation transmission can outperform persistent cooperation transmission. As an example, we report respectively on figure 10 and 9 our simulation results for $PER_{\text{target}} = 10^{-1}, 5 \cdot 10^{-2}$.
As it is shown on figure 9, with $PER_{\text{target}} = 5 \cdot 10^{-2}$, persistent cooperation outperforms non-cooperative transmission over all the considered SNR range since, $PER_{\text{target}} < PER_{\text{cross}}$ for all MCS.
In this case, *hybrid cooperation* outperforms *non-cooperative* and *persistent cooperative* transmission respectively with a gain up to 1.75 dB and 0.75 dB. Relaxing the constraint on the PER_{target} to $PER_{\text{target}} = 10^{-1}$, there are some MCS for which $PER_{\text{target}} > PER_{\text{cross}}$. As a consequence, *non-cooperation* outperforms *persistent cooperation* in same parts of the considered SNR range. Again, *hybrid cooperation* outperforms *non-cooperative* and *persistent cooperative* transmission respectively with a gain up to 1.25 dB and 0.9 dB (see figure 10).
We report hereafter also some simulation results aimed at understanding the average relaying activation ratio χ) - which is the ratio between the number of frames were the relay is active over the total number of transmitted frames - versus the average SNR adopting the proposed *hybrid cooperation* protocol. Results are shown on Fig 11 for $PER_{target} = 10^{-1}$. Two working zones of an AMC mechanism can be distinguished. In the first zone, even if AMC selects the minimum MCS at which the system can operate, we have that $PER > PER_{target}$. Therefore, since PER is large, χ is large too. For such link quality conditions the AMC may decide to avoid transmission since AMC cannot assure the QoS constraints imposed by the upper layers. The second zone starts when MCS selected for transmission assures $PER \leq PER_{target}$. In this zone each saw tooth corresponds to a change of MCS. Our results outline that when AMC can assure a $PER \leq PER_{target}$, χ is very small ($\chi \leq PER_{target}$) since the hybrid cooperation protocol activates the cooperative mode only when direct link transmission is in outage. At the end of the second zone transmission is done at the highest MCS and the system operates at $PER \ll PER_{target}$, with consequent $\chi \ll 1$. Note that, contrary to the cooperative AMC protocol case for which $\chi = 1$ over the whole SNR range, when AMC can assure a $PER \leq PER_{target}$ and the proposed hybrid cooperation protocol is adopted, χ is reduced to the same

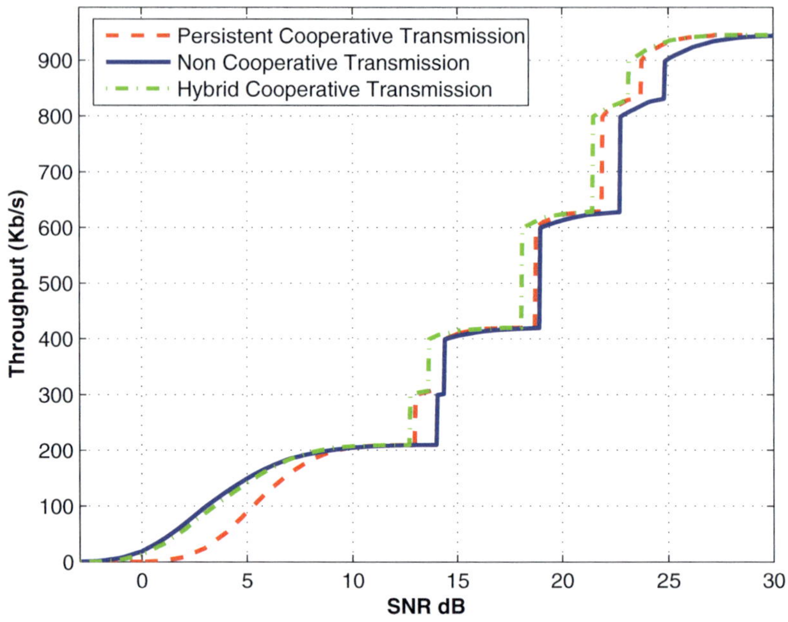

Fig. 9. Cooperative/non-cooperative/hybrid cooperative transmission with $PER_{\text{target}} = 5 \cdot 10^{-2}$

order of magnitude of PER_{target}. Note that the major result in our investigation is reduction of average relaying activation and not the improvement in average system throughput achieved with hybrid cooperative AMC mechanism.

The reduction of average relaying activation ratio achieved with the proposed hybrid AMC protocol presents three main advantages. First, the average power consumed by the active relays is strongly reduced especially when cooperation does not help and consequently cooperation activation results in a waist of relays processing power. Second, the delay caused by the cooperation protocol and consequently the packet delivery delay can be strongly reduced adopting our proposed *hybrid cooperation* protocol. For instance, when direct non-cooperative transmission is not forecasted to be in outage, the destination can immediately send a clear to send (CTS), without waiting for the relay probing process. This is an important attribute for scheduling algorithm with delay QoS constraints. Third, the average computing complexity is reduced by decreasing the number of average operation associated to cooperation.

4.3 An efficient power allocation optimization for hybrid cooperation protocols

In this section we combine the OAF hybrid cooperation protocol presented in section 4.1 with an optimal power allocation algorithm. The goal is to maximize the mutual information of the equivalent cooperative channel via optimal power allocation between the source and the relay. It is well known that the performance of a cooperative scheme is improved by relaying with optimal power values. Hereafter we assume that a maximal overall transmit power is fixed by using, for instance, a suitable power control algorithm in order to minimize co-channel

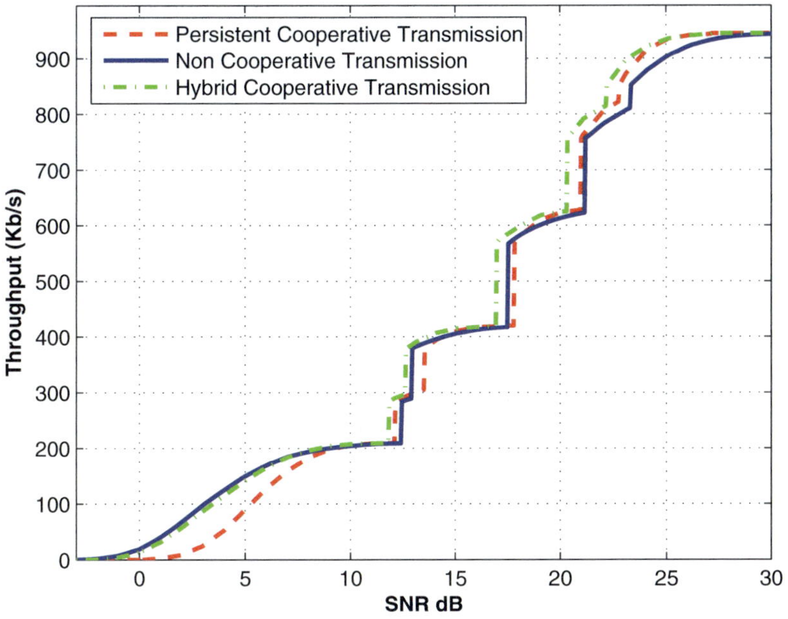

Fig. 10. Cooperative/non-cooperative/hybrid cooperative transmission with $PER_{\text{target}} = 10^{-1}$

interference. The overall total transmitting power should then be optimally shared between the source and the relay. The simplicity of an OAF cooperation scheme leads to an outage probability expression easier to handle than in the NAF case. Basically, we optimize the power allocation by minimizing the outage probability in the high SNR regime.

4.4 Outage probability approximation

First we should find the expression of the outage probability, denoted $\mathbb{P}(\mathcal{O}_c, \mathcal{O}_d)$, and approximate it in the high SNR regime. *Proposition 1:* Let P denotes the total power constraint in the network, $P_s = \alpha P$ and $P_r = (1-\alpha)P$ the fractions of P allocated to the source and the relay, respectively. Let $C_\lambda = \frac{\lambda_g}{\lambda_h}$ and $C_R = \frac{1}{2^R+1}$. Then, the approximation of the outage probability in the high SNR regime is

$$\mathbb{P}(\mathcal{O}_c, \mathcal{O}_d) =$$
$$2(2^R-1)^2(2^R+1)\epsilon^2\, \lambda_f \lambda_h \left(\frac{1-\alpha+\alpha C_\lambda}{\alpha(1-\alpha)}\right)\left(1-\alpha C_R\right)$$

Proof: The following *Lemma* will be used in our proof
Lemma 1: Let δ be positive, and let $r_\delta = \frac{vw}{v+w+\delta}$ where v and w are independent exponential random variables and λ_v and λ_w are, respectively, their parameters. Let $h(\delta)$ be continuous with $h(\delta) \to 0$ as $\delta \to 0$. Then

$$\lim_{\delta \to 0} \frac{1}{h(\delta)}\, \mathbb{P}\{r_\delta < h(\delta)\} = \lambda_v + \lambda_w$$

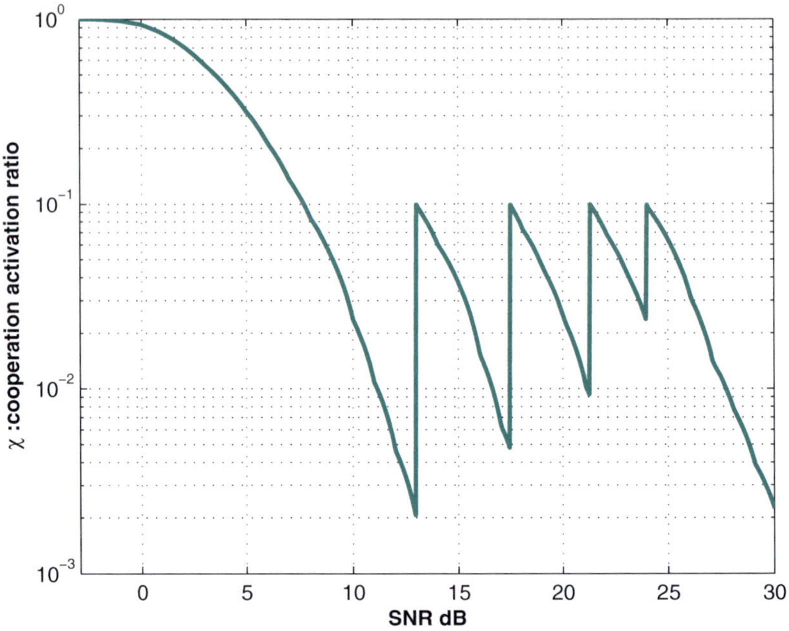

Fig. 11. Average relaying activation ratio for hybrid cooperative transmission with $PER_{\text{target}} = 10^{-1}$

$$\mathbb{P}(\mathcal{O}_c, \mathcal{O}_d) = \mathbb{P}\{\alpha|f|^2 + \frac{\alpha|h|^2(1-\alpha)|g|^2}{\alpha|h|^2 + (1-\alpha)|g|^2 + P^{-1}} < \frac{2^{2R}-1}{P}, \frac{|f|^2}{2} < \frac{2^R-1}{P}\} \tag{2}$$

$$= \mathbb{P}\{u + \frac{vw}{v+w+\epsilon} < (2^{2R}-1)P^{-1}, u < 2\alpha(2^R-1)P^{-1}\} \tag{3}$$

$$= \mathbb{P}\{r_\epsilon < g_1(\epsilon) - u, u < g_2(\epsilon, \alpha)\} \tag{4}$$

$$\mathbb{P}(\mathcal{O}_c, \mathcal{O}_d) = 2(2^R-1)\epsilon^2\lambda_f\left[\frac{\lambda_h}{\alpha} + \frac{\lambda_g}{1-\alpha}\right]\left[(2^{2R}-1) - \alpha(2^R-1)\right] \tag{5}$$

We know that

$$\mathbb{P}(\mathcal{O}_c, \mathcal{O}_d) = \mathbb{P}\{I_c < 2R, I_d < R\}$$

The outage probability can be expressed as in (2), if we define $u = \alpha|f|^2$, $v = \alpha|h|^2$, $w = (1-\alpha)|g|^2$, $\epsilon = P^{-1}$, $g_1(\epsilon) = \frac{(2^{2R}-1)}{P}$, and $g_2(\epsilon, \alpha) = 2\alpha\frac{(2^R-1)}{P}$.
Let λ_u, λ_v and λ_v be the parameters of the exponential random variables u, v and w, respectively. For $i = f, h$, we have

$$\lambda_i = \frac{1}{\alpha\sigma_i^2} = \alpha^{-1}\lambda_i \; and \; \lambda_w = \frac{1}{(1-\alpha)\sigma_g^2} = (1-\alpha)^{-1}\lambda_g$$

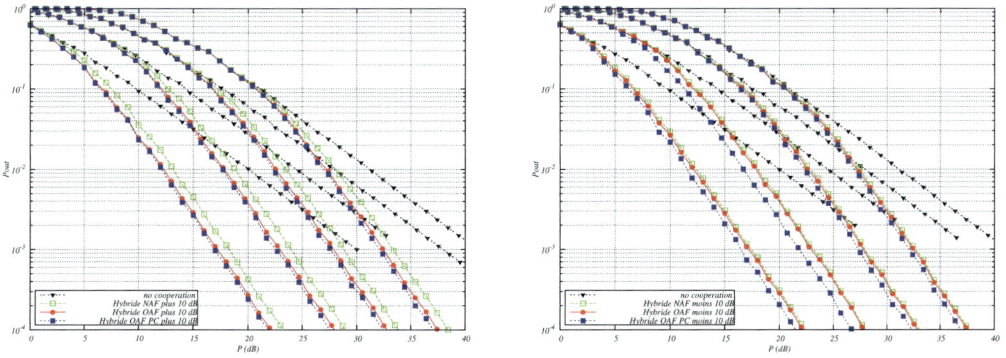

Fig. 12. Outage probabilities for the non-cooperative, Hybrid-NAF, Hybrid-OAF and Hybrid-OAF with power allocation scheme. One relay network. Considered information rates: 1, 2, 3 and 4 BPCU. $C_\lambda = \pm 10$ dB.

Using Lemma 1, we get

$$\begin{aligned}\mathbb{P}(\mathcal{O}_c, \mathcal{O}_d) &= \int_0^{g_2} \mathbb{P}\{r_\epsilon < g_1(\epsilon) - u\} p_u(u) \mathrm{d}u \\ &= \int_0^{g_2} (\lambda_v + \lambda_w)(g_1(\epsilon) - u) p_u(u) \mathrm{d}u.\end{aligned}$$

Knowing the pdf of the exponential variable u, the expression of $\mathbb{P}(\mathcal{O}_c, \mathcal{O}_d)$ is developed (calculation details are omitted due to length constraints). This expression is then approximated in the high SNR regime, using the second order Taylor development of $e^{-a\epsilon}$ when $\epsilon \to 0$, a being positive, which leads to expression (5).

Eventually, define $C_\lambda = \frac{\lambda_g}{\lambda_h}$ and $C_R = \frac{1}{2^R+1}$ which, when substituted in (5), complete the proof.

For a given spectral efficiency R and channels variances, optimizing the power allocation consists in minimizing the outage probability and thus, finding the optimal α, denoted α^*, that verifies

$$(C_\lambda - C_\lambda C_R - 1)\alpha^{*2} + 2\alpha^* - 1 = 0 \tag{6}$$

4.4.1 Simulation results

In order to clarify the impact of the proposed power allocation algorithm we compare non-cooperative, NAF cooperative, hybrid NAF cooperative and hybrid OAF cooperative protocols in two different transmission scenarios. Fist we suppose that both path-loss and shadowing effects are the same between source, relay and destination. This scenario is specified by $C_\lambda = 0$ dB, so that we have $\sigma_h^2 = \sigma_g^2$. In this case α^* is

$$\alpha^* = \frac{1}{1 + \sqrt{1 - C_R}}$$

We observe that minimizing the outage probability leads to almost an equal power allocation between the source and the relay since α^* takes values around 0.5 independently from the transmission spectral efficiency. We evince that, when $C_\lambda = 0$ dB, the algorithm of power

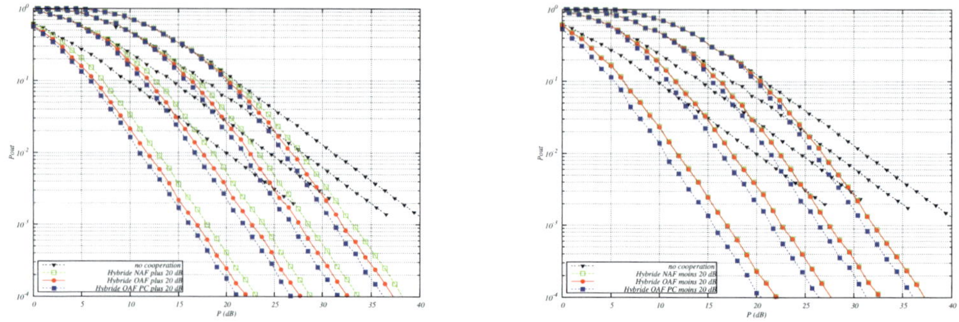

Fig. 13. Outage probabilities for the non-cooperative, Hybrid-NAF, Hybrid-OAF and Hybrid-OAF with power allocation scheme. One relay network. Considered information rates: 1, 2, 3 and 4 BPCU. $C_\lambda = \pm 20$dB.

allocation optimization performs as an equal power allocation $P_s = P_r = P/2$. This is obvious since source-relay and relay-destination links have the same link quality.
As a second scenario, we consider the more realistic case where $C_\lambda \lesseqgtr 0$ dB. Actually, having $C_\lambda \lesseqgtr 0$ dB, we assume that one of the links, source-relay or relay-destination, has a better quality, *i.e.*, $(\sigma_h^2 \lesseqgtr \sigma_g^2)$. Optimizing the power allocation becomes more worthy in this situation since allocating more power to the worst channel helps. In this case, α^* can be derived from (6) as follow:

$$\alpha^* = \frac{1}{1 + \sqrt{C_\lambda (1 - C_R)}}$$

On Figures 12 and 13 we consider the case of $C_\lambda > 0$ dB, having respectively, $C_\lambda = 10$ dB and $C_\lambda = 40$ dB. In this scenario, *e.g.*, the attenuation between source and relay is much smaller than between relay and destination. In this case, if the cooperation is activated by the hybrid cooperation controller, our power optimization allocates a higher fraction of the overall transmit power P to the relay.
A more challenging scenario is when $C_\lambda < 0$ dB or equivalently $\sigma_h^2 < \sigma_g^2$. In this case, an optimal power allocation algorithm can drive to notable performance improvement. Mainly, making reliable the transmission between the source and the relay is imperative since the relay amplifies and then forwards the received signal. That is why our optimization technique allocates, in this case, a higher fraction of P to the source. Simulation results for $C_\lambda = -10$ dB and $C_\lambda = -40$ dB are given on Figures 12 and 13.

5. Conclusion

In this chapter we present an effective scheme to improve the system performance of a cooperative system, reduce cooperation complexity, signalling overhead and cooperation protocol delay, while meeting the QoS constraints from the upper layer. For this reason, we looked for a novel AF cooperative protocol, and its combination with adaptive mechanisms such as AMC and power allocation.
First, we propose a novel cooperation protocol for half-duplex AF cooperative networks. We call this protocol *hybrid cooperation*. We prove by simulation that, NAF hybrid cooperation outperforms both non-cooperative and classical full-cooperative transmission. To evaluate the improvement due to this new strategy, we also propose an *hybrid cooperative AMC*

mechanism, which is the combination of AMC mechanism and *hybrid cooperation* protocol. We show that the advantages of *hybrid cooperative AMC* are twofold. First, its average throughput performance is higher than both AMC combined with non-cooperative and with fixed-cooperation transmission for all values of SNR. This results is benchmarked by our simulation results. Second, the proposed algorithm drives to a reduction of both average power consumed by the active relays and cooperation probing cost. This results in a reduced average packet delivery delay since both throughput performance is improved and cooperation probing delay is strongly reduced. Moreover, we showed how the proposed *hybrid cooperative AMC* mechanism drives to a reduction of cooperation signalling overhead that from a MAC layer point of view, may result in an additional throughput enhancement at the top of the MAC layer.
We further investigate the proposed hybrid AF cooperation protocol. We compared hybrid OAF and hybrid NAF protocols. Imposing a total average power constraint and no power allocation, we showed that the orthogonal strategy (OAF), suboptimal in the case of a classical amplify-and-forward scheme, outperforms both classical NAF cooperative and hybrid NAF schemes. Moreover, we pointed out that from an implementation point of view, the hybrid OAF protocol reduces significantly the cooperation complexity.
Furthermore, we profit of the simplicity of the outage probability expression for the OAF cooperation scheme to derive an optimal power allocation algorithm. The proposed algorithm optimizes the system performance by minimizing the outage probability of the channel at high SNR. We underlined that the need of such an optimization increases with the increasing quality difference within the links (source-relay and relay-destination). Indeed, we succeeded in finding a low complexity algorithm that optimizes the power allocation in the case of a hybrid-OAF schemes.

6. References

E. Calvanese Strinati. *Radio link control for improving the qos of wireless packet transmission*. PhD thesis, Ecole Nationale Supérieure des Télécomunications de Paris, December 2005.

E. Calvanese Strinati, S. Yang, and J-C. Belfiore. Adaptive Modulation and Coding for Hybrid Cooperative Networks. June 2007.

Emilio Calvanese Strinati and Luc Maret, "Performance Evaluation of Hybrid Cooperation Protocol in IEEE 802.16e", *IEEE Vehicular Technology Conference (VTC Spring)*, Singapore, Mai 2008.

Maya Badar and Emilio Calvanese Strinati and Jean-Claude Belfiore, "Optimal Power Allocation for Hybrid Amplify-and-Forward Cooperative Networks", *IEEE Vehicular Technology Conference (VTC Spring)*, Singapore, Mai 2008.

E. Erkip A. Sendonaris and B. Aazhang. User cooperation diversity-part 1: System description. 51:1927–1938, November 2003.

E. Erkip A. Sendonaris and B. Aazhang. User cooperation diversity-part 2: Implementation aspects and performance analysis. 51:1939–1948, November 2003.

D. Gunduz and E. Erkip. Outage Minimization by Opportunistic Cooperation. 2:1436–1442, June 2005.

D. N. Tse J. N. Laneman and G. W. Wornell. Cooperative diversity in wireless networks: Efficient protocols and outage behavior. 50:3062–3080, December 2004.

H. Bölcskei R. U. Nabar and F. W. Kneubühler. Fading relay channels: Performance limits and space-time signal design" algorithm. *ieeeJSAC*, pages 1099–1109, August 2004.

H. El Gamal K. Azarian and P. Schniter. On the achievable diversity-multiplexing tradeoff in half-duplex cooperative channels. 51:4152–4172, December 2005.

S. Yang and J-C. Belfiore. "Optimal space-time codes for the mimo amplify-and-forward cooperative channel". IEEE Trans. Inform. Theory, May. 2006.

Z. Lin and E. Erkip and M. Ghosh. "Adaptive Modulation for Coded Cooperative Systems". pages 615Ű619, June 2005.

M. Lampe and H. Rohling and W. Zirwas, "Misunderstandings about link adaptation for frequency selective fading channels," *IEEE International Symposium on Personal, Indoor, and Mobile Radio Communications* , September 2002.

E. Yazdian and M. R. Pakravan. "Adaptive Modulation Techniques for Cooperative Diversity in Wireless Fading Channels", in proceeding of IEEE PIMRC, September 2006.

M. Hasna and M-S. Alouini. "Optimal power allocation for relayed transmission over rayleigh-fading channels", IEEE Trans. Inform. Theory, 3(6), November. 2004.

Q. Zhang and C. Shao and Y. Wang and P. Zhang and J. Zhang and Z. Zhang. Zhang, "Adaptive optimal transmit power allocation for two-hop non-regenerative wireless relaying systems", Vol 41 :124Ű133, September 2004.

D. P. Reed and A. Bletsas and A. Khisti and A, "Lippman. A simple cooperative diversity method based on network path selection", IEEE Journal on Selected Areas of Communication, 2005.

I. Hammerstrom and A. Wittneben, "On the optimal power allocation for nonregenerative OFDM relay links," in Proc. IEEE Int. Conf. Communications (ICC), June 2006.

Spatial Channel Model Ad Hoc Group (Combined ad-hoc from 3GPP and 3GPP2), "Spatial Channel Model Text Description", *SCM-134,* April 22, 2003.

IEEE Standards Department, "Part 16: Air Interface for Fixed Broadband Wireless Access Systems - Amendment 2: Physical and Medium Access Control Layers for Combined Fixed and Mobile operation in Licensed Bands and Corrigendum 1", New York, IEEE Std 802.16e-2005, Feb. 2006. (Available online at: http://www.ieee.org).

10

Adaptative Rate Issues in the WLAN Environment

Jerome Galtier
Orange Labs
France

1. Introduction

In this chapter, we investigate the problem of mobility in the WLAN environment. While radio conditions are changing, the Congestion Resolution Protocol (CRP) plays a key role in controling the quality of service delivered by the distributed network. We investigate different types of CRP to show the impact of one user to all the other ones. We place our work in an urban context where the users (bus passengers, walkers, vehicule network applications) are using an accessible WLAN (for instance WiFi) network via an access point and interact with one another through the network.
Accessing the network via an access point has become in the last years a more and more popular technique to do some networking at low cost. The reason for it is that the WLAN technologies such as WiFi do not require complex user registration, handovers, downlink/uplink protocol synchronization, or even planification for existing base stations (such as GSM BTS or UMTS Node-B). Of course such transmissions achieve much lower performance profile, but they are often delivered for free or almost for free, for instance simply to attract new clients in cafés or restaurants.
As a result, we come up wih new habits of communications which are not exactly the use for which engineers have designed WiFi for L (and other WLAN networks).

2. Overview of 802.11 modulation techniques

2.1 Techniques employed

In the course of its developpment, the 802.11x family has developped a surprising number of modulation techniques that deeply impact the final performance of the system. We summarize these techniques for a 20 MHz band in the 2.4 GHz frequency area in Tab. 1 (we skip here all the historical modulations that have since been abandonned). All these cards implement backward compatibility, which means that the most recent and sophisticated one also handles previous rates in order to be able to communicate with simpler/older cards. As a result, a new 802.11n card with 4 streams will be able to produce modulations in 44 different modes!
We give in the following some explanations on the different modulation techniques employed for WiFi.

BPSK Binary Phase Shift Keying is a modulation technique that uses the phase of two complementary phases to code the bis 0 or 1. We plot its constellation diagram in Fig. 1.

QPSK Quadrature Phase Shift Keying uses four phases instead of two to code the signal, so that each symbol carries 2 bits instead of 1 for the BPSK.

Protocol (# streams)	Data rate per stream (Mbits/s)	Modulation & Coding
-	1,2	DSSS/BPSK QPSK/Barker seq.
b	5.5,11	DSSS/QPSK/CCK
g	6,9,12,18,24,36,48,54	OFDM/BPSK QPSK QAM/Conv. coding
n (1 st.)	7.2,14.4,21.7,28.9,43.3,57.8,65,72.2	OFDM/MIMO/Conv. coding
n (2 st.)	14.4,28.9,43.3,57.8,86.7,115.6,130,144.4	OFDM/MIMO/Conv. coding
n (3 st.)	21.7,43.3,65,86.7,130,173.3,195,216.7	OFDM/MIMO/Conv. coding
n (4 st.)	28.9,57.8,86.7,115.6,173.3,231.1,260,288.9	OFDM/MIMO/Conv. coding

Table 1. Different rate parameters for 802.11x at 2.4GHz within a 20 MHz band.

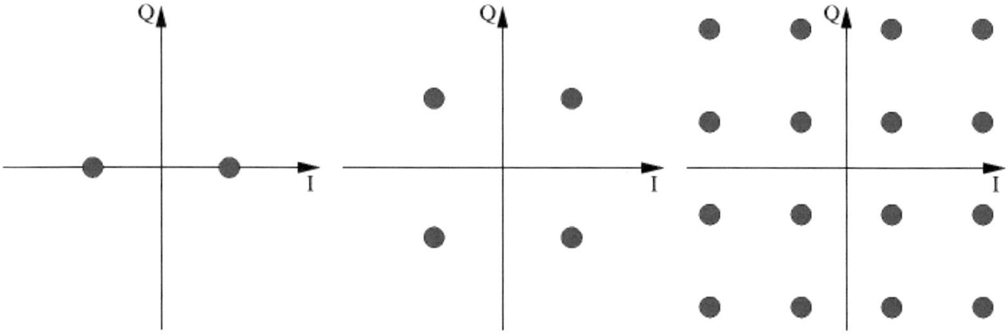

(a) BPSK. (b) QPSK. (c) QAM.

Fig. 1. Constellation diagram for main 802.11 modulation techniques.

QAM Quadrature Amplitude Modulation combines amplitude modulation with phase modulation to carry more information. 16-QAM carries 4 bits, while 64-QAM carries 6 bits.

DSSS The Direct Spread Sequence Spectrum is a modulation technique that uses the whole band (here, 20 MHz) to encode the information via some coding techniques, more precisely Barker codes or CCK in the 802.11 context.

Barker sequences For the 1 Mbit/s coding, the pseudo-random sequence (10110111000) is used to code the "1" symbol, and its complement (01001000111) to code the "0" symbol, in a PSK modulating scheme. The 2 Mbit/s version is obtained by using QPSK modulation instead of PSK.

CCK The CCK (Complementary Code Keying) technique consists in using 16 or 256 different sequences coded in eight chips (QPSK symbols). The 16 or 256 different sequences allow to identify 4 or 8 bits of information.

OFDM The OFDM (Orthogonal Frequency-Division Multiplexing) is a technique that consists in dividing the channel into close sub-carriers to transmit data through these parallel sub-channels. Using orthogonality of signals, this technique allows to reduce significantly the spacing between sub-carriers and therefore improves spectral efficiency.

MIMO The Multiple-Input and Multiple-Output technique consists in using several input antennas and several output antennas in the devices, in order to use spatial diversity and therefore increase the throughput capacity.

2.2 Range of communication

In a paper on adaptativiy and mobility, of course, the range of communication is a crucial parameter. Unfortunately, this very range is very variable depending on radio conditions. We try in this subsection to give a more accurate opinion on that question without falling into two main defaults of the literature on that topic, that would be (1) rely exclusively on simulations or (2) explain the theoretical context without answering the question of range.

Data rate (Mbits/s)	Modulation	Coding rate (R)
6	BPSK	1/2
9	BPSK	3/4
12	QPSK	1/2
18	QPSK	3/4
24	16-QAM	1/2
36	16-QAM	3/4
48	64-QAM	2/3
54	64-QAM	3/4

Table 2. Rate parameters in 802.11a and 802.11g.

We aim to obtain a realistic model for WLAN communications the following way. We take in this section as an example the model of IEEE 802.11g which is a precise, popular industrial context, in which the problem is really accurate. As mentioned in (*Part 11: wireless LAN medium access control (MAC) and physical layer (PHY) specifications*, 1999, chapter 17), the different rates in which such networks operate are the ones of Tab. 2. We can see a continuum of rates that varies from 6 Mbits/s to 54 Mbits/s. However, the radio conditions impact a lot the effective performance of terminals. We do not want to enter too deeply into some specific situations in this paper, but instead we try to extract general enough properties that can be extrapolated to numerous contexts. We need to say that, surprisingly, the theory says that BPSK has exactly the same performance as QPSK, so we will only plot curves for QPSK. We use first the *berfading* function of mathlab, and evaluate the coding gain as $1/R$. This simple implementation gives the plots of Fig. 2, for respectively, (a) a Rayleigh channel of diversity order 2, and (b) a Rice channel of diversity order 3 and K-factor 5.

Although these curves are quite different, they have some characteristics that are kept not only in these two cases, but with a large set of different diversity values ranging from 2 to 10 and, for the Rice channel, K-factors ranging from 1 to 100 or more. We plot in Fig. 3 the same curves, that we normalized by taking the logarithm, substracting the mean of the logarithms in the six cases, and scaling by the standard deviation.

This simple approach matches very well the data that some authors are giving on the range of operations for different configurations, as we plot in Fig. 4, with indoor ranges of Romano (2004), and ranges of *WLAN - 802.11 a,b,g and n* (2008). In (Segkos, 2004, page 93, Fig. 58), again, similar ranges are shown. The conclusion are always the same:

1. within each group of modulation (QPSK, 4-QAM, 64-QAM) the curves are closer one to another, than when one jumps from one group of modulation to another.
2. the closest curves are that of 48 and 54 Mbits/s,
3. the farthest curves are that of 36 and 48 Mbits/s.

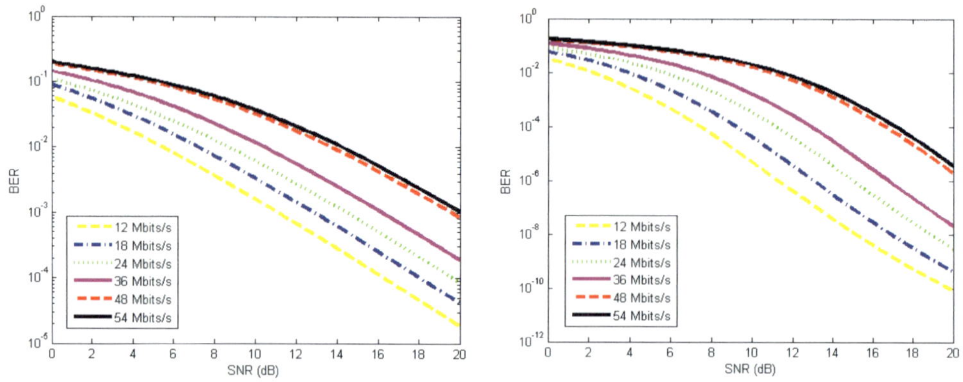

(a) Rayleigh channel of diversity order 2. (b) Rice channel of diversity order 3 and K=5.

Fig. 2. Various analytical SNR to BER behaviors.

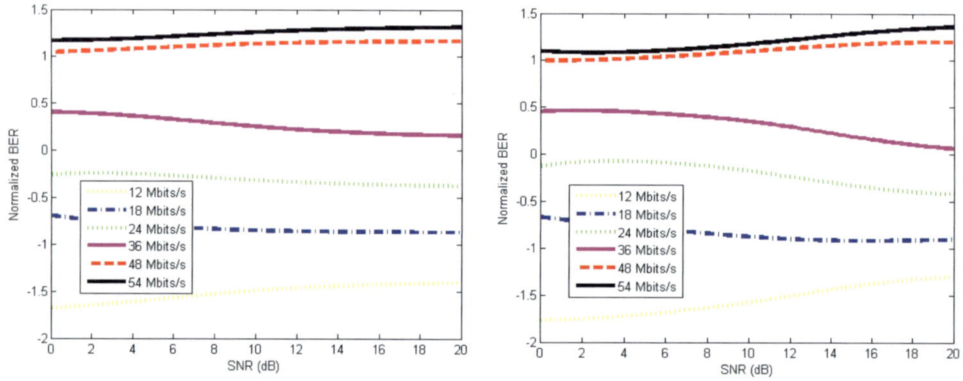

(a) Rayleigh channel of diversity order 2. (b) Rice channel of diversity order 3 and K=5.

Fig. 3. SNR to normalized BER.

Some differences, however, appear. For instance, it sounds like in the simulations and analytic tools (Fig 2 and (Segkos, 2004, page 93, Fig. 58)), the curves of rates 48 and 54 Mbits/s are much closer than they are in the tutorial curves of Fig. 4. We have observed this phenomenon with lots of different parameters for the Rayleigh and Rice channels.

As a result, we can conclude that the behavior of the channel for WLAN networks deeply depends on the type of radio conditions that are experienced. However, all the experiments we have done suggest that we can use a pathloss model where the gain follows a law in $K_r(d/\beta)^\eta$, where K_r depends on the rate r of the connection, d is the distance to the access point, and η the pathloss parameter, depending on radio conditions. Fig. 3 gives a sufficiently precise behavior of all the mechanisms to evaluate the channel.

3. Existing rate adaptation algorithms

There exist several algorithms that are intended to find the optimal rate of communication for WiFi terminals while exploring the channel. In order to adapt the rate of the packet, such

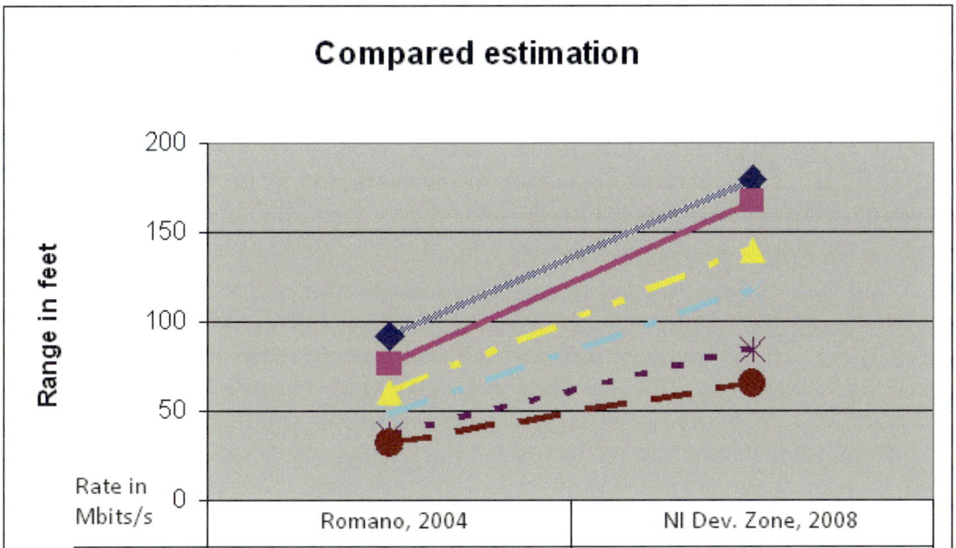

Rate in Mbits/s	Romano, 2004	NI Dev. Zone, 2008
12	92	180
18	76	167
24	60	140
36	48	117
48	36	84
54	32	65

Fig. 4. Comparative given ranges for 802.11g in the litterature.

an algorithm will certainly take into account the fact (or not) that the receiver emitted an acknowledgement at the end of the transmission, since this is mandatory in the legacy mode of IEEE 802.11x.

A very popular approach to that question is that of ARF (*Auto-Rate Fallback*) Kamermann & Monteban (1997). The sender begins to send packets at the minimal rate. After N successes, the sender increases the rate to the next available rate in terms of speed. In case a failure occurs when the new rate is first tried, the rate is fixed to the previous one, and will not be improved again until N successes occur. If, at a given rate, two successive failures occur, then the rate is also decreased, and will also wait N successes before a new try to a faster rate is done.

An improved version, AARF (*Adaptive Auto-Rate Fallback*) Lacage et al. (2004), plays with the value N of successes before a try to faster rate is done. If a first test at a new rate fails, then the system will wait 2N successes to try again. This results in an improvement of the performance of the system.

The TARA scheme (*Throughput-Aware Rate Adaptation*) Ancillotti et al. (2009) combines the information of the Congestion Window (CW) of the MAC of 802.11 with specific parameters to improve the rate mechanism.

Another variant of ARF, ERA (*Effective Rate Adaptation*) Wu & Biaz (2007), tests, in case of collision, a retry at the lowest rate. This retry is used to infer whether the failure is due to a collision or to a radio (SNR) problem. The rate is changed only if the problem is supposed not to be a collision problem, that is, when the retransmission at lower rate is successful.
Indeed, it is not true that all defaults of acknowledgement are due to radio conditions (and accordingly rate of transmission). In fact, in saturation mode, up to 30% of packets are lost because of collisions. This has led to additional research work mainly in four directions:

- Obtain a measurement on the radio conditions. Two main methods are then possible. First, one can assume link symmetry, so that the transmitter will evaluate the signal-to-noise ratio of a packet from the receiver, Pavon & Choi (2003). Second, one can suggest to modify the RTS/CTS mecchanism so that the CTS would send back the received SNR to the receiver (Holland et al. (2001); Saghedi et al. (2002)). These ideas have several drawbacks as mentioned in Ancillotti et al. (2008). The former can only give an assumed SNR, while the latter supposes that both receiver and emitter implement this RTS/CTS modification. But the main inconvenient is that both mechanisms assume the knowledge of a SNR-to-Rate table, which is not easy to obtain and/or update.
- Distinguish physically loss reason. Some algorithms lie on the fact that the physical layer may distinguish packets that are lost due to channel collisions, from packets that are lost because of collisions. This could be for instance implemented by a feature that allows the receiver to say that he could decode the header of the packet (transmitted at minimal rate) but not the payload in itself, see Pang et al. (2005). Of course this solution requires a very specific hardware and nevertheless one cannot be sure that the collision detection feature is fully reliable.
- Test the channel in case of collision by an RTS/CTS. Several mechanisms – J.Kim et al. (2006); Wong et al. (2006) – decide, in case of collision, to send an RTS/CTS to test the channel. Of course, in case the channel comes close to saturation, this has a terrible impact on performance. This has also the terrible side-effect of employing the hidden station procedure (RTS/CTS) for a different purpose, which has several side-effects. Note that this defaults are partially corrected using probabilities in Chen et al. (2007).
- Use Beacon information. In the case where one terminal is connected to an Access Point, some SNR information can be used to have a first idea of the channel quality, see Biaz & Wu (2008).

4. Analytical model

In this section, we make the use of Markov analysis to infer important properties of the rate adaptation algorithm. We model in the following the ARF mechanism, knowing that this model is very popular, and can possibly be extended to alternative approaches. This can be viewed as an extension of the model of Bianchi (2000). It gives an interesting model where, as expected, the rate is connected to the size of the congestion window of the backoff process. Indeed, in IEEE 802.11x, when a station needs to send a packet, it goes through a phase called contention resolution protocol (CRP) that aims at deciding which stations - among the contending ones - will send a packet. In order to do that, it uses two main variables: the contention window (CW) and the backoff parameter (b). The contention window is set at the begining to a minimal value (CWMin) and doubles each time a failure is experienced, to a maximum CWMax. When the maximum is reached, the CW parameter will stay at this value for a fixed number of retries as long as the transmission fails, and then aborts sending and falls to CWMin. In case of success, in all these cases, the next CW is set to CWMin. This typical behavior of CW has been discussed in many ways in the literature (Galtier (2004); Heuse et al. (2005); Ibrahim & Alouf (2006); Ni et al. (2003)), therefore in the following, we will simply use a series of constants $CW_0, \ldots, CW_m$, with CW_0 =CWMin and CW_1 being the value of CW after the first augmentation (2CWMin in the legacy case), and $CW_{m-retries+1} = \cdots = CW_m$ =CWMax. It gives, in the legacy case, the following formula:

$$CW_i = \min(2^i\text{CWMin}, \text{CWMax}).$$

Meanwhile, when a transmission is to be done, the value of b is taken randomly between 0 and CW-1. If b equals zero, the station emits its packet as soon as the channel is available (that is, when the previous transmission is completed). Otherwise, b is decremented and the station waits for a small period (called a minislot) and listen to the channel to see if some other station did not start to transmit. If it is the case, it postpones its own emission to the end of the current transmission, and therefore freezes CW and b. Then, if b is equal to zero, it starts emitting. Otherwise b is decremented and so on.
We describe our model in Fig. 5. In this figure the state s^i_j corresponds to the situation where the rate is j and the congestion window CW_{i-1}. The probability p_j represents the probability that a transmission at rate j fails. One can see in the figure also small states between s^i_0 and s^{i+1}_0. Those states represent the fact that N successful transmissions at rate r_i are necessary to try to send at rate r_{i+1}. The colors of the vertices (or the levels of gray) correspond to the rate of transmission of the state in question. The state s^r_+ represents the case where the current rate is r and at least the last transmission at that rate was successful.

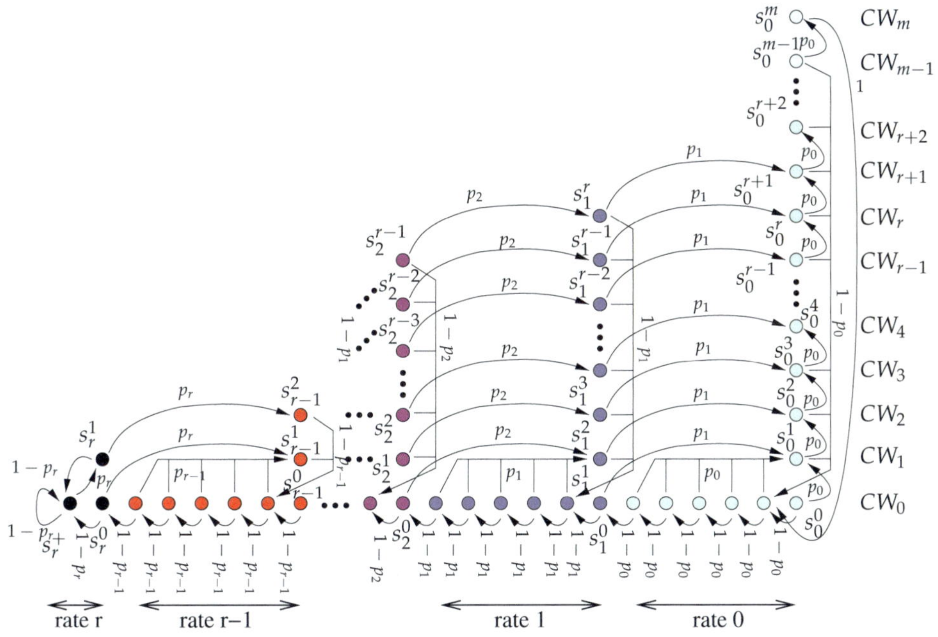

Fig. 5. Model including ARF in the backoff process.

Note that, if we extend the model similarly to Bianchi (2000) to represent the backoff variable, this is a discrete Markov process. Note also that a simpler model also appears in the literature (Singh & Starobinski (2007)) that expresses two models, one for the rates, and one for the ARF internal behavior, and mixes them based on semi-markov properties. Note that our model captures fine properties of backoff behavior, and complex mechanisms of rate improvement.

4.1 Analysis of the left-hand part

Let now design by π_i^j the probability that the transmitter is in state s_j^i. We also note π_+^r the probability that the transmitter is in state s_r^+. The analysis of the left part of Fig 5 gives:

$$\begin{cases} \pi_+^r = (1-p_r)(\pi_+^r + \pi_0^r + \pi_1^r), \\ \pi_1^r = p_r \pi_+^r. \end{cases} \tag{1}$$

And therefore

$$\pi_1^r = \frac{1-p_r}{p_r}\pi_0^r. \tag{2}$$

4.2 Analysis of the intermediate part

Let us now analyze an intermediate level of the model. We can see on Fig. 5 that, for $k \in \{2, \ldots, r\}$,

$$\begin{cases} \pi_{i+1}^{k-1} = p_k \pi_i^k, \text{ for } i \in \{1, \ldots, r+1-k\}, \\ \pi_0^k = (1-p_{k-1})^N \sum_{i=0}^{i=r+2-k} \pi_i^{k-1}, \\ \pi_1^{k-1} = p_k \pi_0^k + p_{k-1}((1-p_{k-1}) + (1-p_{k-1})^2 + \cdots + (1-p_{k-1})^{N-1}) \sum_{i=0}^{i=r+2-k} \pi_i^{k-1}. \end{cases} \tag{3}$$

Rearranging the two last lines of (3) gives

$$\pi_1^{k-1} = \left(p_k - 1 + \frac{1}{(1-p_{k-1})^{N-1}}\right)\pi_0^k \qquad \text{for } k \in \{2, \ldots, r\}. \tag{4}$$

We show by induction that

$$\sum_{i=0}^{i=r+1-k} \pi_i^k = \frac{\pi_0^k}{p_k}. \tag{5}$$

Obviously, using (2), equation (5) is true for $k = r$. Now we suppose that (5) is true for the values $\{k, \ldots, r\}$. Using the first line of (3) and (4) we have

$$\sum_{i=0}^{i=r+2-k} \pi_i^{k-1} = p_k \sum_{i=0}^{i=r+1-k} \pi_i^k + \left(\frac{1}{(1-p_{k-1})^{N-1}} - 1\right)\pi_0^k + \pi_0^{k-1}.$$

Using the induction we have

$$\sum_{i=0}^{i=r+2-k} \pi_i^{k-1} = \frac{1}{(1-p_{k-1})^{N-1}}\pi_0^k + \pi_0^{k-1}.$$

We now replace by the second line of (3) to get:

$$\sum_{i=0}^{i=r+2-k} \pi_i^{k-1} = (1-p_{k-1}) \sum_{i=0}^{i=r+2-k} \pi_i^{k-1} + \pi_0^{k-1}.$$

Hence the result for $k \in \{1,\dots,r\}$ in (5).
If we combine equation (5) and the second line of (3) we have

$$\pi_0^k = \frac{(1-p_{k-1})^N}{p_{k-1}} \pi_0^{k-1}. \tag{6}$$

Now, before evaluating the last stage of the Markov model, we get an analytical expression of any other state π_i^j with $j \geq 1$ in terms of π_0^1. The simplest expression comes from (6):

$$\pi_0^k = \frac{(1-p_1)^N \dots (1-p_{k-1})^N}{p_1 \cdots p_{k-1}} \pi_0^1 \qquad \text{for } k \in \{2,\dots,r\}. \tag{7}$$

Then, using (4), one can deduce

$$\pi_1^k = \left(p_{k+1} - 1 + \frac{1}{(1-p_k)^{N-1}} \right) \frac{(1-p_1)^N \dots (1-p_k)^N}{p_1 \cdots p_k} \pi_0^1 \qquad \text{for } k \in \{1,\dots,r-1\}. \tag{8}$$

Now, let us see a more general case, with $i \geq 2$ and $j \geq 1$.

$$\begin{aligned} \pi_i^j &= \pi_1^{i+j-1} p_2 \dots p_i && \text{using the first line of (3)} \\ &= \left(p_{i+j} - 1 + \tfrac{1}{(1-p_{i+j-1})^{N-1}} \right) p_2 \dots p_i \, \tfrac{(1-p_1)^N \dots (1-p_{i+j-1})^N}{p_1 \dots p_{i+j-1}} \pi_0^1 && \text{using (8), with } i+j \leq r \end{aligned}$$

We then get the following formulas:

$$\begin{aligned} \pi_i^1 &= \frac{(1-p_1)^N \dots (1-p_{i-1})^N}{p_1} \left((1-p_i) - (1-p_{i+1})(1-p_i)^N \right) \pi_0^1, \\ &\text{for } i \in \{2,\dots,r-1\}, \end{aligned} \tag{9}$$

and

$$\begin{aligned} \pi_i^j &= \frac{1}{p_1} \frac{(1-p_1)^N \dots (1-p_{i+j-2})^N}{p_{i+1} \cdots p_{i+j-1}} \left((1-p_{i+j-1}) - (1-p_{i+j})(1-p_{i+j-1})^N \right) \pi_0^1, \\ &\text{for } i \geq 2,\ j \geq 2,\ i+j \leq r. \end{aligned} \tag{10}$$

Of course, the case $i+j = r+1$ remains. It gives:

$$\begin{aligned} \pi_i^{r+1-i} &= \pi_1^r p_2 \dots p_i && \text{using the first line of (3)} \\ &= \tfrac{1-p_r}{p_r} p_2 \dots p_i \, \pi_0^r && \text{using (2)} \\ &= \tfrac{1-p_r}{p_r} \tfrac{p_2 \dots p_i}{p_1 \dots p_{r-1}} \left((1-p_1)^N \dots (1-p_{r-1})^N \right) \pi_0^1 && \text{using (7).} \end{aligned}$$

We distinguish the cases $i = r$, $i = r-1$, and others, and we obtain

$$\pi_r^1 = \frac{1-p_r}{p_1}\left[(1-p_1)^N \dots (1-p_{r-1})^N\right] \pi_0^1, \tag{11}$$

$$\pi_{r-1}^2 = \frac{1-p_r}{p_r}\left[(1-p_1)^N \dots (1-p_{r-1})^N\right] \pi_0^1, \tag{12}$$

$$\pi_i^{r+1-i} = \frac{1-p_r}{p_1 p_r}\frac{(1-p_1)^N \dots (1-p_{r-1})^N}{p_{i+1}\cdots p_{r-1}} \pi_0^1 \qquad \text{for } i \in \{1,\dots,r-2\}. \tag{13}$$

4.3 Analysis of the right-hand part

The analysis of the chain of Fig. 5 gives the following formulas:

$$\begin{cases} \pi_{i+1}^0 = p_0\pi_i^0 \qquad \text{for } i \in \{k+1,\dots,m-1\} \\ \pi_{i+1}^0 = p_0\pi_i^0 + p_1\pi_i^1 \qquad \text{for } i \in \{1,\dots,k\} \\ \pi_0^1 = (1-p_0)^N \sum\limits_{i=0}^{i=m} \pi_i^0 + p_0(1-p_0)^N\pi_m^0 \\ \pi_1^0 = (1-p_0)\left(1-(1-p_0)^{N-1}\right)\sum\limits_{i=0}^{i=m}\pi_i^0 + p_1\pi_0^1 + p_0\pi_0^0 + p_0\left(1-(1-p_0)^{N-1}\right)\pi_m^0 \end{cases} \tag{14}$$

Since there is no entering state for s_0^0, the corresponding probability is a stationary state will verify $\pi_0^0 = 0$. Then, if we denote

$$\pi_{\geq j}^0 = \sum_{i=j}^{i=m} \pi_i^0,$$

we can deduce from the last line of equation (14) the following:

$$\pi_1^0 = \frac{(1-p_0)\left(1-(1-p_0)^{N-1}\right)\pi_{\geq 2}^0 + p_1\pi_0^1 + p_0\left(1-(1-p_0)^{N-1}\right)\pi_m^0}{1-(1-p_0)\left(1-(1-p_0)^{N-1}\right)} \tag{15}$$

We can now make use of the following family of polynomials:

$$\begin{cases} Q_1(X) = 1 - X + X^N, \\ Q_{p+1}(X) = Q_p(X) - (1-X)^p(X-X^N), \qquad \text{for } p \geq 1. \end{cases} \tag{16}$$

We can rewrite this formula as follows for $p \geq 1$

$$\begin{cases} Q_{p+1}(X) = Q_1(X) - ((1-X)+\dots+(1-X)^p)(X-X^N), \\ \qquad = 1 - X + X^N - (1-X)\frac{1-(1-X)^p}{X}(X-X^N), \\ \qquad = 1 - X + X^N - (1-X)(1-(1-X)^p)(1-X^{N-1}). \end{cases}$$

Finally it gives:

$$Q_{p+1}(X) = X^{N-1} + (1-X)^{p+1}(1-X^{N-1}), \qquad \text{for } p \geq 1. \tag{17}$$

We then aim at proving the following formula for $i \in \{2,\dots,r\}$:

$$\begin{aligned} Q_i(1-p_0)\pi_i^0 &= p_0^{i-1}\left((1-p_0)-(1-p_0)^N\right)\pi_{\geq i+1}^0 + p_0^i\left(1-(1-p_0)^{N-1}\right)\pi_m^0 \\ &+\sum_{j=1}^{j=i-2} p_0^{i-j-2}(1-p_0)^N\dots(1-p_j)^N(1-p_{j+1})\pi_0^1 \\ &+p_0^{i-1}p_1\pi_0^1 + p_0^{i-2}(1-p_1)Q_1(1-p_0)\pi_0^1 \\ &-Q_{i-1}(1-p_0)(1-p_1)^N\dots(1-p_{i-1})^N(1-p_i)\pi_0^1 \end{aligned} \quad (18)$$

The case $i = 2$ comes from the second line of equation (14) and equations (15) and (8). The following cases are obtained by induction using again the second line of (14) in conjunction with (9).

Now, using the first line of equation (14) we have

$$\begin{aligned} \pi_{\geq r+1}^0 &= \pi_{r+1}^0 + \pi_{r+2}^0 + \dots + \pi_m^0 \\ &= \pi_{r+1}^0 + p_0\pi_{r+1}^0 + \dots + p_0^{m-r-1}\pi_{r+1}^0 \\ &= \tfrac{1-p_0^{m-r}}{1-p_0}\pi_{r+1}^0. \end{aligned}$$

We then start with the second line, again, of equation (14), that iis $\pi_{r+1}^0 = p_0\pi_r^0 + p_1\pi_r^1$, and combining with equations (18) for $i = r$ and (11) we get:

$$\pi_{r+1}^0 = \left(1+\sum_{j=0}^{j=r-1}\frac{(1-p_0)^N\dots(1-p_j)^N}{p_0^{j+1}}(1-p_{j+1})\right)\frac{p_0^r\pi_0^1}{(1-p_0)^{N-1}}. \quad (19)$$

Naturally, it gives, using now the first line of equation (14),

$$\pi_k^0 = \left(1+\sum_{j=0}^{j=r-1}\frac{(1-p_0)^N\dots(1-p_j)^N}{p_0^{j+1}}(1-p_{j+1})\right)\frac{p_0^{k-1}\pi_0^1}{(1-p_0)^{N-1}} \qquad \text{for } k \in \{r+1,\dots,m\}. \quad (20)$$

Back-tracking further with the help of the second line of equation (14) along with equations (11) and (8), we can write:

$$\begin{aligned} \frac{\pi_k^0}{\pi_0^1} &= \left(1+\sum_{j=0}^{j=k-2}\frac{(1-p_0)^N\dots(1-p_j)^N}{p_0^{j+1}}(1-p_{j+1})\right)\frac{p_0^{k-1}}{(1-p_0)^{N-1}} \\ &-(1-p_1)^N\dots(1-p_{k-1})^N(1-p_k) \qquad \text{for } k \in \{1,\dots,r\}. \end{aligned} \quad (21)$$

We summerize all the different values of the states with respect to π_0^1 in Tab. 3. We come to the conclusion that, even if this model is still analytical, the complexity is much more important than that of Bianchi (2000). All the formulas show the role of the barrier of value $(1-p_k)^N$ to come from rate k to rate $k+1$.

5. Conclusion

In this article, we have opened an approach to study the impact of mobility over WLANs. First, after reviewing different modulation aspects, we have shown some constants that appear in terms of the shape of the area where some rate of communication is likely to operate

Eq.	State	Ratio to π_0^1
(7)	π_0^k $k \in \{2,\dots,r\}$	$\frac{(1-p_1)^N \dots (1-p_{k-1})^N}{p_1 \cdots p_{k-1}}$
(8)	π_1^k $k \in \{1,\dots,r-1\}$	$\left(p_{k+1} - 1 + \frac{1}{(1-p_k)^{N-1}}\right) \frac{(1-p_1)^N \dots (1-p_k)^N}{p_1 \cdots p_k}$
(9)	π_i^1 $i \in \{2,\dots,r-1\}$	$\frac{(1-p_1)^N \dots (1-p_{i-1})^N}{p_1} \left((1-p_i) - (1-p_{i+1})(1-p_i)^N\right)$
(10)	π_i^j $i \geq 2,\ j \geq 2,\ i+j \leq r$	$\frac{1}{p_1} \frac{(1-p_1)^N \dots (1-p_{i+j-2})^N}{p_{i+1} \cdots p_{i+j-1}} ((1-p_{i+j-1}) - (1-p_{i+j})(1-p_{i+j-1})^N)$
(11)	π_r^1	$\frac{1-p_r}{p_1} \left[(1-p_1)^N \dots (1-p_{r-1})^N\right]$
(12)	π_{r-1}^2	$\frac{1-p_r}{p_r} \left[(1-p_1)^N \dots (1-p_{r-1})^N\right]$
(13)	π_i^{r+1-i} $i \in \{1,\dots,r-2\}$	$\frac{1-p_r}{p_1 p_r} \frac{(1-p_1)^N \dots (1-p_{r-1})^N}{p_{i+1} \cdots p_{r-1}}$
(19)	π_{r+1}^0	$\left(1 + \sum_{j=0}^{j=r-1} \frac{(1-p_0)^N \dots (1-p_j)^N}{p_0^{j+1}} (1-p_{j+1})\right) \frac{p_0^r}{(1-p_0)^{N-1}}$
(20)	π_k^0 $k \in \{r+1,\dots,m\}$	$\left(1 + \sum_{j=0}^{j=r-1} \frac{(1-p_0)^N \dots (1-p_j)^N}{p_0^{j+1}} (1-p_{j+1})\right) \frac{p_0^{k-1}}{(1-p_0)^{N-1}}$
(21)	π_k^0 $k \in \{1,\dots,r\}$	$\left(1 + \sum_{j=0}^{j=k-2} \frac{(1-p_0)^N \dots (1-p_j)^N}{p_0^{j+1}} (1-p_{j+1})\right) \frac{p_0^{k-1}}{(1-p_0)^{N-1}}$ $-(1-p_1)^N \dots (1-p_{k-1})^N (1-p_k)$

Table 3. Stationary probabilities for the model of Fig. 5.

well. These figures have been obtained in a very general context, taking into consideration practical and theoretical channel conditions, including Rice and Rayleigh channels.
On top of that, we have described different existing systems for adapting the rate of communication in an unknown medium. We have shown many characteristics and also differences of the approaches.
Finally, we could open a new approach to the difficult, and yet unanswered question of designing a reliable analytical model to explain the behavior of such systems, taking the ARF scheme as a model one. In that case, we were able to highlight an high correlation between the Congestion Window (CW) of the system, and the rate at which packets are emitted. Not only that, but the analysis showed that all the stationary probabilities of the states of this Markov chain can be described with a closed formula. This opens new ways to research in that area and shows that the different mechanisms that have been implemented in the MAC systems of

WLAN cards have strong correlations with one another and therefore have to be redesigned with at least a global understanding of channel access problems (backoff and collisions) and rate adaptation questions.

6. References

Ancillotti, E., Bruno, R. & Conti, M. (2008). Experimentation and performance evaluation of rate adaptation algorithms in wireless mesh networks, *Proc. of 5th ACM PE-WASUN'08*, Vancouvert, BC, Canada, pp. 7–14.

Ancillotti, E., Bruno, R. & Conti, M. (2009). Design and performance evaluation of throughput-aware rate adaptation protocols for ieee 802.11 wireless networks, *Performance Evaluation* **66**: 811–825.

Bianchi, G. (2000). Performance analysis of the IEEE 802.11 distributed coordination function, *IEEE Journal on Selected Areas in Communications* **18**(8): 535–547.

Biaz, S. & Wu, S. (2008). Loss differentiated rate adaptation in wireless networks, *IEEE WCNC 2008*, Las Vegas, NV, USA, pp. 1639–1644.

Chen, X., Qiao, D., Yu, J. & Choi, S. (2007). Probabilistic-based rate adaptation for IEEE 802.11 WLANs, *Globecom'07*, Washington DC, USA, pp. 4904–4908.

Galtier, J. (2004). Optimizing the IEEE 802.11b performance using slow congestion window decrease, *Proccedings of the 16th ITC Specialist Seminar on performance evaluation of wireless and mobile systems*, Antwerpen, pp. 165–176.

Heuse, M., Rousseau, F., Guillier, R. & Duda, A. (2005). Idle sense: An optimal access method for high throughput and fairness in rate diverse wireless LANs, *SIGCOMM*, Philadelphia, USA.

Holland, G., Vaidya, N. & Bahl, P. (2001). A rate-adaptative MAC protocol for multi-hop wireless networks, *MobiCom'2001*, Rome, Italy, pp. 236–251.

Ibrahim, M. & Alouf, S. (2006). Design and analysis of an adaptative backoff algorithm, *Networking*, pp. 184–196.

J.Kim, Kim, S., Choi, S. & Qiao, D. (2006). CARA: collision-aware rate adaptation mechanism of IEEE 802.11 WLANs, *Infocom'2006*, Barcelon, Spain, pp. 1–11.

Kamermann, A. & Monteban, L. (1997). Wave LAN II: a high-performance wireless LAN for the unlicensed band, *Bell Labs Technical Journal* pp. 118–133.

Lacage, M., Manshaei, M. & Turletti, T. (2004). IEEE 802.11 rate adaptation: a practical approach, *MSWiM'04*, pp. 126–134.

Ni, Q., Aad, I., Barakat, C. & Turlletti, T. (2003). Modeling and analysis of slow CW decrease for IEEE 802.11 WLAN, *PIMRC*, Beijing, China.

Pang, Q., Leung, V. & Liew, S. (2005). A rate-adaptation algorithm for IEEE 802.11 WLANs based on MAC layer loss differentiation, *IEEE Broadnets 2005*, Boston, USA, pp. 709–717.

Part 11: wireless LAN medium access control (MAC) and physical layer (PHY) specifications (1999). IEEE Std 802.11a-1999.

Pavon, J. & Choi, S. (2003). Link adaptation strategy for IEEE wlan via received signal strength measurement, *Proc. ICC'03*, Vol. 2, Seattle, WA, USA, pp. 1108–1113.

Romano, P. (2004). The range vs. rate dilemna of (WLANs), EETimes Design.

Saghedi, B., Kanodia, V., Sabharwal, A. & Knightly, E. (2002). Opportunistic media access for multirate Ad Hoc networks, *MobiCom'2002*, Atlanta, Georgia, USA, pp. 24–35.

Segkos, M. (2004). *Advanced techniques to improve the performance of OFDM wireless LAN*, Master's thesis, Naval postgraduate school, Monterey, California.

Singh, A. & Starobinski, D. (2007). A semi markov-based analysis of rate adaptation algorithms in wireless LANs, *IEEE SECON*, pp. 371–380.

WLAN - 802.11 a,b,g and n (2008). NI Developper Zone.

Wong, S., Yang, H., lu, S. & Bharghavan, V. (2006). Robust rate adaptation for 802.11 wireless networks, *MobiCom'2006*, Los Angeles, California, USA, pp. 146–157.

Wu, S. & Biaz, S. (2007). ERA: efficient rate adaptation algorithm with fragmentation, *Technical Report CSSE07-04*, Auburn University.

11

An Overview of DSA via Multi-Channel MAC Protocols

Rodrigo Soulé de Castro, Philippe Godlewski and Philippe Martins
Télécom ParisTech, NMS research group
France

1. Introduction

The development of radio access technologies requires an increasing number of spectrum resources. Unfortunately, spectrum bands are scarce and the development of new wireless communication networks and services are thus more and more challenging ([jia07]). Recent reports indicate that fixed channel allocations result in low efficiency in spectrum utilization because a large portion of the spectrum remains underutilized ([mchenry05]).

One approach capable of dealing with the above problem is Dynamic Spectrum Access (DSA) which allows spectrum sharing. In such an approach, unlicensed users, known as secondary users (SUs), dynamically look for unused spectrum in licensed bands and communicate using "spectrum holes". These idle bands represent spectrum portions assigned to licensed users (known as primary users, PUs) that are not being used at a considered time and location ([timmers07]).

Many researchers have proposed different multi-channel MAC protocols to increase network throughput and reduce interference caused by secondary use of the spectrum. Many of these studies consider Wi-Fi like protocols (or IEEE 802.11 based mechanism).

Cognitive Radios (CR) are a type of radio capable of switching channels and adapting its transmission parameters in real-time ([mitola99]). Common MAC protocols do not provide, in general, mechanisms for channel switching. When having multiple independent channels to be used simultaneously, the need for enhanced Multi-channel MAC protocols becomes paramount. The IEEE 802.11 standard uses a distributed coordination function (DCF), as the fundamental Medium Access Control (MAC) technique. However, the distributed coordinate function, which employs carrier sense multiple access with collision avoidance (CSMA/CA), was not designed to work in a multi-channel environment ([ahmed07]).

Secondary users equipped with a cognitive radio, in a multi-channel environment, may improve the efficiency of spectrum utilization and increase the network throughput.

2. Background

2.1 Secondary use of spectrum

A cognitive radio is an intelligent communication device, which has the ability to adapt its transmission parameters such as channel frequency, modulation and power; based on the interaction with the environment in which it operates ([jia07]).

There are two different approaches of secondary use of spectrum in cognitive radio context. One is in the form of **overlay**, opportunistic usage of idle bands in the primary user's (PU) spectrum by cognitive radios and another in the form of **underlay,** using Ultra Wide Band (UWB) technology ([cabric06]).
The rules in secondary use of frequency spectrum specify that licensed users, known as Primary Users (PUs), have the rights for interference-free communication in certain bands. When these bands are not used by the primary users, they can be used by Secondary Users (SUs). As soon as a primary user starts activity in its channel, the SU has to vacate the channel to avoid interference ([timmers07]). However, a cognitive radio (using a half duplex transceiver) cannot scan the spectrum and transmit simultaneously in the same frequency band. Then, for the protection of primary users, a maximum detection or sensing time must be established. This detection time represents the maximum time of interference, from secondary users, that a primary user can tolerate ([jia07]).

2.2 Rendezvous in multi-channel protocols

In multi-channel MAC protocols, Mobile Stations (MSs) exchange control information to concur on the channel for data transmission in the user plane. Proposed protocols vary in how MSs negotiate the channel to be used for data transmission and the way to solve medium contention; these protocols can be divided according to their principle of operation.
In single rendezvous protocols, the rendezvous between a sender and its receiver can take place on at most one channel at any time, while in Multiple Rendezvous protocols, several rendezvous can take place in different channels simultaneously, thereby mitigating the control channel congestion ([mo07]).
In single rendezvous, three different classes of protocols can be distinguished based on the mechanism of channel negotiation ([sheung07]). The **Dedicated Control Channel** approach, which uses two transceivers (TRx), operates with a single channel only for control packets exchange. In this approach, the MSs always tune one TRx to the control channel to make agreements and be aware of neighbours' negotiations. The other TRx is able to switch channels and is used for data transmission. The **Split Phase** protocol uses only one TRx for control and data packets. In this protocol, time is split into fixed periods of control and data phases. The control phase is used as common control channel to make rendezvous, when control phase ends, MSs switch to their selected channels and begin data transmission. The third class of protocol is named **Common Hopping,** which also has only one TRx for both control and data packets, in this protocol there is no CCCH. MSs hop synchronously through all available channels and pauses hopping when sender and receiver agree on data transmission using their current channel.

2.3 Hidden terminal problem in a single channel environment

Hidden terminal problem occurs when mobile stations cannot detect signal from other MSs by carrier sensing because they do not have a physical connection to each other. Figure 2 illustrates this problem: MS "A" sends a message to MS "B"; "C" cannot detect the signal from "A" since "C" is out of range of "A". For station "C", the channel is idle. When MS "C" sends a message to "B", this message will collide at "B" with the message sent from "A". In this scenario "C" is the hidden node to "A".

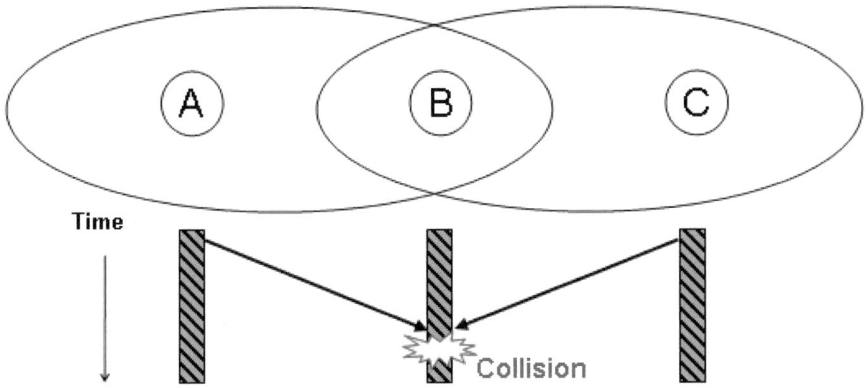

Fig. 2. Hidden terminal problem in a single channel environment

2.4 Virtual carrier sensing using RTS/CTS exchange

To deal with the above problem, the IEEE 802.11 MAC layer uses the Distributed Coordination Function (DCF) mechanism, which employs virtual carrier sensing to solve the hidden terminal problem by using the RTS/CTS mechanism.

In this mechanism, when a mobile station wants to initiate communication, it first sends a **RTS** (Request-To-Send) message and the receiver replies by sending a **CTS** (Clear-To-Send). The RTS and the CTS contains the **NAV** (Network Allocation Vector), which is the expected duration of time that other mobile stations, around the communication pair, must refrain from sending data to avoid collisions.

This procedure can solve the hidden terminal problem in a single channel environment, under the assumption that all mobile stations have the same transmission range. However, the DCF mechanism cannot work well in a multi-channel environment, the reason is because MSs may be transmitting or receiving data packets in different channels, missing the RTS/CTS procedure of the DCF mechanism.

2.5 Multi-channel hidden terminal problem

This problem occurs when mobile stations in the network listen to different channels missing the RTS/CTS procedure.

The Multi-Channel Hidden Terminal Problem is illustrated in figure 3. Initially, mobile station "A" wants to communicate with "B", then "A" sends an A-RTS to "B" on the Common Control Channel (Channel 1). After receiving the A-RTS, MS B selects the Channel 2 to communicate with "A" and sends back an A-CTS, notifying their neighbours that the data channel number 2 has been selected. In a single channel environment the RTS/CTS exchange avoids collisions in the transmission ranges of "A" and "B". However, in multi-channel environments other mobile stations could be involved in communication in different channels when the RTS/CTS procedure took place. That is the case of mobile stations "C" and "D", as they were communicating in channel 3 they did not hear the A-CTS sent by "B". When they finish their communication on Channel 3, mobile stations "C" and "D" switch to Channel 1 and now they select Channel 2 to reinitiate communication. When MS "C" sends the first message to "D", this message will cause collision to mobile station "A" and "B" on Channel 2.

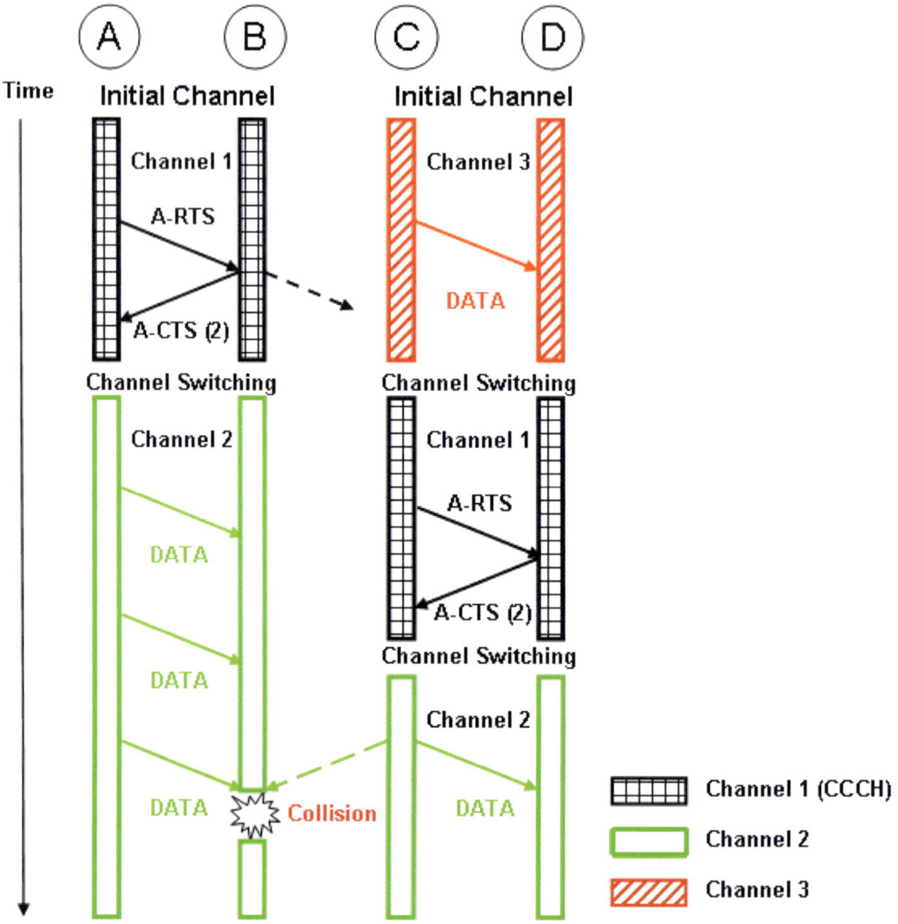

Fig. 3. Hidden terminal problem in Multi-channel protocols (figure inspired from Jungmin So et al. [so04])

One possible solution would be a unique channel or moment in which every MS in the network listens to, thereby, ensuring that the RTS/CTS procedure can be heard by all the MSs, thus avoiding the Multi-Channel Hidden Terminal Problem ([so04]).

3. Multi-channel MAC protocols

3.1 "Comparison of multi channel MAC protocols" [mo07]

[mo07] presents a performance comparison between different multi-channel MAC protocols, single rendezvous protocols (dedicated control channel, common hopping and split phase) and multi rendezvous (parallel rendezvous).

Dedicated Control Channel Approach: This protocol uses 2 TRx per Mobile Station (MS), one is used for control information exchange and the other is able to switch between channels for data transmission. There is no need for synchronization to make rendezvous because the control channel is always tuned by all the MSs in the network. However, this protocol presents two principal problems, the need for 2 TRx and the possibility of control channel bottleneck.

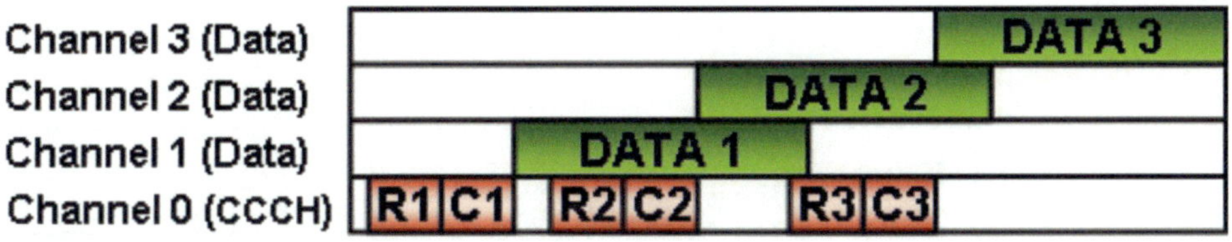

Fig. 4. Dedicated Control Channel Approach (figure inspired from [mo07])

Common Hopping Approach: This protocol uses 1 TRx per Mobile Station (MS); this TRx is able to switch between channels for control information exchange and data transmission. To make rendezvous, MSs hop synchronously over all the channels and pauses its hopping sequence when the agreement between sender and receiver is made. This protocol uses all the channels for data transmission. However, the synchronization among MSs is crucial.

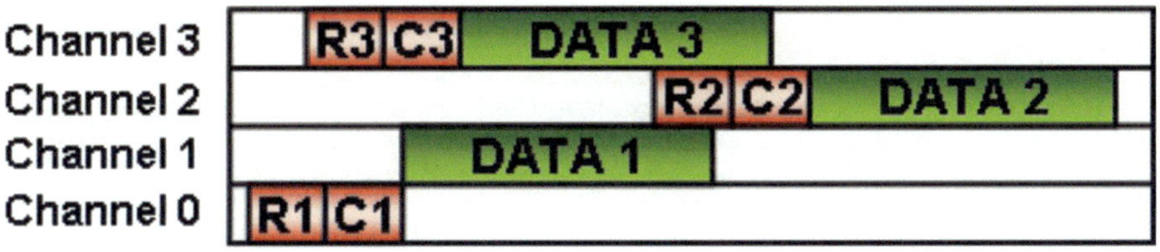

Fig. 5. Common Hopping Approach (figure inspired from [mo07])

Split Phase Approach: This protocol uses 1 TRx per Mobile Station (MS), time is divided into control Phase and Data phase, this division has the objective to ensure that all MSs listen to the control phase, thus avoiding the Multi-Channel Hidden Terminal problem (MCHTP). Two important disadvantages of this protocol are the need for global synchronization and the wasted data channels during the control phase. However, with only one TRx, this protocol solves the MCHTP.

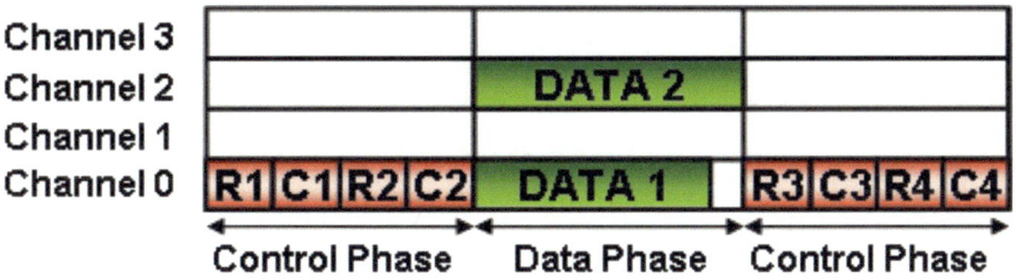

Fig. 6. Split Phase Approach (figure inspired from [mo07])

3.2 "McMAC: A parallel rendezvous multi-channel MAC protocol" [sheung07]

McMAC protocol uses 1 TRx per Mobile Station (MS). At the beginning, a sender chooses a hopping pattern in a pseudo-random way using a seed to generate it, neighbours learn its hopping sequence because is included in all the sender's packets. To make rendezvous, a MS can deviate from its default hopping sequence and hops to the receiver's channel. In this protocol multiples rendezvous can be made in different channels at the same time, thus improving the network throughput and avoiding control channel bottleneck. However, the synchronization and coordination between MSs are essential.

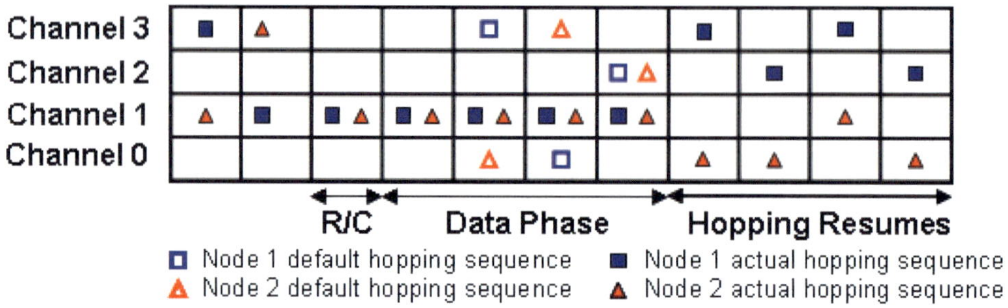

Fig. 7. McMAC protocol (figure inspired from [mo07])

3.3 "SSCH: Slotted Seeded Channel Hopping for capacity improvement in IEEE 802.11 ad-hoc wireless networks" [bahl04]

SSCH protocol uses 1 TRx per Mobile Station (MS). In this protocol, each sender chooses one of the possible hopping patterns generated in a pseudo-random way (one hopping pattern for each available channel). To make rendezvous, a sender must wait until its current hopping pattern intersects with that of the receiver before it can send data. The principal disadvantage of this protocol is the time wasted waiting to coincide with the receiver. However, multiples rendezvous can be made at the same time in different channels and the control channel bottleneck is avoided.

3.4 "Multi-channel MAC for ad hoc networks: handling multi-channel hidden terminals using a single transceiver" [so04]

In MMAC protocol, each MS is equipped with 1 TRx. Time is divided into an alternating periods of control and data phases (split phase). An Ad Hoc Traffic Indication Message (AR), at the start of each control interval, is used to indicate traffic and negotiate channels for utilization during the data interval. A similar approach is used in IEEE 802.11's power saving mechanism (PSM). This scheme uses two new packets which are not used in IEEE 802.11 PSM: the ATIM ACK (AC) and the ATIM-RES (A-RE). These packets inform the neighbourhood nodes of the Sender (S) and Destination (D), of which channels are going to be used during the data exchange. During the control period, named ATIM window, all MSs have to attend the default channel and contend for the available channels. Once reservation is successful, the MSs switch to the reserved channel. With only one TRx this protocol solves the Multi-Channel Hidden Terminal Problem. A Preferred Channel List (PCL) is used to select the best channel based on traffic conditions. In this list all the channels are classified by the status: HIGH, MID, and LOW.

The major drawback of the scheme could be the need for synchronizing beacons, which might be difficult to implement in Ad Hoc networks and the waste of the bandwidth in other channels during the ATIM window (control period). However, with only one TRx this protocol solves the MCHTP.

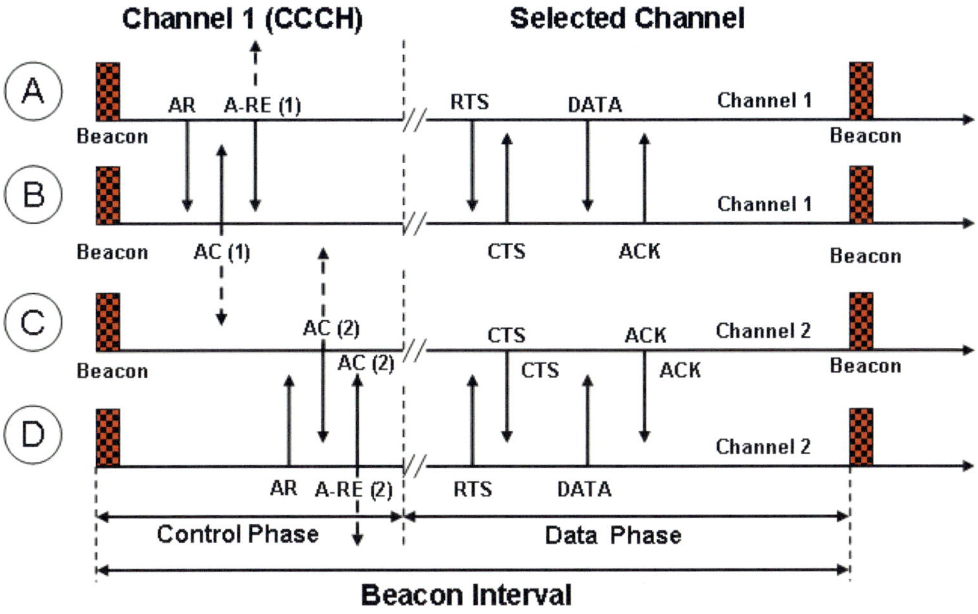

Fig. 8. MMAC protocol (figure inspired from [so04])

3.5 "A distributed multichannel MAC protocol for cognitive radio networks with primary user recognition" [timmers07]

In MMAC-CR protocol, time is split into alternating periods of control and data phase and each user is equipped with 1 TRx. A similar approach is used in IEEE 802.11's power saving mechanism (PSM). This protocol has two data structures: the Spectral Image of Primary users (SIP), which contains the channels used by Primary Users (PUs), and the Secondary users Channel Load (SCL), which is used to select the communication channel in terms of traffic.

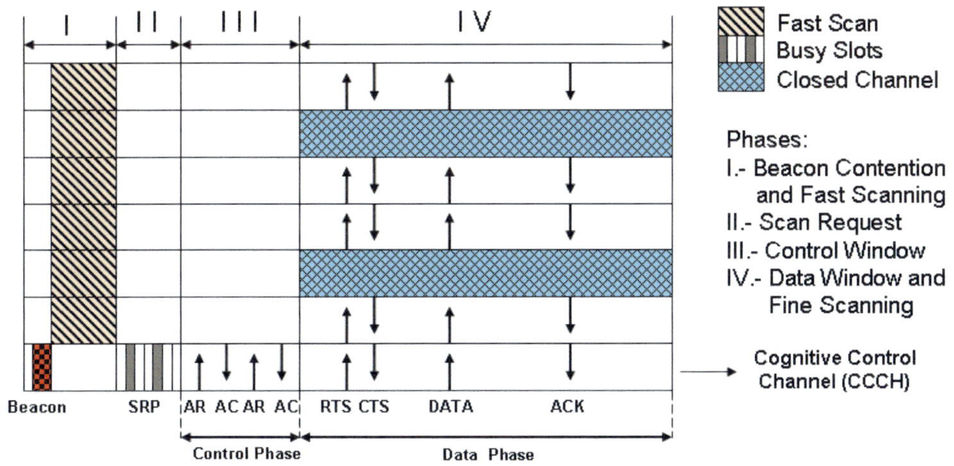

Fig. 9. MMAC-CR protocol (figure inspired from [timmers07])

The proposed protocol is divided into four phases: during phase I, the nodes contend to transmit a beacon and perform a fast scan; this scanning process is used to update the SIP value of the scanned channel. Phase II is used to determine the spectral opportunities by listening to C minislots (each minislot correspond a data channel). Each MS informs the others of the presence of PUs by transmitting a busy signal in the corresponding minislot. In Phase III, using ATIM packets (AR and AC), the channels are negotiated. Phase IV is used for data transmission or fine sensing for idle nodes.

MMAC-CR with only one TRx solves the "Multi-Channel Hidden Terminal Problem". Alternating periods of control and data phases, this protocol avoids the possibility of control channel bottleneck. However, the synchronization and coordination between MSs are essential to make rendezvous which might be difficult to implement in Ad hoc networks.

3.6 "TMMAC: an energy efficient multi-channel MAC protocol for ad hoc networks" [zhang07]

In TMMAC, each user is equipped with 1 TRx; time is divided into control phase (ATIM window) and data phase. The ATIM window size is not fixed and can be adapted based on traffic conditions. The data phase is slotted, only a single data packet can be transmitted or received during each time-slot. The purpose of the control window is twofold, the channel negotiation and the slot negotiation. In the data phase, each node switches to the negotiated channel and uses its respective time slot for packet transmission or reception.

This protocol has the same advantages and disadvantages presented in split phase protocols: the need for global synchronization and the wasted data channels during the control phase. However, with only one TRx, this protocol solves the MCHTP.

3.7 "Hardware-constrained multi-channel cognitive MAC" [jia07]

In HC-MAC, each MS is equipped with 1 TRx. In this protocol, there is no need for global synchronization. To make rendezvous, HC-MAC transfers control packets using a Common Control Channel (CCCH). Time is divided into Contention phase, Sensing phase and Transmission phase and each phase has a RTS/CTS exchange:

1. C-RTS/C-CTS: using the RTS/CST mechanism (cf. IEEE 802.11 DCF mode), a pair of MSs reserves all the channels (CCCH and data channels) for the following two phases (sensing and transmission).
2. After sensing the different data channels, the pair exchanges a S-RTS/S-CTS on the CCCH to mutually inform about channel availability. A set of channels (only one in single Tx case) is then selected.
3. After data transmission on the different selected channels, the communication pair informs the end of transmission by a T-RTS/T-CTS exchange. This allows neighbouring MSs to begin the contention phase with a random back off.

Authors outline two constraints for cognitive radios, sensing and transmission, the former used to optimize the stopping of spectrum sensing and the later used to optimize the spectrum utilized in transmission by secondary users.

The major drawback of this scheme could be that after one communication pair wins the CCCH, using the C-RTS/C-CTS exchange, other mobile stations must defer their sensing and transmission. Then, for a certain time, only one pair uses all available channels and other users must wait for the T-RTS/T-CTS notification to contend again in the control channel.

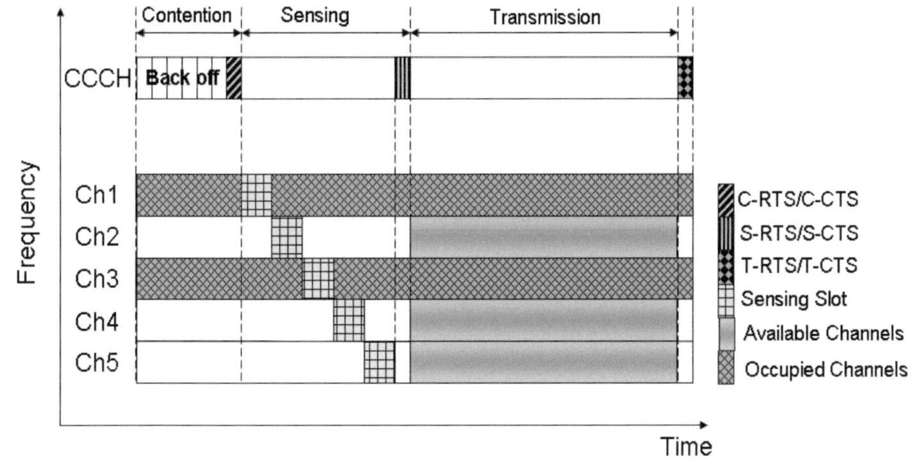

Fig. 10. HC-MAC protocol (figure inspired from [jia07])

3.8 "Distributed coordinated spectrum sharing MAC protocol for cognitive radio" [nan07]

This protocol uses 2 TRx per Mobile Station (MS), one is used for control information exchange and the other is able to switch between channels for data transmission. There is no need for synchronization to make rendezvous because the control channel is always tuned by the MSs. In this protocol, secondary users employ a time slot mechanism for cooperative detection of primary users around the communication pair by using the CHRPT (channel report slots). Each node informs the others about the presence of PUs, in the sender and in the receiver side, by transmitting a busy signal in the corresponding minislot (there is one minislot for each data channel).

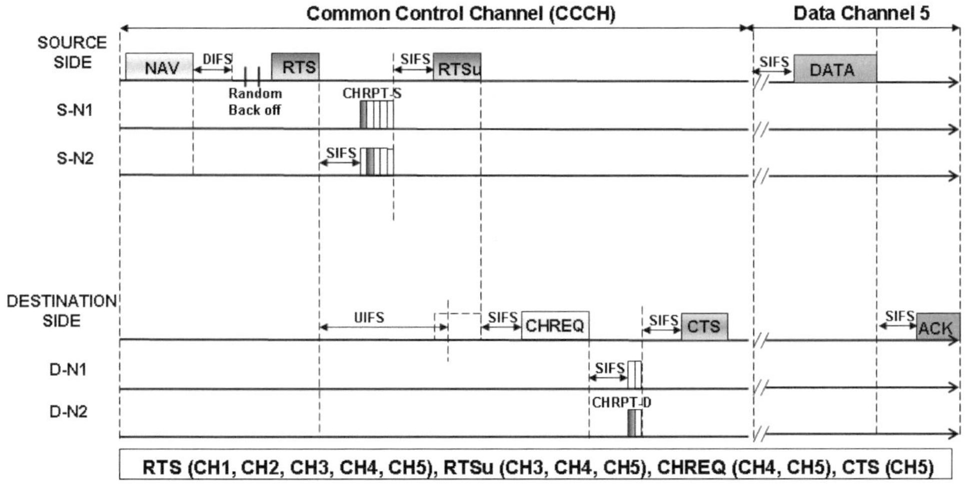

Fig. 11. Procedure of the proposed protocol (figure inspired from [nan07])

The source sends to destination the RTS which includes its available channel list. Neighbour nodes, which hear the RTS, compare the sender list with their own; if they detect a PU occupation in a channel, they reply with a pulse in the specified time slot during CHRPT (signalling occupied channels seen by the neighbours). If necessary, the source update its RTS sending a RTSu. The same mechanism occurs in the destination side. After the RTS reception the destination waits to get the possible RTSu for certain time named UIFS, if the RTSu does not arrives, the destination will handle the first RTS. After the RTS reception, the destination sends to its neighbours the Channel Status Request (CHREQ), which includes the destination available channel list among the listed channels of the source. At the end of channel verification by the destination neighbours, the receiver sends the CTS with the chosen channel.

The major drawbacks of the scheme are the time wasted in channel verification by the neighbours and the need for two TRx. However, this procedure ensures the absence of primary users in the vicinity of the communication pair.

3.9 "Performance of multi channel MAC incorporating opportunistic cooperative diversity" [ahmed07]

In CD-MMAC, time is divided into fixed periods (split phase), each user is equipped with 1 TRx. This protocol uses the same mechanism proposed by So et al. in MMAC ([so04]). The authors of this protocol add the notion of relays between source and destination. Time is divided into fixed-time intervals (control phase and data phase) using beacons, a small window, named ATIM, at the start of each interval is used to indicate traffic and negotiate channels to be used during the data phase. This protocol uses intermediate nodes as relays to increase the probability of transmission success.

This protocol solves the MCHTP with only one TRx. However, two drawbacks of CD-MMAC are the need for global synchronization and the wasted data channels during the control phase.

3.10 "A full duplex multi channel MAC protocol for multi-hop cognitive radio networks" [choi06]

In this protocol, each secondary user is equipped with 3 TRx named: "Receiver, Transmitter and Controller". To communicate, the RECEIVER of the receiving node and the TRANSMITTER of the sending node must be tuned to the same channel.

There is no need for synchronization because the CCCH is always tuned by the MSs using the CONTROLLER. A MS selects an unused frequency band as its home channel (HCh), it tunes its receiver to its HCh and informs the others about its selected channel by broadcast in the control channel. This protocol uses CSMA/CA scheme of IEEE 802.11 DCF mode. With the use of three TRx, MSs can reduce communication delay by transmitting packets while they are receiving. However, the need for 3TRx will increase the overall cost.

3.11 "A multi channel MAC for opportunistic spectrum sharing in cognitive networks" [mishra06]

In AS-MAC (Ad hoc SEC Medium Access Control) protocol, the primary user is a TDMA/FDMA (GSM) cellular network and the secondary user is an Ad hoc network that can decode the control information of GSM system. Sensing the vacant slots, the SU uses the resources left utilized by the primary user, which could be a Base Station (BS) or a Mobile Station (MS). To obtain all the parameters like synchronization, frequency correction and cell information, secondary users decode the beacon channel from the BS. To make rendezvous, this protocol employs RTS/CTS and Reservation (RES) mechanism.

3.12 “Performance evaluation of a medium access control protocol for IEEE 802.11s mesh networks” [benveniste06]

CCC protocol uses 2 TRx per Mobile Station (MS), one is used for control information exchange and the other is able to switch between channels for data transmission. There is no need for global synchronization to make rendezvous because the control channel is always tuned by the MSs. The CCC protocol defines a Common Control Channel (CCCH), over which, mesh nodes will exchange control and management frames, the rest of the channels, called Mesh Traffic (MT) channels, are used to carry the data traffic. Reservations of the various MT channels are made by exchanging control frames on the CCCH.
This protocol has the same advantages and disadvantages presented by the dedicated control channel approach: there is no need for synchronization to make rendezvous. However, this protocol needs two TRx and the possibility of control channel bottleneck exists.

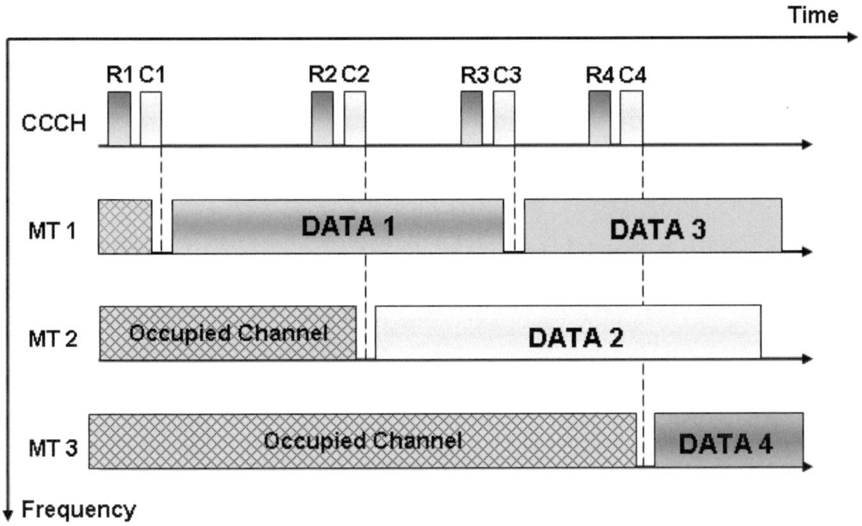

Fig. 12. CCC MAC protocol (figure inspired from [benveniste06])

4. “Os-MAC: an efficient MAC protocol for spectrum-agile wireless networks” [hamdaoui08]

In Os-MAC protocol, each secondary user is equipped with 1 TRx; this protocol uses the IEEE 802.11 DCF mode. This approach seeks to exploit the available spectrum opportunities using MSs coordination. One entity per channel is a "delegate", the delegates are chosen among MSs and makes reports about channel quality. A single ACK notion is used in a "multicast group" named Secondary User Group (SUG).
OS-MAC divides time into periods; each period is named Opportunistic Spectrum Period (OSP). In each OSP, there exist three consecutive phases: Select, Delegate, and Update Phase. In the first phase, each SUG selects the “best” Data Channel (DC) based on traffic conditions

and uses it for communication during the totality of the OSP period. During the second phase, a Delegate Secondary User (DSU) is chosen to represent the data channel during the Update Phase, in which, all DSUs switch to the CCCH to update each other about their channel conditions, mean while, all non-DSUs continue communicating on their DCs.

An important aspect of this protocol is the notion of groups and the Delegate for each DC. This mechanism can improve the channel classification necessary to define the best channel, based in traffic conditions, which could be used for data transmission.

4.1 "Primary Channel Assignment Based MAC (PCAM) a multi channel MAC protocol for multi-hop wireless networks" [pathma04]

In PCAM protocol, each user is equipped with 3 TRx. This scheme eliminates the need for a dedicated control channel that arise the possibility of control channel bottleneck when the traffic increases. In this protocol, a MS selects a frequency band as its primary channel, this will be used as a receiver channel and a secondary channel is used as transmitter while the third TRx is used for transmitting and receiving broadcast messages. PCAM protocol removes the constraints of time synchronization and control channel saturation because the channels are pre-assigned. However, the need for 3 TRx will increase the overall cost and the channel assignment procedure, in this protocol, is not specified.

4.2 "Adaptive MAC protocol for throughput enhancement in cognitive radio networks" [lee08]

In this protocol, each user is equipped with 2 TRx, this protocol proposes two channels, the first one is a WLAN channel which is always available for data transmission; the second one, named "Cognitive channel", is available sporadically. When traffic conditions restrain the use of the cognitive channel, this channel is used for frame errors recovery by transmitting the same information in both channels, known as frequency diversity in MIMO systems; otherwise, the cognitive channel can be used to increase the overall throughput by sending sequential frames using both channels.

The drawback of this scheme could be the need for two TRx. However, this procedure can enhance the overall throughput if the "Cognitive channel" is available.

4.3 "CREAM-MAC: An efficient Cognitive Radio-EnAbled Multi-channel MAC protocol for wireless networks" [su08]

In the Cognitive Radio-EnAbled Multi-channel MAC (CREAM-MAC) protocol, each secondary user is equipped with 1 TRx that can dynamically utilize one or multiple channels to communicate and also has multiple sensors that can detect multiple channels activity simultaneously. This protocol needs neither centralized controllers nor synchronization.

The CREAM-MAC protocol employs a Common Control Channel (CCCH) as the "rendezvous channel". With one TRx, this protocol solves the Multi-Channel Hidden Terminal Problem employing a four-way handshake. These control packets are RTS/CTS and CST/CSR, the RTS/CTS exchange prevents the collisions among the secondary users by reserving the CCCH for channel negotiation. The CST/CSR exchange avoids collisions between secondary and the primary users by allowing secondary users to share sensing information about PU's channel occupation.

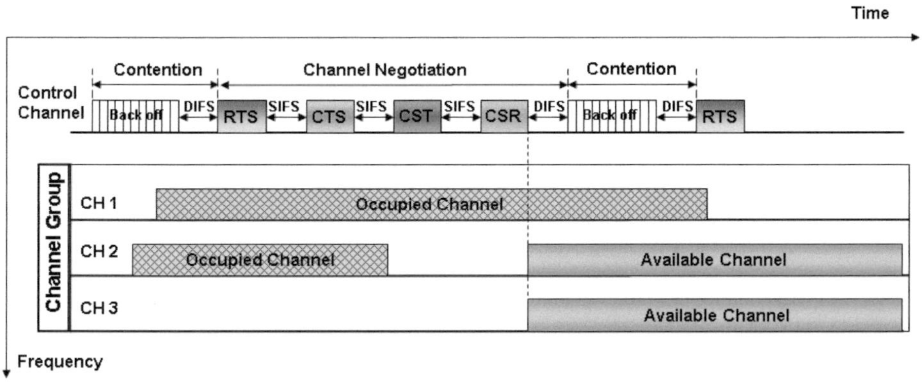

Fig. 13. CREAM-MAC protocol (figure inspired from [su08])

The merit of the CREAM-MAC protocol is the fact that there is no need for global synchronization and with the use of only one TRx and multiple sensors, this protocol solves the MCHTP.

4.4 "Distributed coordination in dynamic spectrum allocation networks" [zhao05]

In this paper, the notion of groups with similar views of spectrum availability is addressed. Each secondary user is equipped with 1 TRx, this protocol employs a voting scheme for selection of a "Coordination Channel" (CCH) for a group and this "user group" is assembled based in similar spectrum channel availabilities.

The CCH is used as the only means to connect secondary users, thus, only members of the same group can directly communicate with each other. To maintain network connectivity "bridge" nodes, located on the edge of each group, must manage at least two different CCH to transfer data packets between groups and connect users with different spectrum perspective.

The advantage of this approach is its possible application in the case of secondary use of the spectrum by WLAN devices in TV white spaces, principally, because the interference condition with primary users is determined by distance.

4.5 "Single-Radio Adaptive Channel Algorithm for spectrum agile wireless ad hoc networks" [ma07]

In the Single-Radio Adaptive Channel (SRAC) algorithm, each secondary user is equipped with 1 TRx. This algorithm proposes an adaptive channelization, where a radio combines multiple fixed channels with minimum bandwidth, named "atomic channels", based on its needs to form a new channel with more bandwidth, thus forming a "Composite channel". In this algorithm there is no need for global synchronization. SRAC also proposes "Cross-channel communication", utilized to enable communications when there are multiple jamming sources and there is no common idle spectrum between the transmitter and the receiver. A node always has a pre-assigned channel for reception, which is well known by its neighbours and will be used to reach that node; this channel can be modified but the selection must follows strict rules to enable future communications.

The merits of this algorithm are the adaptive channelization and the fact that it does need neither CCCH nor synchronization because the MSs have a pre-assigned channel for reception.

4.6 "Cognitive radio system using IEEE 802.11a over UHF TVWS" [ahuja08]

This paper presents a practical implementation of IEEE 802.22 WLAN/TV with Primary and Secondary users. The architecture consists of Cognitive Mobile Stations (CMS) and a Cognitive 802.11 Access Point (CAP), which performs band sensing and available channel determination. The Cognitive Access Point has 1 TRx and 1 Rx for sensing, the Cognitive Mobile Stations are equipped with only 1 TRx. There is no CCCH, the CAP sends a broadcast message to inform all stations about the available channels list and time synchronization. A Geo-location module is used to guarantees that the cognitive radio units will never transmit on a channel that is determined to be within a licensed station's protected contour.

4.7 "Spectrum sharing radios" [cabric06]

This paper proposes the utilization of overlay, opportunistic usage of idle bands, for data transmission and underlay, using UWB technology, for control messages exchange. In this approach, authors propose two different types of control channels. The first one is a low throughput and wide coverage channel, named "Universal Control Channel (UCC)", which is used as a CCCH allowing the co-existence of several Radio Access Technologies (RATs). The second type of channel, named Group Control Channel (GCC), works as "Group Coordination Channel". This channel with high throughput and short coverage allows sensing information exchange, link maintenance and performs channel allocation.
The advantage of the use of UWB Control Channels is that we could have a realistic and reliable Cognitive Control Channel, always free of Primary users, which is one of the principal assumptions in several propositions of Multi-Channel MAC protocols.

5. Conclusions

This chapter presents the main existing multi channel MAC protocols. The merits of several protocols are discussed with regard to different factors: the number of transceivers, the need for synchronization, the need for a common control channel (CCCH) and the different ways to make rendezvous for data transmission. As we showed, each multi-channel MAC protocol faces and resolves differently the various complications that arise in dynamic spectrum access.
In short, Cognitive Radio (CR) technology offers the possibility for additional use of radio spectrum by secondary users. Multiple channel protocols allow dynamic spectrum access (DSA) due to the fact that different rendezvous and data transmissions can be performed on different channels. This type of protocols, compared to others that use a single frequency channel (IEEE 802.11mechanism), may improve spectrum utilization and increase total network throughput.

6. Acronyms

ATIM: Ad hoc Traffic Indication Message
AC: ATIM ACK
AR: ATIM
A-CTS: ATIM CTS (which includes the data channel selection)
A-RE: ATIM Reservation
A-RTS: ATIM RTS (which includes the data channel selection)

CCCH: Common Control Channel
CR: Cognitive Radio
CREAM-MAC: Cognitive Radio-EnAbled Multi-Channel MAC protocol proposed in [su08]
CSR: Channel-State-Receiver
CST: Channel-State-Transmitter
DC: Data Channel
DCF: Distributed Coordination Function {IEEE 802.11}
DSU: Delegate Secondary User
MCHTP: Multi-Channel Hidden Terminal Problem
MC-MAC: Multi-Channel (wireless) MAC
MMAC: Multi-Channel MAC protocol proposed in [so04]
MMAC-CR: Multi-Channel MAC protocol proposed in [timmers07]
MS: Mobile Station
OSMAC: Opportunistic Spectrum Media Access Control proposed in [hamdaoui08]
OSP: Opportunistic Spectrum Period
PN: Primary Network
PSM: Power Saving Mechanism
PCL: Preferred Channel List
PU: Primary User
RAT: Radio Access Technology
SCL: Secondary users Channel Load
SIP: Spectral Image of Primary users
SU: Secondary User
SUG: SU Group
TRx: Transceiver

7. References

[ahmed07] Sabbir Ahmed, Christian Ibars, Aitor del Coso and Abbas Mohammed, Performance of Multi Channel MAC incorporating Opportunistic Cooperative Diversity. in *IEEE Vehicular Technology Conference,* April 2007.

[ahuja08] Ramandeep Ahuja, Robert Corke and Alan Bok, Cognitive Radio System using IEEE 802.11a over UHF TVWS. *New Frontiers in Dynamic Spectrum Access Networks,* 2008.

[bahl04] Paramvir Bahl, Ranveer Chandra and John Dunagan, SSCH: Slotted Seeded Channel Hopping for Capacity improvement in IEEE 802.11 Ad-Hoc Wireless Networks. in *MobiCom* 2004.

[benveniste06] Mathilde Benveniste and Zhifeng Tao, Performance Evaluation of a Medium Access Control Protocol for IEEE 802.11s Mesh Networks, in *Sarnoff Symposium,* 2006.

[cabric06] Danijela Cabric, Ian D. O'Donnell, Mike Shuo-Wei Chen, and Robert W. Brodersen, Spectrum Sharing Radios, in *IEEE Circuits and Systems Magazine*, 2006.

[choi06] Noun Choi, Maulin Patel and S.Venkatesan, A Full Duplex Multi channel MAC Protocol for Multi Hop Cognitive Radio Networks, in *Cognitive Radio Oriented Wireless Networks and Communications Conference*, June 2006.

[hamdaoui08] Bechir Hamdaoui and Kang G. Shin, Os-MAC: An efficient MAC Protocol for Spectrum-Agile Wireless Network, in *IEEE Transactions on Mobile Computing* 2008.

[jia07] Juncheng Jia and Qian Zhang, Hardware-constrained Multi-Channel Cognitive MAC. in *IEEE Global Telecommunications Conference,* November 2007.

[lee08] Byungjoo Lee and Seung Hyong Rhee, Adaptive MAC Protocol for Throughput Enhancement in Cognitive Radio Networks, in: *Information Networking*, 2008.

[mishra06] Amitabh Mishra, A Multi channel MAC for Opportunistic Spectrum Sharing in Cognitive Networks, in *Military Communications Conference*, 2006.

[mitola99] J. Mitola III, Cognitive radio for flexible mobile multimedia communication, in: *Proc. IEEE International Workshop on Mobile Multimedia Communications* (MoMuC) 1999, November 1999, pp. 3–10.

[ma07] Liangping Ma, Chien-Chung Shen, and Bo Ryu, Single-Radio Adaptive Channel Algorithm for Spectrum Agile Wireless Ad Hoc Networks, in: *New Frontiers in Dynamic Spectrum Access Networks*, 2007.

[mchenry05] Mark A; McHenry, NSF, Spectrum Occupancy Measurements, Project Summary, http://www.sharedspectrum.com/ , August 2005

[mo07] Jeonghoon Mo, Hoi-Sheung Wilson So and Jean Walrand, Comparison of Multi channel MAC protocols, in *IEEE Transactions on Mobile Computing* 2007.

[nan07] Hao Nan, Tae-In Hyon and Sang-Jo Yoo, Distributed Coordinated Spectrum Sharing MAC protocol for cognitive radio, 2007.

[pathma04] Jaya Shankar Pathmasuntharam, Amitabha Das and Anil Kumar Gupta, Primary Channel Assignment Based MAC (PCAM) A Multi Channel MAC Protocol for Multi-Hop Wireless Networks, in *Wireless Communications and Networking Conference*, 2004.

[sahin07] Mustafa E. Sahin, Sadia Ahmed, and Hüseyin Arslan. The Roles of Ultra Wideband in Cognitive Networks, in *IEEE International Conference on Ultra-Wideband*, 2007.

[sheung07] Hoi-Sheung Wilson So, Jean Walrand and Jeonghoon Mo, McMAC: A Parallel Rendezvous Multi-Channel MAC protocol. in IEEE *Wireless Communications and Networking Conference*, March 2007.

[so04] Juming So and Nitin Vaidya, Multi-Channel MAC for Ad Hoc Networks: Handling Multi-Channel Hidden Terminals Using A Single Transceiver, in *Proceedings of AMC MobiHoc*, May 2004.

[su08] Hang Su and Xi Zhang, CREAM-MAC: An efficient Cognitive Radio- EnAbled Multi-Channel MAC Protocol for Wireless Networks. in: *International Symposium on a World of Wireless, Mobile and Multimedia Networks*, 2008.

[timmers07] Michael Timmers, Antoine Dejonghe, Liesbet Van der Perre and Francky Catthoor, A Distributed Multichannel MAC Protocol for Cognitive Radio Networks with Primary User Recognition, in *Cognitive Radio Oriented Wireless Networks and Communications,* Aug. 2007.

[zhang07] Jingbin Zhang, Gang Zhou, Chengdu Huang, Sang H. Son and John A. Stankovic, TMMAC: An Energy Efficient Multi-Channel MAC Protocol for Ad Hoc Networks, in *the ICC 2007 proceedings*

[zhao05] Jun Zhao, Haito Zheng and Guang-Hua Yang, Distributed Coordination in Dynamic Spectrum Allocation Networks, in *New Frontiers in Dynamic Spectrum Access Networks*, 2005.

Distance Estimation based on 802.11 RTS/CTS Mechanism for Indoor Localization

Alfonso Bahillo, Patricia Fernández, Javier Prieto, Santiago Mazuelas, Rubén M. Lorenzo and Evaristo J. Abril
University of Valladolid
Spain

1. Introduction

Intense research work is being carried out to design and build localization schemes that can operate in indoor environments where satellite signals typically fail. The objective is to achieve a degree of accuracy, reliability and cost in indoor environments comparable to the well-known Global Navigation Satellite Systems (GNSS) in open areas. These challenging problems are being faced today to fulfill commercial, public safety and military applications (Gustafsson & Gunnarson, 2005; Pahlavan & Krishnamurthy, 2002). In commercial applications for residential and nursing homes there is an increasing need to track people with special needs, such as children and elderly people who are out of regular visual supervision, navigate the blind, and find specific items in warehouses. For public safety and military applications, indoor localization schemes are needed to track inmates in prisons or navigate police officers, fire fighters and soldiers to complete their missions inside buildings.
Among the many indoor technological possibilities that have been considered for indoor localization such as infrared, ultrasonic and artificial vision, radiofrequency based schemes predominate today due to their availability, low-cost and coverage range. Currently, few radiofrequency infrastructures that operate inside buildings are as extensively deployed and used as 802.11. Nowadays, many buildings such as shopping malls, museums, hospitals, airports, etc. are equipped with 802.11 access points (APs). Therefore, it may be practical to use these APs to determine user location in these indoor environments.
Whichever indoor wireless technology is involved, the purpose of localization schemes is to find the unknown position of a mobile station (MS) given a set of measurements called localization metrics. These metrics could be the measured time-of-arrival (TOA) (Golden & Bateman, 2007), angle-of-arrival (AOA) (Seow & Tan, 2008) or received-signal-strength (RSS) (Mazuelas et al., 2009) of the MS's signal at the reference devices or APs. Techniques based on RSS require channel modeling and they are not flexible because they present high variability to environmental changes; even though building and updating a RSS database is much easier in indoor environments than in wide urban areas. The major drawback of pattern recognition techniques still lies in substantial efforts needed in generation and maintenance of the RSS database in view of the fact that the working environment changes constantly. Techniques based on TOA need time synchronization between wireless nodes; and techniques based on AOA require specialized antennas. Furthermore, it is important to point out that as the

measurements of metrics become less reliable, the complexity of the positioning algorithm increases.

In this chapter, the performance of the 802.11 wireless networks for indoor localization is based on the time delay localization metric through round-trip time (RTT) measurements. The challenge is to develop an infrastructure that is inexpensive to design and deploy, complies with frequency regulations, and provides a comprehensive coverage for accurate ranging. RTT is used instead of TOA to avoid the need for time synchronization between wireless nodes. Furthermore, due to the use of an 802.11 infrastructure, the location capabilities will be an added value to the existing connectivity ones. The main characteristic that makes the RTT measurements possible in any 802.11 wireless network is the common protection mechanisms to fully reserve a shared medium, Request To Send/Clear To Send (RTS/CTS) handshake (Bahillo et al., 2009). Therefore, the RTT measurements are obtained by measuring the latency of a series of layer two CTS frames sent by and in response to a corresponding series of RTS frames initiated by the MS that is going to be located. The measuring system is integrated in a Printed Circuit Board (PCB), which is used as additional hardware to the 802.11 adapter from which appropriate signals, such as transmission and receiver pulses of exchange frames, are extracted to quantify the RTT.

The results of RTT measurements in different scenarios are qualitatively consistent because, as it was expected, the delay profile observed shifts as the actual distance between wireless nodes in line-of-sight (LOS) increases following a linear shape. The coefficient of determination is used to measure how much of the original uncertainty in the RTT measurements is explained by the linear model.

Unfortunately, the assumption that a direct sight exists between two wireless nodes in an indoor environment is an oversimplification of reality, where the obstacles usually block the direct path. Known as non-line-of-sight (NLOS), several techniques have emerge to overcome this problem. They can be broadly classified in two groups, techniques which attempt to minimize the contribution of NLOS multipaths (Chen, 1999) or techniques which focus on the identification of NLOS reference devices and discard them for localization (Cong & Zhuang, 2005). However, their reliability remains questionable in an indoor environment with abundant scatterers where almost all reference devices will be in NLOS. In this chapter the PNMC (Prior NLOS Measurements Correction) technique is used to correct the NLOS effect from distance estimates (Mazuelas et al., 2008). This technique manages to introduce the information that actually resides in the NLOS measurements in the localization process.

The chapter is organized as follows. Section 2 presents a method to quantify the time delay between two wireless nodes and proposes a PCB as a measuring system. Section 3 analyzes the best statistical estimator of the time delay assuming a linear regression model to relate that estimator with the actual distance between two wireless nodes in LOS. Section 4 describes the mitigation of the severe NLOS effect on those distance estimates using the PNMC method. Section 5 evaluates the performance of the distance estimation technique in a rich multipath indoor environment, and Section 6 summarizes the main achievements.

2. Time delay quantification

The TOA-based systems measure distance based on an estimate of signal propagation delay between a transmitter and a receiver since, in free space or air, radio signals travel at the constant speed of light. The TOA can be measured by either measuring the phase of received narrowband carrier signal or directly measuring the arrival time of a wideband narrow pulse. However, the challenge for this chapter is to develop a distance estimation system that is inexpensive to design and deploy, complies with 802.11 regulations, and provides a comprehensive coverage for accurate ranging.

That is why in this chapter the performance of the 802.11 wireless networks for indoor localization is based on the time delay localization metric through RTT measurements. By using RTT the need for time synchronization between wireless nodes is avoided which would entail a major increase in the complexity of the location scheme development. Furthermore, due to the use of an 802.11 infrastructure, the location capabilities will be an added value to the existing connectivity ones.

The main characteristic that makes the RTT measurements possible in a 802.11 wireless network is the common protection mechanism to reserve a shared medium, RTS/CTS handshake (Gast, 2002).

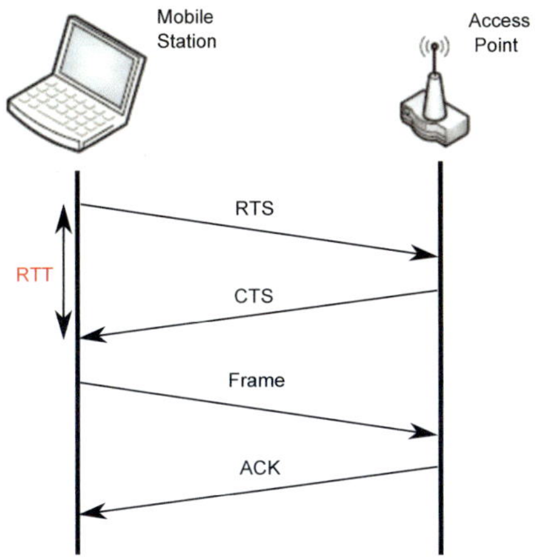

Fig. 1. RTS/CTS handshake.

In wireless communication networks, medium access control (MAC) schemes are used to manage all nodes' access to the shared wireless medium. Due to the randomness of packet arrivals and local competition, it is difficult to completely eliminate packet collisions. Since data packet collisions are costly, researchers proposed to use the RTS/CTS dialogue to reserve the right to channel usage. Assuming a ready node has a frame to send, if it has the RTS/CTS technique activated (see Fig. 1), it initiates the process by sending an RTS frame. The RTS frame serves several purposes; in addition to reserving the radio link for transmission, it silences any station that hear it. If the target station receives an RTS, it responds with a CTS. Like the RTS frame, the CTS frame silences stations in the immediate vicinity. Once RTS/CTS exchange is complete, the mobile station can transmit its frames without worry of interference from any hidden nodes. Hidden nodes beyond the range of the sending station are silenced by the CTS from the receiver. With the use of the RTS/CTS dialogue, it is less likely that data packets will suffer collisions.

Therefore, the RTS/CTS handshake is used to quantify the RTT by measuring the latency of a series of layer two CTS frames sent by and in response to a corresponding series of RTS frames initiated by the MS that is going to be located. The same as acknowledgement (ACK), CTS are considered in the AP the highest priority frames, therefore, the minimum elapsed time in the AP is guarantee when processing these sort of frames.

2.1 Printed circuit board design

In order to quantify the RTT of the RTS/CTS two-frame exchange 802.11 mechanism, appropriate signals from within the WLAN adapter chip set must be selected and accessed by the measuring system. The aim is to extract both transmission pulses and receiver signals in such a way that the RTS frame can be used as the trigger to start the measuring system that would be stopped by the corresponding CTS frame.

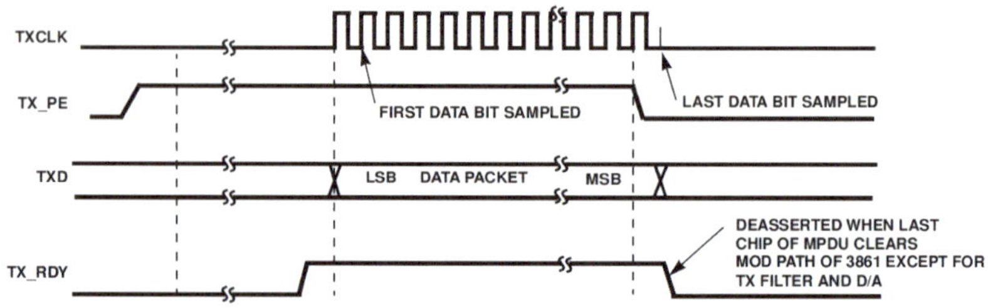

(a) Tx port timing

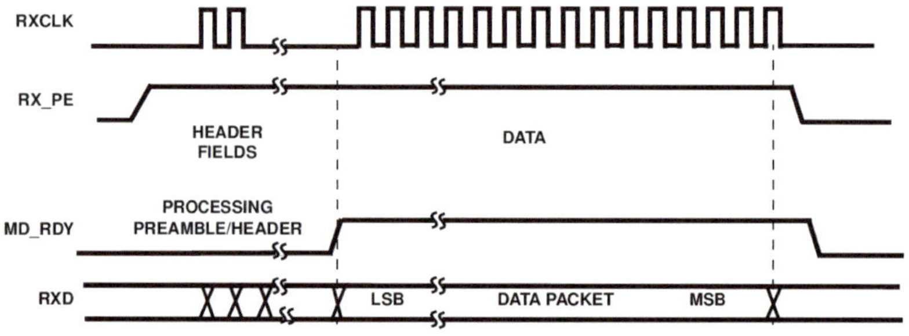

(b) Rx port timing

Fig. 2. Timing behavior of *TX_RDY* and *MD_RDY* signals.

If we access the physical layer of the WLAN adapter, we lose the control of what instant corresponds with which frame sent or received. That is way we access the interface between the physical and MAC layers of the WLAN adapter. This way, it will be easy to associate transmission times and reception times with sent and received frames, respectively. Among the commercial chip sets that work as interface between the physical and MAC layers, the Intersil HFA3861B baseband processor is fully free documented (Intersil, 2002).

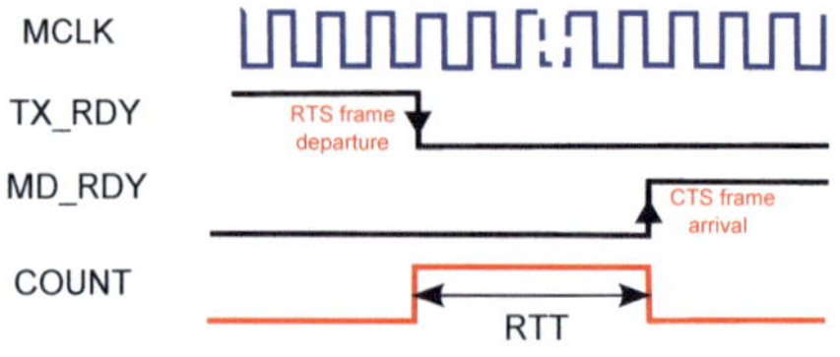

(a) RTT

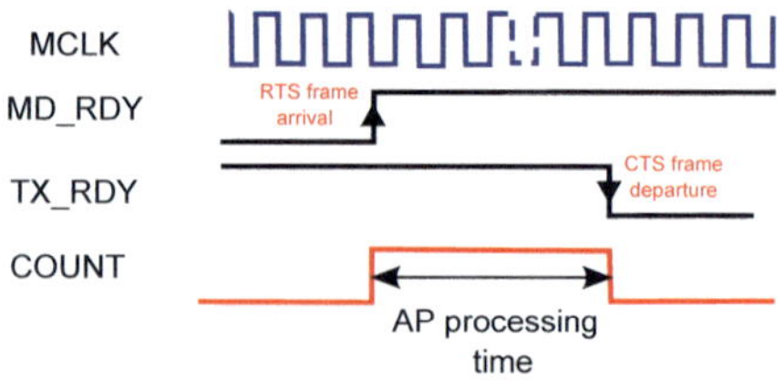

(b) AP processing time

Fig. 3. Timing diagram to mesure the RTT and the processing time of the AP.

From an inspection of the Intersil HFA3861B component pinout diagram, three appropriate leads (and common ground) were identified, *TX_RDY*, *MD_RDY* and *MCLK* (Intersil, 2002).

TX_RDY is an output of the external network processor indicating that preamble and header information has been generated and that the HFA3861B is ready to receive the data packet from the network processor over the *TXD* serial bus (see Fig. 2(a)). *MD_RDY* is an output signal of the network processor, indicating that header data and a data packet are ready to be transferred to the processor (see Fig. 2(b)). *MCLK* is the 44 MHz master clock that governs MAC layer processing. In Fig. 2(a) and 2(b) *TXCLK* and *RXCLK* clocks govern the signals *TX_RDY* and *MD_RDY*, respectively. These signals, *TXCLK* and *RXCLK*, are generated through the master clock *MCLK*. Therefore, as *TX_RDY* and *MD_RDY* signals are synchronized with *TXCLK* and *RXCLK*, respectively, they will be also synchronized with the master clock, *MCLK*.

Thus, in case the RTT is measured, the falling edge of the *TX_RDY* signal is used to start the counter (RTS frame departure) and the rising edge of the *MD_RDY* signal is used to stop it (CTS frame arrival). If the processing time of the AP is wanted to be measured, the rising edge of the *MD_RDY* signal is used to start the counter (RTS frame arrival) and the falling edge of the *TX_RDY* signal is used to stop it (CTS frame departure).

To quantify the RTT a 16-bit counter is used as measuring system and its input is the aforementioned *MCLK* lead, which allows to measure times up to 1.489 ms (2^{16} *MCLK* cycles). The triggers of the counter will be the *TX_RDY* and *MD_RDY* leads. As these triggers are *MCLK* synchronized, the count accuracy will not improve although a higher clock frequency was used.

The counter is integrated in a PCB (see Fig. 4). The PCB is made up of four serial 4-bit counter resulting in a 16-bit counter; one Flip-Flop and one XOR gate to form the RTT signal, managing the *TX_RDY* and *MD_RDY* triggers so that the RTS and CTS frames can be used as triggers to start and stop the measuring system; three 4-bit multiplexers to read the state of the four 4-bit counters; and the corresponding bypass capacitors between power supply and common ground to speed up the PCB commutation times, and to form a low pass filter which will prevent from high frequency disruptions. With the aim of reducing the loop area both for the supply and the signal tracks, top and bottom planes of the PCB were poured of copper, one attached to the power supply lead and the other attached to the common ground. This will minimize the impedance of the return path. In order to control the measuring system, the PCB is governed by the MS through the universal serial bus (USB) port.

The PCB measures the RTT as follows:

1. The MS enables the counters prior to send the RTS frame.
2. The last bit of the sending RTS frame starts the counters.
3. The first bit of the receiving CTS frame stops the counters.
4. Once the RTS/CTS two-frame exchange is completed the MS disables the counters.
5. The MS saves the state of the four 4-bit counters through the multiplexers.

The measuring system proposed has some limitations. First of all, as the *MCLK* that governs the PCB is 44 MHz frequency, the 16-bit counter implemented on the PCB cannot measure RTTs over 1.489 ms, but this time is enough for wireless networks range. Secondly, as a frame coming from other wireless nodes could activate or deactivate the count within the short lapse of time in which the measuring system is enabled, a filter that rejects these undesirable measurements is implemented. Filter limits have been chosen based on previous trials where there were no other wireless nodes interfering. Finally, according to (Bahillo et al., 2009) the elapsed time in the AP, between receiving a RTS frame and sending the corresponding CTS frame, can be assumed to be constant when there are no other processes competing for the AP resources. Obviously, although the CTS frame has the highest priority (Gast, 2002), it could be concurrent RTS frames coming from other MS at the same AP increasing the load of the AP. In that case, if there are not enough APs in range to apply the localization algorithm, the wireless localization system delay increases, but the accuracy is not degraded thanks to the previous filter that rejects the RTT measurements that are out of the expected range.

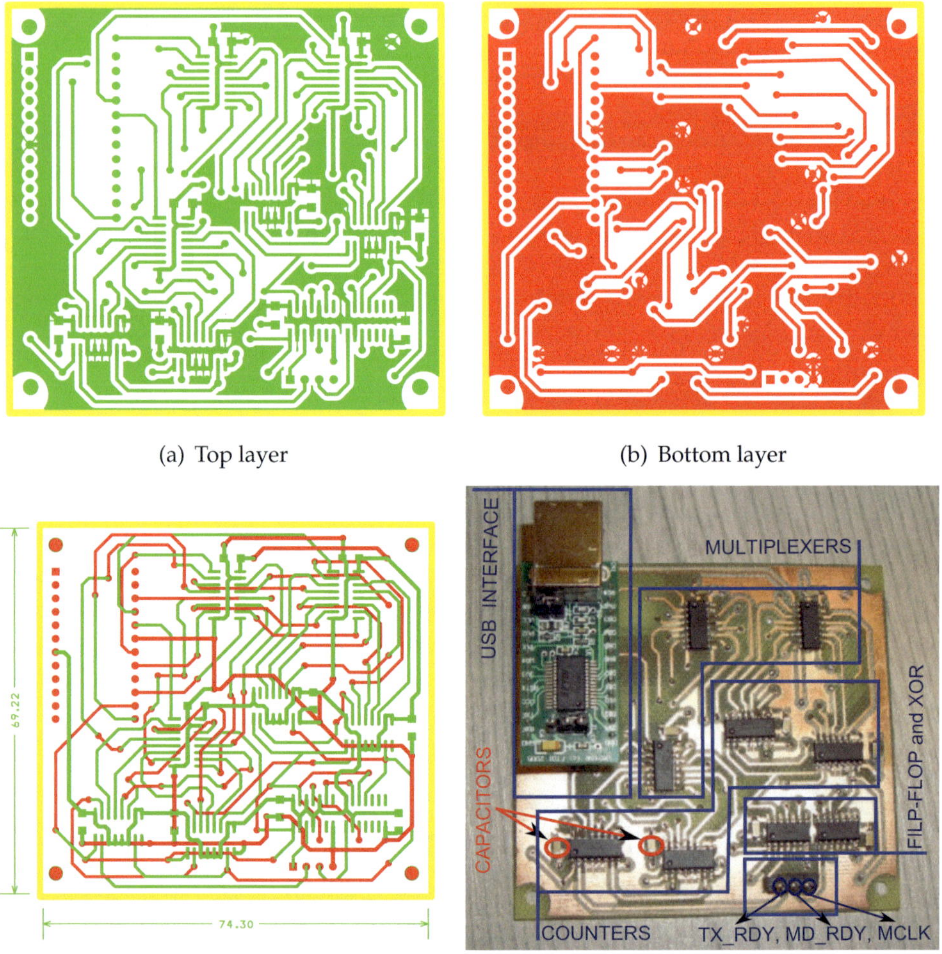

(a) Top layer (b) Bottom layer

(c) Layer connections (d) Printed circuit board

Fig. 4. Printed Circuit Board, 69×74 [mm^2] size, used to measure the RTS/CTS two-frame exchange.

Regardless of the PCB size, the core of the measuring system is the counter, because the other components have to control the measuring system. If the measuring system would be integrated in the WLAN adapter, only the counter component and the driver to control it would be needed.

2.2 Experimental validation

After using the circuit simulation software *Pspice* to check that the PCB design works as expected, it is necessary to check that the PCB connected to the WLAN adapter and managed by the MS is able to measure the time delay when interchanging RTS/CTS frames.
The wireless devices involved in the experimental validation can be found in most wireless networks. They are:

1. A MS, an 802.11b wireless cardbus adapter, specifically a Cisco Aironet AIR-PCM340 with the HFA3861B baseband processor. The wireless adapter has been connected to the computer through a cardbus extender to be able to access to the HFA3861B pinout. This wireless adapter includes two on-board patch antennas with a diversity switch which toggles to and from, and stops when a significant amount of radio frequency power is detected.
2. An AP, a Linksys WRT54GL 802.11b/g. This AP includes two rubber duck omnidirectional antennas in diversity mode that never work at the same time, since diversity circuitry switches to the one with better reception. Rubber duck antennas provide vertical polarization with 360 degrees of coverage in the horizontal plane and 75 degrees in the vertical one. The AP was configured to send a beacon frame each 100 ms at constant power on 802.11 frequency channel 1 (2.412 GHz).

Using these wireless devices, several measurement campaigns were carried out in two different scenarios: an esplanade with a few streetlamps and trees, in the following *exterior*; and the corridor of a building $50 \times 4.3 \times 3.5$ m^3 (length, width and hight) size with wooden and metal doors and a few people walking around, in the following *corridor*. In all scenarios, the two WLAN devices which were involved in the scheme of the experimental setup were always in line-of-sight on a cardboard box 1.5 m high each, in order to guarantee the first Fresnel zone clearance.

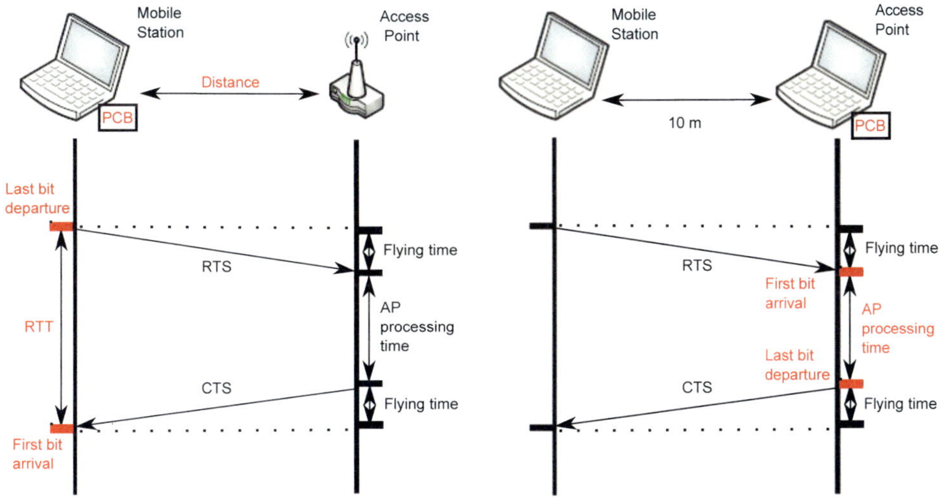

(a) RTT (b) AP processing time

Fig. 5. Experimental setup.

As shown in Fig. 5(a), in case the RTT is measured, the last bit of the RTS frame departure is used to start the counter and the first bit of the CTS frame arrival is used to stop it. If the processing time of the AP is wanted to be measured (see Fig. 5(b)), the first bit of the RTS frame arrival is used to start the counter and the last bit of the CTS frame departure is used to stop it.

2.2.1 RTT measurements

RTT measurements were performed using one Cisco WLAN adapter acting as a MS and one Linksys WRT54GL acting as an AP. The PCB was connected to the WLAN adapter of the MS. Three campaigns of 5000 RTT measurements were conducted at each position for several distances from 0 to 40 m.

Fig. 6 shows the RTT measurements obtained in terms of the number of *MCLK* cycles elapsed at each distance and environment, *exterior* and *corridor*. Each mark in Fig. 6 represents the mean of each group of 150 RTT measurements. The result is qualitatively consistent because, as it was expected, the delay profile observed shifts to the right as the actual distance between the two WLAN nodes increases, and it follows a linear form.

As Fig. 6 shows, although a direct path exists between the MS and the AP in all scenarios and distances, the delay profile observed is spread around 4 *MCLK* cycles. Besides the random behavior of the electronics, there are two main reasons: first of all, the frequency clock that governs the MS and the AP is 44 MHz and 20 MHz, respectively. Secondly, because of the multipath and the scatters of the environment, the direct path is not always the one selected by WLAN adapters. Therefore, to estimate the distance between the MS and the AP, a statistical estimator of the delay profile observed has to be selected. It will be discussed in the following section.

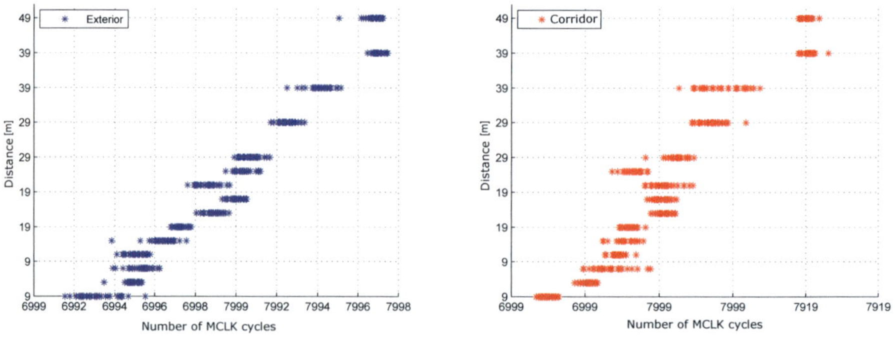

(a) Exterior (b) Corridor

Fig. 6. RTT measurements between two WLAN nodes in LOS at different distances in two different scenarios, *exterior* and *corridor*.

2.2.2 AP processing time measurements

The AP processing time is measured in the two scenarios described above (*exterior* and *corridor*) in order to check that the AP processing time is constant when the RTS/CTS two frame exchange is performed. As the WLAN MAC chip set of the Linksys WRT54GL 802.11b/g is not for public access, in this section the Cisco AIR-PCM340 WLAN adapter is used in AP mode. Therefore, a testing for this AP processing time approach was performed

using two Cisco WLAN adapters, one acting as a MS and the other acting as an AP. The PCB was in the AP. By using a driver designed by ourselves, based on the LORCON library, the MS sends a RTS frame to the AP identified through the MAC address and it waits for the corresponding CTS frame response. When the AP processing time is measured, the measuring system was not disabled between two RTS/CTS frames exchange because the AP does not know, a priori, the time the RTS frames arrival because they are not synchronized. As other frames coming from other WLANs could start or finish the count, the AP processing time measurements have to be filtered a posteriori, because the triggers that starts and finishes the count could not match with our RTS/CTS frames exchange.
As the AP processing time is independent of distance, the measurements were conducted for a distance of 10 m between the two WLAN adapters in the two scenarios, *exterior* and *corridor*.

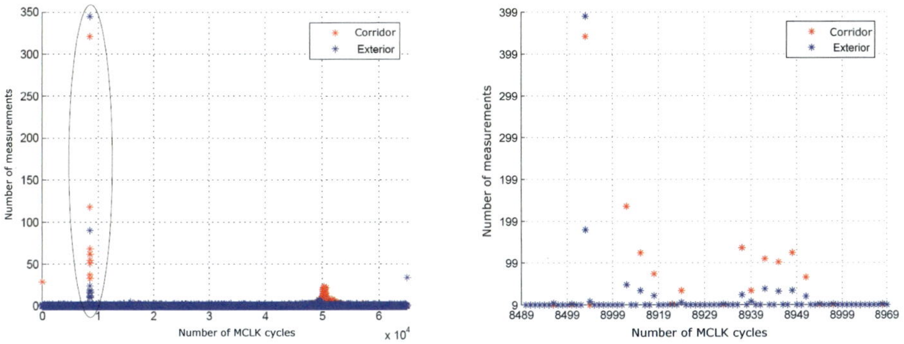

(a) Zoom out (b) Zoom in

Fig. 7. AP processing time measurements in two different scenarios, *exterior* and *corridor*.

As Fig. 7 shows, there are mainly four different behaviors about AP processing time measurements, which are around 0, 8500, 50000 and 65000 *MCLK* cycles. The extreme values, 0 and 65535 *MCLK* cycles, mean that the measuring system has not started the count before reading the counters state and the count has overflow, respectively. Measurements around 50000 *MCLK* cycles are not due to the RTS/CTS frame exchange because this behavior does not appear in the *exterior* environment, where there were not signals coming from other 802.11 devices. Measurements around 8500 *MCLK* cycles do not appear when the MS does not send the RTS frames. Therefore, these measurements, around 8500 *MCLK* cycles, are due to the AP processing time. When a RTS/CTS frame exchange is performed, it is assumed that the AP processing time is roughly constant because more than 50% of measurements, which are around 8500 *MCLK* cycles, were exactly 8494 *MCLK* cycles, although this actual value of the AP processing time is not needed to apply the ranging method we propose in next section.

3. Distance estimation in LOS

According to (Chen & Ling, 2002), the range resolution is determined by the bandwidth of the transmitted signal when RTT measurements are used. High-precision location would require large transmission bandwidths or the use of multiple frequency channels. Furthermore, when using a 44 MHz clock as input of the measuring system to quantify the RTT measurements, the maximum resolution achievable, if only one sample is taken, is hampered by that frequency clock. Moreover, even in a LOS environment the RTT measurements have a random behavior due to the error introduced by the standard noise from electronic devices, that is always present. Therefore, to overcome these limitations several RTT measurements have to be performed at each distance and a representative value, called the location estimator, from this group of RTT measurements has to be selected as the distance estimation. The selection of the location estimator is based on the model that relates the location estimator to the distance that separates the MS and the AP.

The location of a random variable distribution can usually be presented by a single number, the location estimator. In (Olive, 2008), several location estimators of a random variable are analyzed. The mean, median, mode and the scale parameter of the Weibull distribution (scale-W) are examples of the location of a random variable. In this chapter, they have been analyzed and compared as location estimators of the RTT measurements in terms of the coefficient of determination, r^2. This coefficient measures how much of the original uncertainty in the RTT measurements is explained by the model (Weisberg, 2005). In this chapter, a simple linear regression function is assumed to be the model that relates the actual distance between the two wireless nodes involved in RTT measurements to the location estimators in LOS.

Analytically,

$$\widehat{d}_{RTT}^{LOS} = \beta_0 + \widehat{RTT}\beta_1 = d + \epsilon_{LOS}, \tag{1}$$

where, $\widehat{d}_{RTT}^{LOS}$ and d are the estimated and the actual distance between the MS and the AP in LOS, respectively, $\widehat{RTT}$ is the location estimator of RTT measurements, β_0 and β_1 are the intercept and slope of the simple linear regression function, respectively, and ϵ_{LOS} is the error introduced by $\widehat{RTT}$. The error term ϵ_{LOS} has been modeled as a zero-mean Gaussian random variable, because the estimators used are asymptotically Gaussian and a large amount of measurements have been used, so

$$\epsilon_{LOS} \rightsquigarrow N(0, \sigma_{LOS}). \tag{2}$$

In this case, as the expression (1) is a simple linear regression function, r^2 is simply the square of the correlation coefficient, $r^2_{\widehat{RTT},d}$.

The parameters β_0 and β_1 that characterize the simple linear regression function do not depend on the environment where the wireless localization system is going to be deployed, but on the communication system used, i.e. the MS and the AP. These parameters are computed so as to give a best fit of the location estimators to the actual distance. Most commonly, the best fit is evaluated by using the least squares method, but this method is actually not robust in the sense of outlier-resistance. Hence, robust regression has been performed as it is a form of regression analysis designed to circumvent some limitations of least squares estimates for regression models (Olive, 2008).

Assuming LOS between the MS and the AP without any scatter nearby and guaranteeing the first Fresnel zone clearance of the link between both nodes, three campaigns of RTT measurements were conducted for several distances from 0 to 40 m. Fig.8 shows the robust linear regression function which best fits each location estimator to be analyzed. Each location estimator has been computed from each group of 150 RTT measurements at each distance. The different location estimators are analyzed and compared in terms of the coefficient of determination value, r^2.

The mode ($\widehat{RTT}_{md}$) is the value that is most likely to be sampled, thereby it could be a good candidate for the location estimator, but the value that occurs the most frequently in a data

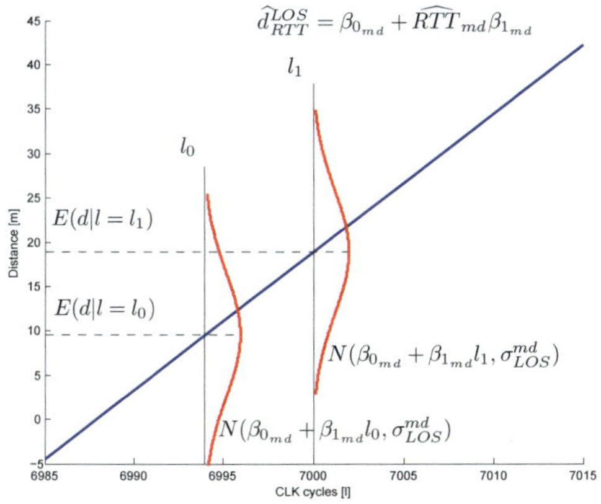

(a) Mode (md)

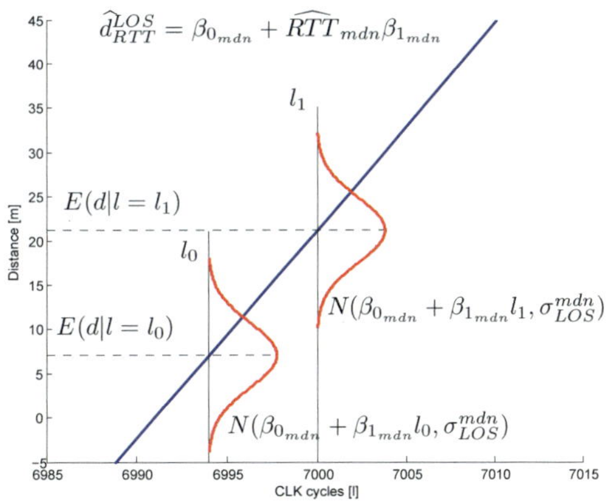

(b) Median (mdn)

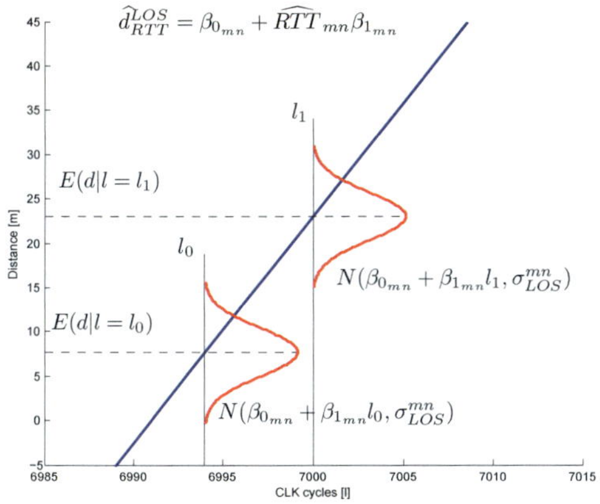

(c) Mean (mn)

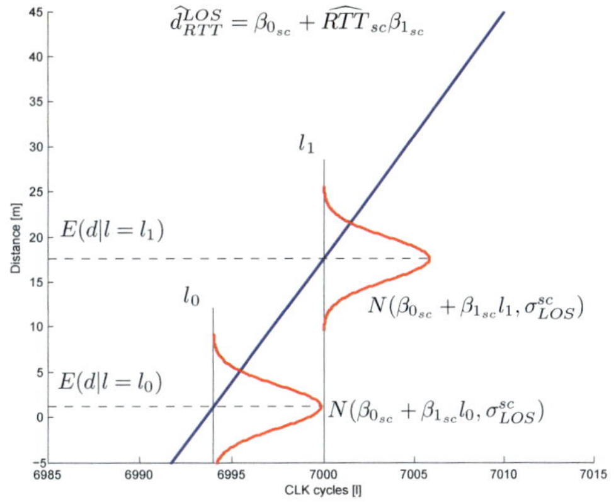

(d) Scale-W (sc)

Fig. 8. Robust linear regression function that best fits each location estimator to be analyzed: the mean, median, mode and scale-W parameter. Where each location estimator is computed from groups of 150 RTT measurements at each distance.

set is a discrete value. Therefore, the resolution achieved, T_{MCLK}, is not enough for indoor localization systems. The same resolution is achieved with the median ($\widehat{RTT}_{mdn}$) as it is a discrete value separating the higher half of a data set. Fig.8 (a) (b) show that the Gaussian distributions that characterize the errors ϵ_{LOS} of the mode and the median are the widest, $\sigma^{md}_{LOS} = 7$ m being $r^2_{md} = 0.64$ and $\sigma^{mdn}_{LOS} = 3.6$ m being $r^2_{mdn} = 0.9$.

The mean ($\widehat{RTT}_{mn}$) is equivalent to the center of gravity of the distribution and it does not take discrete values, thereby the resolution is improved. Although the mean is rather sensitive in the presence of outliers, the use of a robust regression function circumvents this limitation. Fig.8 (c) shows the errors committed when using the mean as location estimator are characterized by a Gaussian, $\sigma^{mn}_{LOS} = 2.7$ m, being $r^2_{mn} = 0.94$, lower than the error commit when the median. However, Fig.8 (d) shows that the best location estimator is the scale-W parameter ($\widehat{RTT}_{sc}$) once Weibull distribution is fitted to the RTT measurements. In this case ϵ_{LOS} is characterized by $\sigma^{sc}_{LOS} = 2.3$ m and $r^2_{sc} = 0.96$. Therefore, the assumption of a linear function as the model that relates RTT measurements to distance is corroborated by a correlation coefficient value close to the unit. This value indicates that the regression line nearly fits the $\widehat{RTT}_{sc}$ perfectly.

There is no phenomenological explanation for choosing the scale-W parameter as location estimator of the RTT measurements set, however this parameter is another kind of a location estimator since the maximum likelihood estimator (MLE) of the scale-W parameter is the Hölder mean (Borwein & Borwein, 1986), a generalized form of the Pythagorean means, taking as parameter the shape parameter of Weibull distribution (for more detail see appendix).

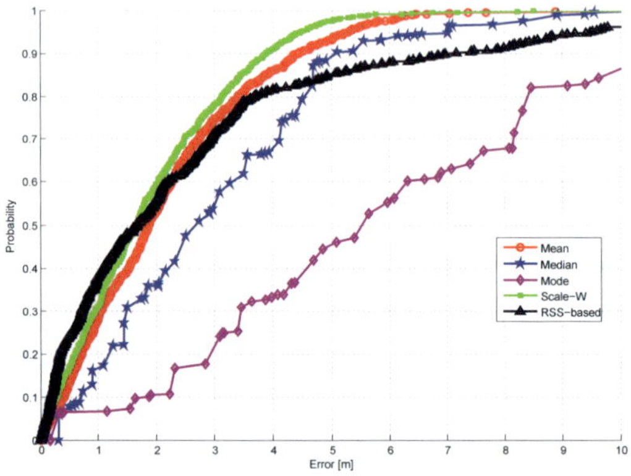

Fig. 9. CDFs of distance errors performed with four different location estimators and a RSS-based method.

Once scale-W parameter is found as the statistical estimator of the RTT measurements that best fits the actual distance when using a simple robust linear regression function as the model that relates the estimator to the distance, its performance is compared to a RSS-based solution to evaluate the goodness of the proposed one. The same two wireless nodes have been used in the same LOS environment. As it is well known the distance between two wireless devices causes an attenuation in the RSS values. This attenuation is known as path-loss and it is modeled to be inversely proportional to the distance between both devices raised to a certain exponent. According to (Mazuelas et al., 2009), the distance between two wireless nodes can be estimated from RSS measurements by

$$\widehat{d}_{RSS} = 10^{\frac{P_{ref}-\overline{P}}{10\alpha}} \tag{3}$$

where $\widehat{d}_{RSS}$ is the estimated distance between the MS and the AP, P_{ref} is the RSS measured in logarithmic units at the reference distance of 1 m, $\overline{P}$ is the average RSS in logarithmic units at the actual distance, and α is the path-loss exponent. According to (IEEE Standard, 2007), for any distance under 20 m in LOS, α is recommended to be 2 while $\alpha = 3.5$ for longer distances. Therefore, having taken this value for the path-loss exponent and from the RSS values measured between both devices, the distance between the two wireless nodes can be estimated by using the expression (3).

Fig.9 shows the cumulative distribution function (CDF) of distance errors. As the mode and the median take discrete values, the CDF has a step-shape with large errors. The mean has a good behavior with an error lower than 2 m on average. However, the scale-W parameter with an error lower than 3 m for a cumulative probability of 80% achieves the best behavior. Also in Fig.9 it can be appreciated that the scale-W parameter outperforms the RSS range based method, specially for cumulative probabilities larger than 50%.

4. Distance estimation in NLOS

The assumption that LOS propagation conditions are present in an indoor environment is an oversimplification of reality. In such environments the transmitted signal could only reach the receiver through reflected, transmitted, diffracted, or scattered paths. Hence, these paths could positively bias the actual distance caused mainly by the blocking of the direct path or due to experiencing a lower propagation speed through obstacles (Allen et al., 2007).

Known as the NLOS problem, this positive bias has been deeply considered through the literature with the aim of mitigating its effect on distance estimates (Mazuelas et al., 2008; Tang et al., 2008; Wylie & Holtzman, 1996), however, in all of them the NLOS is mainly discussed within the cellular networks. Note that such techniques usually assume that the bias for the NLOS range measurements changes over time and has larger variances than LOS range measurements (Güvenç et al., 2008), assumptions that could not be assured in an indoor environment (Yarkoni & Blaunstein, 2006).

In an indoor environment, the easiest method for dealing with NLOS conditions is simply to place APs at additional locations and select those from LOS, however one of the objectives of this chapter is to deploy a wireless localization system in a common and unmodified wireless network. Therefore, in this chapter, the feasibility of the PNMC method presented in (Mazuelas et al., 2008) is analyzed in an indoor environment, taking the PCB proposed in previous section as measuring system, the scale-W as statistical estimator of the RTT and the simple linear regression as the model to relate the scale-W to the actual distance.

The PNMC method relies on the statistical distribution of NLOS errors and on the major variance that NLOS errors present with respect to LOS. The distribution type of NLOS errors depends on the particular environment. Hence, it can follow different statistical distributions such as Gaussian, Exponential, Gamma, etc. (Mazuelas et al., 2008). Regarding the distribution, its parameters can be assumed to be constant in that particular environment. Moreover, those parameters can be obtained before the process of getting distance estimates (Cong & Zhuang, 2005) or directly from the estimated delay spread at that moment (Urrela et al., 2006). In this chapter, those parameters have been obtained beforehand by a campaign of RTT measurements in NLOS.

Let d be the actual distance between the MS and the AP, thus

$$\widehat{d_{RTT}^{NLOS}} = d + \epsilon, \tag{4}$$

where, $\widehat{d_{RTT}^{NLOS}}$ and d are the estimated and the actual distance between the MS and the AP, respectively. The term ϵ denotes the error in the estimation of the distance. This error is the sum of two independent errors, $\epsilon = \epsilon_{LOS} + \epsilon_{NLOS}$, where ϵ_{LOS} describes the noise form electronic errors, while ϵ_{NLOS} is the error due to the lack of direct sight between the MS and the AP. On the one hand, the term ϵ_{LOS} has been evaluated in the previous section and it was found as a zero-mean Gaussian with $\sigma_{LOS} = 2.3$ m. On the other hand, the term ϵ_{NLOS} can follow different statistical distributions. However, regarding the distribution of ϵ_{NLOS}, it can be characterized by its mean and standard deviation. These parameters, as well as the distribution type of ϵ_{NLOS}, depends on the particular environment, but it can be assumed that the NLOS propagation conditions do not change significantly in the time window that contains the record of range measurements, so the mean and standard deviation of ϵ_{NLOS} can be assumed to be constant. Moreover, the parameters that characterize ϵ_{NLOS} can be obtained previously to the localization process by the estimates performed in the environment where the localization system is going to be deployed. For simplicity, the Exponential distribution has been chosen for the term ϵ_{NLOS}. Therefore,

$$\epsilon_{NLOS} \rightsquigarrow Exponential(\lambda) \tag{5}$$

where the λ parameter is fixed previously to the localization process.

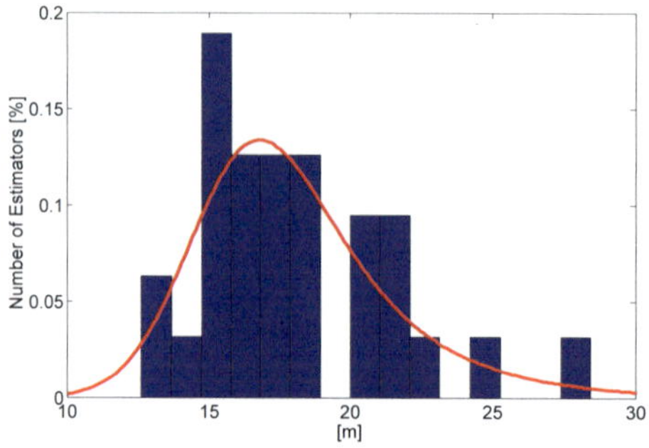

(a) Histogram

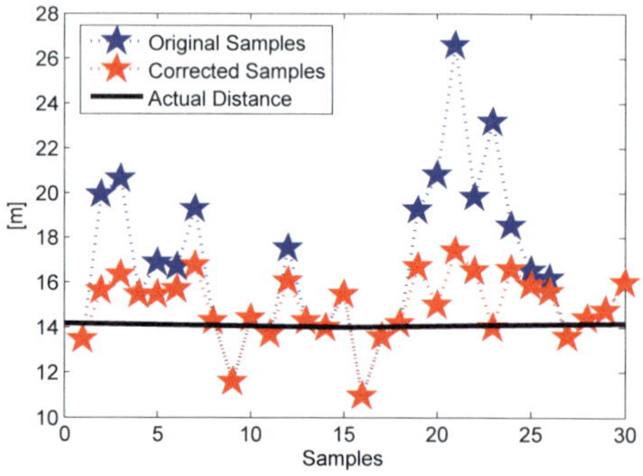

(b) Estimators

Fig. 10. NLOS error correction from a record of distance estimates. (a) Histogram of distance estimates and the probability density function that best fits the data. (b) Record of original and corrected distance estimates after having applied the PNMC method.

In order to show the feasibility of the PNMC technique in an indoor environment, a campaign of measurements in the second floor of the Higher Technical School of Telecommunications Engineering (ETSIT) at the University of Valladolid has been carried out. Specifically, the PNMC technique is applied to the range measurements computed between an AP and a MS 14 m away who is moving 5 m straight perpendicularly to the path that joins the AP and the MS. As $\epsilon = \epsilon_{LOS} + \epsilon_{NLOS}$, the probability density function (PDF) of the term ϵ is the convolution of the Gaussian PDF caused by the ϵ_{LOS} errors and the Exponential PDF caused by the ϵ_{NLOS} errors. Fig.10 (a) shows the histogram of the distance estimates record and the PDF of the term ϵ that best fits these estimates, where the value of the parameter λ that best fits the data is $\lambda = 0.3$ m^{-1}. Once the term ϵ is statistically characterized, the PNMC technique can be applied. Fig.10 (b) shows the result of applying the PNMC technique to the original range measurements computed in a time window equivalent to 5 m walking. In this scenario, the ratio of ϵ_{NLOS} errors from the record of range measurements has been 52%. Subsequently, these NLOS range measurements have been corrected by subtracting the expected NLOS errors for each segment according to the Exponential distribution.

5. Experimental validation

The second floor of the ETSIT as a real indoor environment with several offices, rooms and many people walking around has been the selected scenario to test the wireless localization system's accuracy. The 802.11 wireless network deployed in that building has been used as the one over which the MS communicates with their APs whose positions are previously known. Fig. 11 shows the layout of the south-wets of the second floor of the ETSIT building where positioning tests have been carried out. The route followed by the MS describes a 40×11 m^2 rectangle walking through the middle of the corridors where each pair of continuous positions is separated 0.75 m approximately. The corridors involved in the route are 2 m wide except the widest one that has a width of 4.3 m. As a consequence of the heterogeneous distribution of rooms and offices, and the people walking around, multiple reflection, diffraction or scatter points could appear and alter the signal path. Presumably, although NLOS is always present, multipath will be more noticeable when moving along the narrowest corridors.

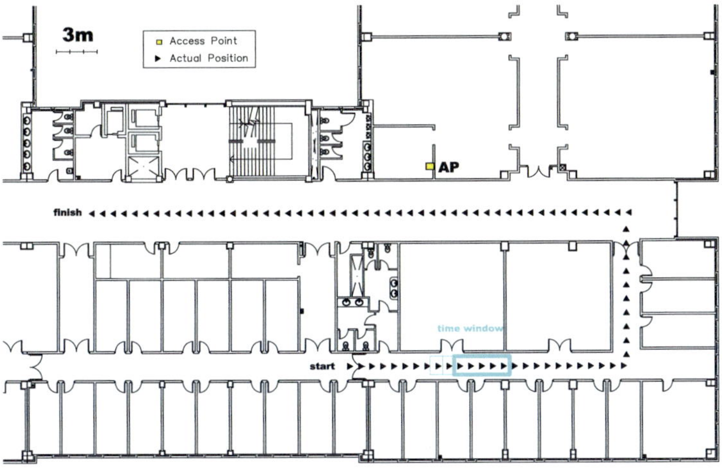

Fig. 11. Indoor environment where RTT measurements have been carried out.

RTT measurements have been performed between an AP fixed in a laboratory and a MS which was moving on the route shown. Any of the positions on the route has a direct sight to the AP, situation that could possibly happen in any indoor environment with high probability. Therefore, the route followed shows different degrees of NLOS instead of LOS and NLOS combinations.

The error introduced by the term NLOS is corrected by using the PNMC technique, where NLOS is observed to be exponentially distributed with λ =0.3 m^{-1} and a time window equivalent to 5 m walking is used. As location estimator, the scale-W parameter has been implemented to reduce the error produced by the term LOS.

Fig. 12(a) shows the actual distance from the MS to the AP at each position through the route shown in Fig. 11. The distance estimate, in red, at each position is shown by using the scale-W as statistical estimator of the RTT, having applied the linear regression model. As it is observed in Fig. 12(a), due to the fact that the positions where the MS is going to be located do not have a direct sight to the AP, the distance estimates are almost always higher than the actual one. Therefore, PNMC method is going to correct distance estimates with severe NLOS. In blue, the distance estimates having applied the PNMC method on the computed distance estimates are shown. They are more similar to those that would be obtained in the absence of severe NLOS propagation.

It can be concluded that although the difference between σ_{LOS} and σ_{NLOS} is not so great, the presence of severe NLOS in the record is detected and corrected. As it is shown in

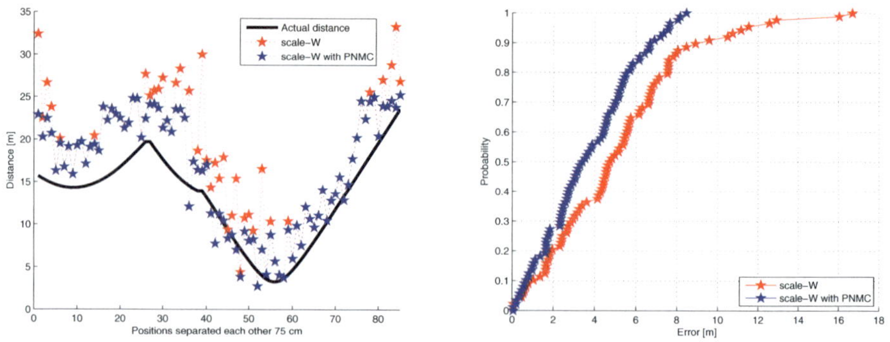

(a) Correction of severe NLOS measurements with PNMC

(b) CDF

Fig. 12. NLOS error mitigation from a record of distance estimates using a window size of 5 m walking. (a)scale-W distance estimates before and after applying PNMC method. (b) Comparison of CDFs errors in distance estimate before and after applying PNMC method.

Fig. 12(b) the improvement of applying the PNMC method can be observed through the CDF of errors in distance estimates. Generally speaking, the distance estimate can be improved on approximately 2 m for cumulative probabilities higher than 30% when applying the PNMC method.

6. Conclusions

The achievable positioning accuracy of traditional wireless localization systems is limited when harsh radio propagation conditions like rich multipath indoor environments are present. In this chapter a novel RTT-based ranging method is proposed over a PCB that performs RTT measurements. The effect of hardware errors has been minimized by choosing the scale-W parameter as RTT estimator. A coefficient of determination value of 0.96 achieved with this estimator in LOS justified the simple linear regression function as the model that relates distance estimates to RTT measurements in LOS. As LOS is not guaranteed in an indoor environment, the accuracy of the proposed localization algorithm has been tested in a rich multipath environment without any NLOS error mitigation technique achieving an error lower than 4 m on average. However, this error is improved after having implemented the PNMC technique to correct NLOS errors. Once reliable RTT-based ranging estimates are obtained, simple geometrical triangulation methods can be used to find the location of the MS (Pahlavan & Krishnamurthy, 2002).

Indoor localization schemes have experienced a flurry of research in recent years. However, there still remain multiple areas of open research that will help systems to meet the requirements of applications that have to operate in indoor propagation environments where GNSS typically fails. These are: i) Interference mitigation: To date, the majority of research effort ignores the effects of interference on time estimation accuracy, and few papers propose robust interference mitigation techniques. ii) Inertial Measurements Units (IMU): the integration of traditional localization metrics, such as TOA, RSS or AOA with IMU information, such as the one reported by accelerometers, gyroscopes and magnetometers, could provide location estimations more precisely and continuously, since IMU-based localization is a beacon-free methodology. iii) Secure ranging: In certain scenarios the localization process may be subject to hostile attacks. While some works have presented secure localization algorithms (see, e.g., (Li et al., 2005; Zhang et al., 2006)), less attention has been paid to secure ranging.

7. Acknowledgment

This research is partially supported by the General Board of Telecommunications of the Council of Public Works from Castilla-León (Spain) and by the spanish national project LEMUR (TIN2009-14114-C04-03).

8. Appendix

8.1 Maximum likelihood estimator of the scale parameter of the Weibull distribution

The scale-W parameter is estimated by using the MLE method and assuming that the shape parameter is known.

The probability density function of a Weibull (two-parameter) random variable x is

$$\begin{aligned} f(x;k,\lambda) &= \frac{k}{\lambda}\left(\frac{x}{\lambda}\right)^{k-1} \cdot e^{-\left(\frac{x}{\lambda}\right)^k} \qquad x \geq 0 \\ &= \frac{k}{\lambda^k} \cdot x^{k-1} \cdot e^{-\left(\frac{x}{\lambda}\right)^k} \qquad x \geq 0 \end{aligned}$$

where $k > 0$ is the shape parameter and $\lambda > 0$ is the scale-W parameter.

Let $X_1, X_2, ...X_n$ be a random sample of random variables with two-parameter Weibull distribution, k and λ. The likelihood function is

$$L(x_1, ...x_n; k, \lambda) = \prod_{i=1}^{n} f(x_i; k, \lambda)$$

Therefore,

$$\begin{aligned} lnL(x_1, ..., x_n; k, \lambda) &= \sum_{i=1}^{n} lnf(x_1, ...x_n; k, \lambda) \\ &= \sum_{i=1}^{n} \left(ln\left(\frac{k}{\lambda}\right) + (k-1) \cdot ln\left(\frac{x_i}{\lambda}\right) - \left(\frac{x_i}{\lambda}\right)^k \right) \\ &= n \cdot ln\left(\frac{k}{\lambda}\right) + (k-1) \cdot \sum_{i=1}^{n} ln\left(\frac{x_i}{\lambda}\right) - \sum_{i=1}^{n}\left(\frac{x_i}{\lambda}\right)^k \\ &= n \cdot (ln(k) - ln(\lambda)) + (k-1) \cdot \left[-n \cdot ln(\lambda) + \sum_{i=1}^{n} ln(x_i)\right] - \sum_{i=1}^{n}\left(\frac{x_i}{\lambda}\right)^k \\ &= n \cdot ln(k) + (k-1) \cdot \sum_{i=1}^{n} ln(x_i) - n \cdot k \cdot ln(\lambda) - \lambda^{-k} \cdot \sum_{i=1}^{n} x_i^k \end{aligned}$$

thus,

$$\frac{\partial lnL}{\partial \lambda} = -n \cdot k \cdot \frac{1}{\lambda} + k \cdot \frac{1}{\lambda^{k+1}} \cdot \sum_{i=1}^{n} x_i^k$$

in order to find the maximum, $\frac{\partial lnL}{\partial \lambda} = 0$ then,

$$0 = -n \cdot k \cdot \frac{1}{\lambda} + k \cdot \frac{1}{\lambda^{k+1}} \cdot \sum_{i=1}^{n} x_i^k$$
$$= \frac{\sum_{i=1}^{n} x_i^k - n \cdot \lambda^k}{\lambda^{k+1}}$$
$$= \sum_{i=1}^{n} x_i^k - n \cdot \lambda^k$$

hence, the MLE of the scale-W parameter

$$\widehat{\lambda} = \left[\frac{1}{n} \sum_{i=1}^{n} x_i^k \right]^{\frac{1}{k}}$$

this expression is known as the generalized mean or Hölder mean.
The Hölder mean is a generalized mean of the form,

$$M_p(x_1, x_2, ..., x_n) = \left[\frac{1}{n} \sum_{i=1}^{n} x_i^p \right]^{1/p} \quad (6)$$

where the parameter p is an affinely extended real number, n is the number of samples and x_i are the samples with $x_i \geq 0$. The Hölder mean is an abstraction of the Pythagorean means which for example includes minimum ($M_{-\infty}$), harmonic mean (M_{-1}), geometric mean (M_0), arithmetic mean (M_1), quadratic mean (M_2), maximum (M_∞), and the MLE of the scale-W parameter (M_k) where k is the shape parameter of Weibull distribution.

9. References

Bahillo, A., Mazuelas, S., Lorenzo, R.M., Fernández, P., Prieto, J. & Abril, E.J. (2009a). Indoor location based on IEEE 802.11 round-trip time measurements with two-step NLOS mitigation. *Progress In Electromagnetics Research B, PIERB*, Vol.2009, No.15, (September 2009) 285-306, ISSN: 1937-6472.

Bahillo, A., Prieto, J., Mazuelas, S., Lorenzo, R.M., Blas, J. & Fernández, P. (2009b). IEEE 802.11 distance estimation based on RTS/CTS two-frame exchange mechanism, *Proceedings of 69th international conference of Vehicular Technologies*, pp. 1-5, ISBN: 978-1-4244-2517-4, Barcelona, June 2009, IEEE VTC, Spain.

Golden, S.A. & Bateman, S.S. (2007). Sensor measurements for wifi location with emphasis on time-of-arrival ranging. *IEEE Transactions on Mobile Computing*, Vol.6, No.10, (October 2007) 1185-1198, ISSN: 1536-1233.

Gustafsson, F. & Gunnarson, F. (2005). Mobile positioning using wireless networks: possibilities and fundamental limitations based on available wireless network measurements. *IEEE Signal Processing Magazine*, Vol.22, No.4, (July 2005) 41-53, ISSN: 1053-5888.

Mazuelas, S., Bahillo, A., Lorenzo, R. M., Fernández, P., Lago, F.A., García, E., Blas, J. & Abril, E. J. (2009). Robust indoor positioning provided by real-time RSSI values in unmodified WLAN networks. *IEEE Journal of Selected Topics in Signal Processing*, Vol.3, No.5, (October 2009) 821-831, ISSN: 1932-4553.

Mazuelas, S., Lago, F.A., Blas, J., Bahillo, A., Fernández, P., Lorenzo, R. M. & Abril, E. J. (2008). Prior NLOS measurements correction for positioning in cellular networks. *IEEE Transactions on Vehicular Technologies*, Vol.58, No.5, (November 2008) 2585-2591, ISSN: 0018-9545.

Morrison, J.D. (2002). IEEE 802.11 wireless local area network security through location authentication. *M.S. Thesis, Naval Postgraduate School Monterey, California.*

Pahlavan, K. & Krishnamurthy, P. (2002). *Principles of wireless networks - A unified approach*, Prentice-Hall Inc., 2nd edition, ISBN: 0-13-093003-2, Upper Saddle River, New Jersey.

Prieto, J., Bahillo, A., Mazuelas, S., Blas, J., Fernández, P. & Lorenzo, R. M. (2008). RTS/CTS mechanism with IEEE 802.11 for indoor location, *Proceedings of the Navigation Conference & Exhibition: Navigation and Location*, pp. 1-5, London, UK, October 2008, NAV & ILA.

Seow, C.K. & Tan, S.Y. (2008). Localization of omni-directional mobile device in multipath environments. *Progress In Electromagnetics Research, PIER*, Vol.85, No.2008, (2008), 323-348, ISSN: 1070-4698.

Soliman, M.S., Morimoto, T. & Kawasaki, Z.I. (2006). Three-dimensional localization system for impulsive noise sources using ultra-wideband digital interferometer technique. *Journal of Electromagnetic Waves and Applications*, Vol.20, No.4, (2006), 515-530, ISSN: 0920-5071.

Gast, M.S. (2002), *802.11 Wireless networks: the definitive guide*, O'Reilly & Associates, Inc., ISBN: 0-596-00183-5, 1005 Gravenstein Highway North, Sebastopol, CA 95472.

Chen, V.C. & Ling, H. (2002), *Time-frequency transforms for radar imaging and signal analysis*, Artech House, Inc., ISBN: 1-58053-288-8, 685 Canton Street, Norwood, MA, 02062.

Olive, D.J. (2008), *Applied robust statistics*, Southern Illinois University, Department of Mathematics, 4408 Carbondale, IL 62901-4408.

Weisberg, S. (2005), *Applied linear regression, 3rd ed.*, John Wiley & Sons, Inc., ISBN: 0-471-66379-4, Hoboken, New Jersey.

Intersil, (2002), *HFA3861B wireless LAN medium access controller*, Intersil Data Sheet.

Borwein, J.M. & Borwein, P.B. (1986), *Pi and the AGM: a study in analytic number theory and computational complexity*, John Wiley & Sons, Inc., ISBN: 0-471-83138-7, USA.

IEEE Standard for Information Technology (2007) *Telecommunications and Information Exchange Between Systems - Local and Metropolitan Area Networks - Specific Requirements - Part 11: Wireless Medium Access Control (MAC) and Physical Layer (PHY) Specifications, IEEE Std 802.11-2007 (Revision of IEEE Std 802.11-1999).*

Allen, B., Dohler, M., Okon, E., Malik, W., Brown, A. & Edwards, D., (2007), *Ultra Wideband Antennas and Propagation for Communications, Radar, and Imaging*, John Wiley & Sons, Inc., ISBN: 0-470-03255-3, West Sussex, UK.

Tang, H., Park, Y. & Qui, T., (2008). NLOS mitigation for TOA location based on a modified deterministic model. *Research Letters in Signal Processing*, Vol.8, No.1, (April 2008) 1-4, ISSN: 1687-6911.

Wylie, M. P. & Holtzman, J., (1996). The non-line of sight problem in mobile location estimation. *Proceedings of the 5th IEEE International Conference on Universal Personal Communications*, Vol.2, pp. 827-831, ISBN: 0-7803-3300-4, Cambridge, October 1996, Mass, USA.

Güvenç I., Chong, C.-C., Watanabe, F. & Inamura, H., (2008). NLOS identification and weighted least-squares localization for UWB systems using multipath channel statistics, *EURASIP Journal on Advances in Signal Processing*, Vol. 2008, (April 2008) 1-14, ISSN: 1110-8657.

Yarkoni, N. & Blaunstein, N., (2006). Prediction of propagation characteristics in indoor radio communication environment. *Progress In Electromagnetics Research*, PIER 59, (2006) 151-174, ISSN: 1070-4698.

Cong, L. & Zhuang, W., (2005). Non-line-of-sight error mitigation in mobile location. *INFOCOM 2004, 23th Annual Joint Conference of the IEEE Computer and Communications Societies*, Vol. 4, pp. 560-572, ISBN: 0-7803-8355-9, Hong-Kong, March 2004.

Urrela, A., Sala, J. & Riba, J., (2006). Average performance analysis of circular and hyperbolic geolocation. *IEEE Transactions on Vehicular Technology*, Vol. 55, (January 2006) 52-66, ISSN: 0018-9545.

Chen, P.-C., (1999). A non-line-of-sight error mitigation algorithm in location estimation. *Proceedings of Wireless Communications and Networking Conference*, Vol. 1, pp. 316-320, ISBN: 0-7803-5668-3, New Orleans, La, USA, September 1999.

Li, Z., Trappe, W., Zhang, Y. & Nath, B., (2005). Robust statistical methods for securing wireless localization in sensor networks, *Proceedings of IEEE Information Processing Sensor Networks*, pp. 91-98, ISBN: 0-7803-9202-7, Los Angeles, California, USA, April 2005.

Zhang, Y., Liu, W., Lou, W. & Fang, Y., (2006). Location-based compromise-tolerant security mechanisms for wireless sensor networks, *IEEE Journal on Selected Areas in Communications*, Vol. 24, (February 2006) 247-260, ISSN: 0733-8716.

13

Data Forwarding in Wireless Relay Networks

Tzu-Ming Lin, Wen-Tsuen Chen and Shiao-Li Tsao
ITRI, NTHU, &NCTU,
Hsinchu,
Taiwan, R.O.C.

1. Introduction

Broadband wireless communication has brought users a number of multimedia services for years. As wireless broadband markets mature, system operators face new problems. First, system capacity is bounded by finite radio resources. Second, service providers increasing spend on network deployments and services provisions. With limited radio resource and increased costs, operators have no choice but to force users to pay higher rates to preserve the same service quality than they had several years ago.

The expectation is that these problems are solved by introducing relay station (RS) into traditional wireless systems. The relaying technology has proven to be a feasible technology to expand system capacity and reduce deployment costs simultaneously. One of the most popular technologies is cooperative communications that boost network throughput significantly by improving the radio quality for transmission and reception (Nosratinia et al., 2004); (Mohr, 2005); (Doppler et al., 2007). Numerical results (Soldani & Dixit, 2008) showed the cost saving benefits when conventional communication networks adopt relay functionality. From this, a commercial cellular system can save more than 56 % on capital expenditures when RSs are deployed.

RS is a wireless communication station that provides relay services for receiving and forwarding radio signals between two stations. In a cellular system, RS is in charge of receiving, decoding, and forwarding data between the Base Station (BS) and the Subscriber Stations (SSs). During the relaying, RS provides additional features to assist data transmissions. For example, (Tao et al., 2007) proposed a novel data forwarding scheme in this environment. The novel data forwarding scheme improved the transmission efficiency by approximately 66 %.

Although data relay shows superior performances in throughput enhancements and cost savings. Multi-hop relaying introduces new issues that impact overall performance. To indicate a routing path in multi-hop circumstances, additional control information should be introduced. The routing indication design for multiple radio links introduces new overhead and may impact transmission efficiency during relay. Imprecise indications may waste RS computation and storage since it does not know what it shall relay and where to forward. Moreover, there may be a case that data destined to others are received by the RS that is not in charge. Imprecise relay indication will not only increase overheads but also lead RS to use its processor and storage inefficiently. In other words, data forwarding with ambiguous control information would increase RS complexity and buffer storage unnecessarily. During

the relaying, RS shall forward data as simple as possible to prevent wasting processing power and storage. This study proposes a burst-switch concept aiming to tackle the issues and provides a simple and efficient data forwarding for wireless relay networks.
The rest of the chapter is organized as follows. First, wireless relaying and a conventional relay system are overviewed in section 2. The proposed new forwarding mechanism is then elaborated in section 3; section 4 presents the evaluation and simulation results for the mentioned issues. At last, conclusions are given in section 5.

2. Background and related works

Relay technology has been investigated for years, and a realistic relay system will be deployed in few years to enhance legacy wireless networks. This section overviews IEEE 802.16 communication system and introduces the relay enhancements. The data forwarding mechanisms adopted in the system are also discussed to address the issues.

2.1 Overview of wireless relay networks

A wireless relay network consists of a BS, one or more RSs, and numbers of SSs. In the network, directly or through the assistances of RSs. BS forwards the downstream data coming from outside network to SS while RSs relay upstream data generated by SSs to BS. Since all the data transmissions within the network are arranged by BS and there are no communications between RSs, the relay network is usually constructed as a tree topology, which is illustrated in figure 1.

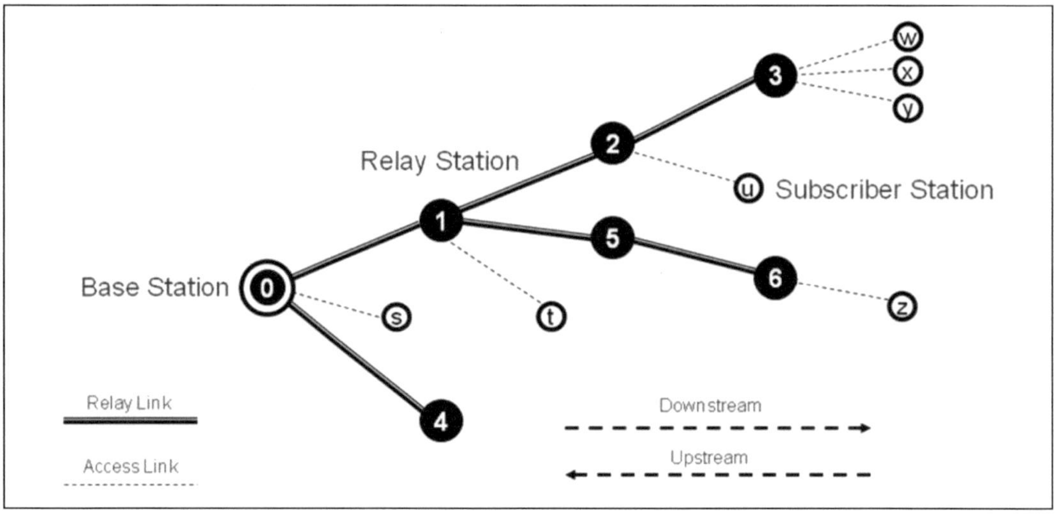

Fig. 1. Wireless Relay Network

There are two types of radio links in this network: relay link and access link. The radio link between a BS and a RS and between two RSs are called relay links, and BS constructs a relay path by multiple relay links. The access link is the radio communication between a SS and its access station, which can be a BS or an access RS. The access RS is a RS attached by a SS and can helps BS for relaying data to the SS. For the example in figure 1, RS_2 is an access RS of SS_u and assists BS_0 to provide relay services for SS_u. BS_0 allocates resources along the relay path between SS_u and itself so that the help on data relay in corresponding relay links.

2.2 IEEE 802.16 and multi-hop relay network

IEEE Std 802.16e™-2005 is one of the most popular wireless broadband networks nowadays, and figure 2 shows the reference model for that system. The system consists of two layers, Medium Access Control (MAC) and Physical (PHY) layers, to handle wireless communications. Packets from TCP/IP layer are translated into MAC Protocol Data Units (MPDUs) and then encoded into a PHY burst. The burst is associated with a MAP Information Element (MAP-IE) that indicates a station for receiving and decoding the burst. After the data process, BS transforms both the burst and the associated MAP-IE into a radio frame and pumps it into wireless medium.

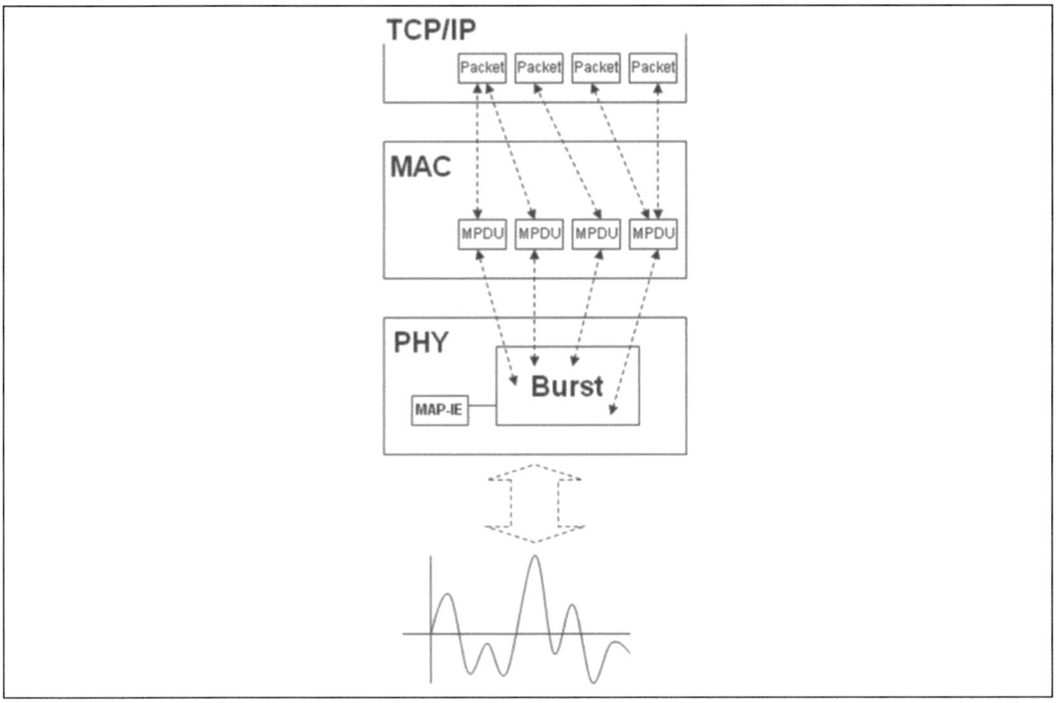

Fig. 2. Data Processing in IEEE 802.16

The overview of 802.16e frame structure is depicted in figure 3. The frame composes two subframes: downlink and uplink subframes, and starts with a synchronization part of preamble and Frame Control Header (FCH). The first part is used for each receiving station synchronize with BS and abstracting the frame. Following the synchronization part, the frame header further includes a downlink MAP (DL-MAP) and an uplink MAP (UL-MAP), which consists of MAP-IEs to indicate the stations where and how to access data bursts. As stated before, each data burst is associated with an MAP-IE, and one or more MPDUs destining to a destination can be concatenated or packed into the burst. With a connection identity (CID) in the MAP-IE, every receiving station locates and receives MPDUs in the desired PHY burst, and has no need to check all the bursts in the frame.

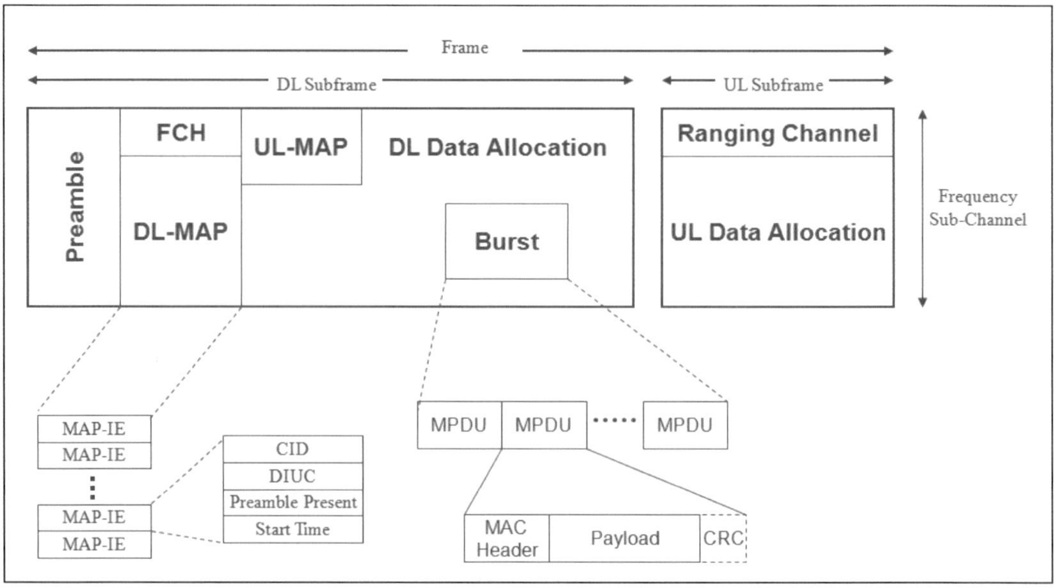

Fig. 3. IEEE 802.16e Frame Structure

The system further specifies an option for disabling MAP to save overheads so that more data can be allocated. In this case, receiving stations should put more efforts to process entire frame since there are no indications in frame header any more. Without MAP indication, the receiving station cannot but store the whole frame to check if there are any desired MPDUs. However, it is inefficient for buffering and checking all the MPDUs in a frame. Although, the operations brought overheads but the problem should not be as serious as that in multi-hop communications. Because of redundant processing and transmissions during relay, multi-hop data forwarding makes the overhead become a severe problem.

2.3 Data forwarding and issues in 802.16j MR network

IEEE working group specifies Multi-hop Relay (MR) support for 802.16e system in IEEE Std 802.16j™-2009. The specification aims to solve the capacity problem and reduce development cost with advanced relay technologies. Efficiency during data relay is also a major concern for implementing RSs. Two data forwarding schemes are specified to facilitate relay functionalities and reduce overheads.

The first forwarding scheme is CID-based transmission, in which RS forwards MPDUs based on the CIDs contained in the MAP-IE or MPDU headers. For saving signalling overheads, relayed MPDUs do not carrier any extra routing information, and are transmitted as in 802.16e conventional system. When MPDUs are relayed, each receiving RS gets CIDs from MAP-IEs and checks associating bursts if there are any data required for further forwarding. The RS discards the burst that is not indicated by the recorded CIDs in its forwarding list. When the CID of the burst is in the forwarding list, the receiving RS forwards the burst to the station in next hop. Besides, there is another implementation that RSs forward MPDUs by identifying CIDs containing in MPDU headers. As mentioned, each RS has to process all the MPDU in receiving frame, and determines the MPDUs for relay. Figure 4 depicts the example for this forwarding scheme based on the relay network in figure 1. RS_1 receives all the packets containing in first frame and stores the MPDUs destining to RS_1, RS_2, RS_3, RS_5, and RS_6. In second frame, RS_2 caches the MPDUs from RS_1 and checks which MPDUs it shall relay. The data for RS_2 and RS_3 are received and only RS_3 data are relayed. After that, RS_3 receives the data by checking MPDU headers and performs relaying for SS_w, SS_y, and SS_y individually. Moreover, the first option for adopting MAP-IE indications can filter unnecessary bursts before looking into every MPDU and RS needs not store all the data in the frame. Otherwise, checking CIDs in MPDU headers would force RSs to receive all the data in a frame. As a result, the using of MAP is a trade-off issue between storage and signalling because enabling MAP prevents RSs to store MPDUs unnecessarily but brings extra control overheads.

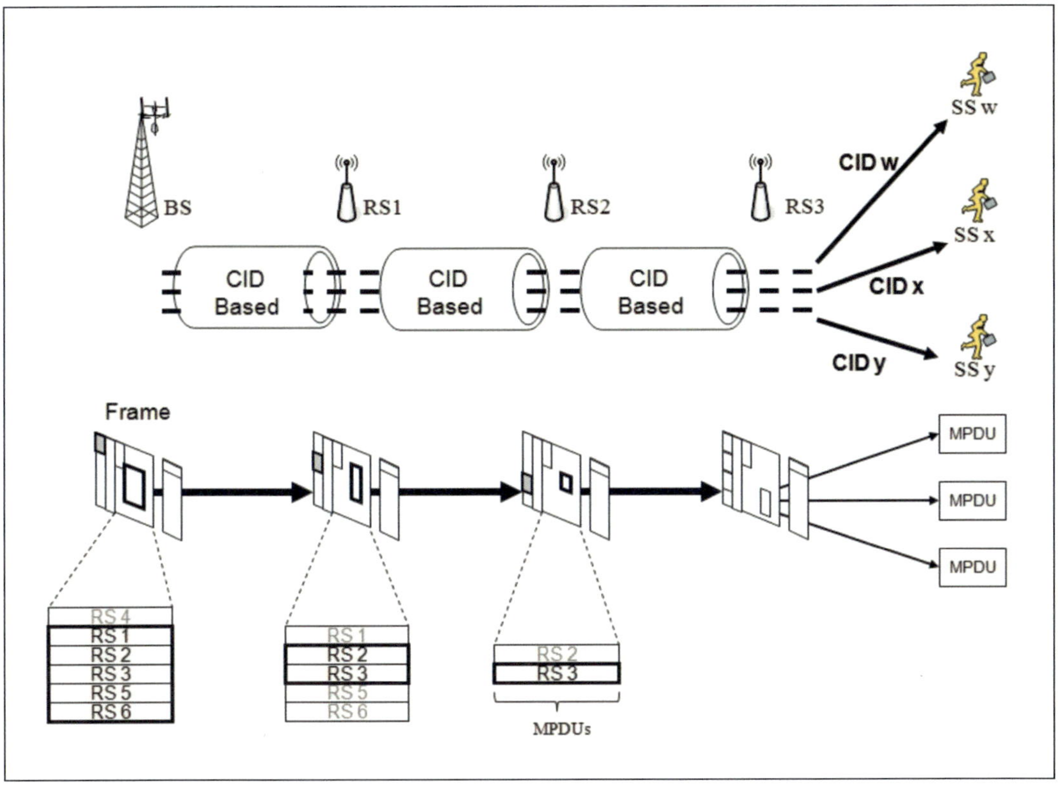

Fig. 4. CID-Based Transmission

The second scheme is tunnel-based transmission, in which BS and access RS encapsulates MPDUs into a Tunnel PDU (T-PDU) and transmits these data through a tunnel in between. Figure 5 shows an example of this scheme. BS, the ingress station of a tunnel, aggregates MPDUs into one or more T-PDUs, and transmits the data to an access RS. The access RS at the egress of tunnel, e.g. RS_3, is responsible for removing tunnel headers and forwarding the decapsulated MPDUs to SSs. Besides, the intermediate RSs along a relay path relay the T-PDU to the tunnel end through the indication of tunnel headers. As CID-based operation, enabling MAP can help RSs to filter unwanted data, and disabling MAP would force RSs to buffer and process all the T-PDUs in the frame. Tunnel header benefits in preventing redundant processes for the group of MPDUs destining to same access RS. However, the extra overhead brought from the tunnel header should be considered. Furthermore, the impact caused by adopting MAP and tunnel headers at the same time shall be also investigated.

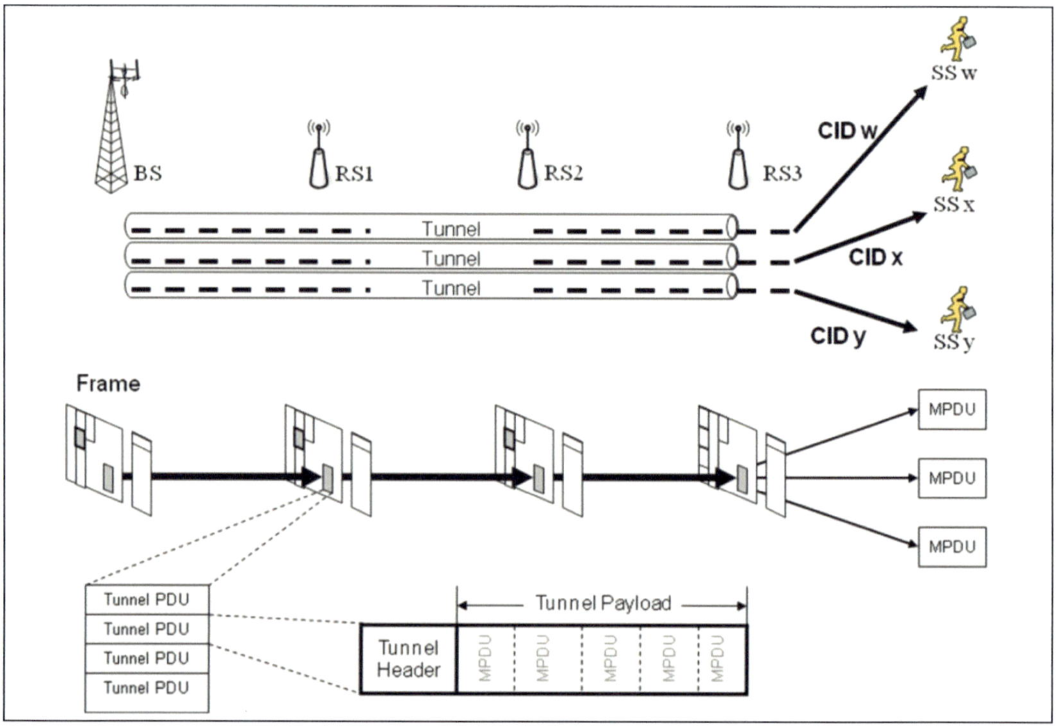

Fig. 5. Tunnel-based Transmission

Comparing these two forwarding schemes, RSs applying tunnels forward data efficiently since RS identifies a MPDU group by a tunnel header in replace of multiple MPDU headers. CID-based scheme provides the forwarding as simple as that in legacy one-hop system, and introduce no extra headers. Although both the two mechanisms can assist data forwarding in multi-hop environments, the overheads caused by excess header processing and unnecessary data buffering shall be discussed in advanced. First, the processing overhead for RSs would rise with the system traffic load. Take the example in figure 1, RS_5 shall process $\boldsymbol{2n}$ tunnel headers in tunnel-based scheme when each RS maintains $\boldsymbol{n}$ tunnels. If $\boldsymbol{m}$ MPDUs are scheduled for each SS in CID-based scheme, $\boldsymbol{4m}$ MPDU headers will be handled by RS_2. In the example, both $\boldsymbol{m}$ and $\boldsymbol{n}$ increase with the traffic load and all the intermediate RSs suffer. The impacts to processing complexity of the two schemes are:

$$C_{Tunnel\_based} = L_{tunnel\_header} \times N_{tunnel} \times N_{RS} \tag{1}$$

$$C_{Station\_based} = L_{MPDU\_header} \times N_{MPDU} \tag{2}$$

where $\boldsymbol{C_{Tunnel\_based}}$ and $\boldsymbol{C_{CID\_based}}$ define the computation in bits, $\boldsymbol{L_{tunnel\_header}}$ and $\boldsymbol{L_{MPDU\_header}}$ denote the lengths of tunnel and normal MPDU headers, $\boldsymbol{N_{tunnel}}$ is the tunnel number that a RS shall handle, N_{RS} is the summation of all the RSs behind a receiving station for a relay path, and N_{MPDU} is the amount of MPDUs that needs to be forwarded. In a stationary relay network, both C_{Tunnel_based} and C_{CID_based} grow with N_{tunnel} and N_{MPDU}, and both the two numbers correspond to the network load. As system load rises, tunnel numbers would increase with provisioned connections. If CID-based forwarding is applied in this case, the amount of processed MPDUs would also grow with the increased connections.

Although tunnel header prevent RSs to process MPDU header redundantly, the issue of inefficient data processing and buffering remains unsolved; every RS needs to store all the T-PDUs or MPDUs in a frame before getting relay information. Some undesired data would still be processed and buffered before being processed. If MAP is applied, RSs could handle fewer data since the bursts for other destinations can be filtered by associating MAP-IEs. However, another issue for excess processing is unsolved because there are two-levels of control information, one for MAP and the other for MPDU or T-PDU headers. No matter which mechanism is applied, the addressed problems cannot be taken over totally. Aiming this, a simple data forwarding mechanism is proposed in this study for relaying data more efficiently.

3. End-to-end burst switch and proposed network model

This study brings out a new concept of switching burst to forward data in the wireless relay network. The burst switch mechanism uses RSs more efficiently by reducing processed control information and buffered data. Besides, a new network model for adopting the concept is also proposed to realize the forwarding process.

3.1 End-to-end burst switch

This literature suggests using an end-to-end burst CID in the associated MAP-IE to forward relayed data, and proposes to relay data with a unique identifier for each relay path in burst level. To reduce carried routing information, data goal to same destination RS are assigned with one burst CID. With the help of the burst CID, intermediate RSs relay bursts without checking the MPDU so as to eliminate the processing in MPDU level. Moreover, checking CIDs in MAP also saves unnecessary processing for the data destining to other destinations.

Before relaying data, BS identifies the access RS and sets up the forwarding path for a SS. As the legacy schemes, a data burst binding a relay path aggregates multiple MPDUs and is transmitted in frames. Since each relay path is identified by a burst CID, each RS along a relay path checks the MAP to locate the relayed burst. For the RS not in the path, it just ignores the data burst after checking the MAP. During the relay process, RSs need not to look into MPDUs to identify the data or routing. The usage of CIDs in this scheme is similar to that used in tunnel-based scheme, but the CID is used in burst level, not in T-PDU level. Such an enhancement saves the overheads for both control and data parts. By referring burst CID in frame header, the RS can decide to receive the associated burst or not. If the burst for the relay path is located, the RS transmits the entire burst toward next RS without processing the data in further. The intermediate RSs do not decapsulate inside data but switch the burst along the relay path until the destination RS receives it. After the burst is switched to the end of relay path, the access RS forwards the data to SSs using individual CIDs in access links as BS does in the conventional one-hop network. Since bursts are identified and switched by the burst CID, the proposed forwarding scheme is so called burst switch.

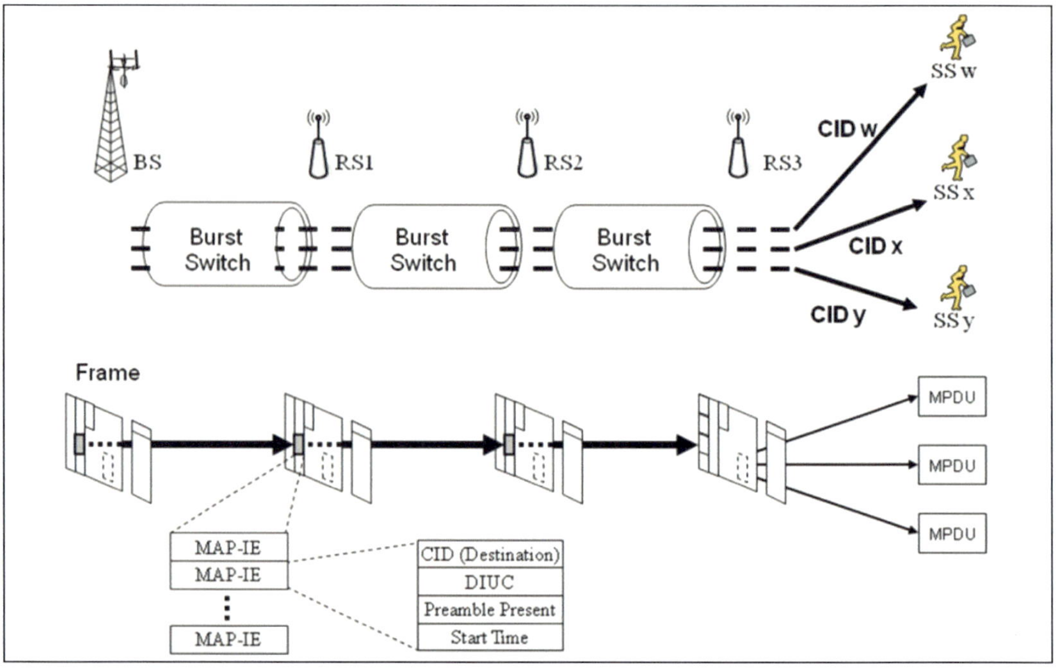

Fig. 6. Burst Switch Transmission

Figure 6 shows the proposed scheme in detail. In the figure, BS establishes a path for switching the burst for SS_w, SS_x, and SS_y, and assigns an end-to-end burst CID for the relay destination, RS_3. When receiving the MAP in first hop, RS_1 and RS_2 check the burst CID and buffer the burst for switching in second hop. After receiving the switched burst, RS_3 decapsulates the burst and processes the MPDUs to see where the destination for the data is. After that, access RS_3 forwards the decoded MPDUs to SS_w, SS_x, and SS_y separately. The computation for this relay process is:

$$C_{Burst\_switch} = L_{MAP-IE} \times N_{RS} \tag{3}$$

Where $\boldsymbol{C_{Burst\_Switch}}$ is the computation cost, $\boldsymbol{L_{MAP\text{-}IE}}$ denotes MAP-IE length and $\boldsymbol{N_{RS}}$ represents the numbers of RSs behind the station. Since $\boldsymbol{N_{RS}}$ is a fixed value as network has been setup, neither the network load nor other traffic factors impact $\boldsymbol{C_{Burst\_Switch}}$. Compared with the tunnel-based and CID-based solutions, the proposed scheme adopts the 32 bits MAP-IE and saves 33 % overhead while the two schemes apply 48 bits T-PDU or MPDU headers. Therefore, it can be expected that $\boldsymbol{C_{Burst\_Switch}} < \boldsymbol{C_{Tunnel\_based}} < \boldsymbol{C_{CID\_based}}$. Moreover, RS with burst switch identifies relay destination before storing data and takes the advantage of storage saving.

3.2 Proposed network model

Before providing relay services, the relay system shall construct a relay network model to determine relay paths for all the SSs. This paragraph provides an overview of the proposed model, in which the burst switch mechanism can be applied. In the network, BS maintains a forwarding table that keeps the CID for every relay path and access link while RSs only record the CIDs for the stations in behind. The CIDs includes the burst CID of access RSs and also the access CIDs of SSs. Figure 6 shows the model for the example of the wireless relay network in figure 1.

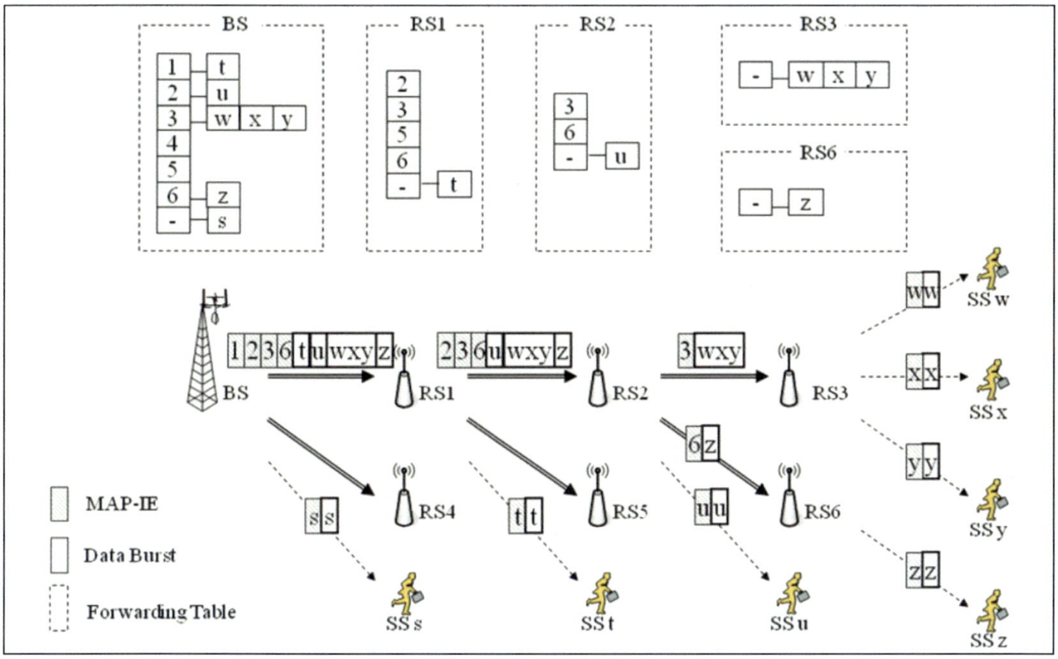

Fig. 7. Proposed Relay Network

In the example, BS can obtain the CIDs for RS_1 to RS_6 and SS_t to SS_z from the table, and RS_2 only holds the CIDs for RS_3 RS_6, and SS_u. When initializing a relay transmission, BS first identifies a receiving SS and the associating access RS from the table. Then, radio resources are allocated for relay links in the relay path. By the MAP indications, intermediate RSs switch burst by checking the CIDs within MAP when receiving a frame. Only when the received burst CID matches the recorded CID, the binding burst is switched to next hop. Intermediate RSs switch the burst until the burst arrivals the relay destination, and the access RS forwards MPDUs to SSs in access links after receiving.

In figure 6, data bursts for all the SSs are ready for relay, and BS assigns burst CID for each burst based on the forwarding table. Bursts for SS_t associates with the CID of RS_1, and bursts for SS_u and SS_z are associated with the CIDs of RS_2 and RS_6 respectively. Because SS_w, SS_x, and SS_y attach the same access RS, data for these SSs are encapsulated together and assigned with the CID of RS_3. When the relaying begins, RS_1 checks CIDs in the MAP and see if there are any bursts associating with the CIDs of SS_u, RS_2, RS_3, and RS_6. Likewise, RS_2 operates similarly, but the difference is switching bursts to RS_3 and RS_6 individually due to separated burst CIDs. After receiving the bursts, RS_3 and RS_6 forward the data to their SSs using the CIDs of SS_w, SS_x, SS_z, and SS_u to complete the relay process.

The major outcomes of the proposed scheme are determined control overhead, simple forwarding, and efficient buffer usage. Unlike tunnel-based and CID-based approaches, processing overheads of proposed method increase with static RS number, not dynamic network traffic load. Moreover, it is simple because each RS identifies desired data by matching CIDs in frame header. Beside, RSs identify relay destination before caching it, and do not store data irrelevantly. To validate the outcomes, next section evaluates the performance of burst switch among various scenarios and compares the results with the other forwarding schemes.

4. Simulation results

This study investigates the issues on data forwarding during relay in terms of process and storage efficiency. Two metrics are introduced for the evaluations. Processing overhead, the first metric, is the total bits that RS shall handle through all the relay process. The second metric, storage overhead, is the memory space in which RS needs to store bursts during relaying. Besides, five forwarding schemes are evaluated in the simulations: (1) *CID_w_MAP*, (2) *CID_w/o_MAP*, (3) *Tunnel_w_MAP*, (4) *Tunnel_w/o_MAP*, and (5) *Burst Switch*. Schemes (1) and (3) represent CID-based and tunnel-based forwarding schemes with MAP support while scheme (2) and (4) denote the relay schemes without MAP indications. For the schemes enabling MAPs, the CID in a MAP-IE directs the station in a hop-by-hop manner, which means RSs swap CIDs for each hop during the relaying. The proposed burst switch scheme employs a burst CID to indicate an end-to-end relay path for each access RS. Table 1 lists detailed parameters, including system and traffic, and the four traffic types defined in 802.16e are also used as inputs to show the performance differences between services.

System Parameters		Traffic Parameters	
Channel Bandwidth	20 MHz	Arrival Time - UGS	30 ms
Frame Duration	5 ms	MPDU Size - UGS	200 bytes
Symbols Per Frame	48	Arrival Time - RTPS	30 ms
FFT Size	2048	MPDU Size - RTPS	1000 bytes
Code Rate	1/2	Arrival Time - NRTPS	100 ms
Modulation	QPSK	MPDU Size - NRTPS	1500 bytes
RS Number	20	Arrival Time - BE	300 ms
SS Number	100	MPDU Size - BE	800 bytes

Table 1. Simulation Parameters

Figure 8 illustrates the statistics of processing overhead for RSs in 2-hop relay case. It is observed that *CID_w/o_MAP* makes severe processing overheads. Moreover, the curve grows dramatically as the number of traffic flows increases. On the contrary, the overheads for the other forwarding schemes rise slightly. The figure tells that disabling MAPs in two-hop case would cause excess overhead for RSs and the results come from processing all the MPDUs in a frame. Although MAP is also disabled in scheme (4), *Tunnel_w/o_MAP* performs best by introducing tunnel header to avoid the drawback. When comparing with scheme (4), *Tunnel_w_MAP* shows a little worse performance since MAPs and tunnel headers are provided redundantly. It is worth to note that *CID_w_MAP* and proposed scheme show identical results in the figure because the CIDs in MAPs both denote the destination RS in two-hop cases.

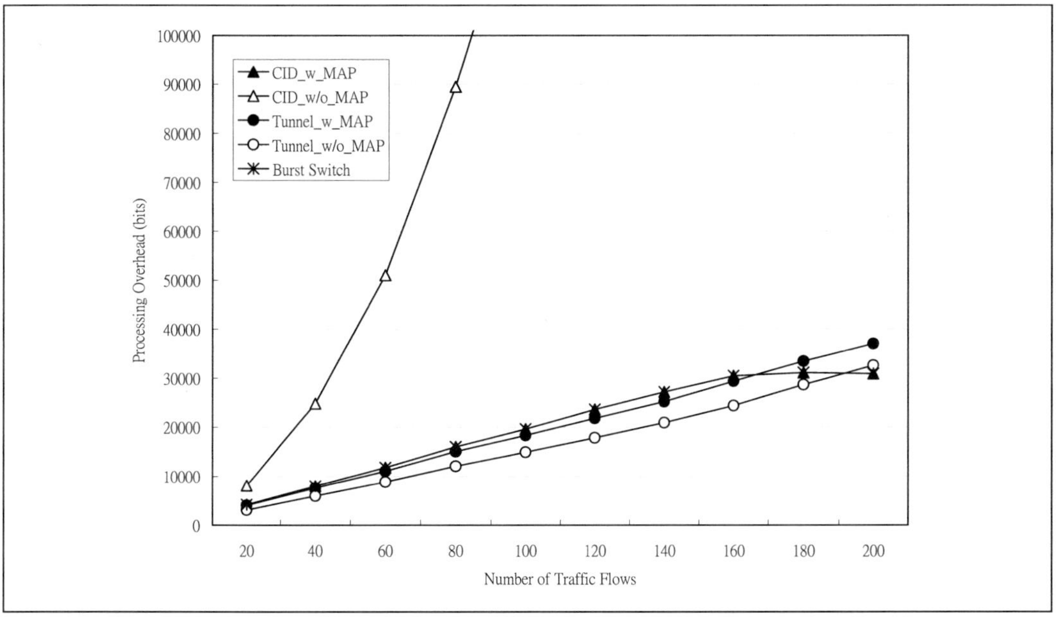

Fig. 8. Processing Overhead in 2-hop Scenario

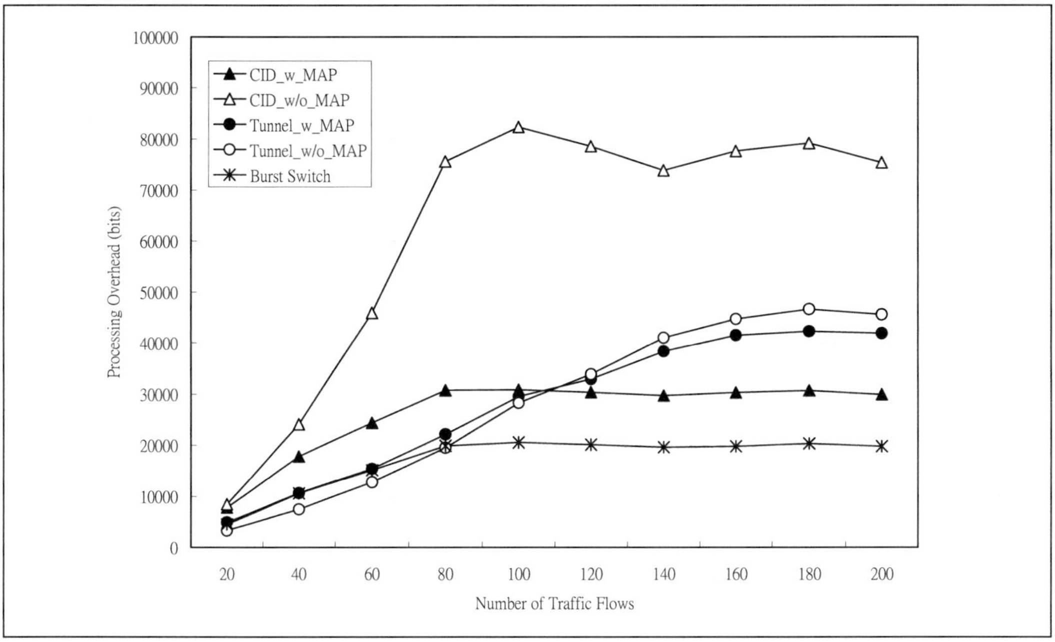

Fig. 9. Processing Overhead in 5-hop Scenario

As the relay hop count increases, RSs shall handle more and more control information in the relaying process, and figure 9 depicts the results in multi-hop relay environments. Generally speaking, the processing overheads of all the schemes increase with traffic load in light traffic load cases (i.e. less than 80 traffic flows). Tunnel-based schemes have less computation overhead than CID-based schemes since tunnel headers can prevent RSs from processing all the MPDUs in a frame. Nevertheless, the curse of tunnel-based schemes also rises as increased traffic in heavy traffic load conditions (i.e. more than 100 traffic flows), and it is because more and more tunnels are established as traffic load increases. Besides, high traffic load also increases the MPDU amount, and, unfortunately, *CID_w/o_MAP* still obtains the worst results. The key worth to be noticed here is that the curve of *CID_w_MAP* and *Birth Switch* scheme rises limitedly in heavy load cases. Compared with the *CID_w_MAP*, RSs employing *Birth Switch* save about 33 % overhead in processing MPDUs, and the figure also shows the outcome. As a consequence, it is obtained that schemes (2), (3), and (4) are impact by system load, and the processing overhead increases with the number of traffic flows. *CID_w_MAP* and the proposed scheme can provide a stable outcome in RS processing, especially when the relay network is loading high. The burst switch concept can save more overheads than *CID_w_MAP*, and is suitable to forward data in multi-hop environments.

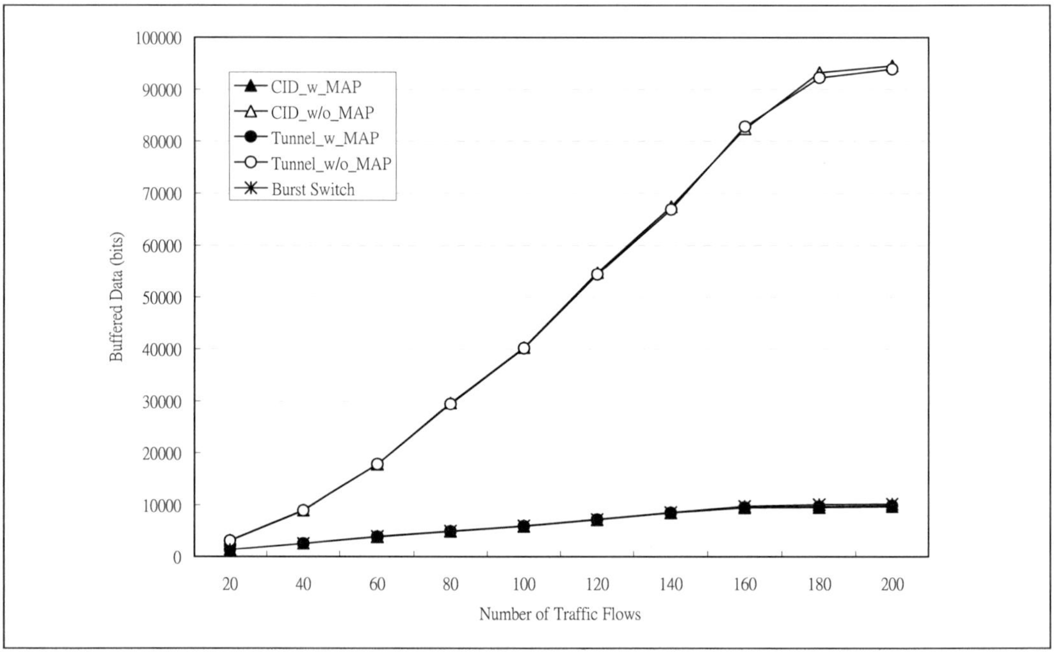

Fig. 10. Storage Overhead in 2-hop Scenario

The second issue is storage overhead, and this study use figure 10 and figure 11 to discuss the evaluations. Both the two simulations confirm the impact resulting from traffic loads. In light load conditions, RSs in 2-hop case do not buffer as much data as those in 5-hop case because the amount of relayed data is small. When the traffic loads high, RSs in both cases have to forward and store more data. Note that the curve becomes saturated for high load cases in figure 10 due to limited radio resource. Moreover, it is easy to find that the schemes without MAP indication, e.g. (2) and (4), cache much more data than the other three schemes do. It is because RSs cannot identify MPDUs before receiving the entire frame, and hence check the MPDUs one by one. On the contrary, RSs in other schemes can filter MPDUs by either frame or tunnel headers so that data would not be buffered inadequately.

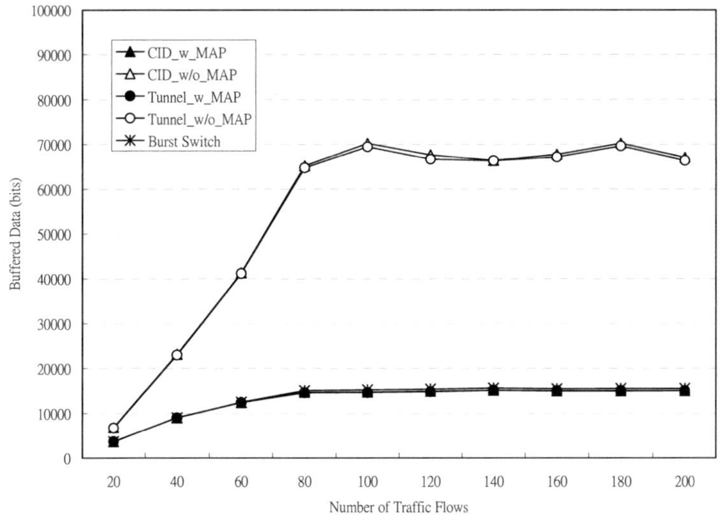

Fig. 11. Storage Overhead in 5-hop Scenario

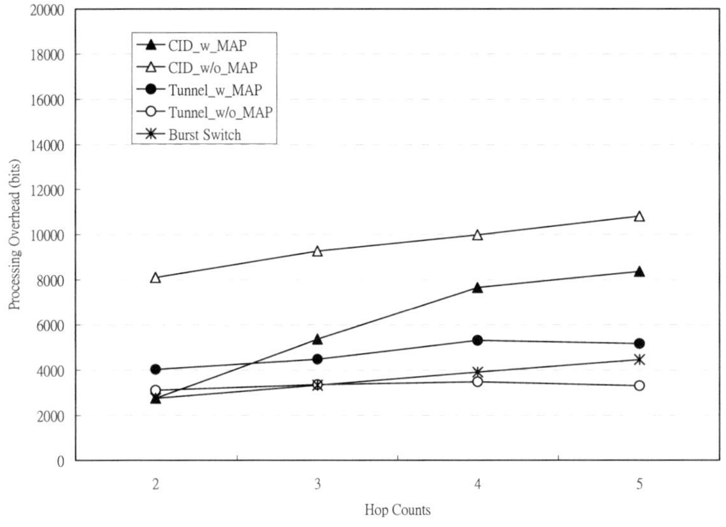

Fig. 12. Processing Overhead in Light Load Cases

This study further evaluates the impact brought from long path lengths. Figure 12 and 13 show the results in light and heavy traffic load conditions respectively. In the simulations of figure 12, the scheme, *CID_w/o_MAP*, performs worst in all the cases. With the MAP assistance, *CID_w _MAP* can perform better than tunnel-based solutions. However, the tunnel solution without MAP outperforms the other tunnel-based scheme because of avoiding extra processing for both MAP and tunnel headers. Furthermore, the RS employing burst switch also avoids redundant and unnecessary processing by burst CIDs so that the proposed scheme also has superior performance in light load conditions.

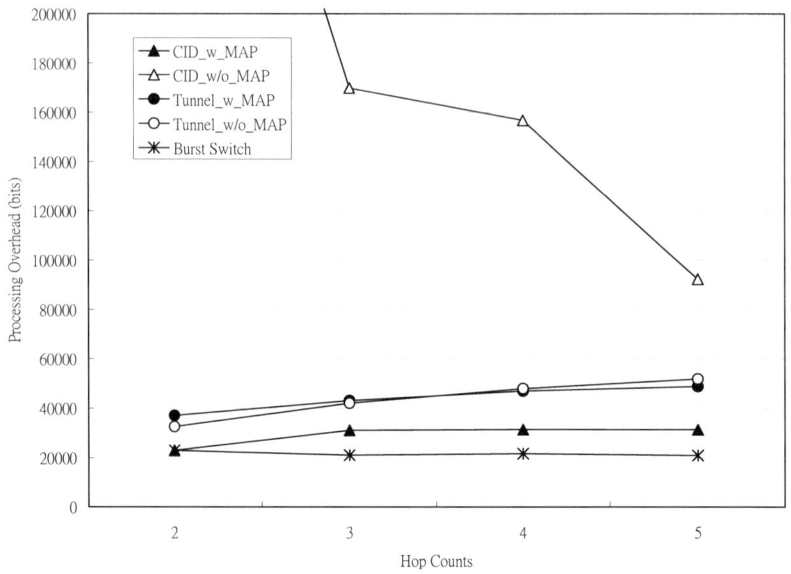

Fig. 13. Processing Overhead in Heavy Load Cases

In figure 13, these forwarding schemes show similar result with that in previous experiment. CID-based forwarding schemes show the frustrated results in heavy load occasions if the network disables MAP support. The other four mechanisms obtain more stable performance in overhead saving when relay hop count increases. Regarding tunnel-based solutions, it is found that enabling MAP does not increase as much overheads as applying tunnel headers because it is tunnel header to dominate the overhead in heavy load conditions. The forwarding schemes without introducing any extra MPDU headers perform better in the simulation. The simulations also suggest the best performance of Burst Switching in multi-hop environments, especially for heavy load cases.

Processing complexity and storage usage are critical issues for RS developments in wireless relay networks. The evaluation results depict overall comparison of the two metrics in figure 14 and figure 15. We can see the performance of overhead saving among the data forwarding schemes. This study uses the results of *Tunnel_w/o_MAP* as basics to show following observations. (1) Without MAP indications, CID-based forwarding scheme performs worst in saving processing and storage overheads, no matter in 2-hop or 5-hop cases. (2) In tunnel-based schemes, enabling MAP indication can reduce the amount of cached data with some penalty from extra MAP-IE processing. (3) Comparing the benefits coming from applying MAP and tunnel headers, *CID_w_MAP* outperforms *Tunnel_w/o_MAP* in storage efficiency since unnecessary MPDU buffering is avoided. (4) *CID_w_MAP* scheme introduce irrelevant processing overhead in 5-hop and light load case although it shows more adequate results than the other three conventional schemes. (5) The proposed burst switch scheme performs better than all other forwarding schemes in both heavy and light traffic load conditions. (6) Considering the performance among different traffic loads, burst switch scheme saves approximately 8 % computation and about 60~73 % storage costs from the results of evaluations.

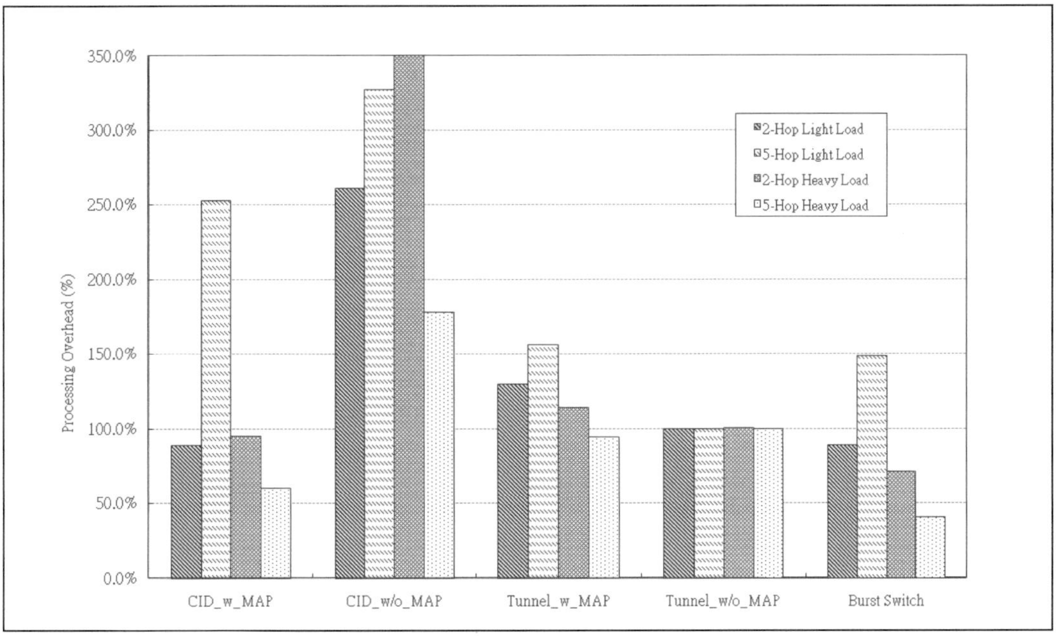

Fig. 14. Overall Comparison for Processing Overhead

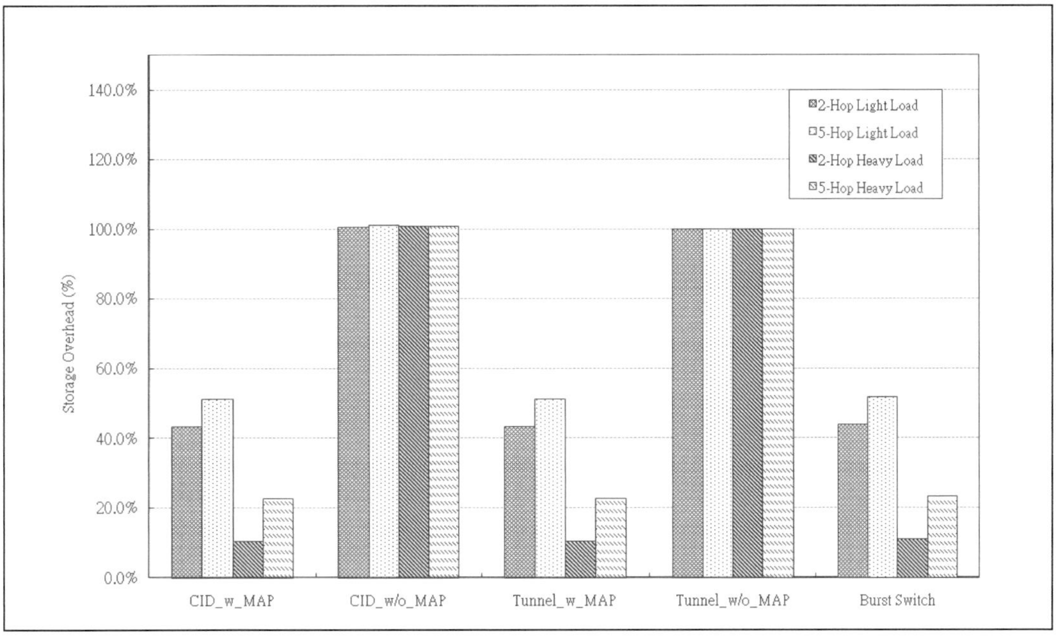

Fig. 15. Overall Comparison for Storage Overhead

Schemes \ Scenario	2-Hop		Multi-Hop	
	Processing	Storage	Processing	Storage
CID_w_MAP	Good	Good	Bad	Good
CID_w/o_MAP	Bad	Bad	Bad	Bad
Tunnel_w_MAP	Good	Good	Bad	Good
Tunnel_w/o_MAP	Relevant	Bad	Relevant	Bad
Burst Switch	Good	Good	Good	Good

Table 2. Evaluation Summary

After comparing all the results from the view point of efficiency in processing and storage, the suggestion for employing data forwarding mechanism can be obtained in TABLE II. In CID-based mechanisms, it is suggested that adopting MAP helps on reducing process and storage overheads. However, both schemes do nothing with the processing problem in multi-hop circumstances. For tunnel-based solutions, network operator shall consider deployment scenario to enable MAP or not since tunnel-based scheme with MAP support performs irrelevantly while disabling MAP for tunnel-based solution causes storage problem in multi-hop cases. After all, *CID_w_MAP*, *Tunnel_w_MAP*, and *Burst Switch* are preferred to be used in 2-hop relay networks since the advantage of low processing and buffering overhead were seen. In multi-hop scenario, all the other forwarding schemes are not suggested because these schemes do not perform as superior as *Burst Switch*. The proposed forwarding can solve all the overhead problems in 2-hop and multi-hop scenarios.

5. Conclusions

In this paper, issues for data forwarding in wireless relay networks are discussed and highlighted. It is found that multi-hop relay has the problem of inefficient processing and buffering by analyzing conventional data forwarding schemes in 802.16j MR system. Through the discussions, introducing tunnel headers can prevent RSs to process data redundantly while reusing MAPs can avoid buffering data unnecessarily. However, none of the convention schemes solve both the problems at the same time.

A new simple forwarding scheme is brought out to tackle the addressed issues. The concept of burst switch is provided with end-to-end relay indications, and RSs can filter data efficiently by the proposed burst CID. Furthermore, data associated with the CID are switched between RSs with simple CID matching. The performance evaluations also prove stable and superior performances of the proposed scheme in various network and traffic environments. Comparing with prior mechanisms, burst switch scheme also shows benefits in both 2-hop and 5-hop cases. As a result, it can conclude that relaying with the proposed method provides simple and efficient data forwarding in wireless multi-hop relay networks.

6. Future research

Apart from the overhead issues, forwarding data in a wireless relay network will suffer from high packet error rate. The issue is also crucial for a wireless multi-hop network since packets during relay will be corrupted due to unreliable wireless medium. It is expected that data forward in multi-hop environments would make the situation worse. Although

Automatic Repeat reQuest (ARQ) mechanism is proposed to maintain the wireless communications reliability, it has been shown that conventional ARQ (Fairhurst & Wood, 2002) cannot overcome the packet error problem efficiently in wireless multi-hop circumstances. Moreover, how a RS participates in error control would impact data forwarding performance directly. Based on error control behaviours, the relay ARQ mechanisms can be classified into three types and are shown in figure 16. The first type is End-to-End Relay ARQ, in which RSs relay packets toward a destination and involve nothing on controlling errors between BS and SSs. The second one is Hop-by-Hop Relay ARQ, and RSs in the scheme trigger feedbacks and retransmissions in per-hop basis. The last type is Two-Link Relay ARQ, which divides an end-to-end transmission into two links and maintains the communications in the links separately. To utilize limited radio resource and provide enhanced relay services, relay ARQ transmissions shall be investigated in the future.

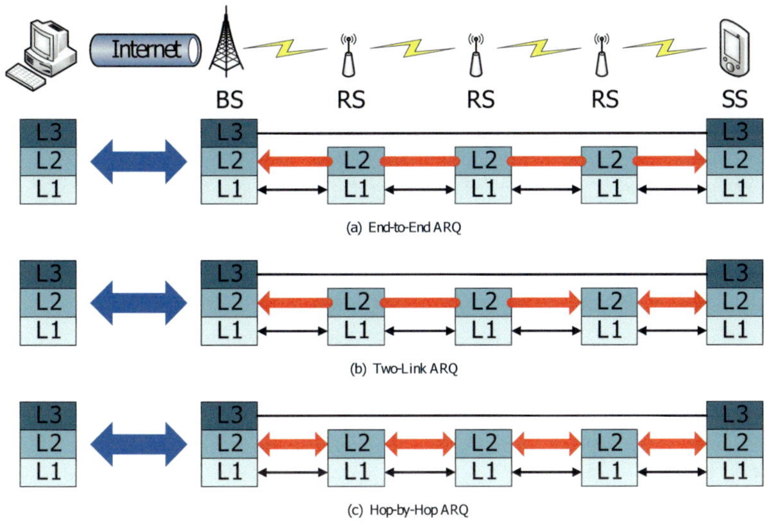

Fig. 16. Wireless Relay ARQ Transmissions

7. References

Nosratinia, A. ; Hunter, T.E. ; Hedayat, A. (2004). Cooperative communication in wireless networks, *IEEE communication magazine*, Vol. 42, Issue 10, Oct. 2004, pp. 74-80.

Mohr, W. (2005). The WINNER (wireless world initiative new radio) project-development of a radio interface for system beyond 3G, *Proceedings of Personal, Indoor and Mobile Radio Communications*, pp. 501-504, July 2005, Berlin, Germany.

Doppler, K. ; Osseiran, A. ; Wodczak, M. ; Rost, P. (2007). On the Integration of Cooperative Relaying into the WINNER System Concept, *Proceedings of Mobile and Wireless Communications Summit 2007*, pp. 1-5, July, 2007, Budapest, Hungary.

Soldani, D. ; Dixit, S.(2008). Wireless relays for broadband access [radio communications series], *IEEE communication magazine*, Vol. 46, Issue 3, March 2008, pp. 58-66.

Tao, Z. ; Teo, K.H. ; Zhang, J.(2007). Aggregation and concatenation in IEEE 802. 16j mobile multihop relay (MMR) networks, *Proceedings of IEEE Mobile WiMAX Symposium*, pp. 85-90, March 2007, Orlando, Florida.

IEEE Std 802.16e™-2005. IEEE Standard for Local and Metropolitan Area Networks - Part 16: Air Interface for Fixed Broadband Wireless Access System, Amendment 2: Physical and Medium Access Control Layers for Combined Fixed and Mobile Operation in Licensed Bands, *IEEE Computer Society and the IEEE Microwave Theory and Techniques Society*, February 2006.

IEEE Std 802.16j™-2009. IEEE Standard for Local and Metropolitan Area Networks - Part 16: Air Interface for Fixed Broadband Wireless Access System, Amendment 1 :Multiple Relay Specification, *IEEE Computer Society and the IEEE Microwave Theory and Techniques Society*, June 2009.

Fairhurst, G.; Wood, L. (2002). Advice to Link Designers on Link Automatic Repeat reQuest (ARQ), RFC3366, IETF Network Working Group, August 2002.

Wang, S.H. ; Yin, H.C. ; Sheu, S.T. ; Chang, C.S. (2009). Efficient Data Foorwarding Schemes for IEEE 802.16j Multi-hop Relay Networks, *Proceedings of Mobile WiMAX Symposium 2009*, pp. 24-29, July 2007, CA, US.

Chun, Nie.; Korakis, T. ; Panwar, S. (2008). A Multi-Hop Polling Service with bandwidth Request Aggregation in IEEE 802.16j Networks, *Proceedings of Vehicular Technology Conference, 2008*, pp. 2172-2176, May 2008, Singapore.

Te, Z. ; Hua, Y. ; (2005). On Link Layer Policies of Data Forwarding over Wireless Relays, *Proceedings of Military Communications Conference*, pp. 2138-2144, Oct. 2005, Atlantic, US.

Wirth, T.; Venkatkumar, V.; Haustein, T.; Schulz, E.; Halfmann, R.; (2009). LTE-Advanced Relaying for Outdoor Range Extension, *Proceedings of Vehicular Technology Conference*, pp. 1-4, September. 2009, Anchorage, Alaska, USA.

Kyungmi, P.; Kang, C.G.; Chang, D.; Song, S.; Ahn, J.; Ihm, J.; (2009).Relay-enhanced Cellular Performance of OFDMA-TDD System for Mobile Wireless Broadband Services, *Proceedings of Computer Communication and Networks*, pp. 430-435, August 2009, Hawaii, US.

Zhao, Z. ;Fang, X. ;Long, Y. ;Hu,Z. ; Zhao, Y. ;Liu. Y. ;Chen, Y. ; Qu, H. Xu, L. ;(2010). Cost Based Local Forwarding Transmission Schemes for Two-Hop Cellular Networks, *Proceedings of Vehicular Technology Conference*, pp. 1-5, May 2010, Teipei, Taiwan.

14

Experiments of In-Vehicle Power Line Communications

Fabienne Nouvel[1], Philippe Tanguy[1], S. Pillement[2] and H.M. Pham[2]
[1]*Laboratory IETR Rennes,*
[2]*University of Rennes 1, Irisa Labs,*
France

1. Introduction

The omnipresence of ECU (electronic control units) in vehicles has lead the automotive industry to face a great challenge in its transition from mechanical engineering towards mechatronical products. The X-by-wire and X-tainment applications involve efficient networks that allow bus sharing while reducing both cabling costs, number of wires and connectors.

This chapter deals with the embedded in-vehicle networks and the use of emerging technologies combining different communication systems like power line communications (PLC) and/or wireless communications and pushing to a dynamic configuration of both networks and ECU.

The ECUs that replace mechanical or hydraulic systems require secure and specific bus for communication. In order to exchange information between sensors and actuators, different networks have been proposed, from low data rate up to high data rate namely LIN, CAN, FlexRay. In section 2, these networks are presented identifying their strengths and possible drawbacks. As a result of using these fieldbuses, the cost of advanced systems should plummet. Furthermore, X-by-wire systems do not depend on conventional mechanical or hydraulic mechanisms. In (Len & Hefferman, 2001), the authors demonstrate the advantages of X-by-wire and embedded networks.

Considering these specific domain embedded networks, we can observe that each solution uses its specific wires and communication system. The growth of the complexity leads to the necessity to commit to a limited set of networks which answers to these multiple applications. An attractive solution to reduce the wires is the power line communication (PLC) using the power lines (12/42V) to transmit both the power and the messages without functional barriers domain. It can answer the vehicles requirements namely cost, decrease of the amount of wires, flexibility and bandwidth. Section 3 is dedicated to PLC systems. Nowadays, this technique is already proposed for domestic uses (Ribeiro et al., 2006). In vehicle PLC seems to be a promising technology and has numerous advantages; it could reduce the weight of wires, the amount of splicing, and simplify cables bundle and the networks between ECU. The background and current studies are first addressed in Section 3. Although high data rate and flexibility obtained for indoor domestic PLC are proven, it is not possible to apply them directly to cars because the geometrical characteristics and wires topologies are totally different. Moreover, the in-vehicle PLC channels are affected by the variable activation schedules of electrical functions, such as brakes, indicators, etc, which produce sharp modifications in the circuit's load impedances over brief time intervals, as presented in (Lienard et al., 2008). To optimize high bit-rate communication, the PLC channel transfer function must be carefully studied because it is frequency selective. Previous studies show that the promising techniques are based on wide bandwidth transmission, such a spread spectrum or orthogonal frequency division multiplex (OFDM). Section 4 provides a description of the experimentations using the existing DC electrical wires. The discussion is framed by describing in particular the area of PLC applications. Results for different cars configurations are presented and demonstrate the feasibility of PLC for automotive.

Section 5 will complete the chapter by describing other alternative solutions for in-vehicle communication based on wireless communications such as ZigBee, Wireless USB, Wifi .... These technologies have been adopted for V2V, R2V and can be extended to in-vehicle communication. Different solutions are proposed, giving a new perspective for ECU communications, both for cars to cars and/or intra-cars communications.

In order to propose more flexibility with these different communication networks, new electronic architecture need to be adopted to reduce the size of ECU. One solution is based on dynamic reconfiguration. Section 6 will analyze this new concept for our embedded system. Reconfigurable systems are already proposed for video driver assistance (Claus & Stechele, 2010). The reconfiguration can increase both safety and flexibility. An ECU can migrate tasks from one node to another. Furthermore, this functionality can be extended to network architecture: according to the channel, the ECU loads, the modem can be dynamically reconfigured to offer seamless communication between ECUs.

2. In-vehicle networks: overview of embedded solutions

Various vehicle buses for different tasks of communications between ECU are used today according to their area of application (Navet, 2008). These embedded networks have both increased the functionality and decreased the amount of wires. However, the usage of different wires for the different networks still has the disadvantage of heavy, complex and expensive. Among these networks, three of them are prevailing:

- the local interconnect network (LIN, the lowest data rate) : proposed by manufacturers, the local interconnect network is used in on-off devices such as car seats, door locks, rain sensors, ...
- the control area network (CAN, medium data rate) developed by BOSCH, is currently the most widely used vehicular network. A typical vehicle can contain two or three separate CAN networks operating at different transmission rates, from 125 Kbps up to a higher-speed at 1 Mbps for more real-time-critical functions.
- the FlexRay is proposed for X-by-Wire applications which require higher data rate(10 Mbps) and safety. FlexRay is a fault-tolerant protocol designed for high-data-rate, advanced-control applications. X-by-wire systems replace the mechanical control systems with electronic component.

Figure 1 illustrates the embedded network architecture. We can observe that these networks have a hierarchical structure.

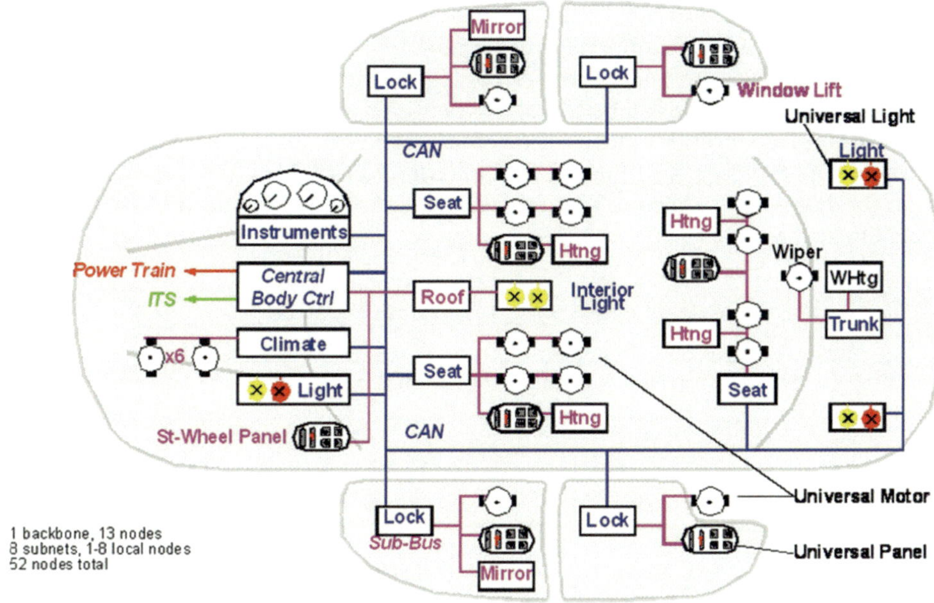

Fig. 1. Network architecture (from http:// www.freescale.com)

2.1 The Local Interconnect Network: LIN

Conceived in 1998, the LIN network (LIN, 2003) is an inexpensive slow and serial bus used for distributed body control electronic systems in vehicle. It enables effective communication for sensors and actuators where bandwidth, speed and versality are not required (i.e. inside mechatronic based subsystems generally made of an ECU and its set of sensors and actuators). LIN is commonly used as a sub bus for CAN and FlexRay. A LIN network is based on one master node and LIN slaves (up to 16 over 40 meters line length).

The master node decides when and which frame shall be transmitted according to the schedule table containing the transmission order. At the moment a frame is scheduled for transmission, the master sends the header inviting a slave node to send its data in response. Any node interested can read a data frame transmitted on the bus. The reliability of LIN is high but it does not have to meet the same levels as CAN. The LIN can be implemented using just a single wire, while CAN needs two. The physical layer (PHY) supports a data rate equal to 20 Kbps (due to electromagnetic limitations) but other transmission supports enabling higher data rates are possible. LIN is widely used in middle range cars but it can not support high data rate as required by devices like portable DVD players or multimedia applications.

2.2 The Control Area Network: CAN

The CAN (CAN, 2009) is a widely communication fieldbus used in automotive and other real time applications. It is a serial communications protocol which efficiently supports distributed realtime control with a middle level of security. In automotive ECUs, sensors, anti-skid-systems, etc. are connected using CAN with bit rates up to 1 Mbps. However, in today's car, CAN is used as an SAE (Society of Automotive Engineers) class C (classification defined in J2056/2 Survey, 1994)) network for real time control in the powertrain and

chassis domains (at 250 or 500 kbps). It is also implemented as an SAE class B network for the electronics in the body domain (at 125 kbps).
In CAN, data are transmitted in frames containing up to 8 bytes of data and a number of control bits. A CAN frame is labelled by an identifier whose numerical value determines the frame priority. Depending on the CAN format (standard CAN 2.0A or extended CAN 2.0B) the size of the identifier is 11 bits (CAN 2.0A) or 29-bits (CAN 2.0B). Between CAN frames sent on the bus, there is also a 3 bit inter-frame space. The standard CAN frame format is depicted in Figure 2.

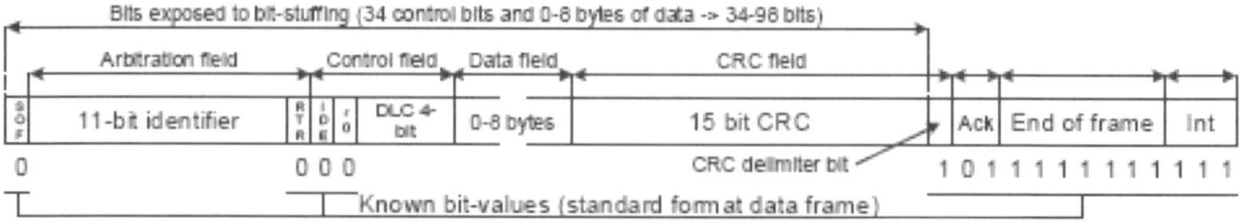

Fig. 2. The CAN frame (from (Nolte, 2006))

Regarding the CAN MAC layer, CAN is a collision-avoidance broadcast bus (CSMA/CA for carrier sense multiple access with collision avoidance), which uses deterministic collision resolution to control access to the bus. It implements a fixed-priority based arbitration mechanism that can provide real time guarantees and that is amenable to timing analysis.
As distributed real time systems become more and more complex, the computing power is steadily growing, and the number of ECUs attached to CAN buses is growing. Thus CAN's maximum speed of 1 Mbps can lead to performance bottlenecks. Hence, methods for increasing the achievable utilisation are needed, e.g., novel analysis methods that allow increased utilisation while guaranteeing timing requirements to be fulfilled, and novel approaches to schedule CAN.

2.3 The FlexRay protocol

X-by-wire systems need fault-tolerant communications with predictable message transmissions and low jitter. This is traditionally solved using TDMA technologies, thanks to their predictable nature. FlexRay (FlexRay Consortium, 2009) is a TDMA communication system developed by a consortium founded in 2000, including both car and semiconductors manufacturers. FlexRay is a fault-tolerant protocol designed for high-data-rate, advanced-control applications. FlexRay is considered by manufacturers as the backbone network for the other networks like CAN or LIN Currently, FlexRay can handle communications at 10 Mbps. An overview of the FlexRay frame format is given in Figure 3. The frame consists of three segments: the header segment, the payload segment (up to 254 bytes of data), and the trailer or CRC segment.
Communication is done in a communication cycle consisting of a static part and a dynamic part, where each of the parts may be empty. The sending slots are represented through the identifier (ID) numbers that are the same on both channels. The sending slots are used deterministically (pre-defined TDMA strategy) in the static part. In the dynamic part there can be differences in the phase on the two channels. Nodes that are connected to both channels send their frames in the static part simultaneously on both channels. An interesting feature of FlexRay is that it can provide scalable dependability i.e., the "ability to operate in configurations that provide various degrees of fault tolerance.".

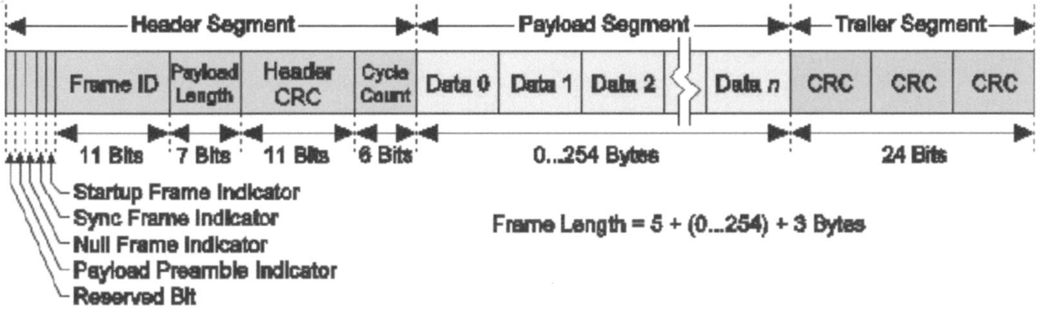

Fig. 3. FlexRay frame (from (FlexRay Consortium, 2009))

However, this network still uses specific wires, which do not achieve compatibility with other networks. So, gateways are necessary to transfer information from ECUs connected to a domain network to ECUs connected to other networks. Those gateways could introduce latency, errors, bottlenecks, and so on.

Other fault-tolerant networks have been developed, namely TTP and TT-CAN, but it seems they are not the best choice for automotive manufacturers due to limited flexibility, high costs and conflicting interests (Nolte, 2006).

Considering these specific embedded networks, we can observe that each solution uses its specific wires and communication system. We can see the wide diversity of the solutions and the necessity to find a limited set of networks which answers to the growing of the multiple applications and requirements. Furthermore, gateways are necessary to switch from one network to another one. This increases the propagation time and does not guarantee real time. One idea to avoid the growth of wires would be to use the PLC technology that is currently developed for indoor AC networks to transmit information over the 12V power distribution (Rubin, 2002). The possible applications of automotive PLC are very wide, extending from low-speed data buses for activating actuators to high-speed multimedia applications.

3. Power Line Communication (PLC)

Many studies are carried out on in-vehicle PLC and focus both on channels, impulsive noises, drivers and protocols. The in-vehicle networks have reduced the number of wires, allowing communication between different typologies of electronic systems (safety devices, entertainment devices, and power train electronics). Additional cost reduction can be accomplished by adopting PLC approach. PLC can be considered to provide the physical layer for serial communications among ECU using for example LIN or CAN transceivers. In this case, the dedicated bus is eliminated. However, PLC can provide both the physical and MAC layers, allowing full compatibility between any ECU. This section considers first the different indoor PLC solutions. Then we focus of in-vehicle PLC driver solutions.

3.1 Indoor PLC

In 2000, a coalition of manufacturers has established a new protocol HomePlug 1.0 that enables the establishment of an Ethernet-IP class network over power line channels (Home Plug V1.0, 2009). The HomePlug process is based on an OFDM technique (Bahai et al., 2004) whose major advantage for the embedded PLC application is to cope with the frequency selectivity of the power line channel caused by the multiple reflections of the loads connected to the power grid and by the coupling to the other cables placed in the same bundle. The modulation is based on 128 subcarriers equally spaced from 4.3 MHz to 25 MHz, in conjunction coding applied before differential encoding. HomePlug uses CSMA/CA protocol to access to the network. Figure 4 represents the PHY frame format.

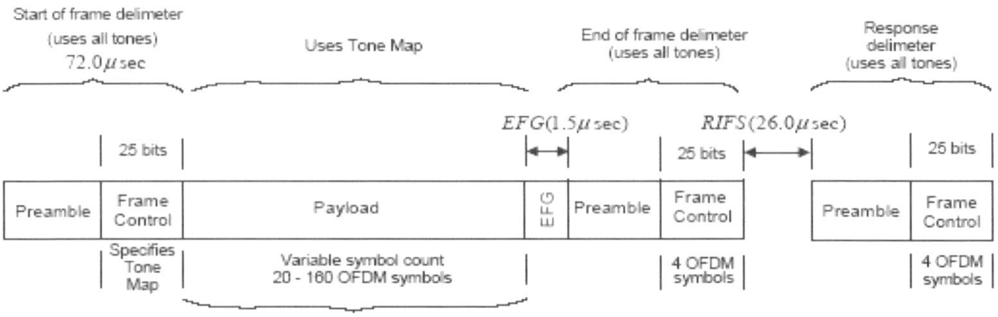

Fig. 4. HomePlug V1.0 PHY frame format (from (HomePlug, 2009))

	Panasonic	HPAV	UPA
Modulation	wavelet OFDM	windowed OFDM	windowed OFDM
Channel coding	RS-CC; LDPC	Parallel-concatenated turbo convolutional code	RS + 4D-TCM concatenation
Mapping	PAM 2-32	QAM 2, 4, 8, 16, 64, 256, 1024	ADPSK 2-1024
FFT/FB size	512 (extendable to 2048)	3072	NC[*1]
Max number of carriers	NC	1536	1536
Sample frequency	62.5 MHz	75 MHz	NC
Frequency band	4-28 MHz 2-28 MHz optional	2-28 MHz	0-30 MHz 0-20 MHz optional
PHY Rate	190 Mbps	200 Mbps	200 Mbps[*5]
Information Rate	NC	150 Mbps	158 Mbps[*5]
Programmable notches	Yes	Yes	Yes
Power Spectral Density	NC	-56 dBm/Hz*[6]	-56 dBm/Hz[*4]
Media Access Method	TDMA-CSMA/CA	TDMA-CSMA/CA	ADTDM[*2]
Hidden Node Avoidance	NC	Yes	Yes
Duration MAC frame	NC	Variable	Variable
MAX number of nodes	64[*3]	NC	64[*4]
Network identifier	Yes	Yes	Yes

Table 1. PHY and MAC layers of current PLC solutions (from (OMEGA, 2008))

More recently, the HomePlug AV (HPAV) has been introduced and will be the second major standard released by the HomePlug Powerline Alliance (Gavette, 2006) (Afkhamie et al.,

2005). The main HomePlug AV's objective is to distribute multi-media content within the house as well as data. The PHY layer still operates in the frequency range of [2 - 28] MHz and provides a 200 Mbps PHY channel rate (150 Mbps net information rate).
Long OFDM symbols with 917 usable carriers (tones) are used in conjunction with a flexible guard interval. Modulation densities from BPSK) to 1024 QPSK are independently applied to each carrier based on the channel characteristics between the transmitter and the receiver. Experimental systems of HPAV have been field tested in houses, suggesting that on average HomePlug AV system achieves 10 times the data rate of a HomePlug 1.0 system.
At the same time, The HD-PLC (High Definition Power Line Communication) (Galli, 2008) solution has been proposed by Panasonic. It is based on a specific OFDM modulation called Wavelet-OFDM which exploits the Wavelet transform combined with 2 to 16 PAM modulations. The Wavelet OFDM achieves highly efficient transmission with characteristics that exceed even FFT-based OFDM systems. Wavelet OFDM features greater speed efficiency and forms a deeper "flexible notch" that prevents interference with shortwave and other broadcasts. No guard interval is included. The MAC layer uses the hybrid TDMA and CSMA/CA protocols synchronized with the AC line cycle. Table 1 summarizes the current PLC solutions.
Todays, the HomePlug Alliance, HD-PLC alliance and the IEEE (IEEE P1901, 2008) are strongly committed to delivering a single mature solution that will be endorsed by the IEEE P1901 work group as the baseline standard. A standard for high speed (over 100 Mbps at the physical layer) communication will be proposed and will use transmission frequencies up to 100 MHz. These PLC solutions could be investigated in an automotive environment.

3.2 In-vehicle PLC

Although high data rate and flexibility obtained for indoor domestic PLC are proven, it is not possible to apply them directly to cars because the geometrical characteristics and wires topologies are totally different. PLC can be considered for the PHY layer only or for the MAC and PHY layers. These two configurations are considered below.
In (Benzi et al., 2008), the authors focus on the issues that need to be addressed when introducing PLC in vehicle. Three main domains need to be covered: the physical (PHY) layer, the data link layer and the performances. In order to answer to them, the properties of the automotive in board PLC supply networks have been investigated (Huck, 2005) (Arabia et al., 2006), (Degardin et al., 2006), (Mohammadi et al., 2009). The results show the insertion losses over the [0-30] MHz bandwidth are about -15 dB and -36 dB in the frequency range [0.500-30] MHz. The noise measurements show an increasing background noise in the frequency ranges [0-100] MHz, especially at frequencies less than 10 MHz, the peaks of the noise could be in the range [-90 dBm/Hz; -40 dBm/Hz]. Varying space between cabling and car body results in changing behaviour of the whole system. A new wiring harness structure is proposed in (Benzi et al., 2008), based on a star structure using active star couplers. The transfer function of this wiring seems to be more flat in the range between 150 and 250 MHz. However this solution needs to re-organize all the harness, which is different from one vehicle to another one.
Considering first the PLC for the PHY layer, a power line communication system over 12 to 42 V wires combining the LIN protocol and a PLC driver has been proposed (De Caro, 2009). In order to avoid interferences between the master and slaves LIN nodes, two different transmission modes have been adopted, one based on BPSK for master to slave

transfers, while slave mode exploits a BASK modulation. The modulation must support data transfer at 10 Kbps, while the accepted conducted emission limits need to be less than 53 dBμv in the [1-30] MHz band. Two carriers have been selected, one at 100 KHz for low power modules and one at 2 MHz for high power modules. Although this LIN and PLC transceiver is an attractive solution, the data rate remains under 10 Kbps that is not convenient for X-by-wire applications.

A similar approach has been proposed for CAN protocol by many authors (Yamar, 2009), (Silva et al., 2009) (Beikirch et al., 2000). The Yamar solution implements CAN and PLC using the DC-BUS technology with different bit rates up to 1.7 Mbps. It uses narrow band channels with a center frequency between [2-12] MHz. The DC-BUS protocol uses the CSMA/CA multiplex mechanism allowing bidirectional communication up to 16 nodes. In addition, this CAN-PLC solution can be used as a redundant channel for the CAN protocol. However, this solution still does not answer to data rate over 10 Mbps.

Additional PLC drivers combining MAC layers have been presented in (Benzi, 2008). The commercial solutions are available for automotive but to our knowledge not implemented yet in vehicles.

More recently, PLC in electric vehicles has been studied in (Bassi et al., 2009). One can think that the requirements of such communication system within an electrical car differ from a fuel car. An experimental setup has been built. It uses 2 ECUs and 2 DCB500 transceivers to modulate the DC-line. The DCB500 transceivers feature PLC communication over DC-line with a bit rate up to 500 Kbps. The conducted and irradiated emissions show substantial compatibility, except for the lower end frequencies (under 1 MHz) where significant peaks are highlighted. In addition, different channel measurements in electric cars have been carried out in (Barmada et al., 2010). Different cases are considered (front to/from rear part) with different vehicle's configuration (position key, battery,...). As for fuel vehicle, the channels are very frequency selective in the [0-30] MHz. We can conclude that the fuel and electric vehicles seem to have similar behaviours in term of frequency channel and noise for PLC applications.

Another solution for PLC is to consider both the MAC and PHY layers. Considering the channel measurements, the candidate techniques for in-vehicle PLC are spread spectrum combined with code division multiple access (CDMA) (Nouvel et all, 1994) and OFDM. OFDM allows high data rate and outperform CDMA performances in term of throughput. and frequency selectivity.

Experimentations using indoor OFDM PLC modems have been carried out and presented in detail in previous studied presented in (Gouret et al., 2006), (Gouret et al., 2007), (Nouvel et al., 2008), (Degardin, 2007) and more recently in (Nouvel et al., 2009A). The results are very promising. Data rate up to 10 Mbps/s can be achieved in the [0-30MHz] bandwidth. The solutions are based on HPAV standards. In (Nouvel et al., 2008) two PLC modems have been tested: SPIDCOM (Spidcom, 2008) and DEVOLO modems. In the SPIDCOM modems, the OFDM modulation is based on 896-carriers from 0 to 30 MHz divided into 7 equal sub-bands. The MAC layer provides a mechanism based on TDMA and CSMA/CA is also available. The PHY and MAC layers are similar to the HPAV ones but differ in some points: number of sub-bands, equalization, and synchronization. With these SPIDCOM modems, an 8 Mbps is achieved with a transmitted power of -50 dBm. With a higher level (-37 dBm), we achieve about 12 Mbps. For multi-media applications, this rate can be sufficient, but decreases rapidly according to the loads. Then measurements have been carried out with

DEVOLO PLC modems. They comply with HPAV and support data speed of up 200 Mbps in a range of 200 meters within a household grid. For intra-car communications, the power supply and the coupling have been modified to take into account the DC channel. Additional measurements are presented in next section. Figure 5 illustrates the spectrum of the transmitted signal over the DC line.

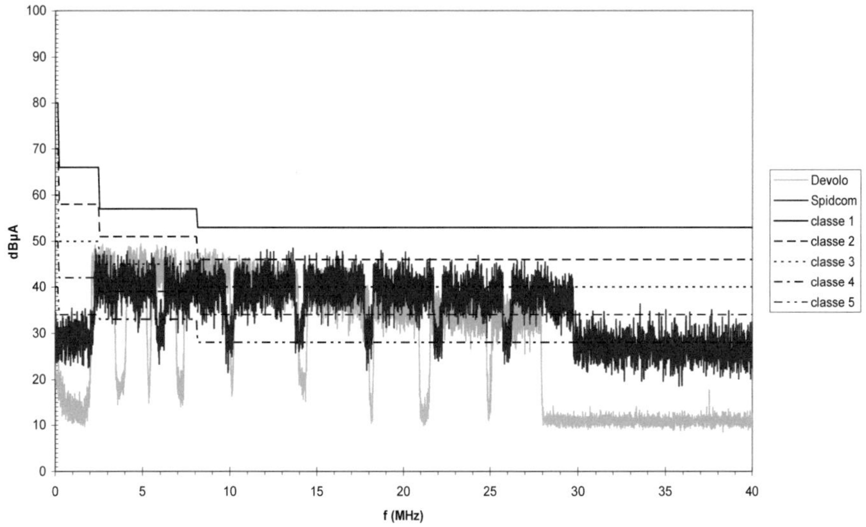

Fig. 5. HPAV and Spidcom spectrum over DC line

Beyond these promising results, the choice of the modulation parameters will be driven by the PLC channels and optimized with regards to the bandwidth, the modulation technique, the coding rate, the guard interval, and so on. This discussion is presented in the next section.

4. In-vehicle measurements

In this section we deal with in-vehicle PLC measurements. In a first time we show some results about real PLC transmissions. Indeed, we have decided to test the feasibility to adapt indoor PLC modems in car. Then, we study in more details the in-vehicle PLC channel with different measurements about the transfer function and the noise. To achieve the capacity of the channel through the cables for PLC, many transfer functions between nodes in the vehicle have been measured. Noises have also been considered.

4.1 In-vehicle PLC transmissions

4.1.1 Data rates measurements testbed

We have tested two indoor PLC modems complying with the standards HPAV and HD-PLC in one car. We have measured throughputs at different points on a gasoline Peugeot 407 SW.

The Figure 6 illustrates the different points used during the throughput measurement. Several scenarios have been used:

1. Car with engine-turned off

2. Car with engine-turned but not moving
3. Car with engine-turned but not moving and effects of lightning, warnings, radio, windscreen wiper, electric windows
4. The car in motion and the effects of the equipments like in 3)

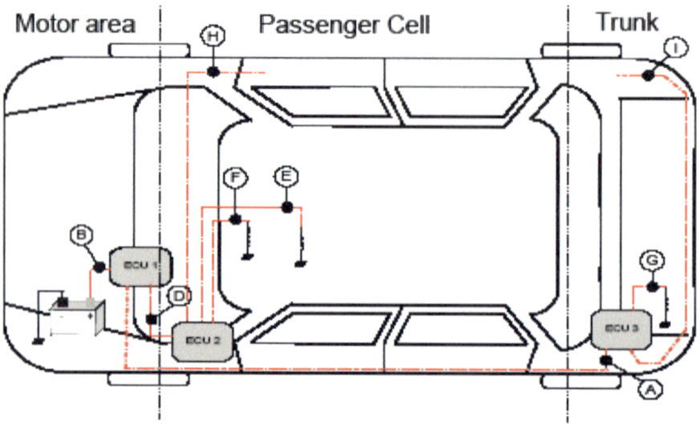

Fig. 6. Measurement scheme: the different uppercases represent the measurement points

The measurements have been achieved with two PLC modems and two computers which have been plugged into the different points shown Figure 6. Therefore, we have measured the TCP throughput between two points with two modems and two PC. The measurement between points A and D has been called path AD. The throughputs are measured associated with the payload ignoring headers. The throughput is also called "Goodput" according the definition in section 3.7 of (Newman, 2009).

4.1.2 Results and discussion

Throughputs for different points have been studied and we can first observe a difference between scenario 1) and the others. Figure 7 to 9 represent the throughput we obtain with the two modems. Throughputs in Figure 7 are higher than 35 Mbps, and in Figures 8 and 9 more than 15 Mbps are achieved for all paths.

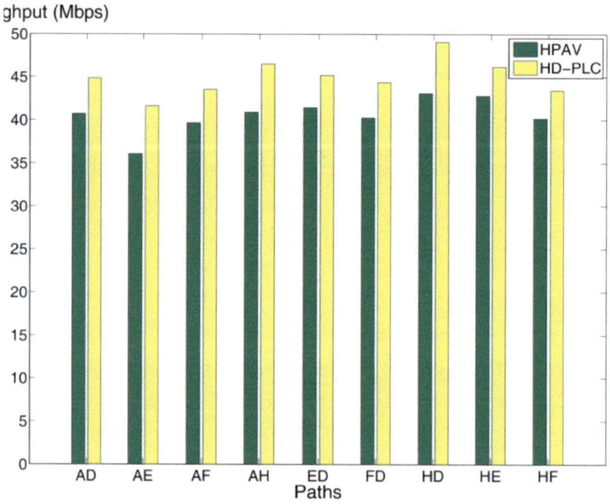

Fig. 7. HPAV and HD-PLC throughputs comparison

For scenario 2), 3) and 4) we remark that the HPAV has the best performances. Moreover, we can observe short variations between the scenarios for the two indoor standards. Furthermore there is a throughput difference according to the path in-vehicle. Indeed, we can see that the path HD has throughput higher than all the others.

Indoor PLC standards have been designed according indoor channel characterization. Moreover, the power level of the transmitted signal has been chosen according the indoor CEM constraints. In fact, to respect the vehicle CEM it has been said in (Degardin et al., 2007) that the power level of transmitted signal should be between -60 dBm/Hz and -80 dBm/Hz. This specific point must be taken into account for next PLC in-vehicle transmission. That's why measurements on several vehicles have been achieved and are discussed in the next subsection.

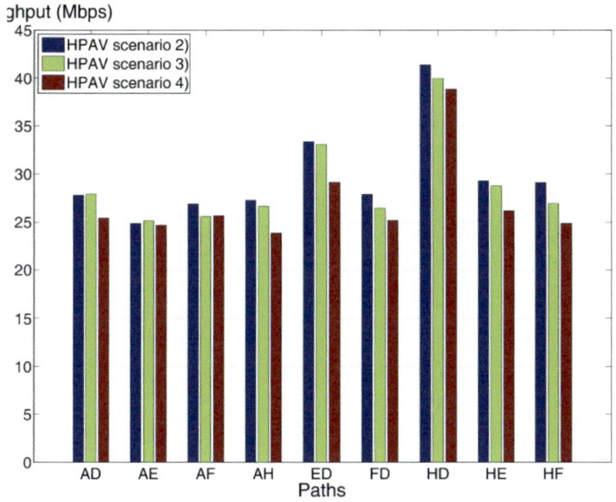

Fig. 8. HPAV throughputs for different paths in-vehicle for scenario 2), 3) and 4)

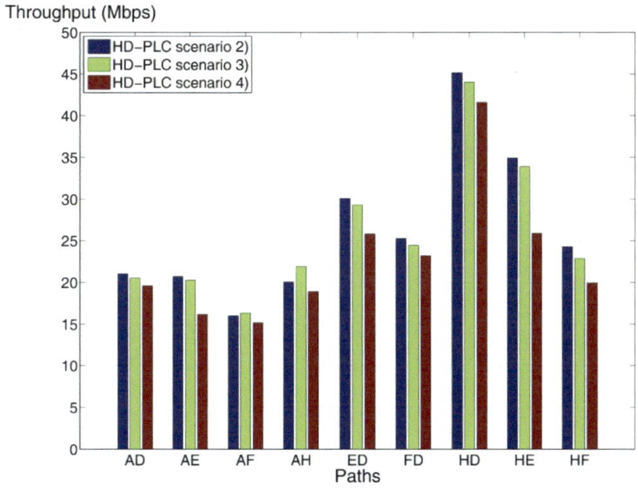

Fig. 9. HD-PLC throughputs for different paths in-vehicle for scenario 2), 3) and 4)

4.2 In-vehicle channel measurements

In order to design a future PLC modem it is necessary to study the PLC in-vehicle channel. Here the transfer function and the background noise is studied.
Additional measurements have been performed on recent vehicles for two classes of paths: front to front and rear to front (Tanguy et all, 2009). Figure 10 and 11 illustrate the results according to our testbed (Figure 6). In order to analyze the DC PLC architectures, additional transfer functions are measured on four different vehicles. The vehicles are classified according to: the number and type of ECUs, the length of wires, the combustion engine.

4.2.1 Measurement testbed

The S-parameters are recorded using a full 4 ports Vector Network Analyzer (VNA) and a PC interfaced to remote the device. We record the S-parameters during about 10 minutes while the car is moving. The S-parameters are recorded about every 10 seconds for the 3 different paths: GF, GH and HD. Compared with the previous subsection we have introduce a new measurement point called G which is for the most of vehicle tested a cigar lighter receptacle. These paths have been chosen in order to analyze the differences between front to front and rear to front.
Regarding the noise, the same points have been considered: G, D, F and H. Two different noise studies have been carried out. The first consists of the measurement of the power spectrum at each point during 10 minutes every 10 seconds with the vehicle moving. The second is a measurement in the time domain. In fact, a digital storage oscilloscope (DSO) has been used to record at each point the signal over the DC line. With this testbed we are able to record two signals at two different points in the same time. Thus, we can observe the level of noise at two different points simultaneously. Finally, the measurements have been performed on a Peugeot 407 SW gasoline and diesel, a Renault Laguna II Estate and a Citroën C3.

4.2.2 Results & discussion

Figure 10 and Figure 11 show an example of time and frequency responses for the three paths GF, GH and HD and for a measurement bandwidth of [1-31] MHz. The impulse responses have been calculated with the inverse Fourier transform of complex parameter S21.

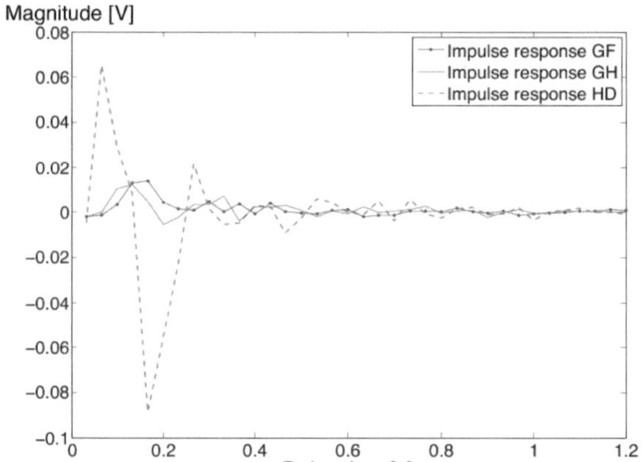

Fig. 10. Impulse response for 3 paths GF,GH,HD on 407SW gasoline

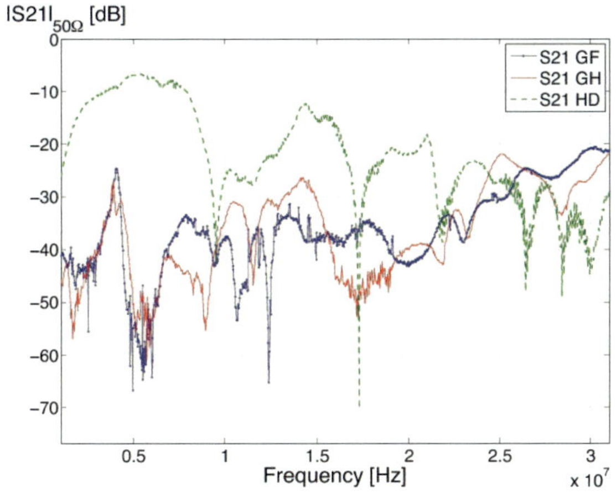

Fig. 11. S21 for 3 paths GF, GH and HD on 407 SW gasoline

		Min	Max	Mean	Std
BC0.9 GF	407 gasoline	391.8 KHz	832.6 KHz	533.8 KHz	89.9 KHZ
	407 diesel	538.7 KHz	881.6 KHz	666 KHz	81.6 KHz
	Laguna II	4.3098 MHz	4.8976 MHz	4.7163 MHz	142.8 KHz
	C3	440.8 KHz	1.3713 MHz	1.1587 MHz	143.9 KHz
BC0.9 GH	407 gasoline	1.3713 MHz	2.1059 MHz	1.7578 MHz	190.3 KHz
	407 diesel	97.9 KHz	1.0775 MHz	748.3 KHz	227.3 KHz
	Laguna II	1.0775 MHz	1.2734 MHz	1.1443 MHz	45.6 KHz
	C3	489.8 KHz	1.5182 MHz	1.0591 MHz	331 KHz
BC0.9 HD	407 gasoline	1.8121 MHz	2.057 MHz	2.006 MHz	40.8 KHz
	407 diesel	685.7 KHz	734.6 KHz	712.6 KHz	24.5 KHz
	Laguna II	685.7 KHz	832.6 KHz	744 KHz	31.9 KHz
	C3	881.6 KHz	1.0775 MHz	995.8 KHz	46.4 KHz

Table 2. Coherence bandwidth (BC0.9) for 3 paths (GF, GH and HD) and for 4 different vehicles

In a previous study on in-vehicle PLC (Lienard et al., 2008) a delay spread under 380 μs and a coherence bandwidth greater than 400 KHz has been found. Moreover, in Table 2, we observe the coherence bandwidths are different from one vehicle to another and from one path to another. This means that the modulation must be adaptive.

Regarding the average attenuation we can also observed differences between the different paths. For example, the Renault Laguna II Estate has a mean average attenuation of 9 dB for the path GF, 31.6 dB for GH and 31.5 for HD. But the 407 SW gasoline has a mean average attenuation of 40.1 dB for the path GF, 40.4 dB for GH and 24.4 for HD. Otherwise, we have a maximum average attenuation of 69.3 dB for the path GH of the 407 SW diesel and a minimum average attenuation of 5.8 dB for the path GF of the Laguna II.

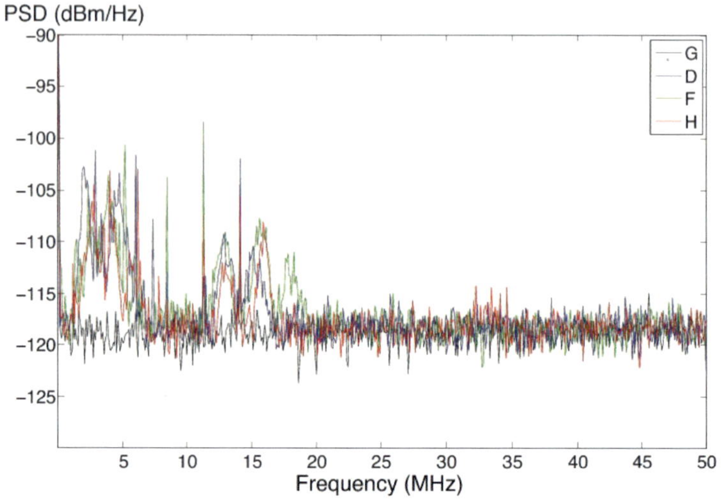

Fig. 12. Noise measured with a spectrum analyser for 4 different paths on a Peugeot 407 SW gasoline

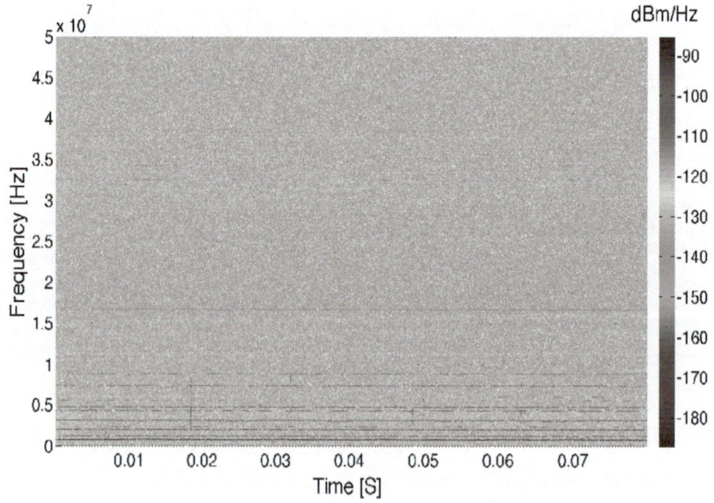

Fig. 13. Spectrogram computed with the DSO recording at point G measured on a Peugeot 407 SW gasoline

To optimize the modulation parameters, we have to consider the noise. Figure 12 represents an example of noise measurement with a spectrum analyzer for 4 different points in a Peugeot 407 SW gasoline. We observe an increase of noise for some frequencies in the bandwidth [0 - 5] MHz. Moreover we can see narrowband noises. Like in (Yabuuchi et al.,

2010) we have applied to noise recordings in-vehicle a time frequency analysis. In Figure 13 we show an example of spectrogram computed with the DSO recording at point G measured on the same vehicle. We have computed the spectrogram with short-time Fourier transform where an Hamming window of length equal to the length of HPAV OFDM symbol (40.96 μs) and an FFT size of 3072 points like in HPAV standard.

In Figure 13 we can observe that in the bandwidth [0 – 5] MHz the noise is constant during the time of the recording. Therefore, in the case of a multi-carriers modulation transmission in the bandwidth [2-30] MHz some subcarriers will be affected during all the transmission time.

We have observed that the average attenuation, the coherence bandwidth and the RMS delay spread are very different according the vehicles, the paths in-vehicle and the paths between vehicles. We verified the capacity for each paths of each vehicles with the parameters of the Table 3 according to

$$C = \Delta f \sum_{0}^{N-1} \log_2(1 + SNR_i) \tag{1}$$

with Δ_f the subcarrier bandwidth, SNR_{i} = (H_{i}^{2}.Pe/Pn) the signal to noise ratio per subcarrier , Pe is the PSD of the emitted signal and Pn is the PSD of the AWGN noise.

Parameters	Values
Fmin	1 MHz
Fmax	31 MHz
Subcarrier	N=1228
FFT/IFFT	3072
Δ_f	24.414 KHz
PSD of noise (Pn)	- 120 dBm/Hz
PSD of signal (Pe)	-60 dBm/Hz

Table 3. Simulation parameters: FFT/IFFT and Δ_f values are parameters used by the HPAV standard

The results show the minimum of the average capacity is about 190 Mbps for the path GH of the Peugeot 407 SW diesel and the maximum is about 507 Mbps for the path GF of the Laguna II. We observed also differences between the paths and the vehicles.

The vehicles have not the same electrical topology. In fact, it depends on car manufacturer, the size of vehicles, the number of ECUs ... Therefore the load on the electrical network, the length of wires and the junctions between cables are different. We have several channels which are different according the paths and the vehicles like we have shown with the coherence bandwidth, the time delay spread, the channel gain and the capacities.

The multicarrier modulation seems to achieve good performances like we have seen during the throughput measurement of HPAV and HD-PLC standards. In this study, only the channel function transfer and the background noise have been studied. The impulsive noise is an other important aspect to take into account (Umehara et al., 2010) and (Degardin et al., 2008) for powerline communication. According to us the MAC/PHY layers must be designed to take into account the differences between vehicles and the differences between paths in-vehicle. Future work will be focus on the integration in a simulator of all the channel measurements (transfer function, background noise, narrowband interference and impulsive noise) in order to optimize the modulation scheme.

5. In-vehicle wireless communications

The interest in wireless networking has grown significantly due to the availability of many wireless products. Looking at in-vehicle communications, more and more portable devices, e.g., mobile phones, laptop computers and DVD player can exploit the possibility of interconnection with the vehicle. Wireless communication could be an attractive solution to reduce the number of cables and disturbances in cars. We have reviewed potential wireless solutions, specifically two of them in (Nouvel et al., 2009A). We have performed tests similar to PLC tests in order to qualify the channel in the 2.4 GHz band. Data rate measurements show it is possible to achieve more than 10 Mbps/s in the vehicle, using also OFDM technology. Additional studies have been carried out in (Nolte et al., 2009). The authors in (Zhang et al., 2009) have conducted measurements in the [0.5 - 16] GHz band. One can observe the different delay profile, different clusters, different paths and the impact of passengers. Due to lake of space, it is not possible to describe all the measurements. And we invite the interested readers to look at the papers and chapters.

6. From static to dynamic ECU and communication networks

Taking into account all these networks, from specific network up to PLC or wireless combined with the constraint of flexibility and security, one attractive idea is to be able to switch from one network to another one, without additional cost. If the main communication fails, the ECU (modem) can switch to the secondary protocol and continue to run. Reconfigurable architectures based on FPGA may offer very flexible links inside a vehicle. A dynamically reconfigurable system allows changing parts of its logic resources without disturbing the functioning of the remaining circuit. This property can applied for networks, in order to allow changing from one protocol to another one according to the channel behaviour, errors, load, etc. This section will discuss about this new concept and demonstrates how it can be integrated in vehicle.

Certain modern FPGAs offer dynamic and partial reconfiguration (DPR - Dynamically and Partially Reconfigurable) capability that allows to change dynamically one portion of the FPGA without affecting the rest of the circuit. Currently, the Xilinx Virtex FPGAs (Xilinx, Inc, 2008) are the only commercially available circuits supporting the DPR paradigm and large applications implementation. Internal structure of a Xilinx Virtex5 is presented in Figure 1. The main resources dispatched in the FPGA matrices are slices, DSP blocks (DSP48E), memory blocks (BRAM), input/output (IO) banks, and Clock Management Tiles (CMTs) as well as the reconfiguration interfaces, so called ICAP. Slices are the smallest configurable elements constituted of LUTs (Look-Up Table), registers and logic gates. DSP blocks offer a powerful set of processing elements for data applications.

The dynamic reconfiguration takes place in Partially Reconfigurable Region (PRR) which can be partially reconfigured independently. Designing a dynamically self-reconfigurable system always require the declaration of PRRs. A PRR is implemented statically despite the fact that its content is dynamic. Thus, at runtime, dynamic reconfiguration can only take place into the PRR. Communications between a dynamic task and its static environment is assured through the bus macro interfaces. Bus macros are also specified statically.

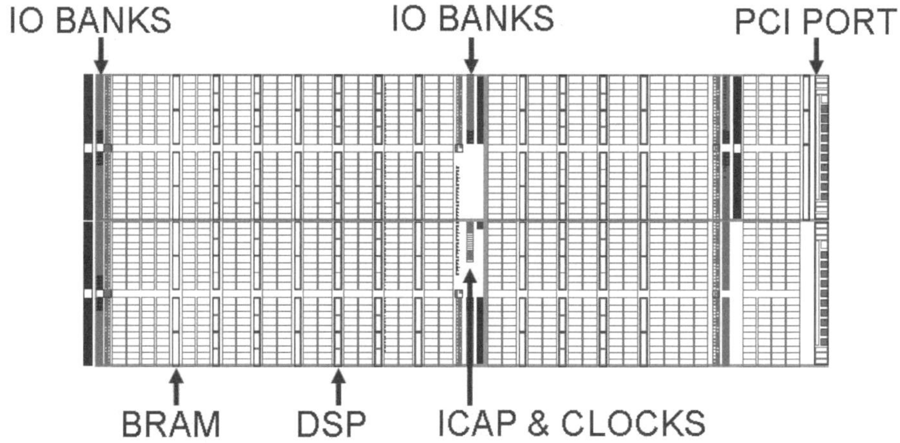

Fig. 14. View of the Virtex5 5VSX50T captured from Xilinx PlanAhead design tool

The FPGA fabric is partitioned into one static logic and one or more partially reconfigurable regions (PRRs). This fabric partitioning enables reconfiguration of a single PRR without system interruption (the static region and other PRRs continue execution while only the reconfigured PRR halts). Each PRR has a related partial bitstream and the reconfiguration process can be done by sending this partial bitstream to the reconfiguration port. In modern FPGAs, the reconfiguration is stored in SRAM based memory, leading to a weakness from a reliability point of view.

Modern FPGAs, besides customary high-density reconfigurable resources, offer the designers the possibilities of implementing programmable processors having features of Commercial Off-The-Shelf (COTS) components (no need to modify processor architecture or application software). Processors play the role of processing units, and one particular is used as coordination units in the embedded system. Besides, processors are in charge of collecting the data from peripherals and from the memory, process the data and send them to the memory and to the peripherals. Also, processors manage the memory and initialize the peripherals. Xilinx FPGA devices include two categories of processors: the hardcore embedded processor (PowerPC) (Xilinx, Inc, 2004) and softcore processors (MicroBlaze, PicoBlaze) (Xilinx, Inc, 2009). Hardcore embedded processors are hard-wired on the FPGA die and their number is limited on each device. On the other hand, softcore processors use reconfigurable resources, so the number that can be actually implemented depends on the device size only. The MicroBlaze tasks can be classified into 2 types: hardware tasks and software tasks. Hardware tasks are peripherals connecting to MicroBlaze. Software tasks are software programs running inside MicroBlazes. Generally, hardware tasks are designed using High Description Language HDL like VHDL, Verilog and software tasks are programmed using C. Regardless of their design methods they are presented in our system in compiled forms of binary files called bitstreams. A bitstream is the set of binary data describing the circuits implemented on the FPGA, or in PRRs (partial bitstream). Shorten term "bitstream 1" will refer to all the bitstreams of FPGA1, idem with "bitstream 2" for FPGA2, ….

The processor software context is a set of information needed to uniquely define the state of the processor at a given moment. It could include the states of the processor registers, the cache, the memory, etc. Saving and restoring all relevant values allow for processor context switching and error recovery. The softcore processor MicroBlaze context is represented by the 32-bit values of 32 General Purpose Registers and two Special Registers: the Program Counter (PC) and the Machine Status Register (MSR).

A MicroBlaze task migration consists in migrating hardware task, software task and restoring the software context. Hardware task migration requires the appropriate peripheral to be added using dynamic reconfiguration. Software context is also migrated by dynamic reconfiguration. And copying the saved software context into the related MicroBlaze program memory does the software context recovery process.

Due to their flexibility, FPGAs are attractive for mission-critical embedded applications like automobile domain, but their reliability could be insufficient unless some fault-tolerance techniques capable of mitigating soft errors are used. Dynamic partial reconfiguration provide not only the flexibility in both hardware and software, but also further solutions dealing with reliability problem in critical domains. The dynamic reconfiguration allows the reloading of the defected module to the correct state and the re-execution of the attributed tasks. It cans also re-distribute defected tasks in the faulty module to other processing units in the system.

We present here the feasibility of integrating dynamic reconfiguration features into automotive-aimed applications in which certain fault-tolerance degrees should be maintained. In case of a fault occurrence, the system must be capable of react in real-time to ensure the safety for driver as well as pedestrian. The reaction in this case can be the fast fault detection and correction by loading the original configuration to put the faulty module to the state as at start-up. It can also be the critical task migration from the defected module to another module.

To define a new embedded automotive platform based on reconfigurable architecture, in CIFAER (CIFAER, 2008) we advocate for the use of Radio Frequency and Power-Line Communication for intra-vehicle communications (Nouvel et al., 2008). The communication can be switched from one to the other by dynamically reconfiguring a defined communication zone on an FPGA. These two modes offer very flexible links inside a vehicle. Figure 15 shows the fault-tolerant multi-FPGA platform. The system consists of four FPGAs connected together using two Ethernet communication (in future development one will be based on PLC interface, while the other will be constructed on RF connections). The first network is routed via a network switch, the other network form a ring topology for the fault-tolerance purpose. The Ethernet protocol is built by Ethernet controller as MicroBlaze hardware peripherals and LightWeight IP (LightWeight) as the software library. The lwIP is an open-source stack using TCP/IP protocol, which can be easily adapted to PLC and wireless modem. Each FPGA contain a fault-tolerant dynamic multi-processors system, consisting of several MicroBlaze (Figure 16). Further details about this system architecture, called FT-DyMPSoC, as well as the fault-tolerance schemes implemented can be found in (Pham et al., 2009) and (Pham et al., 2010) for interested readers.

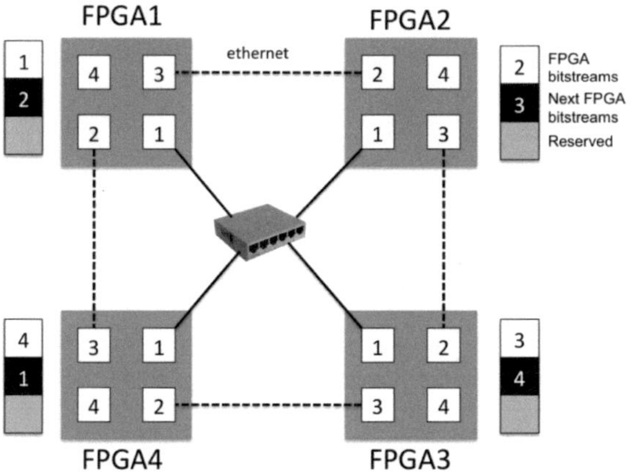

Fig. 15. Fault-tolerant multi-FPGA platform. Two communications networks are supported for reliability purpose

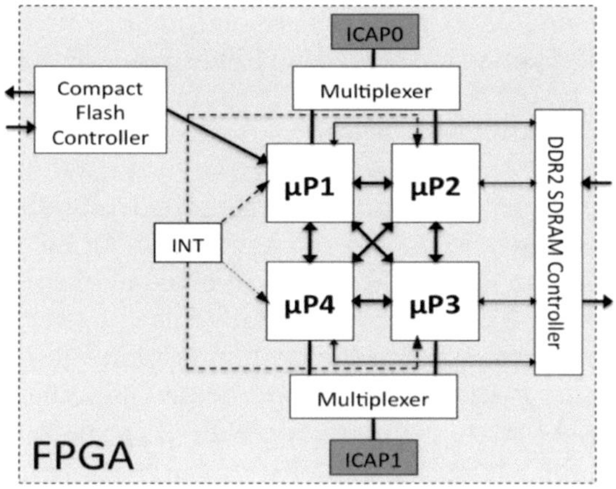

Fig. 16. FT-DyMPSoC Architecture. Included in each FPGA this architecture insert fault-tolerant mitigation schemes

On the overall system, each FPGA is interfaced with a memory that can be accessed by all the processors inside the same FPGA. This memory is partitioned into three segments (Figure 15):

- One for saving all the bitstreams and the software contexts of all the processors of this particular FPGA.
- One for saving all the bitstreams of the next FPGA in the ring network.
- One reserved and used in case of failure occurrence in the system. This segment helps to transfer the bitstreams and contexts between different FPGAs.

The memory segmentation guarantees the existence of at least one copy of all the bitstreams over the whole network.

As we can see in Figure 17, the bitstream of each FPGA is present in its local memory and also in the local memory of the previous FPGA in the ring topology. For example, FPGA1 stores its own bitstream 1 and and the bitstream 2, FPGA2 stores bitstream 2 and bitstream 3… These copies will be used in case of system failure, and permit fast context switching.

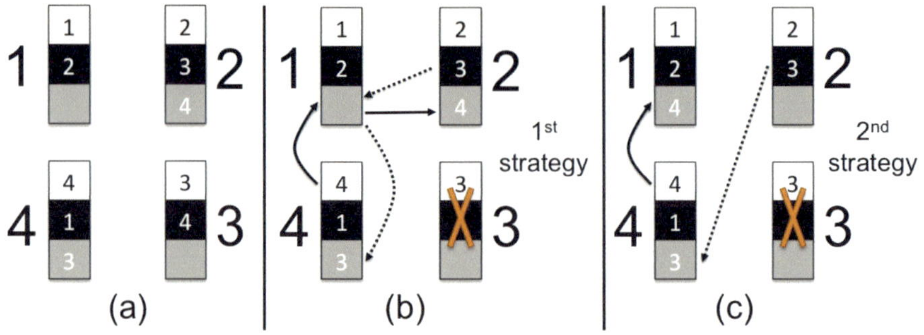

Fig. 17. Fault recovery strategies. Once the faulty FPGA is identified, the copies of the bitstreams are exchanged in order to keep a valid copy of all the configurations

The fault-tolerance degrees are maintained at two levels in the system. The Intra-FPGA level corresponds to the fault-tolerance strategy inside each FPGA, and is related to the design of the FT-DyMPSoC system. The fault-mitigation strategy is realized using the connection matrices algorithm (Pham, 2009), and fault are mitigated by using dynamic reconfiguration at the processors level. The second level called Inter-FPGA level corresponds to the overall system presented in Figure 15. To detect error in the overall network, all the FPGAs exchange frequently among them detection frames. These frames contain the software contexts of the four MicroBlazes of each FPGAs. On one hand, this helps detecting error in the network. On the other hand, including the contexts within the detection frame will help to resume the tasks of a faulty FPGA on another FPGA. During the exchange if the contexts of one FPGA (i.e. FPGA3 in Figure 17) are not received by the others circuits, the FPGA3 is declared faulty. There are 2 possibilities: the MicroBlaze 1 (supporting the interface to the network) of FPGA3 is faulty, causing the communication lost of this FPGA, or the whole FPGA3 is faulty. In order to distinguish these 2 possibilities, the secondary ethernet links is used. FPGA2 and FPGA4 try to communicate with MicroBlaze 2 and 3 of FPGA3. If these communications fails, the whole FPGA3 is declared defected, if not, only the MicroBlaze 1 is defected.

If only one MicroBlaze inside one FPGA fails, we can manage this error thanks to dynamic reconfiguration of this processor or by using task migration within the MPSoC system. The error is managed at the FPGA level. If the whole FPGA fails the task migration concerns the overall circuit. In this case the task of the FPGA3 needs to be dispatched across the remaining circuits. If the system cannot manage all the tasks with one missing FPGA priority needs to be defined and used to maintain critical services for example. In this case, arbitration on the running tasks needs to be executed, and reconfiguration of the remaining FPGA is launched.

If one FPGA is lost, we need to maintain the two bitstreams copy stored in the faulty FPGA. For example, if the FPGA3 is lost (Figure 17), the copies of bitstream 3 and 4 are inaccessible requiring a clone of bitstream 3 and bitstream 4. We propose here 2 strategies delivering the bitstream 3 and 4 to other FPGAs.

1. The first strategy uses only the secondary communication media. We need to use FPGA1 reserved segment as intermediate medium. First the bitstream 4 is copied from FPGA 4 to FPGA1 reserved segment, then to FPGA2. Afterwards, bitstream 3 is copied from FPGA2 to FPGA1, then to FPGA4.

2. The second strategy requires both communication media. Bitstream 4 is copied from FPGA4 to FPGA1 using direct Ethernet link. Simultaneously, bitstream 3 is copied from FPGA2 to FPGA4 using the primary Ethernet via the switch.

In case the Ethernet switch fails, all the primary Ethernet connections are defected; This leads to a connection loss between all the FPGAS. At this moment all circuits switch to the ring topology. The second network will then ensure proper operation of the overall system. The use of redundancy of the network, coupled with the new dynamically reconfigurable paradigm permits to construct highly reliable system.

7. Conclusion

In this chapter, an initial foreseeable solution with PLC has been presented to allow communications and interoperability between embedded applications with different requirements. PLC network answers both to cost, flexibility and throughput requirements. Future work should be devoted to optimize both PLC modulations and ECU architectures in order to minimize the number of cables and ECU etc. This implies rethinking the DC bundles as rethinking the implementation of networks as independent domain.

Furthermore, it is possible to build a reconfigurable ECU for both application and communication. This new concept will allow combining different network technologies. It will answer to fault tolerance constraints, required in X-by-wire applications

This work has been carried out by the CIFAER project, supported by the ANR and by the French Premium Cars competitiveness Cluster ID4car.

8. References

Afkhamie, K.H.; Katar, S.; Yonge, L. & Newman R. (2005). An overview of the upcoming HomePlug AV standard, *Proceedings of the IEEE International Symposium on Power Line Communications and Its Applications*, pp. 400–404, 0-7803-8844-5, 6-8 Apr. 2005, Vancouver, BC, Canada.

Arabia, E.; Ciofi, C.; Consoli, A.; Merlino, R. & Testa A. (2006). Electromechanical Actuators for Automotive Applications Exploiting Power Line Communication, *Proceedings of SPEEDAM*, pp. 909-914, 1-4244-0193-3, Taormina, 26-26 May 2006

Bahai, A.; Saltzberg, R. & Ergen, M (2004). *MultiCarrier Digital Communications*, ISBN: (HB) 0-387-22575-7, Springer NewYork

Barmada, S.; Raugi, M.; Tucchi, M. & Zheng, T. (2010). Powerline communication in a full electric vehicle, *Proceedings of IEEE International Symposium on Power Line Communications and Its Applications*, pp. 331-336, 978-1-4244-5009-1, Rio de Janeiro , 28-31 March 2010

Bassi, E.; Benzi, F.; Almeida, L. & Nolte, T. (2009). Powerline communication in electric vehicles, *Electric Machines and Drives Conference*, pp. 1749–1753, 978-1-4244-4251-5, Miami, 3-6 May 2009

Beikirch, H. & Voss, M. (2000). CAN-transceiver for field-bus power line communication, *Proceedings of the International Symposium on Power Line Communications and Its Applications*, Limerick, pp. 257–264, April 2000

Benzi, T.; Facchinetti, T. Nolte & Almeida L. (2008). Towards the powerline alternative in automotive applications, *Proceedings of Factory Communication Systems*, pp. 259-262, 978-1-4244-2349-1, Dresden, 21-23 May 2008

CAN. (2009), Retrieved from official web site http://www.can-cia.org/ (2009)

CIFAER (2008), Communications Intra-véhicule et Architecture Embarquée Reconfigurable, French ANR Project 2008-2011. Available at web site www.agence-nationale-recherche.fr/AAPProjetsOuverts

Claus, C.; Stechele W. (2010), AutoVision - Reconfigurable Hardware Acceleration for Video-Based Driver Assistance, *In: Platzner, Teich, Wehn (Editors): Dynamically Reconfigurable Systems*, ISBN 978-90-481-3484-7, Springer, 2010

De Caro S. & Testa, A. (2009). A Power Line Communication approach for body electronics modules, *Proceeding of Power Electronics and Applications*, pp. 1-10, 978-1-4244-4432-8, Barcelona, 8-10 September 2009

Degardin, V.; Laly, P.; Liénard, M. & Degauque, P. (2006). Impulsive Noise on In-Vehicle Power Lines: Characterization and Impact on Communication Performance, *Proceedings of IEEE International Power line Communications and Its Applications*, pp. 222-226, 1-4244-0113-5, Orlando, 26-29 March 2006.

Degardin, V.; Lienard, M.; Degauque, P. & Laly, P. (2007). Performances of the Homeplug PHY layer in the context of in-vehicle powerline communications, *Proceedings of Power Line Communications and Its Applications*, pp. 93-97, 1-4244-1090-8, Pisa, 26-28 March 2007

FlexRay Consortium. (2009). FlexRay Communication System, Protocol Specification, Version 2.0. Retrieved from: http://www.flexray.com

Galli, S.; Koga, H. & Kodama, N. (2008). Advance signal Processing for PLCs: Wavelet OFDM, *Proceedings of Power Line Communications and Its Applications*, pp. 187-192, 978-1-4244-1975-3, Jeju city, Jeju Island, 2-4 April 2008

Gavette, S. & al. (2006). HomePlug AV Technology Overview, *Technical report. Sharp Laboratories of America*. Retrieved fromdownload.microsoft.com/download.

Gouret, W.; Nouvel, F. & El Zein, G. (2006). Additional Network Using Automotive Powerline Communication, *Proceedings of International Conference on Intelligent Transport Systems Telecommunications*, pp. 1087-1092, 0-7803-9587-5, Chengdu, 26-29 March 2006

Gouret, W.; Nouvel, F. & El Zein, G. (2007), High Data Rate Network Using Automotive Power Line, *Proceedings of International Conference on Intelligent Transport Systems Telecommunications*, pp. 1-4, 1-4244-1178-5, Sophia Antipolis, 26-28 March 2007

IEEE P1901 (2008), IEEE P1901 Working group. Retrieved from web official web site http://grouper.ieee.org/groups/1901/

Home Plug (2009). HomePlug Powerline Alliance. Retrieved from official web site www.homeplug.org

Huck, T., Schirmer, J. & Dostert, K. (2005). Tutorial about the implementation of a vehicular high speed communication system. in *IEEE International Power line Communications and Its Applications IEEE ISPLC* , pp 162-166, 6-8 April, 2005.

J2056/2 Survey. (1994). J2056/2 survey of known protocols. *In SAE Handbook*. Warrendale, PA: Soc. Automotive Eng. (SAE), vol. 2.

Leen, G; Hefferman, D. (2001). Vehicles without wires. Computing & Control Engineering Journal, Volume. N° 12(Issue 5), October (pp. 205-21).

Lienard, M.; Carrion, M.; Degardin, V. & Degauque, P. (2008). Modeling and Analysis on In-vehicle power line communication channels, *Proceeding of IEEE Transaction on Vehicular Technology*, vol 57, N°2, pp. 670-679, 0018-9545

LIN Consortium. (2003). LIN Specification Package, Revision 2.0.Retrieved from http://www.lin~subbus.org/(2009)

Mohammadi, M.; Lampe L.; Lok, M.; Mirabbasi, S.; Mirvakili, M.; Rosales, R. & Van Veen, P. (2009). Measurement study and transmission for in-vehicle power line communication, *Proceedings of IEEE Power Line Communications and Its Applications*, pp. 73–78, 978-1-4244-3790-0, Dresden, 29-March -1 April 2009

Navet (2008), *The Automotive Embedded Systems Handbook, Industrial Information Technology series*, Taylor and Francis , CRC Press, ISBN 978-0849380266, 2008.

Newman, D. (2009). Benchmarking Terminology for Firewall Performance, RFC 2647.

Nolte,T. (2006). Share-Driven Scheduling of Embedded Networks. *University, Dissertation, May 2006. Department of Computer Science and Engineering Mälardalen* University Västeras, Sweden, Printed by Arkitektkopia, Västeras, Sweden Distribution , 2006

Nolte,T. & Hansson H.(2009). Wireless Automotive Communications, Internal *Report* available at http://ant.comm.ccu.edu.tw/course/97_ITS/1_HW1/0.Wireless%20Automotive%20Communications.pdf ,2009

Nouvel, F.; El Zein, G. & Citerne, J. (1994). Code division multiple access for an automotive area network over power-lines, *Proceedings of IEEE Vehicular Technology Conference*, pp. 525–529, 0-7803-1927-3, Stockholm, 8-10 June 1994.

Nouvel, F. & Maziéro, P. (2008). X-by-Wire and intra-car communications: power line and/or wireless solutions, *Proceedings of International Conference on Intelligent Transport Systems Telecommunications*, pp. 443-448, 978-1-4244-2857-1, Phuket, 24 October 2008.

Nouvel, F; Gouret, W; Maziéro (2009a), Automotive Network Architecture for ECUs Communications, *in Automotive Informatics and Communicative Systems: Principals in Vehicular Networks*", 28 pp, 2009, ISBN 978-160566338-8, 2009

Nouvel F.; Tanguy, P. & Maziero P. (2009b), What is about next high speed power line communication systems for in -vehicle networks, *Proceedings of ICICS*, pp. 533-537, 978-1-4244-4656-8, Macau, 8-10 December 2009

OMEGA (2008), Deliverable D3.1 State of the art, application scenario and specific requirements for PLC, *OMEGA ICT-213311 Project*, FP7, Available on web site http://www.ict-omega.eu/fileadmin/documents/deliverables/Omega_D3.1.pdf

Pham, H.-M.; Pillement S. & Demigny, D. (2009) A Fault-Tolerant Layer For Dynamically Reconfigurable Multi-Processor System-on-Chip, *in Proc. Int. Conf. on ReConFigurable Computing and FPGAs*, pp. 284–289, Cancun, Mexico, Dec. 2009.

Pham, H.-M.; Pillement, S. & Demigny, D. (2010) Evaluation of Fault-Mitigation Schemes for Fault-Tolerant Dynamic MPSoC, *in Proc. Int. Conf. on Field Programmable Logic and Applications*, 2010

Rubin, A. (2002). Implementing automotive protocols for communications over noisy battery power lines, *Proceedings of the IEEE Conference on Conv. Elect Electron. Eng.*, pp. 306, 1 December 2002

Ribeiro. et al. (2006) Power Line communications : a promising communication system paragidm for last miles and meters applications, *Telecommunications : Advances and trends in transmissions*, pp. 134-154 , ISBN 85-98876-18-6.

Silva, P.; Almeida, L.; Caprini, D.; Facchinetti, T.; Benzi, F. & Nolte, T. (2009). Experiments on timing aspects of DC-powerline communications, *Proceedings of IEEE international Conference on Emerging Technologies & Factory Automation*, pp. 1674-1677, 978-1-4244-2727-7, Palma de Mallorca, Spain, September 22-25 2009

Spidcom (2008). Spidcom Inc., Retrieved from official web Official web site http://www.spidcom.com.

Tanguy P.; Nouvel, F. & Maziéro, P. (2009b), Power Line Communication standards for in-vehicle networks, pp. 533-537, 978-1-4244-5347-4/09, Lilles, 20-22 October 2009

Tanguy, P.; Nouvel, F. (2010). In-Vehicle PLC Simulator Based on Channel Measurements, *in next Proceedings of International Conference on Intelligent Transport Systems Telecommunications*, Kyoto, 9-11 November 2010.

Umehara, D.; Morikura, M.; Hisada, T.; Ishiko S. & Satoshi, H. (2010). Statistical Impulse Detection of In-Vehicle Power Line Noise Using Hidden Markov Model, *Proceedings of Power Line Communications and Its Applications*, pp. 341-346, March 2010.

Valéo (2006). Valéo Inc., Electrical and Electronic Distribution Systems: Focus on Power Line Communication. Retrieved from official web http://www.valeo.com

Xilinx, Inc (2004). PowerPC 405 Processor Block Reference Guide, 2004

Xilinx, Inc (2008). Early Access Partial Reconfiguration User Guide UG208, September 2008.

Xilinx, Inc (2009). MicroBlaze Processor Reference Guide UG081 (v10.3), 2009

Yamar, 2009, http://www.yamar.com, last access 20.09.2009

Yabuuchi, Y.; Umehara, D.; Morikura, M.; Hisada, T.; Ishiko ,S. & Horihata, S. (2010). Measurement and Analysis of Impulsive Noise on In-Vehicle Power Lines, *Proceedings of Power Line Communications and Its Applications*, pp. 325-330, March 2010.

Zhang, N.; Zhu, X.; Liu, L.; Yu, C.; Zhang, Y.; Dong, Y.; Zhang, H.; Kuai, Z. & Hong, W. 2009. Measurement and characterization of wideband channel for in-vehicle environment. In *Proceedings of the 4th international Conference on Radio and Wireless Symposium* (San Diego, CA, USA, January 18 - 22, 2009). IEEE Press, Piscataway, NJ, 183-186.

15

Kinesthetic Cues that Lead the Way

Tomohiro Amemiya
NTT Communication Science Laboratories
Japan

1. Introduction

Wayfinding is of vital importance to pedestrians walking in unfamiliar areas. Generally, pedestrians rely on directional information, street names, and landmarks [Bradley & Dunlop, 2005]. Recently, many mobile devices, such as mobile smart phones, can provide us with detailed digital maps, global positioning information, and navigational information. These location-based data and services are usually presented on visual displays. However, the visual displays in mobile devices are very small, which makes it hard to see and use the data. With the increasing complexity of information, and the variety of contexts of its use, it becomes important to consider how other non-visual sensory channels, such as audition and touch, can be used to communicate necessary and timely information to users. Additionally, there are a number of user groups, such as visually impaired people and the emergency services, who also require non-visual access to geographical data.

Kinesthetic stimulation, such as that for pulling or pushing the hand, has the potential to be more intuitive and expressive than cutaneous stimulation, such as rumbling vibration, in conveying direction information because force feedback devices can indicate a one-dimension direction directly. Although a substantial number of force feedback devices have been developed in the last twenty years, most of them use either mechanical linkage to establish a fulcrum relative to the ground (Massie & Salisbury, 1994), use a huge air compressor (Suzuki et al., 2002; Gurocak et al., 2003), or require wearing a heavy device (Hirose et al., 2001). Physical constraints mean that none of them can be used in portable information devices. Some portable "torque" displays have been proposed, based on the gyro effect (Yano et al., 2003) or angular momentum change (Tanaka et al., 2001) have been proposed; however, they can produce neither a constant force nor a translational force without also producing a reaction force; they can generate only a transient rotational force since they use a change in angular momentum. Recently, there have been a number of proposals for generating both constant and directional forces without an external fulcrum by using two oblique motors whose velocity and phase are controlled (Nakamura & Fukui, 2007), by shifting the center-of-mass of a device dynamically to simulate kinesthetic inertia (Swindells et al., 2003), and by producing an air pressure field with airborne ultrasound (Iwamoto et al., 2008).

In contrast, our idea is to exploit the characteristics of human perception to devise a new force perception method for portable information devices that can generate a translation force sensation with a long duration (Amemiya et al. 2005; Amemiya and Maeda 2009). The method uses an asymmetric oscillation, where brief intense pulses of acceleration alternate with longer periods of low-amplitude recovery. Although the net acceleration is zero, humans perceive a sustained force sensation in the direction of the pulses. This is attributed to the nonlinear relationship between perceived acceleration and physical acceleration. We built a handheld prototype that generates periodic motion through asymmetric acceleration, in which asymmetric oscillation is generated by a swinging slider-crank mechanism. Our previous findings indicated that the pulse frequency determines the effective generation of the kinesthetic illusion of being pulled. In this chapter, we present a new hybrid configuration comprising a swinging slider-crank mechanism and a cam mechanism as an approach to fabricating a smaller force feedback system for portable information devices and describe an experiment in which we conducted an empirically examined turn-by-turn navigation with the device used by pedestrians with visual impairments. The results show the device intuitively conveys turning instructions and has potential to be used by untrained users.

2. Pseudo-attraction force

2.1 Haptic sensory illusion

The study of illusions can provide valuable insights into not only human perceptual mechanisms but also the design of new human interfaces. To generate a sustained translational force without grounding, we focused on the characteristics of human perception, which until now have been neglected or inadequately implemented in haptic devices. Although we human beings always interact with the world through human sensors and effectors, the perceived world is not identical to the physical world. For instance, different spectra can elicit the same color in human perception. When we watch television, the images on TV (a combination of RGB colors) we see are different from what we see through windows, i.e., a natural image is a composition of all wavelengths of light. Furthermore, animation actually consists of a series of still pictures in a flip book. Different stimuli can produce almost the same percept, which, though it may seem strange, is normal for humans. Since some illusions are very stable independent of individual variation, we can apply those illusions in practice, such as in designing human interfaces, if we can figure out ways to convert them to subjectively equivalent percepts. Hayward has pointed out that illusions are at the basis of virtually all technological displays (Hayward 2008), mentioning this also includes haptic interfaces.

2.2 Principle

The kinaesthetic illusion of being pulled or pushed, discovered by the authors (Amemiya et al. 2005), can be used to design haptic interfaces. Using different acceleration patterns for two directions to create a perceived force imbalance, the method exploits the characteristics of human perception to generate a force sensation and thereby produce the sensation of directional pushing or pulling. Specifically, a quicker acceleration (stronger force) is generated for a very brief time in the desired direction, while a slower acceleration (weaker force) is generated over a longer period in the opposite direction. The internal human haptic sensors do not detect the slower acceleration (weaker force) , so the original position of the mass is washed out. The result is that the user is tricked into perceiving a unidirectional force. This force sensation can be made continuous by repeating the motions. If the acceleration patterns are well considered and designed, a kinesthetic illusion of being pulled can be created because of this nonlinearity.

2.3 Requirements

There are still many aspects of the manifestation of the kinaesthetic illusion of the pseudo-attraction force that are not well understood, but putative mechanisms have been accumulating. We know that no directional force is felt if the mass is merely moved back and forth, but that using different acceleration patterns for the two directions to create a perceived force imbalance produces the perception of a pseudo-attraction force (Amemiya & Maeda, 2009). The frequency of the oscillation plays an important role in eliciting the perception of a pseudo-attraction force. Oscillations with high frequency might create a continuous force sensation, but previous experimental results have shown that the performance decreases steadily at frequencies over ten cycles per second (Amemiya et al., 2008). In contrast, oscillations with low frequency tend to be perceived as a discrete knocked sensation. If we wish to create a sustained rather than a discrete force sensation, such as the sensation of being pulled continuously, the frequency should be in the five to ten cycles per second range. In addition, changes in the gross weight and the weight of the reciprocating mass affects the perceived force sensation. The threshold of the ratio of the gross weight and the weight of the reciprocating mass is 16%, which is a rough standard for effective force perception in the developed prototype (Amemiya & Maeda, 2009).

3. Hardware design

Our first prototype used a swinging-block slider-crank mechanism to create an asymmetric oscillation [Fig. 1(a)]. In the mechanism, a circular motion of constant speed (crank OB) is transformed into a curvilinear motion since a swinging linkage BC slides and turns around point A. The end point on the curvilinear motion (point C) is connected with a rod (point D), which slides along a linear slider with asymmetric acceleration back-and-forth. Because of the length of the linkages, especially linkage CD (rod), the overall length of the mechanism tends to be large at about 175 mm (Amemiya and Maeda 2008; 2009). Figure 1(b) shows the new mechanism, which is the equivalent mechanism of the previous one but with the length of linkage CD decreased to virtually zero. As in the previous prototype, a circular motion of constant speed (crank OB) is transformed into a curvilinear motion by a swinging-block slider-crank mechanism. The difference is that the end point on the curvilinear motion (point C) slides along a grooved cam, whose shape is a circular arc with a radius of CD, with point D as the center of the arc. This produces a reciprocating motion with asymmetric oscillation.

The mechanisms in Fig. 1 have a single DOF (degree of freedom). We previously developed a prototype of a two-dimensional force display by having one module based on the slider-crank mechanism mounted on a turntable. The direction of the force display module was set by driving the turntable with a belt drive system. Turntable rotation, however, took considerable time, which meant that immediacy was lost. To overcome the problem of turntable rotation, we adopted the summation of linearly independent force vectors. We then fabricated a new 2-DOF prototype to generate a force sensation in at least eight cardinal directions by the summation of linearly independent force vectors. To create the force display, we stacked four layers, each containing a single module and two of which were orthogonal. By combining the force vectors generated by each module, the force display can create a force sensation on a two-dimensional plane more quickly than the turntable approach. The force display has the potential to create a force sensation in any arbitrary direction on a two-dimensional plane if the amplitude of the force vector can be changed. The asymmetric oscillation ($F(t)$) is given by

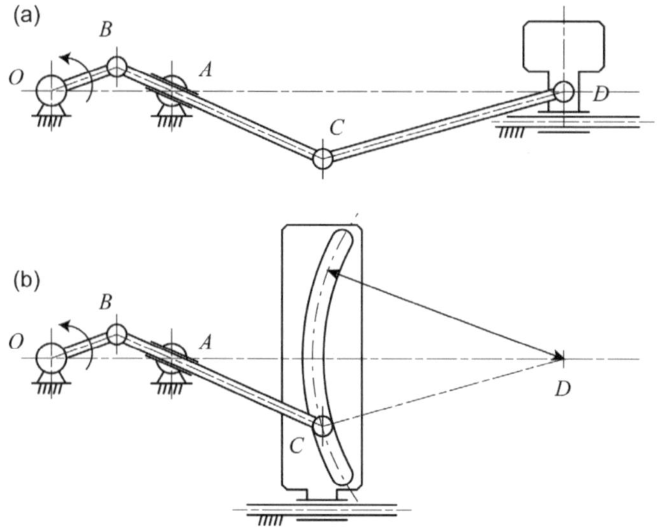

Fig. 1. Mechanisms for generating asymmetric oscillation in first prototype (a) and new one (b). The two mechanisms move identically

$$F(t) = \sum_{j=1}^{n} m_j \frac{d^2 x_j(t)}{dt^2} e_j \tag{1}$$

where m_j is the weight in module j, n is the number of modules, and d^2x_j/dt^2 is the acceleration generated by module j. The acceleration d^2x_j/dt^2 is given by the second derivative with respect to time of the motion of the weight x_j. The equation for the motion of the weight in module j is

$$x_j(t) = l_1 cos\omega_j t + \mu_j (d - l_1 cos\omega_j t) + \sqrt{l_3 - \{l_1(\mu_j - 1)\sin\omega_j t\}^2} \tag{2}$$

where

$$\mu_j = \frac{l_2}{\sqrt{l_1^2 + d^2 - 2l_1 d \cos\omega_j t}} \tag{3}$$

and $x_j(t)$ = OD, d = OA, l_1 = OB, l_2 = BC, l_3 = CD, and $\boldsymbol{\omega}_j t$ = AOB in Fig. 1. The $\boldsymbol{\omega}_j$ is the constant angular velocity, and t is time. In the prototype, d = 28 mm, l_1 = 15 mm, l_2 = 60 mm, l_3 = 70 mm, and n = 4, and the unit vectors are $\langle e_j, e_{j+1} \rangle = 0$, $||e_j|| = 1$. All m_j and $\boldsymbol{\omega}_j$ values are identical, with m_j = 40 g, $\boldsymbol{\omega}_j/2\pi$= 5 Hz.

In the developed 2-DOF prototype, the output shaft of each motor (DC 6.0 V, 2232R006S; Faulhaber) is mounted in a roller made of nylon. The roller drives the crank wheel by friction. The external diameter of the motor roller is 4 mm, and that of crank wheel is 44 mm. The reduction ratio is basically 1:11, but it changes slightly as a result of changes in factors such as temperature or pressure. The prototype weighs approximately 430 g. The diameter of the base is the same as that of a compact disc (i.e., 120 mm). The prototype is 36-mm thick. Each weight on the slider is equipped with a photo-interrupter (PM-R24; SUNX Ltd.) to detect its position. The speed of each crank is stabilized at five counts per second by closed feedback loop control (P control) with a microprocessor so that the signal intervals of the photo-interrupters are close to 200 ms. When combining two orthogonal force vectors, the phases of the cranks are synchronized by another closed feedback loop (PID control) so that the onset intervals of the photo-interrupters are close to zero.

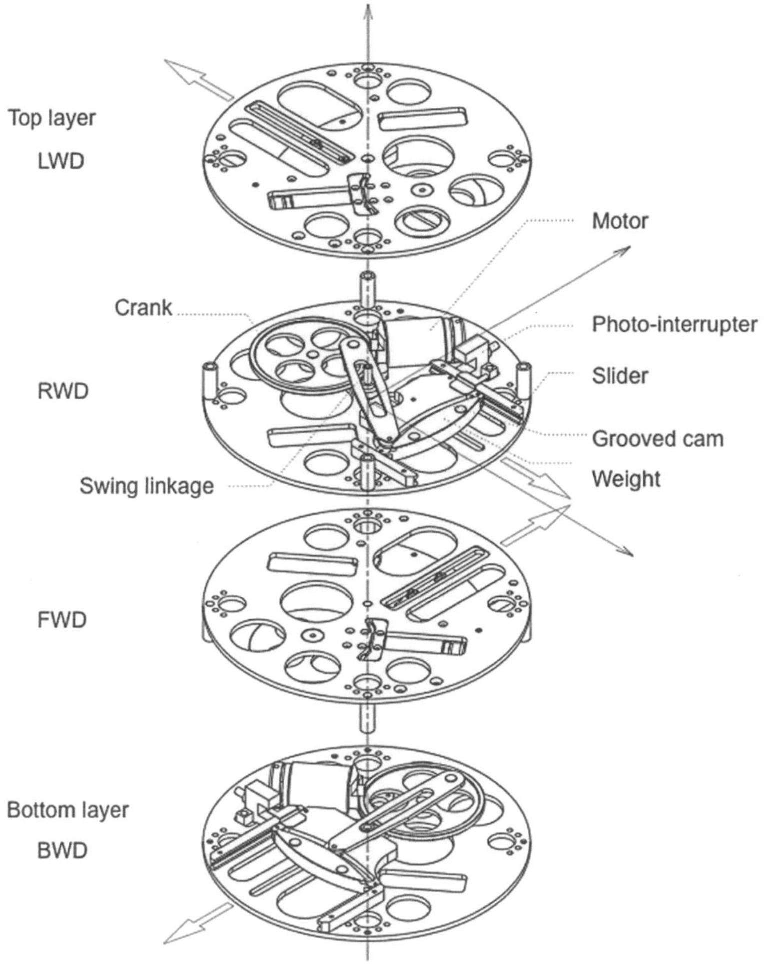

Fig. 2. Structure of the new prototype for generating pseudo-attraction force on a two-dimensional plane

4. Hardware evaluation

Figure 3 shows the measured acceleration profile generated by the device at five counts per second. Acceleration values were calculated from the position data of each weight, which were acquired with a laser sensor (Keyence Inc., LK-G150, 10 kHz sampling) with the bottom of the device fixed to a base. The acceleration profile of the top layer differed slightly from the others, due its distance from the fixed base. The effect of oscillation was augmented by the principle of leverage, leading to some degree of measurement error. The acceleration amplitude reached about 50% of the theoretical acceleration peak. We assume that the friction drive transmitted less torque than the previous gear drive. This clarified that there is a trade-off between the torque transmission efficiency and noise level when we select the friction drive or the gear drive.

Figure 4 shows examples of the response profiles of phase synchronization. The onset intervals between pairs consisting of two orthogonal modules were acquired from the photo-interrupters when the bottom of the device was fixed to the base. An onset interval of zero means that the two orthogonal modules are synchronized. The phases were synchronized within five seconds, which showed that the force display created a force sensation in the eight cardinal directions on a two-dimensional plane.

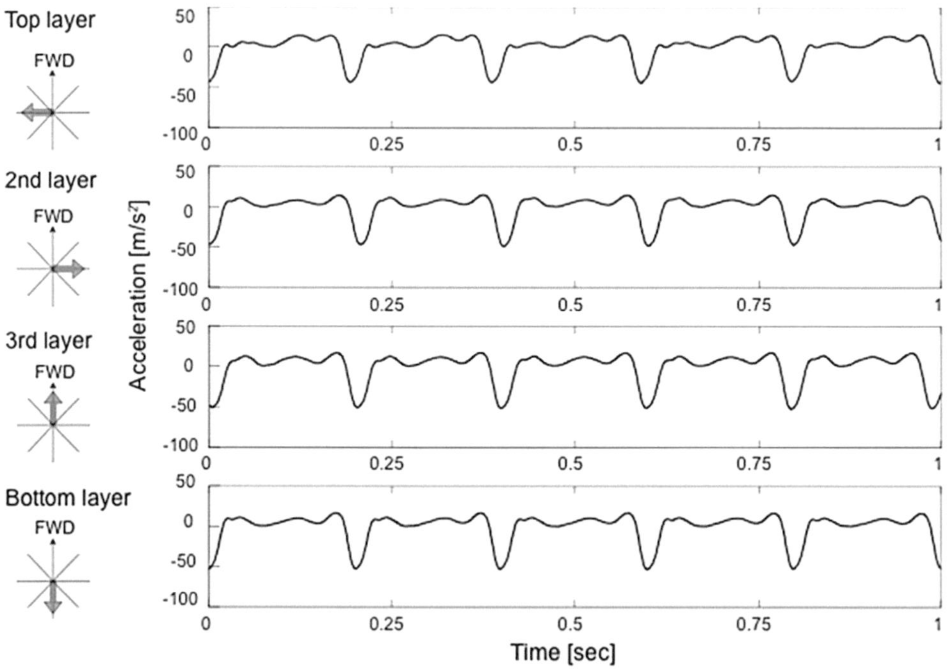

Fig. 3. Acceleration profile calculated from the position data measured with a laser sensor. The data were processed by a seventh order LPF Butterworth filter with a cut-off frequency of 100 Hz

To drive the crank, the first prototype used a pinion gear and crown gears, whose axes were relatively displaced. The relative displacement of the gear axes caused gear noise, which annoyed the users. In fact, many people, including people who are blind, who have held the previous prototype, have complained about the noise. The sound pressure level of the noise generated by the previous prototype was measured in an anechoic room at NTT Communication Science Laboratories. A sound level meter (Rion, Inc., NL-31 Class 1) was used to measure the noise with the A-weighted sound pressure level (SPL). The sound level meter was fixed to a tripod at a height of 1.0 m from the ground and 30 cm from the prototype. The measured noise SPL showed that with the gear drive in the prior prototype, the noise level exceeded 60 dB(A), whereas environmental noise level was 15 dB(A). In contrast, the new prototype uses friction drive, and its noise level at frequencies of 3, 5, 10 counts per second does not exceed 50 dB(A), which shows that the friction drive emits much less noise.

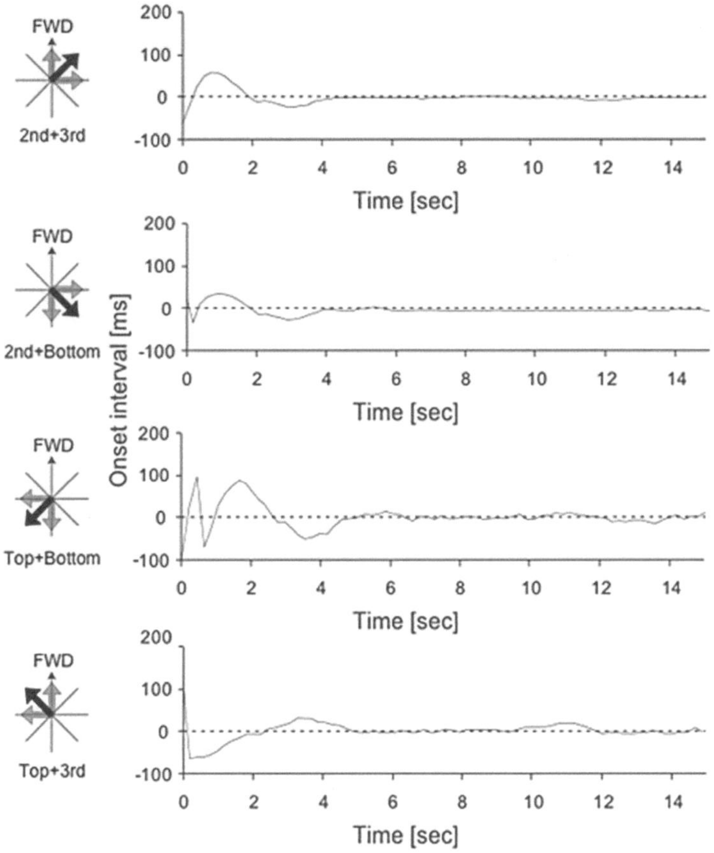

Fig. 4. Examples of measured phase synchronization responses for combinations of each module. The onset intervals obtained by the photo-interrupters were controlled to be close to zero by changing the angular velocity of the motor in the force display

5. User studies

We performed two user studies to investigate the applicability of the force display to pedestrian navigation. One was a psychophysical pilot study, which revealed that people who held the force display clearly sensed a directed force. The other was a navigation experiment related to predefined route guidance, in which people with visual impairments used the force display.

5.1 Directional perception

Five people without visual impairments (three right-handed men and two right-handed women, aged 22-34 years, 25.4±5.3), who were paid volunteers, participated in the psychophysical pilot study. None of the participants had any previous experience with force display prototypes. They were required to reply with one of eight cardinal directions after a five-second oscillating stimulus (north was defined as 0 degrees and the forward direction, and east as 90 degrees and right) without any prior training. Each participant experienced eight conditions × five trials, for a total of 40 trials (randomized). Visual and auditory effects were suppressed by having the participants wear eye-masks and earmuffs. The results showed that the force display provided a well-perceived directed force sensation (pseudo-attraction force sensation) even though the measured acceleration peak was only half the expected value, and all responses from participants were almost identical to the direction of stimuli.

5.2 Pedestrian navigation

For the empirical experiment (Fig. 5), to verify the navigation system, we recruited people with visual impairments who were untrained. We measured the time required to complete a walking task, the level of ease of use and, the users' expectations of the guidance on a subjective scale, and collected user feedback. We built a human-size maze in the gymnasium of the Kyoto Prefectural School for the Visually Impaired, Japan. Since the streets and avenues in Kyoto are mainly laid out in a grid, the routes of the maze were designed in a checkerboard shaped.

Fig. 5. Overview of kinesthetic navigation for pedestrians with visual impairments. After location data received from a satellite system or embedded sensors and orientation data had been collected and processed, a haptic cue of direction was presented to the pedestrian

5.2.1 Method

There were twenty-three visually impaired participants, 19 males and 4 females, who were volunteers from the Kyoto Prefectural School for the Visually Impaired. Ages ranged from 17 to 62 years (30.0±14.7). Thirteen are totally blind, and the other ten are partially sighted. The participants all reported that they had no irregularities with their hands in terms of tactile perception at the time of the experiment, and they were all untrained. The research protocol was approved by local ethics committees. All participants provided written informed consent prior to testing. No participants underwent any prior training. Participants who usually used a cane when walking held a cane with one hand and the haptic direction indicator with the other. Others held the haptic direction indicator with the dominant hand. They carried a small shoulder bag (less than 600 g), which contained a notebook computer (OQO model 02; OQO Inc.), a custom-build circuit with a microprocessor (PIC18F252; Microchip Tech. Inc.), and a 12-V battery (ENAX Corp.). Bluetooth 2.1 allowed the portable computer to communicate with the remote computer (ThinkPad X60s; Lenovo Corp.) within 100 meters. Since the global positioning system (GPS) did not work inside the gym with satisfactory accuracy, we used nine infrared sensors installed at the junctions of the maze as a local positioning system; they were connected to the remote computer. To produce white noise, a clip-shaped music player (iPod shuffle 2nd generation; Apple Inc.) and noise-cancelling headphones (Quiet Comfort 3; Bose Corp.) were used. To measure the yaw angle of the force display, a motion sensor (MDP-A3U9S; NEC TOKIN Corp.) was attached to it. The human-sized experimental labyrinth was formed with a series of foam panels (1,800 mm × 900 mm) occupying a space of 9 m × 15 m. The pads were soft enough not to cause injury in collisions. The route and journey duration were automatically logged by the system with infrared sensors placed at the corners of the walls, captured by digital video cameras from the second floor (about 4 m from the ground), and manually noted by the experimenters.

A participant holding the haptic directional indicator was brought to one of three different departure points in the labyrinth. An experimenter walked behind the participant to ensure his/her safety. First, the force display presented a sensation of pushing forward. The participant then started walking from the departure point. The participant was guided by the force display to turn left or right at a certain turning point, and there were totally nine possible turning points. At that point, the infrared sensor detected the arrival of the participant. In response, the remote computer connected to the infrared sensors sent the turn instruction to the notebook computer in the participant's bag. The direction of the force vector (go straight; turn left or right) was initially determined by the predefined route at each turning point and automatically updated to one of the eight cardinal directions according to the orientation of the participant. In a similar way, they were then guided to the second, third, and fourth turning points and finally to the destination point. Experimenters indicated that the destination was reached after the participants had arrived at the destination point.

The force display was always "on" during the navigation. If the participant made a wrong turn (and was about to begin walking the wrong route), the experimenter changed the direction of the haptic stimuli manually to return the participant to the predefined route and sent them via the remote computer. The same haptic stimuli for the eight cardinal directions were used for the revision. Nevertheless, if this correction failed or the participants did not notice the stimulus change, the experimenter walking behind intervened by touching their backs, giving verbal information about the correct turn and taking note of any incorrect actions. Note that the participants were made aware of a wrong turn in an interruption but not in a revision. The participants had to go back one block to return to the correct path when there was an interruption.

The turn-by-turn navigation task consisted of walking predefined routes in the human-size maze under two auditory information conditions (audio masked and audio unmasked). Under the audio-masked condition, all the audio information was masked by noise-cancelling headphones with white noise. Under the audio-unmasked condition, the participants were able to hear ambient sounds, but no artificial sounds were presented. We expected that people with visual impairments would utilize ambient audio information to identify obstacles or walls in front of them. Since the foam panels along the route were 900 mm high, the audio cues would be different from those provided when walking along an ordinary hallway. However, the wooden wall forming part of the building was over five meters high, and it provided a clear acoustic echo that assisted localization. Each participant completed one trial per one condition. All predefined routes were determined so that the departure and destination points were incongruent, and there were four turns. The routes were different for the two conditions and were randomly selected to reduce the learning effect, for example to prevent the distance to a turn being remembered or guessed. The participants were instructed to walk as fast and as accurately as possible. All participants were invited to complete our two-item questionnaire and to comment freely about what they felt during the task after the experiment. The statements were presented in a different randomized order for each participant. Each statement was rated by the participants on a seven-point Likert scale, with -3 meaning `totally disagree' and +3 `totally agree'. The questions were:

Q1. The guidance was easy to understand.
Q2. I expect it would be useful in disaster situations.

Our intention with Q1 was to gain some insight into the usability of the force display in the experiment for users with visual impairments. The aim with Q2 was to gain some insight into the feasibility of using it during disasters, such as in heavy smoke (visual cues unavailable) or when there is a loud siren noise (audio cues unavailable). In Q2, the experimenter explained that a *disaster* meant a typical situation that lacked some audiovisual information, and which required quick and safe directional navigation. These questions focused on obtaining a subjective rating of the understandability and reliability of the force display and the navigation system.

5.2.2 Results

The experimental results show that our proposed system successfully enabled the participants to follow predefined routes. Twenty-one of twenty-three participants (i.e., 91%) successfully walked from the entry point to the endpoint with four turns under both conditions. Figure 6 shows examples of walking trajectories. Note that trials revised by the experimenter were also counted as successes, whereas trials interrupted by the experimenter were counted as failures. The same two participants failed to complete the tasks under both conditions. The experimenter intervened twice under the audio-masked condition, and three times under the audio unmasked condition. No other participants required intervention during the navigation task. Specifically, one of those two participants always seemed to judge left stimuli as right stimuli, and the other interpreted right stimuli as left stimuli. This tendency appeared to prevent them from returning to the predefined route even if the stimuli were changed manually. Seven participants under the audio-masked condition and six under the audio- unmasked condition failed to perceive the force sensation indicating a turn. Two of them could not recover the route at all. The other participants were able to recover the route when the force display was driven by the experimenter. Under the audio-masked condition, five participants made five mistakes. Under the audio- unmasked condition, four made five mistakes. The average time required to recover from an incorrect turn was about five seconds, and the longest time was about ten seconds.

Almost all the participants rated both statements highly. The medians of the questionnaire results were +2 and +2. The quartile ranges were 1 and 1. No outliers were observed. Moreover, we analyzed the questionnaire results very conservatively by considering only the high scores (+2 or +3) for each statement as indicating the usefulness; in other words, the seven-point questionnaire responses were converted into a binary response (`high score' or `not high score'). The scientific motivation for this is that we wanted to perform an analysis that only took account of people who felt strongly about the usefulness of this force-based navigation. By chance alone, the probability of a high score is 2/7. The ratings for the two statements have frequencies that are much higher than would be expected by chance (21 and 18 respectively out of 23, with corresponding $p<.0001$ and .0001, using the binomial distribution).

Some participants commented after performing the task that they felt the device to be very useful and made it easy to comprehend the direction. A negative comment was that it was hard to keep the force display horizontal and to maintain the direction indicated by the display for a long period of time. Another negative comment from one participant was that his hand felt numb because of the vibration and weight of the force display.

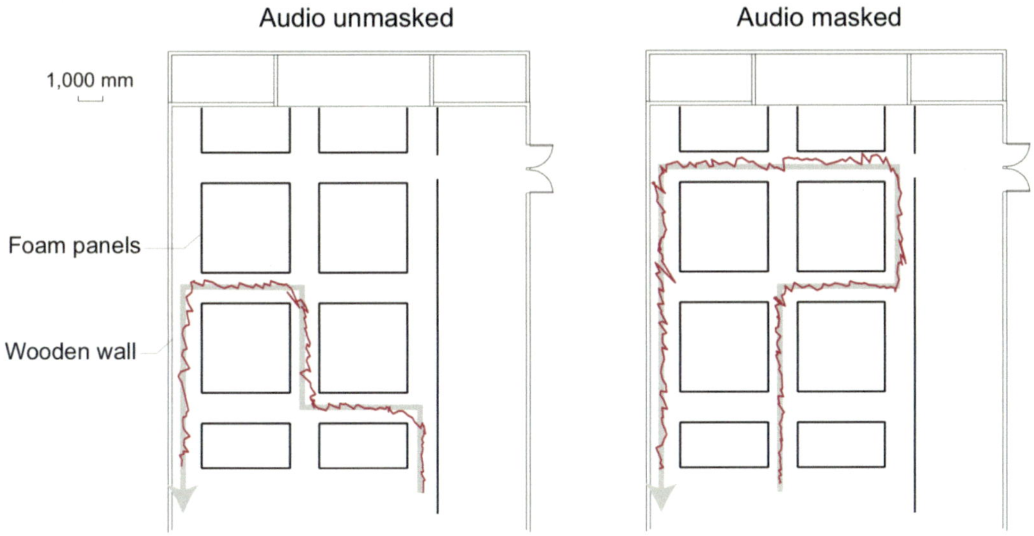

Fig. 6. Examples of a visually impaired participant's walking trajectories. Even if auditory information was masked, they could follow predefined routes

6. Discussion and design implications

The results provide clear evidence of the usefulness of force-based navigation for pedestrians with visual impairments. Some participants collided with the walls because the developed navigation system had no function for the recovery to a straight path from a meandering walk. Due to the lack of such a function, the participants' direction tended to diverge from the center line after they had turned a corner. To avoid such collisions, haptic signals for mid-course corrections are required. There is a trade-off between increasing the number of haptic signals and intuitive comprehension (or mental load), which is a version of the haptic icon problem (Enriquez and MacLean 2008). Nonetheless, this should be considered in future implementations. Our proposed system would also benefit from information displays for other sensory modalities. For instance, adding an audio navigation aid would make it relatively easily to provide complex and semantically rich information, such as categories of landmarks and street names. This semantic richness and complexity is very difficult to achieve with just a force display. Ross and Blasch pointed out that the combination of speech (auditory) and tapping (tactile) information would be useful as orientation aids (Ross and Blasch 2000). Future work will include investigating the effect of such additional meaningful audio information such as street names, and landmarks.

Our haptic navigation system can easily be expanded to support sighted pedestrians. In particular, travellers with a different mother tongue visiting an area would reap significant benefits because haptic instruction is nonverbal. The proposed system would also enable users with reading disabilities to travel independently. Iconic symbols such as left or right arrows would be sufficiently clear and accurate to indicate directions. However, haptic cues not only indicate direction (at the sensory phase) but also make bodies move directly (at the motor phase), which requires less response time. Therefore, haptic cues would be more suitable for navigation applications. Future work will include the integration of satellite navigation instruction, which would expand the available area of our technique.

Manually changing the force vector assisted participants who made a wrong turn to return to the original route. Since the haptic stimuli were identical for the automatic and manual operations, one of the main reasons manual intervention was required was the incomplete algorithm for changing the direction of the force vector. The algorithm depended on the accuracy and reliability of the local position sensors (infrared sensors placed at the maze corners). The detection onset for the local position sensors and turning a corner varied as the walking pace changed.

Our experimental setup has some limitations with regard to generalizing the findings for all areas of pedestrian navigation since we only examined turn-by-turn navigation task in orthogonal-crossed route. Since most streets and avenues, with the exception of certain cities such as Kyoto, are not laid out in a grid, a navigation task in relation to complicated routes must be the next step. In addition, we have to clarify whether the same angular resolution (i.e., the eight cardinal directions) is sufficient for arbitrary routes, in particular when avoiding obstacles. Pielot et al. have reported deficiencies in the human perception of body orientation when small angles are involved (Pielot et al. 2008). Our previous study also showed that there is a systematic error with respect to the human perception of force orientation among people with visual impairments when they are not moving, which is less than 15 degrees when the force vector is provided in the eight cardinal directions (Amemiya 2009). A dynamic exploration of the force vector would be useful when walking complicated routes, which is similar to the guidance a human or a guide dog might provide and depends on the active or passive perception of haptic feedback while walking. Therefore, the systematic error could be minimized with a closed feedback loop, which could lead to precise routing. However, debate continues about the best way to understand the user's orientation for dynamic exploration, and about the most suitable part of the body to which to attach the orientation sensors. Moreover, we must consider whether the force display should be held in the hand, carried in a bag, or worn.

In the experiment, the force display always generated force during navigation, because in a pilot study we received feedback that the absence of stimuli made the participants anxious. However, it is well known that continued vibration tends to cause perceptual fatigue (i.e., adaptation) if it is presented for too long, as with all other sensations (Coren et al. 2003). Nevertheless, there were no differences in the number of collisions with the walls between the L-shaped walking paths of the same trial: one is from the entry point to the second turn, the other is from the second to the last turn to the exit point. The average numbers of collisions were 0.10 and 0.12 (audio unmasked condition), and 0.07 and 0.09 (audio masked), respectively. The average lengths of the former and latter paths were 11.2 meters and 13.2 meters (audio unmasked), and 9.4 meters and 14.2 meters (audio masked), respectively. In addition, all but one of the participants reported that they did not feel any subjective perceptual difference between the start and the end as mentioned above. We conjecture that the effect of vibrotactile adaptation may not appear during short-term navigation, since the asymmetrically oscillating stimuli involve not only cutaneous but also proprioceptive sensations. The muscle spindles or the Golgi tendon organs, which are receptors generating the proprioceptive sensation, are slowly adapting units, while the Pacinian corpuscles, receptors that detect skin vibration, are rapidly adapting units.

Many of the participants made similar positive comments. The questionnaire rating clearly revealed high confidence levels and high expectations. Their medians and quartiles confirmed that the force sensation was well perceived. Responses to Q1 indicated that the participants were aware of the direction information and found the information provided by our system to be very intuitive. Those to Q2 confirmed that many of the participants realized the importance of force feedback in emergency navigation. The participants also commented on the quietness of the force display. They commented that the noise level of the force display would be acceptable in daily life, such as in public spaces. People with visual impairments sometimes use the information provided by acoustic echoes to gain awareness of the environment. The proposed device was so quiet that this information could still be used.

It is crucial to miniaturize and lighten the force display so that it can be carried more easily. However, the amplitude of the kinesthetic stimuli, which should be large enough to be perceived, is proportional to the mass of the weight and the amplitude of acceleration. The trade-off between the mass of the weight and strength of perception limits the amount by which the weight can be reduced. However, the size of the force display could be reduced by using other mechanisms to generate similar asymmetric oscillation, such as linear actuators. We speculate that a miniaturized version of the force display could be embedded in canes for people with visual impairment. It is true that some people feel that no device should be attached to the cane, but only a relatively small amount of retraining would be needed.

7. Related work

Recent research has addressed a range of issues concerning pedestrian navigation aid. Pedestrian navigation systems on mobile devices, such as mobile phones with a satellite positioning function, that employ different sensory channels (i.e., visual and/or audio channels) are being increasingly developed. Since people with visual impairments cannot use vision-based navigation aids, many handheld auditory-feedback devices for people with visual impairments have been developed, such as Talking Signs (Crandall et al. 1999) or similar acoustic information output devices (Loomis et al. 2005). Although such audio interfaces help users move in the right direction by providing sound cues, they can be problematic when they conflict with other sounds or speech around the users, making it difficult for them to distinguish and interpret the sounds generated by the system (Wilson et al. 2007). In addition, pedestrians with visual impairments often rely on information contained within the ambient sounds for navigation purposes. Wearing headphones prevents them from hearing these ambient sounds thereby making navigation less safe. Furthermore, auditory information cannot be used in noisy situations, such as on busy city streets.

Tactile interaction may help overcome such navigation issues for people with visual impairments. It has been reported that tactile interaction can effectively assist pedestrians with visual impairments when crossing the street (Ross and Blasch 2000). As tactile-based navigational aids, vibrotactile stimulation systems that use several vibrators in the shape of a cap (Cassinelli et al. 2006), rings (Amemiya et al. 2004), a vest (Erp et al. 2005), belt (Tan et al. 2003; Heuten et al. 2008), and glove (Zelek et al. 2003) have been proposed. Unfortunately, these tactile approaches require that users learn how to convert stimuli to information. This is not intuitive and requires training since the tactile stimuli employed are basically non-directional.

8. Acknowledgements

We thank Dr. Ichiro Kawabuchi for his technical assistance. We also thank Hisashi Sugiyama, the staff of Kyoto City Fire Department, and the staff of the Kyoto Prefectural School for the Visually Impaired for their kind cooperation. This study was supported by Nippon Telegraph and Telephone Corporation and was also partially supported by the sponsorship of the Fire Defence Agency, Japan.

9. References

Amemiya, T.; Ando, H. & T. Maeda T. (2005). Virtual force display: Direction guidance using asymmetric acceleration via periodic translational motion, *Proceedings of World Haptics Conference*, IEEE Computer Society, pp. 619-622.

Amemiya, T.; Ando, H. & T. Maeda T. (2008). Lead-Me interface for pulling sensation in hand-held devices, *ACM Transactions on Applied Perception*, Vol. 5, No. 3, pp. 1-17.

Amemiya, T. & Maeda, T. (2008). Asymmetric oscillation distorts the perceived heaviness of handheld objects, *IEEE Transactions on Haptics*, Vol. 1, No. 1, pp. 9-18.

Amemiya, T. & Maeda, T. (2009). Directional force sensation by asymmetric oscillation from a double-layer slider-crank mechanism, *Journal Computing Information Science in Engineering*, Vol. 9, No. 1, 011001.

Amemiya, T.; Maeda, T. & Ando, H. (2009). Location-free Haptic Interaction for Large-Area Social Applications, *Personal and Ubiquitous Computing*, Vol. 13, No. 5, pp. 379-386, Springer.

Amemiya, T. & Sugiyama, H. (2009). Haptic Handheld Wayfinder with Pseudo-Attraction Force for Pedestrians with Visual Impairments, *Proceedings of 11th ACM Conference on Computers and Accessibility (ASSETS 2009)*, Pittsburgh, PA, pp. 107-114.

Amemiya, T. & Sugiyama, H. (2010). Orienting Kinesthetically: A Haptic Handheld Wayfinder for People with Visual Impairments, *ACM Transactions on Accessible Computing*, Vol. 3, No. 2, pp. 1-23.

Amemiya, T.; Yamashita, J.; Hirota, K. & Hirose, M. (2004). Virtual Leading Blocks for the Deaf-Blind: A Real-Time Way-Finder by Verbal-Nonverbal Hybrid Interface and High-Density RFID Tag Space, *In Proc. of IEEE Virtual Reality Conference 2004 (VR 2004)*, pp. 165-172.

Bradley, A. & Dunlop, D. (2005). An experimental investigation into wayfinding directions for visually impaired people, *Personal Ubiquitous Computing*, Vol. 9, No. 6, pp. 395-403.

Cassinelli, A.; Reynolds, C. & Ishikawa, M. (2006). Augmenting spatial awareness with haptic radar. *In Proc. International Conference on Wearable Computing*. IEEE Computer Society, pp. 61-64.

Coren, S.; Ward, L. M. & Enns, J. T. (2003). Sensation and Perception. John Wiley and Sons, Inc.

Crandall, W.; Brabyn, J.; Bentzen, B. & Myers, L. (1999). Remote infrared signage evaluation for transit stations and intersections. *Journal of Rehabilitation Research and Development* Vol. 36, pp. 341-355.

Enriquez, M. & MacLean, K. (2008). The role of choice in longitudinal recall of meaningful tactile signals. *In Proc. of 16th IEEE Symposium on Haptic interfaces for virtual environment and teleoperator systems*. pp. 49-56.

Erp, J. B. F. V.; Veen, H. A. H. C. V.; Jansen, C. & Dobbins, T. (2005). Waypoint navigation with a vibrotactile waist belt. *ACM Transactions on Applied Perception*, Vol. 2, No. 2, pp. 106-117.

Foulke, E. (1996). The roles of perception and cognition in controlling the mobility task. *International Symposium on Orientation and Mobility*.

Golledge, R. G. (1992). Place recognition and wayfinding: making sense of space. *Geoforum* Vol. 23, No. 2, pp. 199-214.

Gurocak, H.; Jayaram, S.; Parrish, B. & Jayaram U. (2003). Weight sensation in virtual environments using a haptic device with air jets, *Journal of Computing and Information Science in Engineering*, Vol. 3, No. 2. ASME, pp. 130-135.

Hayward, V. (2008). A Brief Taxonomy of Tactile Illusions and Demonstrations That Can Be Done In a Hardware Store, *Brain Research Bulletin*, Vol. 75, pp. 742-752.

Heuten, W.; Henze, N.; Boll, S. & Pielot, M. (2008). Tactile wayfinder: a non-visual support system for wayfinding. *In NordiCHI. ACM International Conference Proceeding Series*, Vol. 358. ACM Press, pp. 172-181.

Hirose, M.; Hirota, K.; Ogi, T.; Yano, H.; Kakehi, N.; Saito, M. & Nakashige, M. (2001). HapticGEAR: The Development of a Wearable Force Display System for Immersive Projection Displays, *Proceedings of Virtual Reality 2001 Conference*, pp. 123–130.

Hoshi ,T.; Takahashi, M.; Iwamoto, T. & Shinoda, H. (2010). Noncontact Tactile Display Based on Radiation Pressure of Airborne Ultrasound, *IEEE Transactions on Haptics*, Vol. 3, No .3, pp. 155-165.

Loomis, J.; Marston, J.; Golledge, R. & Klatzky, R. (2005). Personal guidance system for people with visual impairment: A comparison of spatial displays for route guidance. *Journal of Visual Impairment and Blindness* Vol. 8, No. 5, pp. 61-64.

Massie, T. & Salisbury, J. K. (1994). The phantom haptic interface: A device for probing virtual objects, *Proceedings of the ASME Winter Annual Meeting, Symposium on Haptic Interfaces for Virtual Environment and Teleoperator Systems*, Vol. 55-1, pp. 295-300.

Nakamura, N. & Fukui, Y. (2007). Development of Fingertip Type Non-grounding Force Feedback Display, *Proceedings of World Haptics Conference 2007*, pp. 582-583.

Pielot, M., Henze, N., Heuten, W., & Boll, S. (2008). Evaluation of continuous direction encoding with tactile belts. *In Proc. the 3rd international workshop on Haptic and Audio Interaction Design*, Springer, LNCS, pp. 1-10.

Richard, C. & Cutkosky, M. (1997). Contact Force Perception with an Ungrounded Haptic Interface, *Proceedings of the ASME Dynamic Systems and Control Division*, pp. 181-187.

Ross, D. & Blasch, B. (2000). Wearable interfaces for orientation and wayfinding. In Proc. ACM Conference on Assistive Technologies. ACM Press, pp. 193-200.

Suzuki, Y.; Kobayashi, M. & Ishibashi, S. (2002). Design of force feedback utilizing air pressure toward untethered human interface, *Proceedings of CHI '02 Extended Abstracts on Human Factors in Computing Systems*. ACM Press, 2002, pp. 808-809.

Swindells, C.; Unden, A. & Sang, T. (2003). TorqueBAR: an ngrounded haptic feedback device. *Proceedings of the 5th international conference on multimodal interfaces*. ACM Press, pp. 52-59.

Tan, H. Z.; Gray, R., Young, J. J. & Traylor, R. (2003). A haptic back display for attentional and directional cueing. Haptics-e: *The Electronic Journal of Haptics Research* Vol. 3, No. 1.

Tanaka, Y.; Masataka, S.; Yuka, K.; Fukui, Y.; Yamashita, J. & Nakamura, N. (2001). Mobile torque display and haptic characteristics of human palm. *Proceedings of 11th international conference on augmented tele-existence*, pp. 115-120.

Wilson, J.; Walker, B.; Lindsay, J.; Cambias, C. & Dellaert, F. (2007). Swan: System for wearable audio navigation. *In Proc. International Conference on Wearable Computing*. IEEE Computer Society, pp. 91-98.

Yano, H.; Yoshie, M. & Iwata, H. (2003). Development of a nongrounded haptic interface using the gyro effect, *Proceedings of 11th international symposium on Haptic Interfaces for Virtual Environment and Teleoperator Systems*. IEEE Computer Society, pp. 32-39.

Zelek, J. S.; Bromley, S.; Asmar, D. & Thompson, D. (2003). A haptic glove as a tactile-vision sensory substitution for wayfinding. *Journal of Visual Impairment and Blindness* Vol. 97, No. 10, pp. 621-632.

Part 2

Transmission Technologies and Propagation

16

Technological Trends of Antennas in Cars

John R. Ojha, René Marklein and Ian Widjaja
Germany

1. Introduction

Antennas have become a commonplace in automotive applications. These are broadly classified as wire and patch antennas which are used in cars for inter-vehicle communication. Besides its use in the automotive sector, these antennas are also used as arrays in the aviation sector e.g. fuselage integrated microstrip phased antenna arrays. These wire and patch antennas can either be modeled analytically e.g. using the Green's function, derived from Eigen functions or numerically using various approaches e.g. MoM, FDTD, FEM etc. Besides the common usage of wire and patch antennas of various shapes, integrated antennas are also widely used. Antennas starting from the traditional monopole antenna followed by patch antennas on car roof tops and mesh antennas on car windscreens will be discussed in this chapter.

2. Figures of merit

This section lists and explains some salient figures of merit of antennas. The input impedance and the radiated fields (near and far) are termed as the primary figures of merit since they form the basis on which other secondary figures of merit such as VSWR, bandwidth, and directivity etc. are determined. Section 2.1 elaborates on the primary figures of merit viz. input impedance. Section 2.2 explains some secondary figures of merit which are obtained from the input impedance. The theory of how the effective radiating power is calculated from the far-field gain patterns is explained in section 2.3.

2.1 Input impedance

The input impedance Z_{in} is defined as the impedance presented by an antenna at its input terminals a – b, as shown in Fig. 1. In other words, the input impedance of an antenna is the ratio of the voltage to the current or the ratio of the electric to the magnetic field measured at the input terminals (feeding point). The input impedance of an antenna is expressed in terms of its real and imaginary parts as

$$Z_{in} = R_{in} + jX_{in}, \tag{1}$$

where Z_{in} is the antenna impedance at the input terminals a – b,
R_{in} is the antenna resistance at the input terminals a – b,
and X_{in} is the antenna reactance at the input terminals a – b.

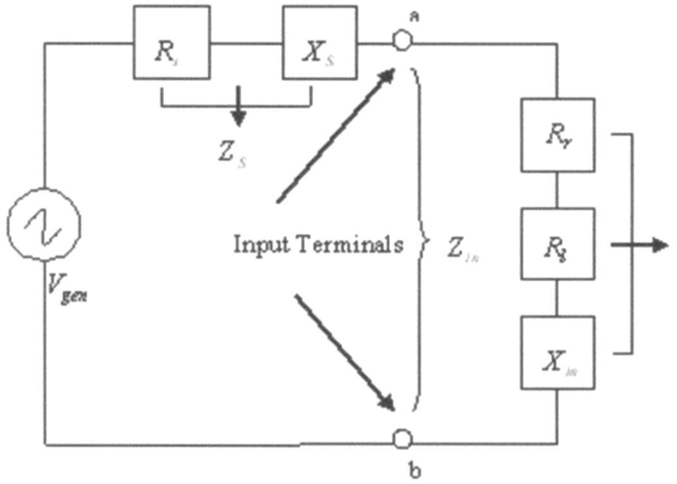

Fig. 1. Block diagram of a transmitting antenna

The imaginary part X_{in} of the input impedance represents the power stored in the near field region of the antenna. The resistive part R_{in} of the input impedance consists of two components, the radiation resistance R_r and the loss resistance R_l. The power associated with the radiation resistance R_r is the power actually radiated by the antenna and the loss resistance R_l represents the dielectric or conducting losses resulting in power dissipation.
The input impedance is of great importance in wire and patch antennas and is therefore discussed here. The input impedance is used as a foreboding of unwanted radiation for EMC related aspects especially in the automotive sector. However, in the case of antennas, the input impedance with the source impedance is used as an intermediate parameter for determining the *S11* parameter, return loss, Voltage Standing Wave Ratio (*VSWR*), and bandwidth. This is explained in more detail in section 2.2, where the matching characteristics of a patch antenna and its bandwidth are explained.

2.2 Reflection coefficient / S11 / VSWR / return loss

Antennas are commonly used in various type of smart antenna systems. In order for any given antenna to operate efficiently, the maximum transfer of power must take place between the feeding system and the antenna. Maximum power transfer can take place only when the input impedance of the antenna (Z_{in}) is matched to that of the feeding source impedance (Z_S). According to the maximum power transfer theorem, maximum power can be transferred only if the impedance of the source is a complex conjugate of the impedance of the antenna under consideration and vice-versa. If this condition for matching is not satisfied, then some of the power may be reflected back. This is expressed as

$$VSWR = \frac{1+|\Gamma|}{1-|\Gamma|}, \tag{2}$$

with

$$\Gamma = \frac{V_r}{V_i} = \frac{Z_{in} - Z_S}{Z_{in} + Z_S}, \tag{3}$$

where Γ is called the reflection coefficient, V_r is the amplitude of the reflected wave, and V_i is the amplitude of the incident wave. The *VSWR* is basically a measure of the impedance mismatch between the feeding system and the antenna. The higher the *VSWR*, the greater is the mismatch. The minimum possible value of *VSWR* is unity and this corresponds to a perfect match. The return losses (*RL*), obtained from equations (2) and (3), indicate the amount of power that is transferred to the load or the amount of power reflected back. In the case of a microstrip-line-fed antenna, where the source and the transmission line characteristic impedance or the transmission line and the antenna edge impedance do not match, waves are reflected. The superposition of the incident and reflected waves leads to the formation of standing waves. Hence the *RL* is a parameter similar to the *VSWR* to indicate how well the matching is between the feeding system, the transmission lines, and the antenna. The *RL* is

$$RL = -20\log|\Gamma| \text{ (dB)}. \tag{4}$$

To obtain perfect matching between the feeding system and the antenna, $\Gamma = 0$ is required and therefore, from equation (4), RL = infinity. In such a case no power is reflected back. Similarly at $\Gamma = 1$, RL = 0 dB, implies that all incident power is reflected. For practical applications, a *VSWR* of 2 is acceptable and this corresponds to a return loss of 9.54 dB. Usually return losses ranging from 10 dB to 12 dB are acceptable.

The bandwidth could be defined in terms of its Voltage Standing Wave Ratio (*VSWR*) or input impedance variation with frequency. The *VSWR* or impedance bandwidth of an antenna is defined as the frequency range over which it is matched with that of the feed line within specified limits. The *BW* of an antenna is inversely proportional to its quality factor *Q* and is expressed as

$$BW = \frac{VSWR - 1}{Q\sqrt{VSWR}}. \tag{5}$$

The bandwidth is usually specified as the frequency range over which the *VSWR* is less than 2 (which corresponds to a return loss of 9.5 dB or 11 % reflected power). Sometimes for stringent applications, the *VSWR* requirement is specified to be less than 1.5 (which corresponds to a return loss of 14 dB or 4 % reflected power). In the case of a patch antenna, the input impedance with the source impedance is used as an intermediate parameter for determining the *S11* parameter (a measure of the reflection coefficient Γ), return loss, Voltage Standing Wave Ratio (*VSWR*), and bandwidth. The return loss is expressed in dB in terms of *S11* as the negation of the return loss. The bandwidth can also be defined in terms of the antenna's radiation parameters such as gain, half power beam width, and side-lobe levels within specified limits.

2.3 Effective radiating power

For every other antenna, the directivity is defined as the ratio of the radiation intensity in a given direction from the antenna to the radiation intensity U_0 averaged over all directions. If the direction is not specified, the direction of maximum radiation intensity is implied.

Hence mathematically the directivity is

$$D_0 = \frac{U_{\max}}{U_o} = \frac{4\pi U_{\max}}{P_{rad}}, \tag{6}$$

where U_{max}, P_{rad} are the maximum radiation intensity and total radiated power, expressed in Watts / solid angle and Watts respectively.
The antenna gain is directly associated with the directivity of an antenna and is therefore associated with only the main lobe. The term K is the radiation efficiency expressed in terms of the conduction efficiency K_c and dielectric efficiency K_d as

$$K = K_c K_d \,, \tag{7}$$

Gain and directivity extraction are based on the source power. Let us assume that P_t is the source power and P_v are some losses in the structure (e.g. dielectric losses), then a power P_t $P_r = P_t - P_v$ will be radiated. The directivity (as compared to an isotropic point source) is then defined as

$$D = 4\pi R2 * (S_s / P_r), \tag{8}$$

where $S_s = (1/2)(|E_\vartheta|2 + |E_\phi|2 \; / \; Z_{F0})$
Z_{F0} denotes the wave impedance of the surrounding medium.
From the equation the gain is extracted from the directivity as

$$G = K \cdot D \;, \tag{9}$$

where G is the gain and D is the directivity. (For an antenna with 100% efficiency, $K = 1$.)
The far field gain is determined from the electric far-field components E_θ and E_ϕ and the source power. The electric field components E_θ and E_ϕ are calculated from the surface electric current densities. The effective radiating power is extracted from the gain by removing the effect of the losses in the form of metallic or /and dielectric losses.

$$\mathit{Effective\,Radiating\,Power} = \mathit{Gain} - \mathit{Power\,loss} \tag{10}$$

3. Numerical approaches for determining figures of merit

The numerical analysis e.g. MoM can be carried out either in the spectral or in the time domain. A patch antenna comprising metallic and dielectric parts with a feeding pin or microstrip line is solved using the traditional MoM by decomposing the antenna as

- discretized surface parts
- wire parts
- attachment node of the wire to the surface element.

Metallic surfaces contain different basis functions as shown in Fig. 2. The MoM uses surface currents to model a patch antenna. In the case of ideal conductors, the boundary condition of $E_{tan} = 0$ is applied.
The most commonly used basis functions for line currents through wires are stair case functions, triangular basis functions, or sine functions. The MoM code uses triangular basis functions. In contrast to wires, two-dimensional basis functions are employed for surfaces. The current density vectors have two-directional components along the surface. Figure 2 shows the overlapping of so-called hat functions on triangular patches. An integral equation is formulated for the unknown currents on the microstrip patches, the feeding wire / feeding transmission line, and their images with respect to the ground plane. The integral equations are transformed into algebraic equations that can be easily solved using a

computer. This method takes into account the fringing fields outside the physical boundary of the two-dimensional patch, thus providing a more exact solution. The coupling impedances Z_{ik} are computed in accordance with the electric field integral equation.

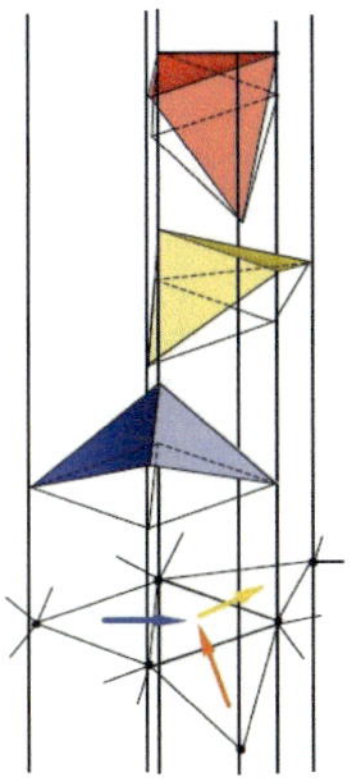

Fig. 2. Hat basis functions on discretised triangular elements on patches

The MoM uses either surface-current layers or volume polarization to model the dielectric slab. In the case of dielectric materials we have to consider 2 boundary conditions

$$\vec{n} \times \vec{E}_1 = \vec{n} \times \vec{E}_2 , \quad (11)$$

$$\vec{n} \times \vec{H}_1 = \vec{n} \times \vec{H}_2 . \quad (12)$$

The traditional full-model applied in the MoM code uses a surface-current approach which is categorised as

- double electric current layer approach or
- single magnetic and electric current layer approach.

4. Various type of antennas

Various type of antennas are described here. Antennas e.g. the conventional monopole, which is of historical importance is still widely used due to its simplicity in construction. The following sections deal with technological trends with respect to the monopole family of antennas as well as patch antennas.

4.1 Wire antennas (monopole antenna)

Monopole antennas are commonly used in automotive applications where range is important. A brief description of how a monopole antenna is characterised will be illustrated e.g. a monopole antenna is suitably placed on a car and then meshed effectively for numerical simulation. These antennas are also very easy to design and tune simply by slightly varying the length. It is assumed the antenna is a quarter wavelength long, which is typical of monopole antennas in the UHF band. The radiation characteristics are linearly polarized, either horizontally or vertically, depending on antenna orientation. Radiation resistance of a quarter wave monopole is approximately 37Ω, and does not vary much with presence or absence of ground plane. The radiation resistance of monopole antennas is length dependent. Resonance of a quarter-wavelength monopole occurs when its length is slightly less than a quarter-wavelength. The appropriate length for a quarter-wave monopole at 433.92MHz would be 2808 ÷ 433.92 = 6.47 inches. Sophisticated antenna measurements are generally not necessary unless a highly optimized design is desired. This makes the monopole very popular and easy to apply. The bandwidth of the antenna can either be broadened by providing an LC circuit or by providing a parasitic element near the wire part connected to the source. Fig. 3 shows a simple sketch of the traditional monopole antenna.

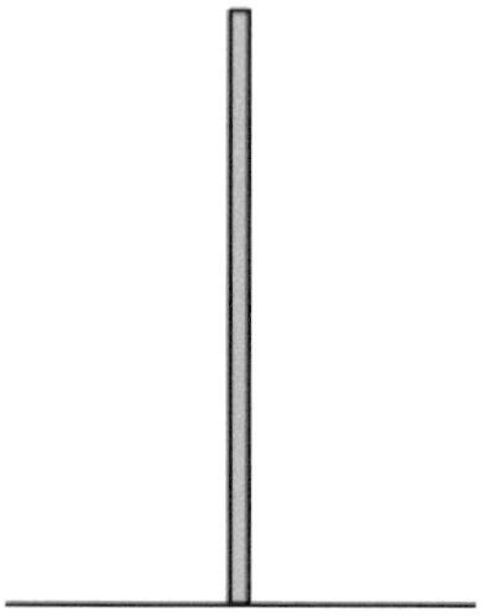

Fig. 3. The traditional monopole antenna

Some salient features are

- To increase the resonant frequency, decrease the monopole height.
- To increase the bandwidth, increase the wire thickness. Variation in wire thickness will have a small effect on the resonant frequency of the antenna. The resonant frequency of the antenna should be corrected for by adjusting the length.
- To decrease the impedance variation versus frequency, increase the element size.

4.2 Monopole antenna with sleeve

Monopole antennas have problems of low bandwidths. The aim of this section is to show a scheme to broaden the bandwidth by providing a sleeve as shown in Fig. 4. The cylindrical sleeve acts as a parasitic element. The advantage of the monopole antenna with the provision of a sleeve is clear from Fig. 5. If the diameter of the wire is not large a wire can still be used instead of cylinder.

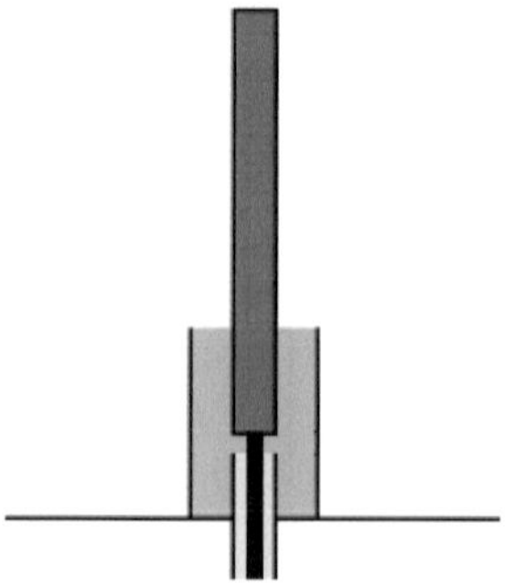

Fig. 4. Monopole antenna with sleeve

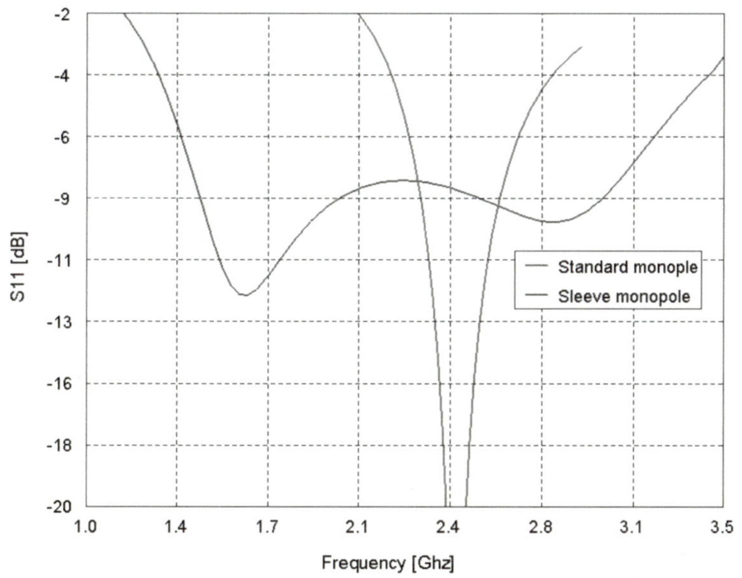

Fig. 5. Comparison of sleeve monopole antenna and the traditional monopole antenna

The design approach is to adjust the exterior dimensions of the antenna to achieve pattern stability and then to use the region within the sleeve for impedance matching [Poggio et al.].

- To increase the operating frequency, decrease the monopole height.
- To increase the bandwidth, increase the wire thickness. (Note that changes in wire thickness will have a small effect on the operating frequency of the antenna. This should be corrected for by adjusting the length according to the previous guideline).
- To decrease the impedance variation versus frequency, increase the element diameter.

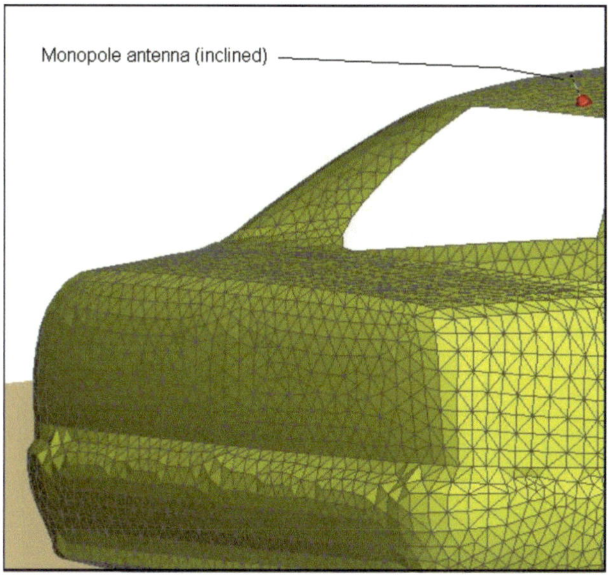

Fig. 6. Monopole antenna (inclined) mounted on a car

Fig 5 shows the characteristics of a traditional monopole antenna on an infinite ground plane. The far-field gain, antenna efficiency, and matching characteristics change with change in location of a monopole antenna in positions A, B, and C shown in Fig. 7. Fig 8 shows variation in the far- field gain patterns for change in the antenna location. There is also a variation in the far-field gain, shown in Fig. 9 when the monopole antenna is upright and inclined. In today's world the antenna is mounted inclined on a car as shown Fig. 6 and Fig. 7 (scheme D). The determination of antenna efficiency and matching characteristics (*VSWR*) is left as an exercise to the reader.

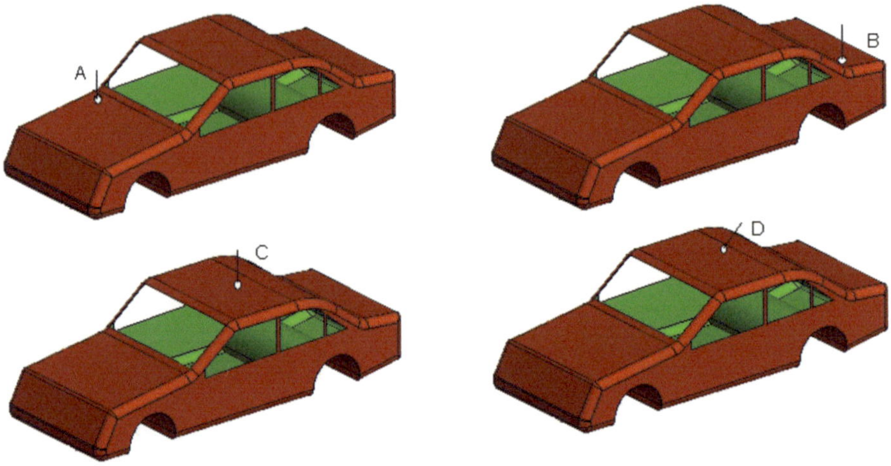

Fig. 7. Monopole antennas mounted at various locations

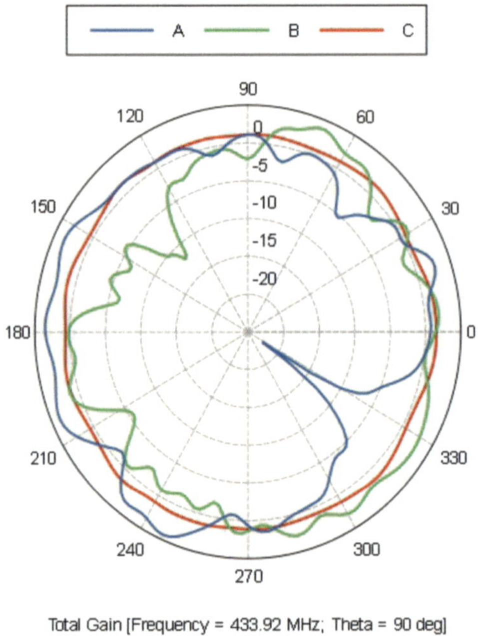

Fig. 8. Far-field gain patterns of antennas at various locations

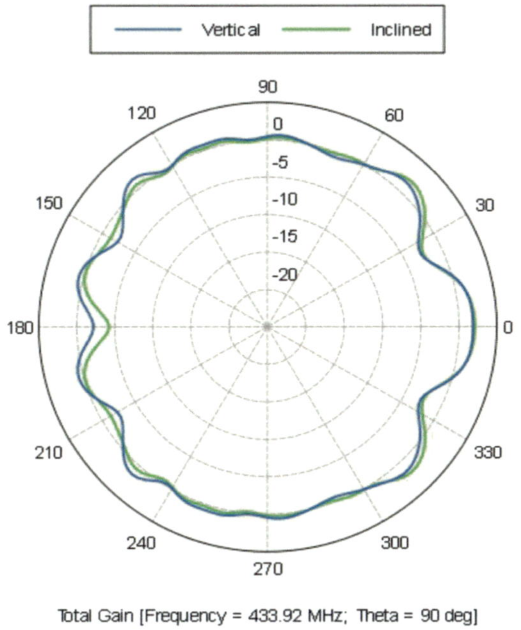

Fig. 9. Far-field gain patterns of antennas at orientations (vertical/tilted)

4.3 Patch antennas

The most common patch antennas in today's world are primitives such as squares, triangles, etc, metallised on a substrate backed by a ground plane. The next section gives a brief overview of a rectangular, a circular, and an elliptical patch antenna.

4.3.1 Rectangular patch antenna

From the cavity model point of analysis, the wave numbers $k_{x'}$, $k_{y'}$, k_z in the corresponding x' , y' , z' directions are

$$\left\{\begin{array}{l} k_x = \left(\dfrac{m\pi}{L}\right), m = 0,1,2,..... \\ k_y = \left(\dfrac{n\pi}{W}\right), n = 0,1,2,..... \\ k_z = \left(\dfrac{p\pi}{H}\right), p = 0,1,2,.... \end{array}\right\}, \tag{13}$$

where m, n, p represent the number of half-cycle field variations along the x, y, and z directions respectively. The primed cylindrical co-ordinates x' , y' , z' are used to represent the field within the cavity. The resonant frequency for such a patch or cavity is

$$(f_r)_{mnp} = \frac{1}{2\pi\sqrt{\mu\varepsilon}}\sqrt{\left(\frac{m\pi}{L}\right)^2 + \left(\frac{n\pi}{W}\right)^2 + \left(\frac{p\pi}{H}\right)^2}, \tag{14}$$

where *W, L, H* represent the width, length and height of the patch antenna. Since the substrate height *H* is very small $(H << \lambda_0)$ the electric field along the z direction is assumed constant and hence $p = 0$ and $k_z = 0$ and consequently the last term in equation (14) disappears.
Some design guidelines for a rectangular patch antenna shown in Fig. 10 are

- To increase (decrease) the resonant frequency, decrease (increase) the patch length.
- To increase bandwidth, increase the substrate height and/or decrease the substrate permittivity (this will also affect resonant frequency and the impedance).
- The bandwidth may be increased (decreased) by increasing (decreasing) the patch width.
- To increase (decrease) the input impedance decrease (increase) the pin inset.

Note: Antennas on very thin substrates have high copper-losses, while thicker and higher permittivity substrates may lead to performance degradation due to surface waves and feed-pin impedance. The maximum impedance that can be realised is governed by the impedance seen at the edge of the patch. The minimum realisable impedance is zero, at the centre of the patch. However, the practical minimum is governed by the rapid impedance variation as the centre is approached. A typical patch antenna similar in nature is mounted on a car as shown in Fig. 11 with no substrate.

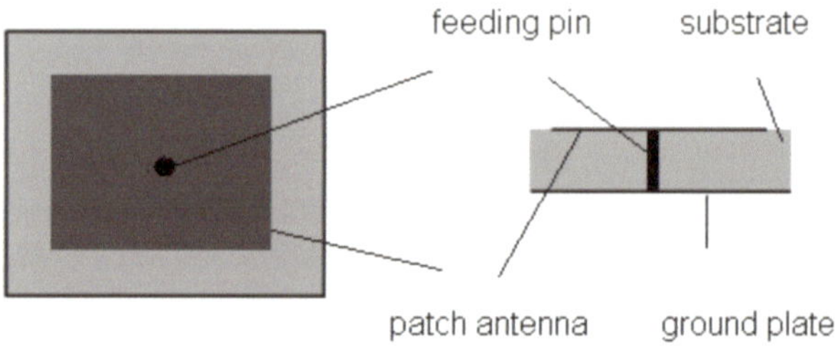

Fig. 10. Square/ rectangular patch antenna

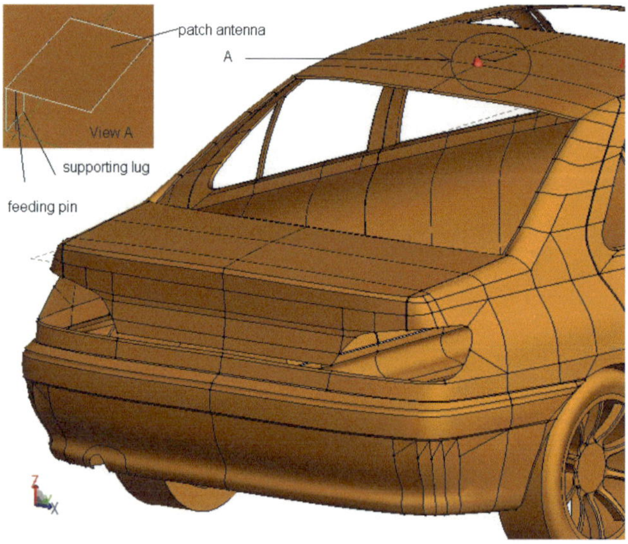

Fig. 11. Patch antenna (no substrate) mounted on a car

4.3.2 Circular patch antenna

From the cavity model point of analysis, the wave numbers k_ρ, k_ϕ, k_z in the corresponding ρ', ϕ', z' directions are

$$\begin{Bmatrix} k_\rho = \left(\frac{\chi'_{mn}}{A'}\right), m = 0,1,2,..... \\ k_\phi = \quad 0, \quad n = 1,2,3..... \\ k_z = \left(\frac{p\pi}{H}\right), p = 0,1,2,.... \end{Bmatrix}, \tag{15}$$

where m, n, p represent the number of half-cycle field variations along the ρ, ϕ, and z directions respectively. The primed cylindrical coordinates ρ', ϕ', z' are used to represent the field within the cavity. Taking into account the condition $k_z = 0$, as in the case of the rectangular structure, the resonant frequency for such a circular patch is

$$(f_r)_{mnp} = \frac{1}{2\pi\sqrt{\mu\varepsilon}}\left(\frac{\chi'_{mn}}{A'}\right), \tag{16}$$

where A' represents the radius of the disk and χ'_{mn} represents the zeros of the derivatives of the Bessel function $J_m(x)$ whose values are given in table 1.

Mode (m,n)	χ'_{mn}
(1,1) or 1st mode	1.8412
(2,1) or 2nd mode	3.0542
(0,1) or 3rd mode	3.8318
(3,0) or 4th mode	4.2012

Table 1. Zeros of the derivatives of $J_m(x) = 0$ at mode (m, n) of order m at the nth zero cross-over point

The use of electrically thick substrates in designs will have degraded matching due to increased feed pin inductance.

- Increasing the patch's diameter will decrease the resonant frequency and vice versa.
- Increasing the substrate height will increase the bandwidth, but will decrease the resonant frequency slightly.
- Increasing the substrate height will also result in a more inductive reactance due to the feed pin.
- To increase/decrease the input impedance, increase/decrease the feed offset.
- The circular patch antenna may be fine tuned for both impedance and centre frequency by the use of trimming stubs, as for the rectangular patch.

Note: Antennas on very thin substrates have high copper losses, while thicker and higher permittivity substrates may lead to performance degradation due to surface waves and feed-pin impedance. The maximum impedance that can be realised is governed by the impedance seen at the edge of the patch. The minimum realisable impedance is zero, at the centre of the patch. However, the practical minimum is governed by the rapid impedance variation as the centre is approached. The patch antenna is on the x-y plane.

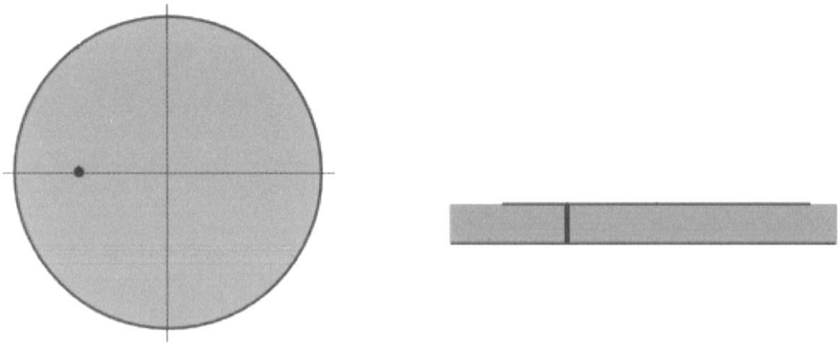

Fig. 12. Circular patch antenna

4.3.3 Elliptical patch antenna

Elliptical antennas are used for single-fed circular polarized antennas, especially in automotive applications. These antennas are characterized analytically making use of the Mathieu function in the case of elliptical antennas. A circular patch antenna could also be used however 2 feeds are necessary with the physical angle and electrical angle displaced by 90 degrees, namely

feed 1: y = 0 and V = 1 at phase angle = 0 degrees.
feed 2: x = 0 and V = 1 at phase angle = 90 degrees.

A typical substrate would have an ε_r of 2.48 and a substrate height approximately 1.5% of a free-space wavelength.

- To increase the operating frequency, reduce the patch dimensions while keeping the ratio of the major to the minor ellipse axes constant.
- To improve the axial ratio at the centre frequency, increase or decrease the ratio of the major to the minor ellipse axes.
- To increase the bandwidth, try increasing the substrate height and/or decreasing ε_r.
- To increase/decrease the input impedance, the feed offset should be increased/decreased.

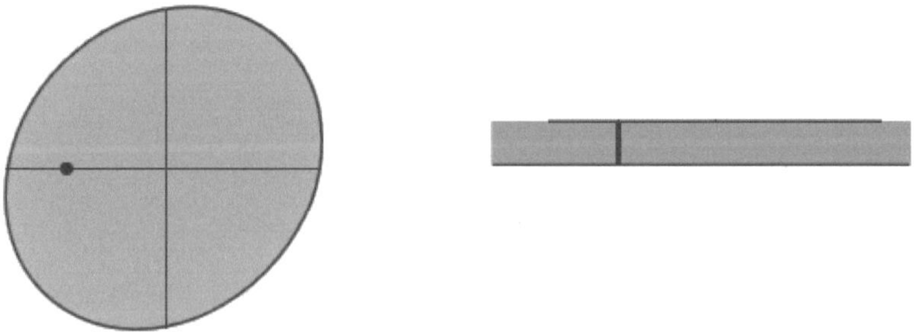

Fig. 13. Elliptical patch antenna

Note: Antennas on very thin substrates have high copper losses, while thicker and higher permittivity substrates may lead to performance degradation due to surface waves and feed-pin inductance. The maximum impedance that can be realised is governed by the impedance seen at the edge of the patch. The minimum realisable impedance is zero, at the centre of the patch. However, the practical minimum is governed by the rapid impedance variation as the centre is approached. Furthermore the best performance is achieved when the ratio of the minor axis to the major axis is almost unity. As in section 4.2.1 and 4.2.2 the antenna is on the x-y plane.

Properties of patch antennas of rectangular and circular geometries on planar surfaces were listed briefly. Such patch antennas also exist on cylindrical and spherical surfaces. Other patch antenna shapes (besides rectangular, circular and elliptical) widely used are triangular and annular in nature. One of the most widely used triangular shaped patch antennas is the bow-tie antenna. Annular antennas are used in applications where a broader bandwidth is required. In some cases, the inner radius of the annulus is short circuited.

5. Design guidelines for patch antenna arrays

For a given center frequency and substrate relative permittivity, the substrate height should not exceed 5% of the wavelength in the medium. The following guidelines are a must for designing a patch antenna and its arrays fed by microstrip lines.

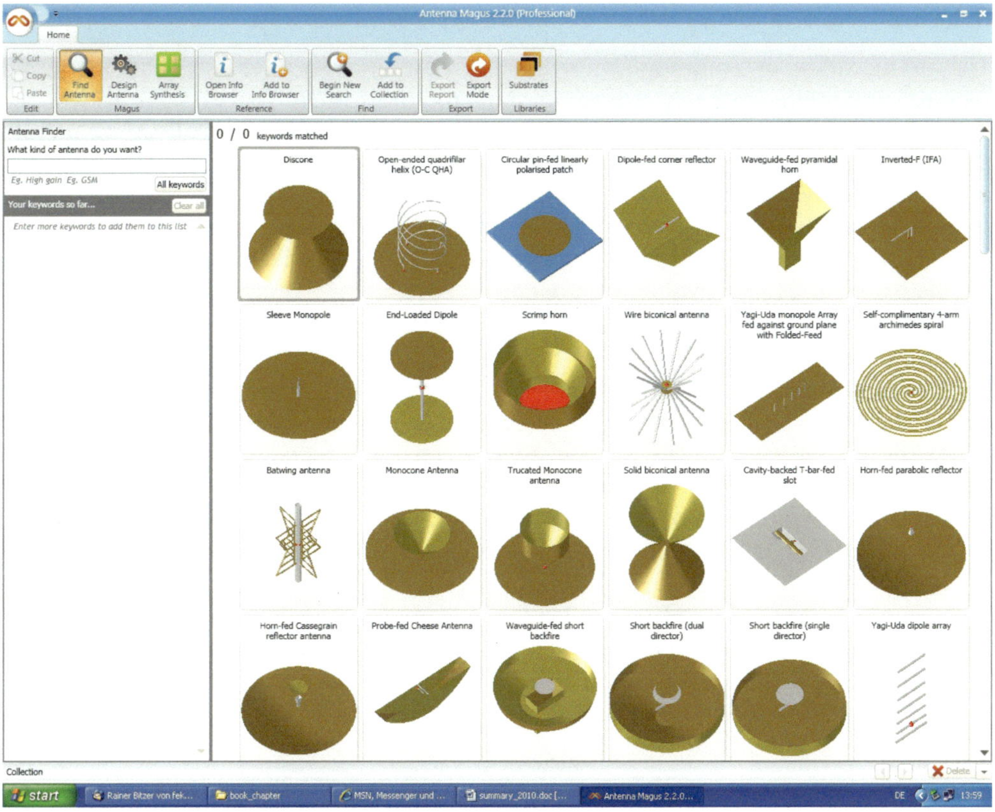

Fig. 14. Automotive patch antenna selected from Antenna Magus

- The length of the patches may be changed to shift the resonances of the centre fundamental frequency of the individual patch elements. The resonant input resistance of a single patch can be decreased by increasing the width of the patch. This is acceptable as long as the ratio of the patch width to patch length (W/L) does not exceed 2 since the aperture efficiency of a single patch begins to drop, as W/L increases beyond 2.

- To increase bandwidth, increase the substrate height and/or decrease the substrate permittivity (this will also affect resonant frequency and the impedance matching).
- To increase the input impedance, decrease the width of the feed lines attached directly to the patches as well as the width of the lines attached to the port. The characteristic impedance of the quarter-wave sections should then be chosen as the geometric mean of half the impedance of the feed lines attached to the patches and the impedance of the port lines.

Antenna Magus (see Fig. 14) is a software tool that helps choose the appropriate antenna for a given application and estimates the S11 / VSWR and the far field gain characteristics.
Caution: Antennas on very thin substrates have high copper-losses, while thicker and higher permittivity substrates may lead to performance degradation due to surface waves. Although arrays are not directly used in cars, they are used in base stations for car to car communication.

6. Modelling of a strip / mesh antenna on a windscreen

The proliferation of communication devices that are required in modern automobiles, require automobile designers to include more and more antennas into their vehicle designs. Requirements include FM/AM antennas, TV antennas, etc. Aesthetically speaking, this is a problem that can only be overcome by including such antennas into vehicle designs in unobtrusive ways. A prominent modern development is to include these antennas into the windscreens of a vehicle. These windscreens include multiple layers of glass and wiring that form the antenna. As with other antenna designs, engineers require the ability to simulate new designs to evaluate many antenna operating characteristics, including:

- Efficiency
- Impedance bandwidth
- Far-field radiation characteristics

FEKO includes a solution method based on the MoM that can be used for rigorous analysis of windscreen antennas. The method meshes only the metallic antenna elements, so the resource requirements that are devoted to modelling of the dielectric layers of the glass is almost negligible. Features of the method include:

- Boundaries of the dielectric interfaces between different layers of glass are accurately accounted for.
- Coupling between closely spaced antenna elements are taken into account.
- Finite size glass antennas can be integrated into a full car model.
- Curvature and rotation of the window is considered.

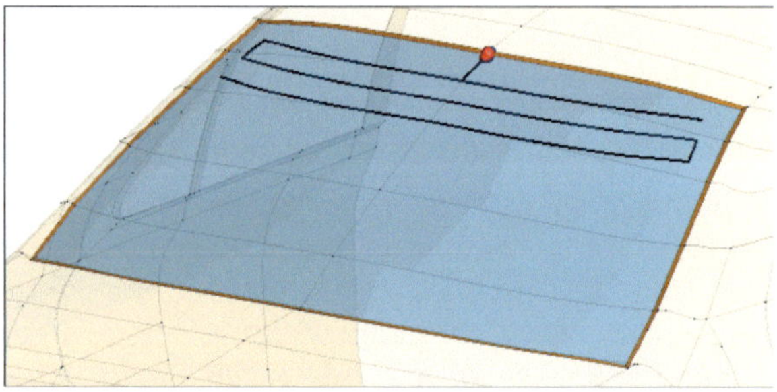

Fig. 15. Integrated windscreen antennas

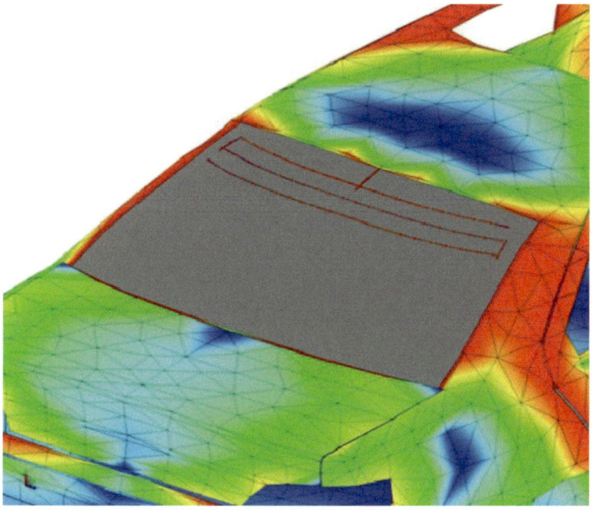

Fig. 16. Current distribution due to integrated windscreen antennas

The windscreen can consist of one or more layers and the different layers do not have to be meshed and thus simulation time is greatly reduced when compared to conventional methods. Fig. 15 and Fig. 16 show a 3D representation of the car and windscreen being simulated e.g. for current distribution, the input impedance / S11, etc. Besides the use of integrated antennas on windscreens, these are also integrated to car tires, mirrors, and bumpers for collision avoidance at the 76 GHz band.

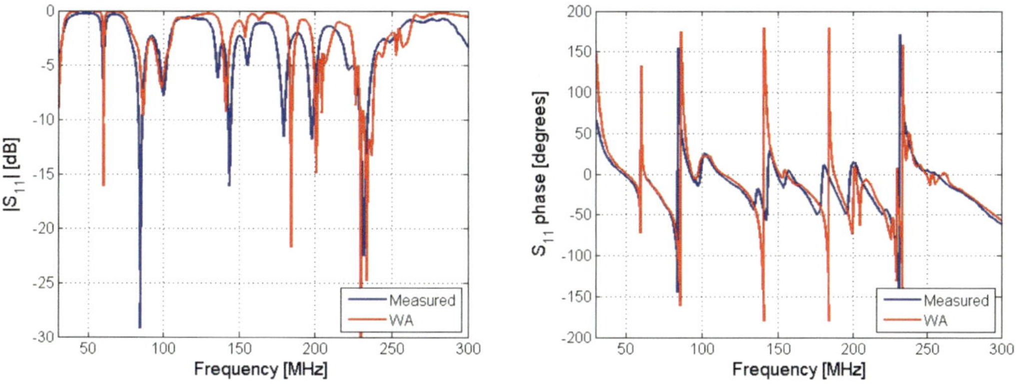

Fig. 17. Comparison of Antenna reflection co-efficient (simulation / measured results)

The currents are calculated

- Based on MoM solution with the incorporation of planar green function
- Full consideration of:
  - Boundaries between dielectric layers of glass
  - Coupling between closely spaced antenna elements
  - Curved/rotated windscreens
  - Multiple windscreens

Using the aforesaid approach only the metallic parts need to be meshed and not the dielectric parts of the windscreen elements of a car. Alternately the dielectric material i.e. the windscreen can be modelled using various methods e.g. FEM. However this approach is more time consuming as even the windscreen has to be meshed.
Results show that the planar Green's function approach (windscreen analysis - WA) is in good agreement with the measured results. Fig.17 shows a fairly good agreement between the simulation and the measured results.

7. References

Balanis, C. A. (2005). *Antenna Theory, Analysis and Design, Wiley & Sons,* Artech House, ISBN 978-0471603528, USA.

Kumar, G. & Ray, K.P. (2003). *Broadband Microstrip Antennas,* Artech House, ISBN 978-1580532440, USA.

Garg, R. (2001). *CAD for Microstrip Antennas Design Handbook,* Artech House, ISBN, Artech House

Sainati, R. A. (1996). *CAD for Microstrip Antennas for Wireless Applications,* Artech House, Publisher, ISBN 978-0890065624, Boston.

Bancroft, R. (1996). Understanding Electromagnetic Scattering Using the Moment Method – A Practical Approach, Artech House, ISBN 978-0890068595, Boston.

Refer to FEKO by using the following information:
Author: EM Software & Systems - S.A. (Pty) Ltd
Title: FEKO (www.feko.info)
Suite: (the suite number reported by FEKO)
Publisher: EM Software & Systems - S.A. (Pty) Ltd
Address: PO Box 1354, Stellenbosch, 7599, South Africa

Refer to Antenna Magus by using the following information:
Author: Magus (Pty) Ltd
Title: Antenna Magus (www.antennamagus.com)
Version: (the version number reported by Antenna Magus)
Publisher: Magus (Pty) Ltd
Address: PO Box 1354, Stellenbosch, 7599, South Africa

17

Link Layer Coding for DVB-S2 Interactive Satellite Services to Trains

Ho-Jin Lee[1], Pansoo Kim[1], Balazs Matuz[2], Gianluigi Liva[2], Cristina Parraga Niebla[2], Nuria Riera Dıaz[2] and Sandro Scalise[2]
[1]*Broadcasting and communication convergence research division, ETRI, 161 Gajeong-dong Yuseong-gu Daejeon,*
[2]*Institute of Communications and Navigation, German Aerospace Center (DLR), 82234 Wessling,*
[1]*Republic of Korea*
[2]*Germany*

1. Introduction

The growing number of railway passengers represents an appealing market for multimedia services. Satellites could be used to fulfill these demands due to the large coverage area and the low cost of associated terrestrial infrastructure. However, transmissions to mobile users through satellite links always pose a big challenge, especially since line of sight connection is frequently interrupted by obstacles between the satellite and the mobile receiver. The railroad satellite channel (RSC) in particular suffers from severe fadings that can be described using a combined statistical/deterministic model. In this paper, we will focus on the DVB-S2 [1] forward link providing service to high-speed trains. Additional protection of the data on link layer (LL) has been taken into account to mitigate the fading effects. The LL coding scheme investigated in this paper is based on the adoption of an erasure correcting code whose symbols are packets of constant size. Examples of erasure correcting codes applied in satellite communication systems can be found in [2]–[4]. The effort for the LL code design is mainly focused on the mitigation of the fade events due electrical trellises or power arches (PA) that are placed aside the tracks in order to provide the electric power to the trains along many railways. Such events are frequent and nearly periodic. In [5] it has already been shown that without a proper mitigation technique they would lead to an unacceptable quality of service. The rest of the paper is organized as follows. In Section II. we will provide an overview of the railroad satellite channel, focusing on the effect of electrical trellises on the received signal power level. In Section III the overall system architecture is described. Some insights on the link layer code design are provided as well. Section IV shows a performance comparison between the proposed link-layer coding approach and an enhancement of the DVB-S2 physical layer (PHY layer) through a long inter-frame interleaver. Moreover, a further, simplified model for the railroad satellite channel is introduced to give a basic understanding of the performance for the different solutions. Concluding remarks follow in Section V.

2. Railroad satellite channel model

An appropriate model for the propagation channel in a railway environment can be derived using the land mobile satellite channel (LMSC) as a reference scenario [6]: in the first instance it is sufficient to characterize the channel behavior by two different states, i.e. a line of sight (LOS) state with relatively high received signal power and a non line of sight (NLOS) state where the signal is shadowed or blocked by objects in the vicinity of the receiver. In the former state, the received signal is composed of a direct and a multipath component, with the instantaneous received signal power S obeying a Ricean probability density function:

$$p_{Rice}(S) = c \cdot \exp\left(-c(S+1)\right) \cdot I_0\left(2c\sqrt{S}\right).$$

Here, c denotes the so-called Rice factor, i.e. the direct-tomultipath signal power ratio and I_0 is the modified Bessel function of order zero. In the NLOS state, with no direct signal path present, the signal power shows Rayleigh behavior around a short-term mean value S_0 with the PDF described by:

$$p_{Rayl}\left(S|S_0\right) = \frac{1}{S_0}\exp\left(-S/S_0\right).$$

For the short-term mean S_0 a lognormal distribution is assumed:

$$p_{LN}\left(S_0\right) = \frac{10}{\sqrt{2\pi}\sigma_{dB}\ln 10} \cdot \frac{1}{S_0}\exp\left[-\frac{\left(10\log S_0 - \mu_{dB}\right)^2}{2\sigma_{dB}^2}\right],$$

with μ_{dB} describing the average power level (in dB) and σ_{dB} the variance of the power level (in dB^2) due to large scale fading. The railroad satellite channel has some peculiarities that have not been modeled properly by the previous description. Measurement campaigns show that a constant attenuation of 2-3 dB is introduced by catenaries above the tracks. Also, long fades occur mainly due to structures like bridges or tunnels, and shorter but periodic ones that are caused by several metallic obstacles along the railroad. Among them there are posts (with or without brackets), electrical trellises or arches spanning over the tracks, but they will be simply referred to as power arches for the rest of this paper. In the sequel, we will restrict ourselves on a RSC model corresponding to the LMSC in the LOS case superimposed by short deep fades ascribed to the power arches. Since these fades are nearly-periodic (and thus deterministic), they are not suitable for a statistical characterization. In turn, the modeling approach proposed in [5], here recalled for sake of clarity, will be adopted. The attenuation introduced by the above-mentioned power arches can be accurately described using the knife-edge diffraction theory. The knife-edge attenuation describes the ratio between the received electro-magnetic field E_D in presence of an obstacle and the received field under free space conditions E_0. For an object of two finite dimensions, it can be represented as sum of two diffracted signal components:

$$\frac{E_D}{E_0} = \frac{1+j}{2}\left(\frac{G(\alpha_1)}{G_{MAX}}\int_{v}^{\infty} e^{-j\frac{\pi}{2}t^2}dt + \frac{G(\alpha_2)}{G_{MAX}}\int_{-\infty}^{v-\frac{d\cdot v}{h}} e^{-j\frac{\pi}{2}t^2}dt\right),$$

where d is the width of the obstacle, h the height above LOS and $G(\alpha)/G_{MAX}$ denotes the radiation pattern of the directive antenna. Moreover, the Fresnel parameter v can be calculated out of h, the wavelength λ and the distance d_1 between the receiver and the object, as well as the distance d_2 between the object and the satellite according to:

$$v = h\sqrt{\frac{2(d_1+d_2)}{\lambda d_1 d_2}}.$$

Following that, the railroad satellite channel which is the basis for all further investigations looks like depicted in Figure 1. It comprises Ricean fading (with a Rice factor of 18 dB), as well as periodic deep fades as a result of equally spaced power arches aside the railway. The worse scenario, where additional NLOS is present according to the LMSC model, will not be considered in the sequel.

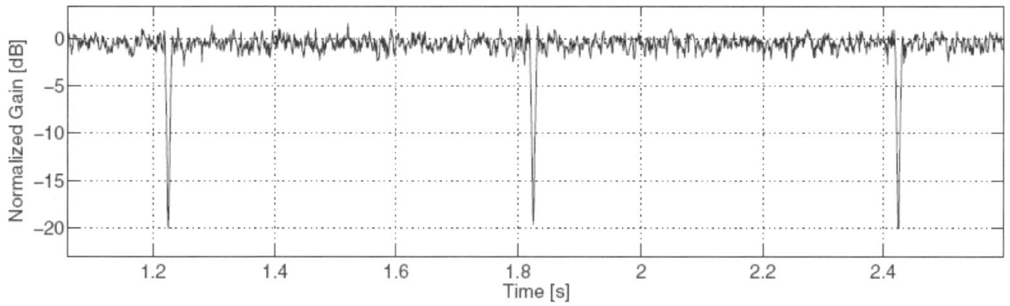

Fig. 1. RSC realization with a power arch distance of 50m at a speed of 300km/h; periodic deep fades occur every time the train passes by a PA

3. System description

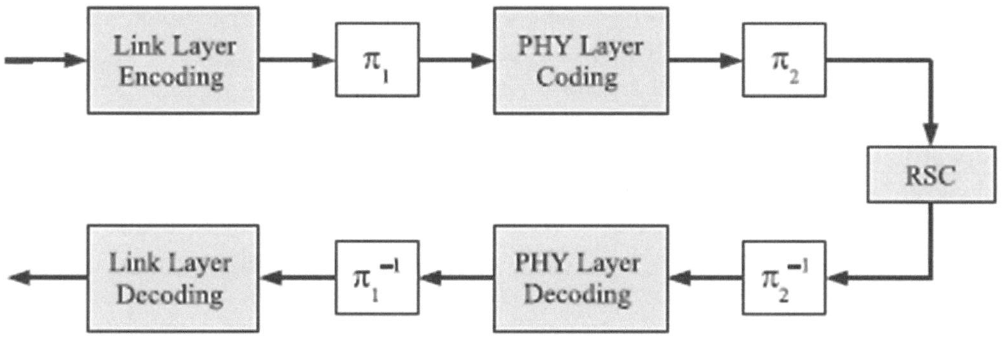

Fig. 2. Transmission chain used for the simulation

Our approach to mitigate the impairments of the RSC is based on link layer coding. For illustration purposes, a simplified block diagram of the link is depicted in Figure 2. In our case LL coding is applied on MPEG-TS packets. The systematic link layer encoder receives at its input K MPEGTS packets and produces at its output N packets, referred to as LL codeword. Due to the systematic nature of the code, this codeword consist of the K input MPEG-TS packets followed by M = N − K parity packets. On the receiver side, the decoder takes care of recovering lost MPEG-TS packets. Note that in the link layer coding framework the code works on an erasure channel, where the erased units (i.e., the LL codeword symbols) are whole packets. In the context of this paper, such feature is guaranteed by the error detection performed after physical layer decoding on each MPEGTS/parity packet. In other words, the PHY decoder attempts to protect individual MPEG-TS packets, whereas the LL decoder is meant to recover lost MPEG-TS packets. The LL decoder design is highly facilitated by the underling erasure channel, permitting for some kind of codes the adoption of software based decoding up to several tens of Mbytes per second. To allow this appealing feature, our investigation is focused on the adoption of low-density parity-check (LDPC) codes [7] as erasure correcting codes. LDPC codes provide capacity approaching performance on many communication channels [8], and in the framework of LDPC codes some astonishing erasure correcting codes have been developed [9]–[11]. The LDPC codes adopted for the simulations belong to the family of the so-called irregular repeat-accumulate (IRA) codes [12]. IRA codes allow simple efficient encoding while keeping nearcapacity performance. The design of the IRA codes have been optimized through extrinsic information transfer (EXIT) analysis [13]–[15], constraining the parity-check matrix of the code to a block-circulant form that would also permit a simple hardware decoder design. Further performance enhancements with similar encoding complexities are expected by adopting more sophisticated IRA-like designs [16] [17]. After link layer coding is done, the MPEG-TS packets within each LL codeword are interleaved (denoted by π_1 in Figure 2) to break up channel correlations. For our investigations we limited the maximum length of the link layer codeword in a way that it spans over to 200 ms. For example, taking into account a symbol rate of 27.5 MBaud, QPSK modulation and physical layer code rate r = 1/2, the LL codeword length would be N = 3400 packets. This constraint has been introduced to avoid long delays which could affect real-time applications. The stream of MPEG-TS/parity packets is then forwarded to a DVB-S2 transmitter, which takes care of the physical layer coding through the serial concatenation of a BCH (Bose, Ray-Chaudhuri, Hocquenghem) code and LDPC code according to the DVB-S2 standard (for details see [1]). The physical layer codeword size corresponds to the large frame size of the DVB-S2 standard (64800 bits). Before forwarding the data to radio frequency (RF) frontend, we allow a further (optional) physical layer inter-frame block interleaver π_2 that permutes the bits among several frames. For the sake of comparison, in the following we will consider also the scenario where the LL coding block is disabled. In such case, the diversity necessary to overcome the short periodic fade events will be provided by the inter-frame interleaver only. However, the interleaver latency will be constrained to be lower than 200 ms. Furthermore, to keep the comparison fair, physical layer code rate will be lowered in a way that the overall efficiency of the two systems is the same.

4. Outcomes

In section IV, we provide some numerical results obtained through Monte Carlo simulations on the RSC described in Section II. The analysis is focused on the case of LOS conditions with superimposed power arches. The performance is depicted in terms of MPEG-TS packet error rate (PER) versus signal-to-noise ratio (SNR) E_s / N_0. Here, E_s denotes the energy per modulated symbol and N_0 the one-sided noise power spectral density. For the LL coded solution, the results are shown in terms of residual MPEG-TS packet error rate at the output of the LL decoder vs. E_s / N_0. Besides the PA width and the train velocity that have a high impact on the fade duration, also other factors have to be taken into account for the simulations, such as the current latitude and the traveling direction of the train. To simplify the simulations it is advisable to determine an effective power arch width that takes into account these factors. For a PA width of 30 cm and a latitude of 38° geometric considerations yield to an effective PA width of roughly 87cm. Considering train speeds from 30 km/h to 150 km/h and a north-south traveling direction, the resulting fade durations range from ~100 ms to ~20 ms. The distance between two subsequent PA is constantly set to 50 m.

4.1 Simulation results

Assuming only PHY layer coding with a code rate of 1/4 combined with physical layer interleavers of different lengths we obtain the plots in Figure 3 (at 30 km/h on the left, at 150 km/h on the right). For both speeds, intermediate error floors arise at error rates proportional to the interleaver duration (in the charts, the performance with interleaver lengths of 200 ms, 100 ms, 50 ms and no interleaving are depicted). However, at high speeds the rate 1/4 code in combination with a sufficiently long interleaver is able to overcome the floor, but the steepness of the curve in the subsequent waterfall region remains quite poor. For low speeds the error floors remain, for all the investigated interleaver lengths and also for relatively high SNRs. The outcomes for joint physical/link layer coding with rate 1/2 codes on both layers and with a LL interleaver duration of 200 ms combined with different PHY layer interleaver durations are shown in Figure 4 (at 30 km/h on the left, at 150 km/h

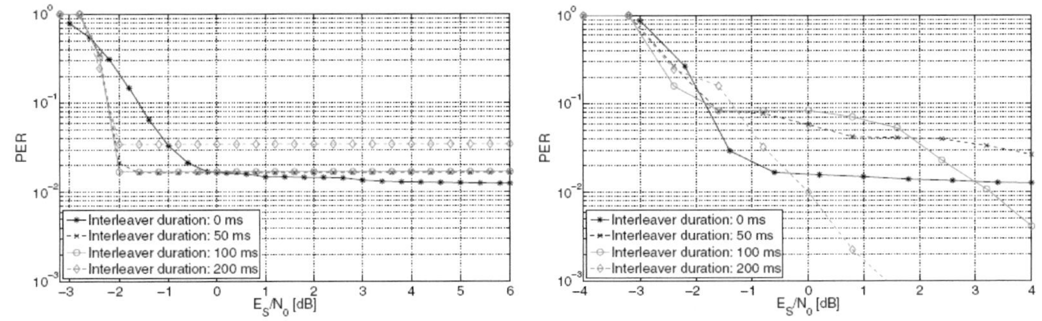

Fig. 3. PER vs Es/No with only physical layer protection(DVB-S2, QPSK, r=1/4) and different interleaving depths. Overall spectral efficiency of 0.5 bps/Hz. Speed of 30km/h (left) and 150km/h (right)

on the right). The best results can be achieved by using no physical interleaver at all. In this case the LL code is able to overcome the errorfloor at both speeds and ensures a steep slope of the PER curve in the waterfall region. Compared to plain physical layer coding with inter-frame interleaving, joint physical/link layer coding clearly shows an improvement of performance. Note that for joint PHY/LL coding both interleavers π_1 and π_2 are synchronized, so that the overall delay that is experienced due to interleaving is equivalent to the maximum of the delays introduced by π_1 and π_2 (in our case not more than 200 ms).

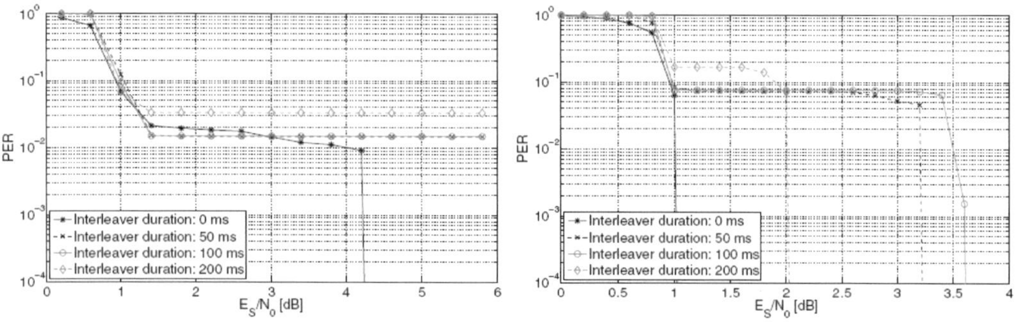

Fig. 4. PER vs Es/No with joint physical layer protection(DVB-S2, QPSK, r=1/2) and link layer Protections(link layer LDPC code, R=1/2) and different interleaving depths. Speed of 30km/h (left) and 150km/h (right). Overall spectral efficiency of 0.5bps/Hz

4.2 Insights on the use of PHY layer interleavers on the RSC

Fig. 5. Main concept of the AWGN/EC-AWGN channel model for the RSC channel

The behavior of a coding scheme including physical layer coding and long inter-frame interleavers on the LOS channel with superimposed blockages can be easily understood by splitting the contributions to the packet error probability P_e (that is the stochastic equivalent to the simulated PER) into two parts, the LOS error rate and the blockage error rate. The LOS condition is referred as the good state (*G*). To simplify the analysis in such state, the channel characteristic is approximated by an AWGN channel (recall the high Rice factor in LOS). The blockage condition ascribed to PAs is referred to as the bad state (*B*). Here, the channel is basically a bursty erasure channel. Assuming interleaving windows longer than a

PA fade duration (thus, spreading the erasures on a wider duration), the PHY layer decoder deals with a combination of an erasure channel with AWGN (EC-AWGN). This state spans over a whole interleaver window. As an illustration the overall channel model is depicted in Figure 5. Denoting by X the channel state random variable, the stationary probabilities of being in the good/bad state are given by

$$P_G = \frac{T_G}{T_B + T_G} \text{ and } P_B = \frac{T_B}{T_B + T_G}$$

where $T_G(T_B)$ represents the time spent in the good (bad) state. Assuming periodic fade events due to the power arches, T_B shall be replaced by the interleaver length L, expressed in seconds, or by the fade duration T_f, in case no interleaving is applied. For sake of simplicity, let's summarize such parameter as Δ_B. The sum $T_B + T_G$ has to be replaced by the power arch periodicity τ, while T_G becomes the time interval between two interleaving windows affected by consecutive power arch fades, Δ_G. Let's define $\Pr\{E|X=G\}$ and $\Pr\{E|X=B\}$ as the packet error conditional probabilities given the good (bad) state, where E denotes the packet error event. The error probability P_e can be therefore expressed as

$$\begin{aligned} P_e &= P_G \cdot \Pr\{E|X=G\} + P_B \cdot \Pr\{E|X=B\} \\ &= \frac{\Delta_G}{\Delta_G + \Delta_B} \Pr\{E|X=G\} + \frac{\Delta_B}{\Delta_G + \Delta_B} \Pr\{E|X=B\}. \end{aligned} \tag{1}$$

Note that, due to the parameters chosen for the simulations, $\Delta_B \sim 10^{-2} \cdot \Delta_G$. Consequently, P_B is in the order of magnitude of 10^{-2}. Thus, at low SNRs, where the error rates are high even in LOS conditions, the first term in (1) dominates the summation. At high SNRs, the error rate $\Pr\{E|X=G\}$ in LOS condition quickly decreases. The second term becomes therefore dominant. The performance curve of the system in such conditions can be composed in a two-fold fashion:

1. The error probability $\Pr\{E|X=G\}$ in LOS conditions is evaluated numerically down to error rates that are negligible respect to P_B. With the current scenario of $P_B \sim 10^{-2}$ the simulation can be stopped once $\Pr\{E|X=G\}$ approaches 10^{-3}
2. As a second step the probability $\Pr\{E|X=B\}$ for the EC-AWGN channel has to be computed. In case no interleaving is applied, Δ_B is equal to T_f. In this interval $\Pr\{E|X=G\}$ can be reasonably set to 1. Otherwise, the problem of computing $\Pr\{E|X=B\}$ reduces to the performance evaluation of the channel code on the ECAWGN, where, assuming random interleaving, each bit soft-value is erased with a probability ε , with $\varepsilon = T_f / L$ in the interval $\Delta_B = L$.

The two error probabilities are then combined on the same chart following equation (1). This is exemplified in Figure 6. The LL code employed for the simulation is a short (2048,1024) LDPC code. The impact of the physical layer interleaver length becomes therefore quite clear: large interleavers lower the erasure rate (recall that $\varepsilon = T_f / L$) of the codeword bits in blockage conditions, increasing the steepness of the $\Pr\{E|X=B\}$ curve. At the same time, high values of L rise up the intermediate floor, at the error rate given by $P_B = L/\tau$. This is compliance with the results presented in charts 3 and 4: the higher the interleaver length, the higher the intermediate error floor. As it can be seen, there are some exceptions, since the same error floor arises for the 50 ms and 100 ms interleavers. This is due to the fact that PA fades affect two interleaver windows for the 50 ms case (i.e., the interleaving window has a length which comparable to the fade duration). In case of a scheme employing a link level code on top of the physical layer, it is often advisable to abstain from the use of physical layer interleavers to keep the intermediate error floors as low as possible (especially at low train speeds). The packet recovery task in presence of the PA fades is then left to the LL code.

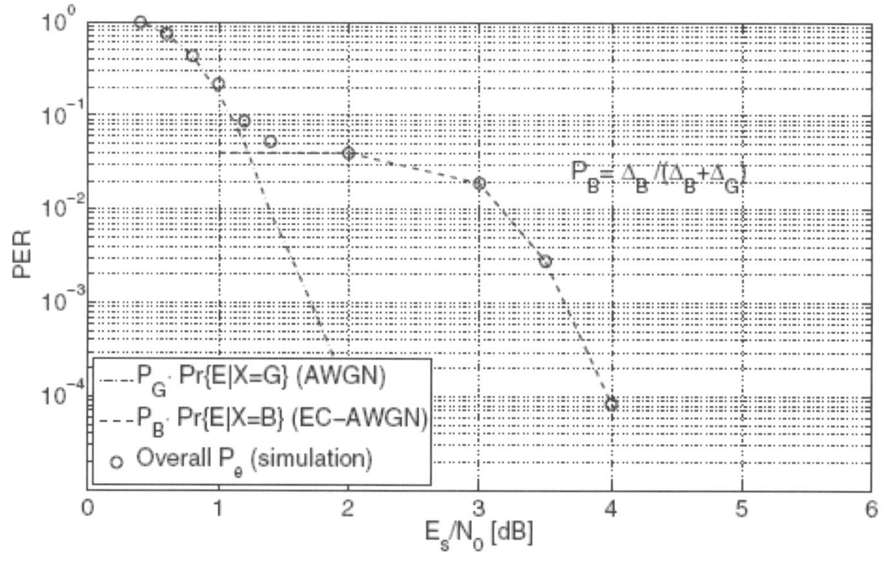

Fig. 6. Packet error rate for the simplified RSC (hybrid AWGN/EC-AWGN model)

5. Concluding remarks

The main purpose of this work was to draw a comparison between physical layer coding and interleaving and the innovative approach of joint physical/link layer coding for the railroad satellite channel. It was shown that the system performance can be highly improved for this type of channel by splitting redundancy on different layers. Employing link layer coding shows some performance advantages with respect to the use of long physical layer interleavers, especially in case of frequent short blockages. A simplified model for the LOS railroad satellite channel with superimposed periodic fades was introduced, with focus on the performance of a scheme employing physical layer coding enhanced by long inter-frame interleavers. The proposed model allows a precise calculation of the arising error floors, as well as simple explanation of the system behavior for different interleaver lengths and train velocities. This knowledge turns out to be very helpful for the code design.

6. Acknowledgments

This work was supported by the ETRI-DLR Collaborative Research under the name of "Communication Technologies for Satellite Broadband Mobile based on DVB-S2/RCS". The work has been presented in part at VTC 2008 spring[18].

7. References

[1] Digital Video Broadcasting (DVB): Second generation framing structure,channel coding and modulation systems for Broadcasting, Interactive Services, News Gathering and other broadband satellite applications, ETSI Std. EN 302 307, 2004.

[2] C. Di, H. Ernst, E. Paolini, S. Coletto, and M. Chiani, Low-density parity-check codes for the transport layer of satellite broadcast, in *Proc. AIAA International Communications Satellite Systems Conference(ICSSC 2005)*, Rome, Italy, Sep. 2005.

[3] M. Chiani, G. Liva, and E. Paolini, Investigation of long erasure codes for space communication protocols, CCSDS, Rome, Tech. Rep., June 2006, spring Meeting.

[4] E. Paolini, G. Liva, M. Chiani, and G. Calzolari, Tornado-like codes: a new appealing chance for space applications protocols?, in *3rd European Space Agency Workshop on Tracking, Telemetry and Command Systems for Space Applications, TTC 2004*, Sep. 2004.

[5] S. Scalise, R. Mura, and V. Mignone, Air Interfaces for Satellite Based Digital TV Broadcasting in the Railway Environment, in *IEEE Transactions on Broadcasting*, IEEE, Ed., vol. 52, no. 2, June 2006, pp. 158–166.

[6] E. Lutz, M. Werner, and A. Jahn, *Satellite Systems for Personal andBroadband Communications*. Springer Verlag, 2000.

[7] R. G. Gallager, *Low-Density Parity-Check Codes*. Cambridge, MA: M.I.T. Press, 1963.

[8] T. Richardson and R. Urbanke, The capacity of low-density parity check codes under message-passing decoding, *IEEE Trans. Inform Theory*, vol. 47, 2001.

[9] M. Luby, LT-codes, in *Proc. of the ACM Symposium on Foundations of Computer Science (FOCS)*, 2002.

[10] E. Paolini, M. Fossorier, and M. Chiani, Analysis of doubly-generalized LDPC codes with random component codes for the binary erasure channel, in *Proc. of Allerton Conf. on Communications, Control and Computing*, Monticello, USA, Sep. 2006.

[11] A. Shokrollahi, Raptor codes, *IEEE Transactions on Information Theory*, vol. 52, no. 6, pp. 2551-2567, Jun. 2006.

[12] H. Jin, A. Khandekar, and R. McEliece, Irregular repeat-accumulate codes, in *Proc. International Symposium on Turbo codes and Related Topics*, Sep. 2000, pp. 1–8.

[13] S. ten Brink, Convergence behavior of iteratively decoded parallel concatenated codes, *IEEE Trans. Commun.*, vol. 49, pp. 1727–1737, Oct. 2001.

[14] G. Liva and M. Chiani, Protograph LDPC codes design based on EXIT analysis, in *Proc. IEEE Globecomm*, Nov. 2007.

[15] G. Liva, S. Song, L. Lan, Y. Zhang, W. Ryan, and S. Lin, Design of LDPC codes: A survey and new results, *J. Comm. Software and Systems*, Sep. 2006.

[16] A. Abbasfar, K. Yao, and D. Disvalar, Accumulate repeat accumulate codes, in *Proc. IEEE Globecomm*, Dallas, Texas, Nov. 2004.
[17] G. Liva, E. Paolini, and M. Chiani, Simple reconfigurable low-density parity-check codes, *IEEE Comm.Letters*, vol. 9, pp. 258–260, March 2005.
[18] B. Matus, Link Layer Coding for DVB-S2 Interactive Satellite Services to Trains, in *Proc. IEEE VTC*, Sigapore, May. 2008

18

Mobility Aspects of Physical Layer in Future Generation Wireless Networks

Asad Mehmood and Abbas Mohammed
Blekinge Institute of Technology Karlskrona
Sweden

1. Introduction

The demand from social market for high speed broadband communications over wireless media is pushing the requirements of both the mobile and fixed networks. The past decade has witnessed tremendous advancement in the blooming development of mobile communications including mobile-to-mobile and mobile-to-fixed networks. Wireless fixed and cellular networks of future generation will need to support new protocols, standards and architecture leading to all IP-based networks. Different systems like digital video broadcasting (DVB) via satellites have great success commercially as they provide ubiquitous coverage and serve large number of users with high signal quality. Satellite communications have proven to be attractive means to provide communication services such as broadband communications (3G services), surveillance, remote monitoring, intelligent transportation systems, navigation, traffic warnings and location-based information etc. to fixed and mobile users. However, to meet the growing demands of mass market integration of satellites and terrestrial networks seems to be inevitable for future generation wireless networks.

Due to technology advances and growing traffic demands, communication systems must evolve to completely new systems or within themselves in order to provide broadband services in a safe and efficient way. While enhancements continue to be made to leverage the maximum performance from currently deployed systems, there is a bound to the level to which further improvements will be effective. If the only purpose were to deliver superior performance, then this in itself would be relatively easy to accomplish. The added complexity is that such superior performance must be delivered through systems which are cheaper from installation and maintenance prospect. Users have experienced an incredible reduction in telecommunications charges and they now anticipate receiving higher quality communication services at low cost. Therefore, in deciding the subsequent standardization step, there must be a dual approach: in search of substantial performance enhancement but at reduced cost. Long Term Evolution (LTE) is that next step and will be the basis on which future mobile telecommunications systems will be built. LTE is the first cellular communication system optimized from the outset to support packet-switched data services, within which packetized voice communications are just one part.

In case of highly mobile scenarios, the effects of signal blockages and Doppler shifts introduce more burdens on the receiver demodulator. The signals blockage is prominent in the case of land mobile communications as compared to satellite communications. In deciding the technologies to comprise in LTE, one of the key concerns is the trade-off between cost of implementation and practical advantage. Fundamental to this assessment, therefore, has been an enhanced understanding different scenarios of the radio propagation environment in which LTE will be deployed and used.

The organization of the chapter is as follows. In section 2, different mobility aspects related to the physical layer of future generation mobile communication networks are discussed. Section 3 discusses the propagation scenarios in which LTE will be deployed. Section 4 describes space-time processing techniques to enhance the system performance. In section 5 LTE system's performance is evaluated at different mobile speeds. Finally, section 6 concludes the chapter.

2. Physical layer aspects

The high data rate multimedia broadcast/multicast services at cheap rates with appropriate quality-of-service (QoS), fast handoff techniques and wide area seamless mobility pave the way for future generation wireless communications. Wireless network operators require different schemes for including new services to take benefits from new access technologies. Fundamental to these strategies is to incorporate mobility that can bring unique advantages to mobile users. In response to these requirements, the wireless industry is foreseen to shift toward LTE and world wide interoperability of microwave access (WiMAX) technologies to be able to support cost effectively the capacity required by mobile operators to meet mass market demands of data services (Motorola, 2010). LTE must be able to provide superior performance compared to the existing wireless network infrastructures which suffer from cell-edge performance, spectral efficiency and desired QoS to end users. In order to provide high data rates with high QoS in already crowded spectrum, LTE is susceptible to different impairments: noise and interference. Therefore to mitigate these propagation impairments, efficient and robust techniques need to be adapted to take full benefits of the technology. A thoughtful design of physical layer aspects to mitigate these propagation impairments and improve the system performance is thus crucial for successful operation and support of the desired QoS.

2.1 Objectives of physical layer

The objectives of LTE physical layer are the significant increase in peak data rates up to 100 Mb/s in downlink and 50 Mb/s in uplink within 20 MHz spectrum leading to spectrum efficiency of 5 Mb/s, increased cell-edge performance maintains site locations as in Wide Band Code Division Multiple Access (WCDMA), reduced user and control plane latency to less than 10 ms and less than 100 ms, respectively (Kliazovich1, et al.). LTE will be able to provide interactive real-time services such as high quality video/audio conferencing and multiplayer gaming with mobility support for up to 350 km/h or even up to 500 km/h and reduced operation cost. It also provides a scalable bandwidth 1.25/2.5/5/10/20 MHz in order to allow flexible technology to coexist with other standards, 2 to 4 times improved spectrum efficiency the one in Release 6 HSPA to permit operators to accommodate increased number of customers within their existing and future spectrum allocation with a reduced cost of delivery per bit, low power consumption and acceptable system and terminal complexity. The system should be optimized for low mobile speed but also support high mobile speed as well. In this section we will discuss some of the features included in LTE physical layer to mitigate propagation impairments.

Scalable OFDMA: Multiple access schemes are used in multi-user communications to provide on-demand data rates to users by sharing the available resources in available finite bandwidth. The orthogonal frequency division multiple access (OFDMA) is used as multiple access scheme in the downlink and single carrier frequency division multiple access (SC-FDMA) is used in the uplink. OFDMA is OFDM based multiple access technique used for LTE to facilitate the exploitation of multi-user diversity, frequency diversity and flexible users scheduling to enhance the system capacity in challenging multi-user communications with wide range of applications, data rates and QoS requirements. The flexible structure of OFDMA allows efficient implementation of space-time processing techniques, e.g., multiple-input multiple-output (MIMO) with reasonable complexity. The scalable bandwidth with different FFT sizes and dynamic subcarrier allocation allows the efficient use of spectrum in different regional regulations for mobile applications.

Frame Structure and Transmission Modes: LTE supports two types of frame structures: type1 frame structure which is designed for frequency division duplex (FDD) and is valid for both half duplex and full duplex FDD modes. Type 1 radio frame has a duration 10 ms and consists of 20 slots each of 0.5 ms. A sub-frame comprises two slots, thus one radio frame has 10 sub-frames. In FDD mode, half of the sub-frames are available for downlink and the other half are available for uplink transmission in each 10 ms interval, where downlink and uplink transmission are separated in the frequency domain (3GPP, 2008). Type 2 frame structure is applicable for time division duplex mode (TDD). The radio frame is composed of two identical half-frames having duration of 5 ms. Each half-frame is further divided into 5 sub-frames having duration of 1 ms. Two slots of length 0.5 ms constitute a sub-frame which is not special sub-frame. The special type of sub-frame is composed of three fields Downlink Pilot Timeslot (DwPTS), GP (Guard Period) and Uplink Pilot Timeslot (UpPTS). Seven uplink-downlink configurations are supported with both types (10 ms and 5 ms) of downlink-to-uplink switch-point periodicity. In 5 ms downlink-to-uplink switch-point periodicity, special type of sub-frames are used in both half-frames but it is not the case in 10 ms downlink-to-uplink switch-point periodicity, special frame is used instead of are used only in first half-frame. For downlink transmission sub-frames 0, 5 and DwPTS are always reserved. UpPTS and the sub-frame next to the special sub-frame are always reserved for uplink communication (3GPP, 2009).

Mobility Support: One of the features of LTE is appropriate physical layer design to facilitate users at high vehicular speeds to support delay sensitive applications (e.g., VOIP) with appropriate QoS. The physical layer features such as power control, hybrid automatic repeat request (HARQ), sub-channelization and pilot structure are used to mitigate the fluctuations in the received signal caused by channel fast fading. In addition, link adaptation technique is used to adjust system parameters according to channel dynamics, i.e, to select appropriate parameters under available propagation conditions. This permits to optimize the spectral and power sources of the system under poor propagation conditions.

Advanced Antenna Techniques: Multiple antenna systems based on space-time processing algorithms have brought great benefits to wireless communications by exploiting the spatial domain to use the resources in efficient way. Advanced antenna techniques such as diversity techniques, spatial multiplexing and beamforming are employed to create independent multiple parallel channels which result in overall system improvement in terms of link reliability, high capacity, extended coverage and reduced transmitted power. LTE uses advanced antennas techniques in both single-user and multi-user MIMO cases.

Link Adaptation and Channel Coding: Link adaptation is used to adjust the system parameters in time varying propagation conditions to facilitate users at different data rates. Thus link adaptation scheme is very closely related to channel coding schemes used for forward error correction (Sesia, et al. 2009). LTE schedules down link data transmission and selects modulation and coding schemes based on the feedback information in terms of signal-to-interference plus noise ratio (SINR) provided by channel quality indicator (CQI) in uplink direction. The LTE specifications define the signalling between user terminal and eNodeB for link adaptation and switching between different modulation schemes and coding rates that depend on several factors including cell throughput and required QoS.

Scheduling and Quality-of-Service: The purpose of scheduling is to manage the resources in uplink and downlink channels while maintaining the desired QoS according to user expectations. In LTE eNodeB performs this operation. The principle of scheduling algorithm is to allocate the resources and transmission powers in order to optimize certain set of parameters such as throughput, user spectral efficiency, average delay and outage probability. The LTE MAC layer can support large number of users with desired QoS.

3. Radio propagation models

From the beginning of wireless communications there is a high demand for realistic mobile fading channels. The reason for this importance is that efficient channel models are essential for the analysis, design, and deployment of communication systems for reliable transfer of information between two parties. Realistic channel models are also significant for testing, parameter optimization and performance evolution of communication systems. The performance and complexity of signal processing algorithms, transceiver designs and smart antennas etc., employed in future mobile communication systems, are highly dependent on design methods used to model mobile fading channels. Therefore, correct knowledge of mobile fading channels is a central prerequisite for the design of wireless communication systems (Rappaport, 1996; Ibnkahla, 2005; Ojanpera, et al., 2001).

The difficulties in modeling the wireless channel are due to complex propagation processes. A transmitted signal arrives at the receiver through different propagation mechanisms as shown in Figure 1. The propagation mechanisms involve the following basic mechanisms: i) free space or line of sight (LOS) propagation ii) specular reflection due to interaction of electromagnetic waves with plane and smooth surfaces which have large dimensions as compared to the wavelength of interacting electromagnetic waves iii) Diffraction caused by bending of electromagnetic waves around corners of buildings iv) Diffusion or scattering due to contacts with objects having irregular surfaces or shapes with sizes of the order of wavelength v) Transmission through objects which cause partial absorption of energy (Oestges, et al., 2007; Rappaport, 1996). It is significant here to note that the level of information about the environment a channel model must provide is highly dependent on the category of communication system under assessment. To predict the performance of narrowband receivers, classical channel models which provide information about signal power level distributions and Doppler shifts of the received signals, may be sufficient. The advanced technologies (e.g., UMTS and LTE) build on the typical understanding of Doppler spread and fading also incorporate new concepts such as time delay spread, direction of departures (DOD), direction of arrivals (DOA) and adaptive antenna geometry (Ibnkahla, 2005). The presence of multipaths (multiple scattered paths) with different delays, attenuations, DOD and DOA gives rise to highly complex multipath propagation channel. Figure 2 illustrates power delay profile (PDP) of a multipath channel with three distinct paths.

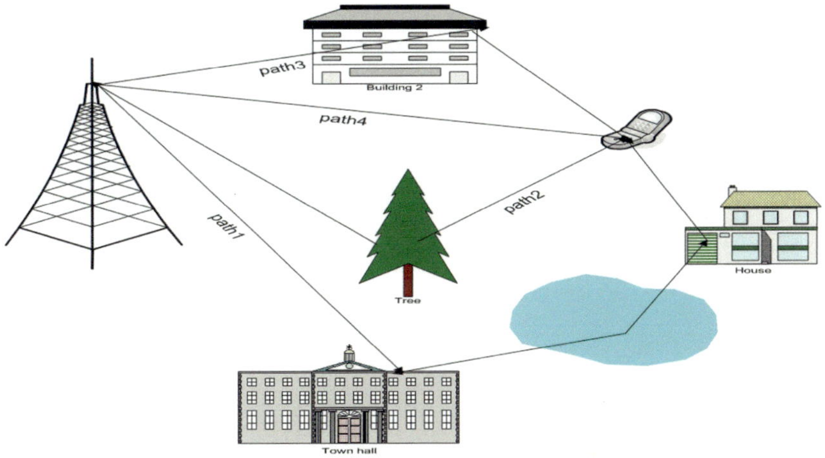

Fig. 1. Signal propagation through different paths showing multipath propagation phenomena

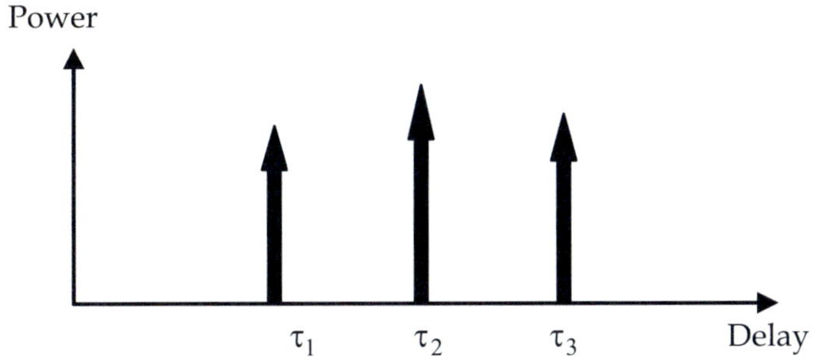

Fig. 2. Power delay profile of a multipath channel

3.1 Propagation aspects and parameters

The behaviour of a multipath channel needs to be characterized in order to model the channel. The concepts of Doppler spread, coherence time, delay spread and coherence bandwidth are used to describe various aspects of the multipath channel.

3.1.1 Delay spread

To measure the performance capabilities of a wireless channel, the time dispersion or multipath delay spread related to small scale fading of the channel needs to be calculated in a convenient way. One simple measure of delay spread is the overall extent of path delays called the excess delay spread. This is an appropriate way because different channels with the same excess delay can exhibit different power profiles which have more or less impact on the performance of the system under consideration. A more efficient method to determine channel delay spread is the root mean square (rms) delay spread (τ_{rms}) which is a statistical measure and gives the spread of delayed components about the mean value of the channel power delay profile. Mathematically, rms delay spread can be described as second central moment of the channel power delay profile (Rappaport, 1996) which is written as:

$$\tau_{rms} = \frac{\sqrt{\sum_{n=0}^{N-1} P_n(\tau_n - \tau_m)^2}}{\sum_{n=0}^{N-1} P_n} \quad (1)$$

where, $\tau_m = \frac{\sum_{n=0}^{N-1} P_n \tau_n}{\sum_{n=0}^{N-1} P_n}$ is the mean excess delay.

3.1.2 Coherence bandwidth

When the channel behaviour is studied in frequency domain then coherence bandwidth Δf_c is of concern. The frequency band, in which the amplitudes of all frequency components of the transmitted signal are correlated, i.e., with equal gains and linear phases, is known as coherence bandwidth of that channel (Ibnkahla, 2005). The channel behaviour remains invariant over this bandwidth. The coherence bandwidth varies in inverse proportion to the delay spread. A multipath channel can be categorized as frequency flat fading or frequency selective fading in the following way.

Frequency flat fading: A channel is referred to as frequency flat if the coherence bandwidth $\Delta f_c >> B$, where B is the signal bandwidth. In this case frequency components of the signal will experience the same amount of fading.

Frequency selective fading: A channel is referred to as frequency selective if the coherence bandwidth $\Delta f_c \le B$. In this case different frequency components will undergo different amount of fading. The channel acts as a filter since the channel coherence bandwidth is less than the signal bandwidth; hence frequency selective fading takes place (Fleury, 1996).

3.1.3 Doppler spread

The Doppler spread arises due to the motion of mobile terminal. Due to the motion of mobile terminal through standing wave the amplitude, phase and filtering applied to the transmitted signal vary with time according to the mobile speed (Cavers, 2002). For an unmodulated carrier, the output is time varying and has non-zero spectral width which is Doppler spread. For a single path between the mobile terminal and the base station, there will be zero Doppler spread with a simple shift of the carrier frequency (i.e., Doppler frequency shift) at the base station. The Doppler frequency depends on the angle of movement of the mobile terminal relative to the base station.

3.1.4 Coherence time

The time over which the characteristics of a channel do not change significantly is termed as coherence time. The reciprocal of the Doppler shift is described as the coherence time of the channel. Mathematically we can describe coherence time as:

$$T_c = \frac{1}{2\pi v_{rms}} \quad (2)$$

where v_{rms} is root mean square vale of Doppler spread.

The coherence time is related to the power control schemes, error correction and interleaving schemes and to the design of channel estimation techniques at the receiver.

4. Standard channel models

Standard channel models can be developed by setting up frame work for generic channel models and finding set of parameters that need to be determined for the description of the channel. Another method is to set up measurement campaigns and extracting numerical values of parameters and their statistical distributions (Meinilä, et al., 2004).
When designing LTE, different requirements are considered: user equipment (UE) and base station (BS) performance requirements which are crucial part of LTE standards, Radio Resource Management (RRM) requirements to ensure that the available resources are used in an efficient way to provide end users the desired quality of service, the RF performance requirements to facilitate the existence of LTE with other systems (e.g., 2G/3G) systems (Holma, et al., 2009). The standard channel models play a vital role in the assessment of these requirements. In the following section, some standard channel models are discussed which are used in the design and evaluation of the UMTS-LTE system.

4.1 SISO, SIMO and MISO channel models

COST projects, Advanced TDMA (ATDMA) Mobile Access, UMTS Code Division Testbed (CODIT) conducted extensive measurement campaigns to create datasets for SISO, SIMO and MISO channel modeling and these efforts form the basis for ITU channel models which are used in the development and implementation of the third generation mobile communication systems (Sesia, et al., 2009). COST stands for the "European Co-operation in the Field of Scientific and Technical Research". Several Cost efforts were dedicated to the field of wireless communications, especially radio propagation modeling, COST 207 for the development of Second Generation of Mobile Communications (GSM), COST 231 for GSM extension and Third Generation systems, COST 259 "Flexible personalized wireless communications (1996-2000)" and COST 273 "Towards mobile broadband multimedia networks (2001-2005)". These projects developed channel models based on extensive measurement campaigns including directional characteristics of radio propagation (Cost 259 and Cost 273) in macro, micro and picocells and are appropriate for simulations with smart antennas and MIMO systems. These channel models form the basis of ITU standards for channel models of Beyond 3G systems. Detailed study of COST projects can be found in (Molisch, et al., 2006; Corria, 2001).
The research projects ATDMA and CODIT were dedicated to wideband channel modelling specifically channel modelling for 3rd generation systems and the corresponding radio environments. The wideband channel models have been developed within CODIT using physical-statistical channel modelling approach while stored channel measurements are used in ATDMA which are complex impulse responses for different radio environments. The details of these projects can be found in (Ojanpera, et al., 2001).

4.2 ITU multipath channel models

The ITU standard multipath channel models proposed by ITU (ITU-R, 1997) used for the development of 3G 'IMT-2000' group of radio access systems are basically similar in structure to the 3GPP multipath channel models. The aim of these channel models is to develop standards that help system designers and network planners for system designs and performance verification. Instead of defining propagation models for all possible environments, ITU proposed a set of test environments in (ITU-R, 1997) that adequately span the all possible operating environments and user mobility. In this chapter we use ITU standard channel models for pedestrian and vehicular environments.

4.2.1 ITU Pedestrian-A, B

In both Pedestrian-A and Pedestrian-B channel models the mobile speed is considered to be 3 km/h. For Pedestrian models the base stations with low antennas height are situated outdoors while the pedestrian users are located inside buildings or in open areas. Fading can follow Rayleigh or Rician distribution depending upon the location of the user. The number of taps in case of Pedestrian-A model is 3 while Pedestrian-B has 6 taps. The average powers and relative delays for the taps of multipath channels based on ITU recommendations are given in Table 1 (ITU-R, 1997).

4.2.2 ITU Vehicular-A (V-30, V-120 and V-350)

The vehicular environment is categorized by large macro cells with higher capacity, limited spectrum and large transmit power. The received signal is composed of multipath reflections without LOS component. The received signal power level decreases with distance for which path loss exponent varies between 3 and 5 in the case of urban and suburban areas. In rural areas path loss may be lower than previous while in mountainous areas, neglecting the path blockage, a path loss attenuation exponent closer to 2 may be appropriate.

For vehicular environments, the ITU vehicular-A channel models consider the mobile speeds of 30 km/h, 120 km/h and 350 km/h. The propagation scenarios for LTE with speeds from 120 km/h to 350 km/h are also defined in (Ericsson, et al., 2007) to model high speed scenarios (e.g., high speed train scenario at speed 350km/h). The maximum carrier frequency over all frequency bands is f=2690 MHz and the Doppler shift at speed v=350 km/h is 900 Hz. The average powers and relative delays for the taps of multipath channels based on ITU recommendations are given in Table 2 (ITU-R, 1997).

Tap No	Pedestrian-A		Pedestrian-B		Doppler Spectrum
	Relative Delay (ns)	Average Power(dB)	Relative Delay (ns)	Average Power(dB)	
1	0	0	0	0	Classical
2	110	-9.7	200	-0.9	Classical
3	190	-19.2	800	-4.9	Classical
4	410	-22.8	1200	-8	Classical
5	NA	NA	2300	-7.8	Classical
6	NA	NA	3700	-23.9	Classical

Table 1. Average Powers and Relative Delays of ITU multipath Pedestrian-A and Pedestrian-B cases

	Tap No					
Average Power(dB)	0	-1.0	-9.0	-10.0	-15.0	-20.0
Relative Delay(ns)	0	310	710	1090	1730	2510

Table 2. Average Powers and Relative Delays for ITU Vehicular-A Test Environment.

5. Multiple antenna techniques

Broadly, multiple antenna techniques utilize multiple antennas at the transmitter or/and receiver in combination with adaptive signal processing to provide smart antenna array processing, diversity combining or spatial multiplexing in a wireless system (Dahlman, et al., 2007; Salwa, et al., 2007). Previously, in conventional single antenna systems the exploited dimensions are only time and frequency whereas multiple antenna systems exploit an additional spatial dimension. The utilization of spatial dimension with multiple antenna techniques fulfils the requirements of LTE; improved coverage (possibility for larger cells), improved system capacity (more user/cell), QoS and targeted date rates are attained by using multiple antenna techniques as described in (3 GPP, 2008). Multiple antenna techniques are an integrated part of LTE specifications because some requirements such as user peak data rates cannot be achieved without the utilization of multiple antenna techniques.

The radio link is influenced by the multipath fading phenomena due to constructive and destructive interferences at the receiver. By applying multiple antennas at the transmitter or at the receiver, multiple radio paths are established between each transmitting and receiving antenna. In this way dissimilar paths will experience uncorrelated fading. To have uncorrelated fading paths, the relative location of antennas in the multiple antenna configurations should be distant from each other. Alternatively, for correlated fading (instantaneous fading) antenna arrays should be closely separated. Whether uncorrelated fading or correlated fading is required depends on what is to be attained with the multiple antenna configurations (diversity, beamforming, or spatial multiplexing) (Dahlman, et al., 2007). Generally, multiple antenna techniques can be divided into three categories (schemes) depending on their benefits: spatial diversity, beamforming and spatial multiplexing which will be discussed further in the following sections.

5.1 Spatial diversity

Conventionally, multiple antennas are exercised to achieve increased diversity to encounter the effects of instantaneous fading on the signal propagating through the multipath channel. The basic principle behind spatial diversity is that each transmitter and receiver antenna pair establishes a single path from the transmitter to the receiver to provide multiple copies of the transmitted signal to obtain an improved BER performance (Zheng, et al., 2003). In order to achieve large gains with multiple antennas there should be low fading correlation between the transmitting and the receiving antennas. Low value of correlation can be achieved when inter-antenna spacing is kept large. Hence it is difficult to place multiple antennas on a mobile device due size restrictions depending upon the operating carrier frequency. An alternative solution is to use antenna arrays with cross polarizations, i.e., antenna arrays with orthogonal polarizations. The number of uncorrelated branches (paths) available at the transmitter or at the receiver refers to the diversity order and the increase in diversity order exponentially decreases with the probability of losing the signal. To achieve spatial diversity for the enhancement of converge or link robustness multiple antennas can be used either at the transmitter side or at the receiver side. We will discuss both transmit diversity where multiple antennas are used at the transmitter (MISO-multiple-input signal-output), and receive diversity using multiple receive antenna (SIMO signal-input multiple-output). On the other hand, MIMO channel provides diversity as well as additional degree of freedom for communication.

5.2 Transmit diversity

The transmit diversity scheme relies on the use of $N_t \geq 2$ antennas at the transmitter side in combination with pre-coding in order to achieve spatial diversity when transmitting a single data stream (Furht, et al., 2009; Jankiraman, 2004). Usually transmit diversity necessitates the absolute channel information at the transmitter but it becomes feasible to implement transmit diversity without the knowledge of the channel with space-time block coding (Jankiraman, 2004). The simplest transmit diversity technique is Alamouti space-time coding (STC) scheme (Alamouti, 1998). Transmit diversity configuration is illustrated in Figure 3.
The use of transmit diversity is common in the downlink of cellular systems because it is easier and cheaper to install multiple antennas at base station than to put multiple antennas on every handheld device. In transmit diversity to combat instantaneous fading and to achieve considerable gain in instantaneous SNR, the receiver is being provided with multiple copies of the transmitted signal. Hence transmit diversity is applied to achieve extended converge and better link quality when the users experience hostile channel conditions.
In LTE, transmit diversity is defined only for 2 and 4 transmit antennas and these antennas usually need to be uncorrelated to take full advantage of the diversity gain.
LTE physical layer supports both open loop and closed loop diversity schemes. In open loop scheme channel state information (CSI) is not required at the transmitter, consequently multiple antennas cannot provide beamforming and only diversity gain can be achieved. On the other hand, closed loop scheme does not entail channel state information (CSI) at the transmitter and it provides both spatial diversity and beamforming as well.
By employing cyclic delay diversity and space frequency block coding, open loop transmit diversity can be accomplished in LTE. In addition, LTE also implements close loop transmit diversity schemes such as beamforming.

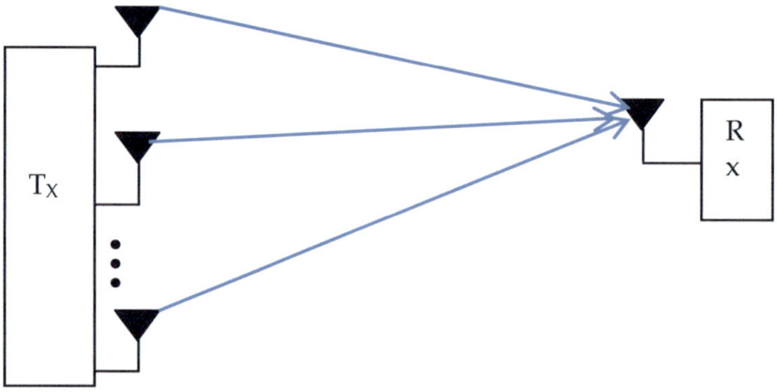

Fig. 3. Transmit diversity configuration

5.3 Space-Frequency Block Coding (SFBC)

In LTE, transmit diversity is implemented by using Space-Frequency Block Coding (SFBC). SFBC is a frequency domain adaptation of the renowned Space-Time Block Coding (STBC) where encoding is done in antenna/frequency domains rather than in antenna/time domains. STBC is also recognized as Alamouti coding (Rahman, et al.). Thus, SFBC is merely appropriate to OFDM and other frequency domain based transmission schemes.
The advantage of SFBC over STBC is that in SFBC coding is done across the subcarriers within the interval of OFDM symbol while STBC applies coding across the number of OFDM symbols equivalent to number of transmit antennas (Rahman, et al.). The implementation of STBC is not clear-cut in LTE as it operates on the pairs of adjacent symbols in time domain while in LTE the number of available OFDM symbols in a sub-frame is often odd. The operation of SFBC is carried out on pair of complex valued modulation symbols. Hence, each pair of modulation symbols are mapped directly to OFDM subcarriers of first antenna while mapping of each pair of symbols to corresponding subcarriers of second antenna are reversely ordered, complex conjugated and signed reversed as shown in Figure 4.
For appropriate reception, mobile unit should be notified about SFBC transmission and linear operation has to be applied to the received signal. The dissimilarity between CDD and SFBC lies in how pairs of symbols are mapped to the second antenna. Contrarily to CDD, SFBC grants diversity on modulation symbol level while CDD must rely on channel coding in combination with frequency domain interleaving to provide diversity in the case of OFDM.

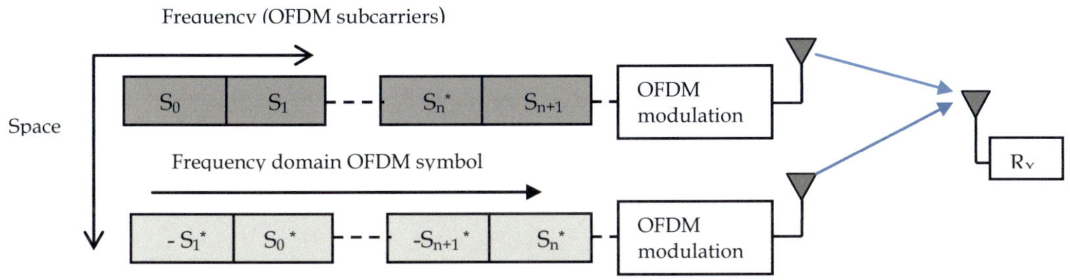

Fig. 4. Space-Frequency Block Coding SFBC assuming two antennas

The symbols transmitted from two transmitted antennas on every pair of neighboring subcarriers are characterized in (Sesia, et al., 2009) as:

$$X = \begin{bmatrix} x^{(0)}(1) & x^{1}(1) \\ x^{(0)}(2) & x^{1}(2) \end{bmatrix} \tag{3}$$

where $X^{(P)}(K)$ denotes the symbols transmitted from antenna port 'p' on the kth subcarrier. The received symbol can be expressed as:

$$y = Hs + n \tag{4}$$

$$\begin{bmatrix} y_0 \\ y_1^* \end{bmatrix} = \begin{bmatrix} h_{00} & -h_{01} \\ h_{11}^* & h_{10}^* \end{bmatrix} \begin{bmatrix} S_0 \\ S_1^* \end{bmatrix} + \begin{bmatrix} n_0 \\ n_1^* \end{bmatrix} \tag{5}$$

where h_{ij} is the channel response for symbol i transmitted from antenna j, and n is the additive white Gaussian noise.

6. Performance comparison of channel estimation schemes

We simulate LTE down link using the SISO system with the parameters given in the specifications (3GPP, 2009). The system bandwidth selected is 15 MHz with the numbers of subcarriers 1536 out of which 900 subcarriers are used and the remaining are zero padded. The sub frame duration is 0.5 ms which leads to a frame length of 1 sec. This corresponds to a sampling frequency of 23.04 MHz or sampling interval of 43.4 ns. A cyclic prefix of length 127 (selected from specification which is extended CP) is inserted among data subcarriers to render the effects of multipath channel which completely removes inter-symbol-interference (ISI) and inter-carrier-interference (ICI). In simulating the SISO system, only one port of an antenna is considered and this antenna port is treated as a physical antenna. We consider one OFDM symbol of size 900 subcarriers and the reference symbols which are (total numbers of reference symbols are 150) distributed among data subcarriers according to specifications (3GPP, 2009) transmitted from the antenna during one time slot. The constellation mappings employed in our work are QPSK, 16 QAM and 64 QAM.

The channel models used in the simulation are ITU channel models (ITU-R, 1997). At the receiver end we used regularized LS and LMMSE estimation methods for the channel estimation. All channel taps are considered independent with equal energy distribution. In addition, frequency domain linear equalization is carried out on the received data symbols. The performance of the system is evaluated by calculating the bit error rates using ITU channel models with different modulation schemes.

The designed simulator is flexible to use. A scalable bandwidth is used, i.e., there is option for using bandwidths of 5 MHz, 10 MHz, 15 MHz and 20 MHz. In addition, cyclic prefixes of different lengths specified in (3GPP, 2009) can be easily selected in the simulation of the system. We used single port of antenna which is taken as physical antenna however changes can be easily made to include two ports antenna.

The performance of LTE transceiver is shown in terms of curves representing BER against SNR values and is compared with AWGN for different channel models. Figures 5 and 6 show BER versus SNR for LMMSE and LS channel estimations, respectively, for different ITU channel models using QPSK modulation. From these figures, it can be seen that LMMSE channel estimation gives better performance than LS channel estimation. Figures 7 and 8 show BER plots for ITU channel models using 16QAM modulation format. It is seen that by increasing the modulation order, the system performance degrades as compared to QPSK modulation. This is due to the fact that higher modulations schemes are more sensitive to channel estimation errors and delay spreads. For 16QAM, LMMSE still have superior performance as compared to LS estimation but its performance also diminishes in environments with high mobile speeds (Doppler spread) and large delay spreads. The LS estimation gives poor performance for higher modulation schemes. Some interpolation techniques can be employed to mitigate ISI effects which can enhance system performance.

Figure 9 illustrates the performance of transceiver for ITU vehicular-A channel model using multiple antennas. The SISO system is also shown for comparision purposes.

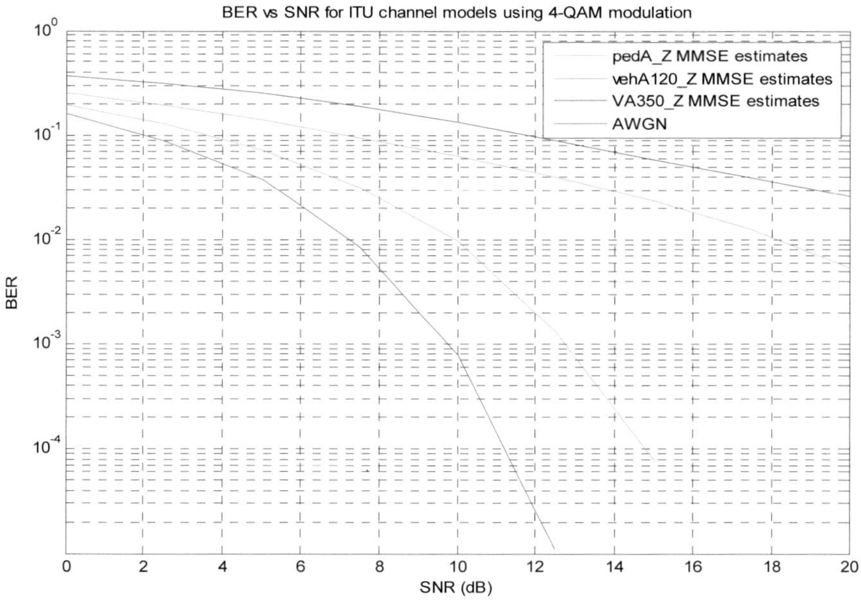

Fig. 5. BER performance of LTE transceiver for different channels using QPSK modulation and LMMSE channel estimation

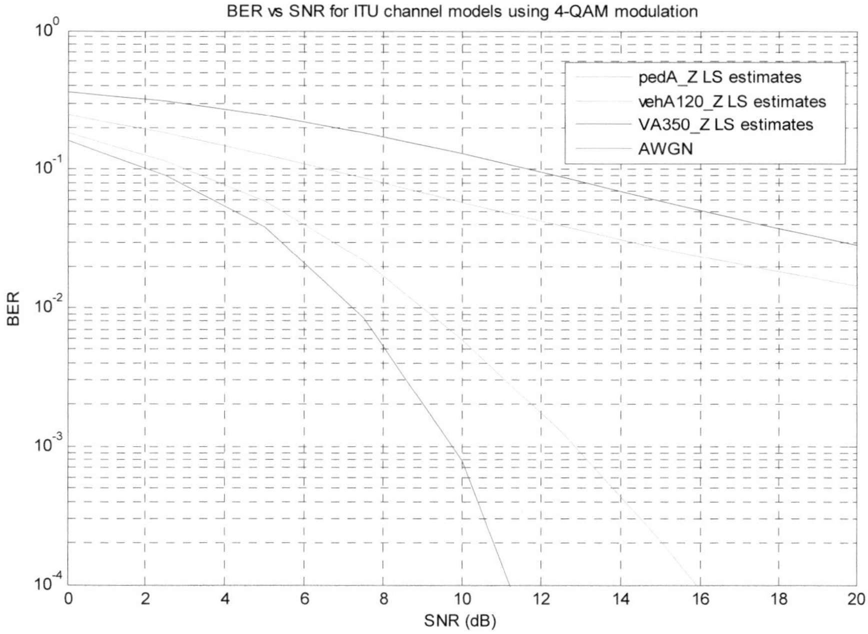

Fig. 6. BER performance of LTE transceiver for different channel models using QPSK modulation and LS channel estimation

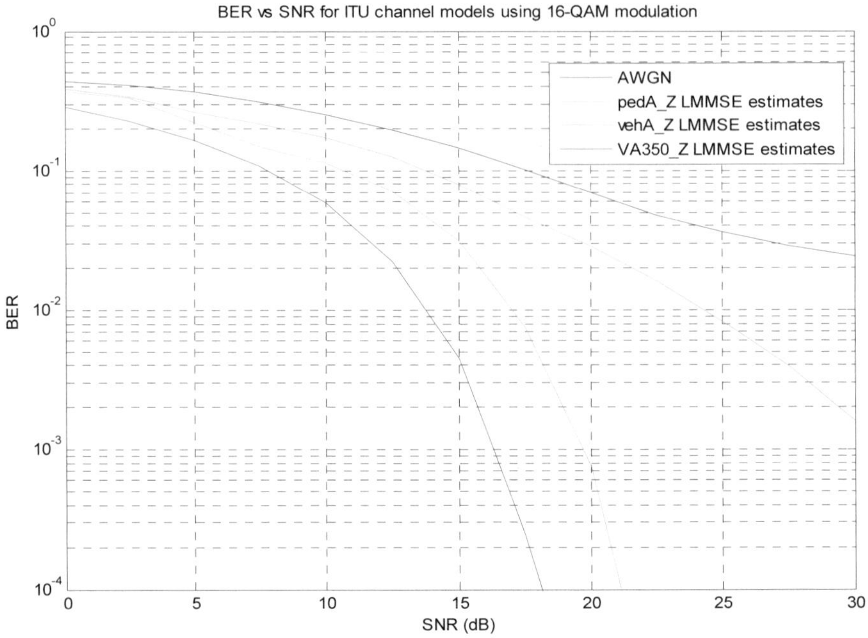

Fig. 7. BER performance of LTE transceiver for different channel models using 16 QAM modulation and LMMSE channel estimation

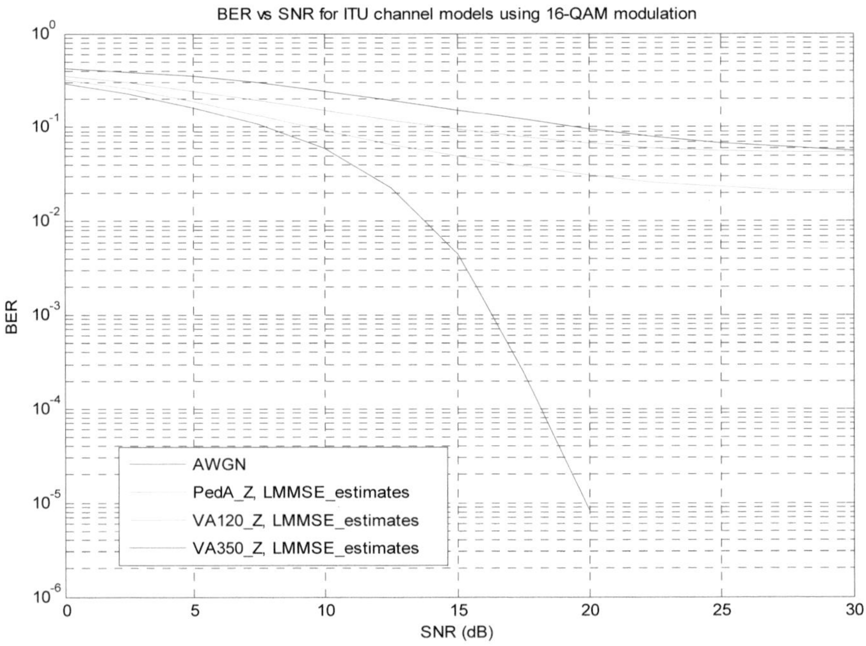

Fig. 8. BER performance of LTE transceiver with multiple antennas for different ITU channel models using 16 QAM modulation and LS channel estimation

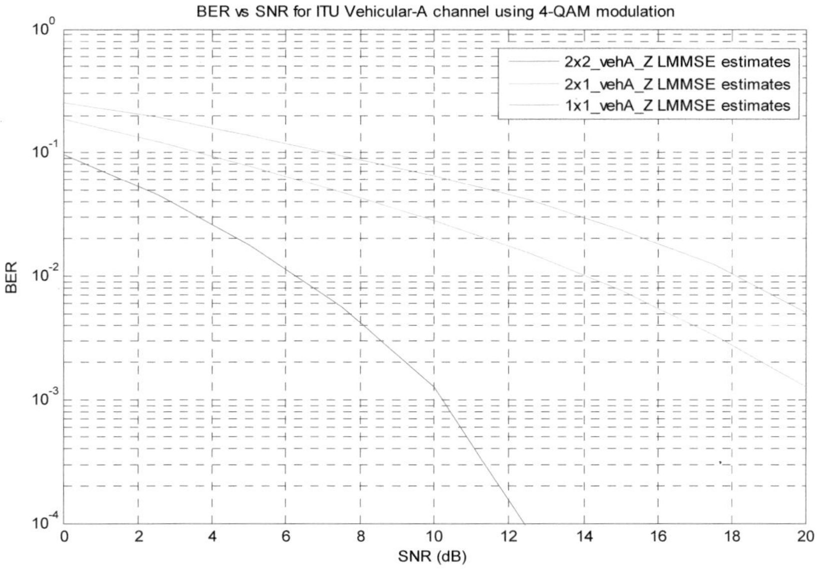

Fig. 9. BER performance of LTE transceiver with multiple antennas for ITU Vehicular-A channel model using 4-QAM modulation and LMMSE channel estimation

7. Conclusions

This chapter illustrates the physical layer aspects of future generation mobile communication systems. Proper knowledge of propagation impairments and channel models is necessary for the design and performance assessment of advanced transceiver techniques employed to establish reliable communication links in future generation mobile communication systems.

The results have been presented by means of simulations. The performance is evaluated in terms of BER and SER and the obtained results are compared with theoretical values. The LS estimator is simple and suitable for high SNR values; however its performance degrades with higher constellation mappings for high mobile speeds. On the other hand, LMMSE estimator is computationally complex and requires a priori knowledge of noise variance but its performance is superior to LS estimates for higher modulation schemes and large delay spreads. The performance of future generation mobile communication systems will be highly dependent on different factors including operating frequency, elevation angles, geographic location, climate etc.

8. References

3GPP (2008). TR 25.913:Requirements for Evolved UTRA (E-UTRA) and Evolved UTRAN (E-UTRAN). Release 8, V 8.0.0.0.

3 GPP (2008). Overview of 3GPP Release 8: Summery of all Release 8 Features, V0.0.3.

3GPP (2009). Physical Channels and Modulation. TR 36.211 V8.7.0, Release 8.

Alamouti, S. M. (1998). A Simple Transmit Diversity Technique for Wireless Communication. *IEEE Journal Select. Areas Communications*, pp. 1451-1458.

Correia, L., M. (2001). Wireless Flexible Personalized Communications (COST 259 Report), John Wiley & Sons, Chichester, UK.

Cavers, J. K. (2002). Mobile Channel Characteristics. Kluwer Academic Publishers, New York, Boston, Dordrecht, London, Moscow.

Dahlman, E., Parkvall, S., Sköld, J., & Beming, P. (2007). 3G Evolution: HSPA and LTE for Mobile Broadband, Elsevier Ltd.

Ericsson, Nokia, Motorola, and Rohde & Schwarz. (2007). R4-070572: Proposal for LTE Channel Models. www. gpp.org, 3GPP TSG RAN WG4, meeting 43, Kobe, Japan.

Fleury, B. H. (1996). An Uncertainty Relation for WSS Processes and Its Application to WSSUS Systems. *IEEE Transactions on Communications*, 44(12):1632–1634.

Furht, B., & Ahson, S. A. (2009). Long Term Evolution: 3GPP LTE radio and cellular technology, published by Taylor & Francis Group, LLC.

Holma, H., & Toskala, A. (2009). LTE for UMTS: OFDMA and SC-FDMA Base Band Radio Access. John Wiley & ISBN 9780470994016 (H/B) John Wiley & Sons Ltd.

Ibnkahla, M. (2005). Signal Processing for Mobile Communications. CRC Press, New York Washington, D.C.

ITU-R (1997). M.1225. International Telecommunication Union: Guidelines for evaluation of radio transmission technologies for IMT-2000.

Jankiraman, M. (2004). Space-Time Codes and MIMO Systems. Artech House Boston, London.

Kliazovich1, D., Granelli, F., Redana, S., & Riato, N. (2007). Cross-Layer Error Control Optimization in 3G LTE," *IEEE Global Telecommunication Conference*, Trento.

Meinilä, J., Jämsä, T., Kyösti, P., Laselva, D., El-Sallabi, H., Salo, J., Schneider, C., & Baum, D. (2004). IST-2003- 507581 WINNER: Determination of Propagation, IST-2003- 507581 WINNER.

Molisch, A., F., Asplund, H., Heddergott, R., Steinbauer, M., & Zwick, T. (2006). The COST 259 vol.5, no. directional- channel model A-I: overview and methodology. *IEEE Transactions on Wireless Communications*, 12, pp. 3421-3433.

Motrola. (2008). Inter-technology Mobility: Enabling Mobility between LTE and other Access Technologies.

Oestges, C., & Clercks, B. (2008). MIMO Wireless Communications: From Real-World Propagation to Space-Time Code Design, Elsevier Publishers.

Ojanpera, T., & Prasad, R. (2001). WCDMA: Towards IP Mobility and Mobile Internet. Artech House Publishers, Boston, London.

Rahman, M. I., Marchetti, N., Das, S. S., Fitzek, F., & Prasad, R. Combining Orthogonal Space-Frequency Block Coding and Spatial Multiplexing in MIMO-OFDM System. for TeleInFrastruktur (CTiF), Aalborg University, Denmark.

Rappaport, T. (1996). Communications, Principles and Practice. Prentice-Hall, Englewood Cliffs, NJ, USA.

Salwa, A. A., Thiagarajah, S. (2007). A Review on MIMO Antennas Employing Diversity Techniques, Proceedings of the International Conference on Electrical Engineering and Informatics Institute Technology Bandung, Indonesia.

Sesia, S., Toufik, I., & Bakkar, M. (2009). LTE- The UMTS Long Term Evolution, John Wiley and Sons, Ltd, First Edition.

Zheng, L., & Tse, D. N. C. (2003). Diversity and Multiplexing: A Fundamental Tradeoff in Multiple-Antenna Channels. *IEEE Transactions on Information Theory.*

19

Verifying 3G License Coverage Requirements

Claes Beckman
Center for RF-Measurement Technology, University of Gävle, and
Center for Wireless Systems, Wireless@KTH, Royal Institute of Technology,
Sweden

1. Introduction

In the beginning of the 21'st century, the 3rd generation mobile phone systems, 3G, were introduced all around the world. In most countries, spectrum for this technology was allocated through some kind of licensing procedure. In Europe, the prevailing approach was to allocate spectrum through auctions, a process which led to a situation where the European operators found themselves committed to pay a staggering 130Bilion Euros for their 3G licenses.
However, in most European countries, the fee was not the only obligation put on the licensee: A coverage, "roll-out" requirement was in many cases also connected to the license (Northstream, 2002). Typically, these coverage requirements required that the licensees cover a certain area at a certain point in time after that the licenses had been awarded.
In order for the regulators to verify that the licensees had met the coverage requirement and, hence, complied with the regulation, a method for coverage verification was needed. Such methods have therefore since then been developed by several European regulators (e.g. PTS 2004; ECC 2007). In this book chapter we describe some general underlying consideration for the verification of radio coverage in UMTS systems and in particular we describe the Swedish methodology developed by the Swedish Telecom regulator Post & Telestyrelsen (PTS).

2. Licensing of 3G in Sweden

In 2001, the Swedish Telecom regulator Post & Telestyrelsen (PTS) granted four licenses for the operation of third generation mobile phone systems (PTS 2001). In contrast to most other European countries, the Swedish licenses were granted through a beauty contest. When acquiring the licenses, the licensees committed themselves to build networks that covered a population of 8.860.000 inhabitants. This requirement implied that each operator would cover some 99.98% of the Swedish population (as counted for in 1996). However, in order to support the roll out, the regulator allowed the operators to build their networks in a combination of self owned sites in the major cities (30% of population) and shared sites in the countryside (70%) (Beckman and Smith, 2005). The roll-out of these 3G networks was delayed several times and the coverage requirements somewhat modified, but in 2007 all Swedish operators reported that they complied with the license requirements. Today Sweden is unique in that more than 98% of the population and 48% of the of the national territory (170.000 km^2) has 3G service coverage (PTS 2008).
In contrast to many other European countries, the original Swedish 3G license defined coverage by specifying a field strength requirement to be measured outdoors on the primary

common pilot channel, CPICH. The assumption was that depending on the environment and the average building penetration pathloss, the pilot signal strength can be related to a particular data service (rate) indoors.
In the original Swedish license requirement the operators where obliged to provide an out-door signal strength that in the 3GPP release 99 standard of the UMTS system (3GPP 2000) corresponded to an in-door packet switched data services, of 384 kbps in downlink and 144kbps in uplink. These requirements were then translated into a field strength for the signal received from the base station. In the original license agreement coverage requirement was that when measured outdoors at a height of 1.7m above ground over 5MHz, the field strength on the CPICH should be at least 58 dBμV/m with an area probability of 95% (PTS 2001).
The design of the Swedish measurement method is previously described in a number of papers, e.g. PTS 2004; PTS 2004 II; Beckman et al 2006; Beckman et al 2008.

3. General considerations

To verify coverage one needs to develop a practical test procedure for measuring field strength. The verification can then easily be performed e.g. in a drive test (PTS 2004; ECC 2007). However, designing such test presents a number of challenges:

- A requirement can be given for a particular field strength measured on the common pilot channel. However, in the UMTS systems the power to be allocated to the CPICH is not given by the standard or by the regulator
- There is no given relation between pilot power and service. In the Swedish license requirements it was assumed that an outdoor signal strength of 58dBμV/m on the CPICH in practise relates to a downlink service indoors of 384 kbps and an uplink service of 144kbps (PTS 2001). However, building penetration path loss varies in different environment. Hence, field strength requirement must vary accordingly.
- A license is typically given for area and population coverage while a drive test only measures along a linear route. In order to convert measurement data from drive testing to a probability of coverage for a given area with a certain population, one needs a statistical model based on population density and geography.

4. The primary Common Pilot Channel

The Universal Mobile Telephony System (UMTS) is a 3G systems specified by the Third Generation Partnership Project organization (3GPP 2002). It has a radio interface based on a code division multiple access scheme, cdma, and 5MHz wide radio channels. Since the radio channel is somewhat wider than previous cdma systems it is referred to as: "wideband" cdma or WCDMA.
The primary Common Pilot Channel, CPICH, is one of many codes in the WCDMA common downlink pilot channel (Holma and Toskala 2002). It is a control channel mainly used for handovers. It does not have a fixed power allocated to it so it is principle not related to any service in either the up- or down-link.

4.1 Allocating power to the CPICH

In theory it is possible to allocate anything between 0% and 100% of the available power to the CPICH. In practice the allocated power has a lower bound which can be derived as follows (PTS 2004)

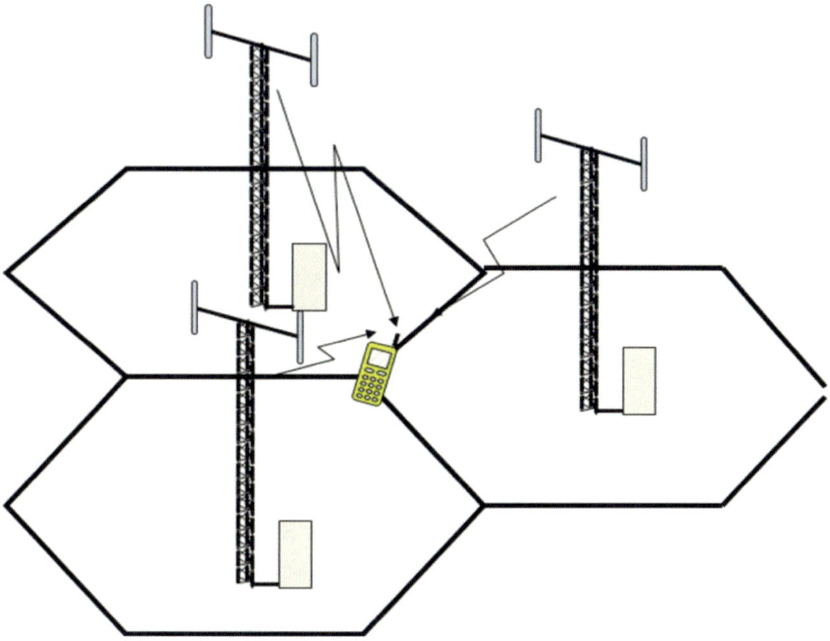

Fig. 1. Illustration of the downlink interference situation at the border between two cells

In order to initiate a soft handover at the border between two cells (Fig. 1.), a cell's pilot must be detected when an adjacent cell's pilot is 5 dB stronger. The required E_b/N_0 on the primary CPICH on the downlink is approximately 10 dB (3GPP 2002). The processing gain on the pilot is 10*log(3840/12.2) = 25 dB which means that the minimum output power for the pilot is approximately: 5 + 10 - 25= -10dB (10%) compared to the total output power from the base station. A worse case scenario is of course when the mobile is at the intersection of 3 cells. The interference level would then of course be doubled. Allocating between 10% and 20% of the available power in the radio channel is also often suggested in industry literature (PTS 2004 II). However, it is in the interest of the operators not to increase the pilot power unnecessarily since raising the pilot power will mean that less power is available for services.

4.2 Relationship between pilot power and services

As described above, there is no given relation between pilot power and services. Still, the regulator needs to have measurable criteria:

First of all one needs to consider what measure is most suitable. By tradition regulators uses prefers to measure the signal strength in e.g. dBμV/m. The main reason for this is that this parameter is easy to measure in a drive test and is independent on frequency and antenna gain. The relationship between signal strength E (as measured in dBμV/m) and signal power P (as measured in dBm) can we written as:

$$P = E - 20log10f - 77.219 + G, \quad (1)$$

where f is the frequency given in MHz and G the antenna gain given in dBi.

Assuming that 10% of the available power is allocated to the primary CPICH and that the building penetration path loss is known, it is now possible to estimate the pilot power needed to provide the required services for different environments by calculating the link budgets (Holma and Toskala 2002).

The base station has typically 10-20 W (40-43 dBm) output power available, while the mobile unit has 0.125 W (21 dBm). The Noise Factor of the base station is typically ~4 dB compared to ~7 dB for the mobile receiver. Antenna diversity is implemented at the base station for the uplink and therefore approximately 4-5 dB lower Eb/N_0 than required in the downlink. Still the downlink has a 10-15 dB path loss advantage over the uplink in a symmetrical service. In case of asymmetrical load (higher bitrates in the downlink than in the uplink), the 10-15 dB advantage reduces to around 5-10 dB (assuming 384 kbits/s downlink and 144 kbit/s uplink).

Uplink coverage can be improved by introducing Tower Mounted Low Noise Amplifiers, i.e. an amplifier directly after the antenna. The gain of this is that the feeder losses in the uplink can be ignored (expect for a short jumper cable between the antenna and the amplifier), and that the TMA often has a better Noise Factor (NF) than the base station (1.5–2 dB compared to 4-5 dB). TMA is widely used by the operators to improve coverage in rural areas.

Mobile a terminals are used in a variety of environments, but to a large extent they are used indoors. The signal is thus being attenuated as it has to propagate through the walls or windows of the building where the user is located. Therefore, the link budget needs to include a margin for the penetration loss in case service is planned for indoor users.

It is evident that a single penetration loss value will not apply to all environments. In rural areas, people often live in small houses that have thin walls and windows in different directions, thus giving a lower penetration loss. In Sweden single family house are mainly constructed out of wood, while multi family and multistory buildings are normally made of concrete.

In the Swedish example, the following guidelines for building attenuation was suggested:

1. In rural areas, single family houses (11 dB attenuation)
2. In suburban areas, single family houses and semi detached homes (11dB)
3. In urban areas, concrete houses with large separation (16dB)
4. In dense urban areas, concrete houses with small separation (20dB)

The link budget calculations are summarized in Tables 1 and 2. Calculations are done for the four different scenarios mentioned above, with and without tower mounted low noise amplifiers, TMA, in rural, and for packet switched uplink and downlink data rates of 144kbps and 384kbps, respectively. The input data is in accordance with the 3GPP UMTS release 99 standard (3GPP 2001) and Holma and Toskala (2002).

In literature link budgets normally includes a margin for the "log normal fading", which can be described statistically, to arrive at a maximum path loss that can be used for radio planning purposes. When comparing the above link budget with the license requirements, it is important to understand that the margin for the statistical variation of the measured signal in the outdoor environment is already

4.3 Coverage criteria

In Table 3 the main results of the link budget calculations are presented. As can be seen, in all cases it is the up-link that limits the service performance. However, the CPICH signal strength required in order to be able to provide the respective services varies in different environments. The pilot signal strength requirement of 58dBµV/m set out in the Swedish license seems to be ~ 7dB too low in dense urban and ~8dB too strict in rural environments.

Environment		Dense	Urban	Suburban	Rural	Rural TMA	Rural TMA
Service UL	kbit/s	144	144	144	144	144	64
Max mobile transmit power	dBm	21	21	21	21	21	21
Mobile Antenna Gain	dBi	0	0	0	0	0	0
Body loss	dB	0	0	0	0	0	0
EIRP	dBm	21	21	21	21	21	21
Thermal Noise	dBm/Hz	-174	-174	-174	-174	-174	-174
Noise Figure	dB	4	4	4	4	2	2
Noise Density	dBm/Hz	-170,0	-170,0	-170,0	-170,0	-172,0	-172,0
Noise Power	dBm	-104,2	-104,2	-104,2	-104,2	-106,2	-106,2
Interference Margin	dB	3	3	3	1	1	1
Receiver interference Power	dBm	-104,2	-104,2	-104,2	-110,0	-112,0	-112,0
noise + interference	dBm	-101,2	-101,2	-101,2	-103,2	-105,2	-105,2
Processing Gain	dB	14,3	14,3	14,3	14,3	14,3	17,8
Required Eb/No	dB	1,5	1,5	1,5	2	2	2
Receiver Sensitivity	dBm	-113,9	-113,9	-113,9	-115,4	-117,4	-120,9
Base station antenna gain	dBi	18	18	18	18	18	18
Cable loss	dB	4	4	4	4	1	1
Max Path Loss	dB	148,9	148,9	148,9	150,4	155,4	158,9
Fast fading margin	dB	4	4	4	4	4	4
Max Fading Path Loss	dB	144,9	144,9	144,9	146,4	151,4	154,9
Average Penetration Loss	dB	20	16	11	11	11	11
Max outdoor UL Path loss	dB	124,9	128,9	133,9	135,4	140,4	143,9

Table 1. Uplink link budgets for different services and environment used for the calculation of the Swedish license requieremnts

What is then a sufficient pilot strength criteria in order to determine whether an area is covered or not with 3G? What is evident from the above link budget calculations is that the signal strength requirements needs to be set differently for different environments. In Table 3, the modified Swedish CPICH requirements for different environments are summarized (PTS 2004 II).

Environment		Dense	Urban	Suburban	Rural	Rural TMA
Service DL	kbit/s	384	384	384	384	384
Total available Power	dBm	43	43	43	43	43
Cable Loss	dB	4	4	4	4	4
Antenna Gain	dBi	18	18	18	18	18
Transmitter total ERP	dBm	57	57	57	57	57
Max Service Power %		25%	25%	25%	50%	50%
Max Service ERP	dBm	51,0	51,0	51,0	54,0	54,0
Thermal Noise	dBm/Hz	-174	-174	-174	-174	-174
NF	dB	7	7	7	7	7
Noise Density	dBm/Hz	-167	-167	-167	-167	-167
Noise Power	dBm	-101,2	-101,2	-101,2	-101,2	-101,2
Processing Gain	dB	10,0	10,0	10,0	10,0	10,0
Required Eb/No	dB	6	6	6	6	6
Receiver Sensitivity	dBm	-105,2	-105,2	-105,2	-105,2	-105,2
Base station antenna gain	dBi	18	18	18	18	18
Cable loss	dB	4	4	4	4	4
Max Path Loss	dB	156,1	156,1	156,1	159,1	159,1
Fast fading margin	dB	4	4	4	4	4
Max Fading Path Loss	dB	152,1	152,1	152,1	155,1	155,1
Average Penetration Loss	dB	20	16	11	11	11
Max outdoor DL Path loss	dB	132,1	136,1	141,1	144,1	144,1

Table 2. Downlink link budgets for different services and environment used for the calculation of the Swedish license requieremnts

Environment	Limiting Link	Required CPICH [dBµV/m]	Modified Swedish Requirements[dBµV/m]
Dense Urban	UL	65.1	58
Urban	UL	61.1	58
Suburban	UL	56.1	52
Rural	UL	54.6	52
Rural TMA withTMA	UL	49.6	50

Table 3. Summary of the Swedish link budget calculations and modified CPICH requirements

5. Measurement set-up

The method used to verify the operators networks needs for obvious reasons to be accurate but also well accepted. The traditional way of performing radio coverage measurements is by conducting drive tests with a vehicle upon the roof of which antennas are mounted. The signal is then sampled, measured and stored on equipment carried inside.

Fig. 2. Photo of the measurement car including the antenna solution with an extra disc as ground plane, used by the Swedish regulator, PTS

5.1 Instrumentation

The measurement system needs by necessity be able to simultaneously detect several control channels from several base station. The reason for that is that when the measurement is performed in urban environments the receiver will detect several base stations. In sub-urban or rural areas, it is of importance to be able to carefully measure at least two base staion control channels during (soft) handowver.

The standard way of doing this is to perform a so called Top N measurement. The measurement instrument then measures the scrambling codes transmitted on each detected CPICH. In a "Top N" measurement, the system scans for all 512 scrambling codes and returns the "N" strongest. In the Swedish measurement the top 6 scrambling codes were detected and measure but N can typically be any number between 1 and 32.

Since the receiver is often in a scattering environment, much of the power is not in the direct path but rather in the "echos" reflected from surrounding buildings and objects. The test instrument therefore needs to support "aggregate power" measurements where the receiver integrates over all reflected signals stronger than a certain level, (e.g. -17 below the maximum power, PTS 2004)
In addition to this it is o course of importance to keep track of the position of each sample, using a GPS, and store the all the acquired data. Table 4 below gives a list of the instrument chosen for the Swedish test. In Fig. 3, the test receivers and the operator display are shown.

Instrumentation	**Equipment**
Test receivers:	4 * Agilent E6455C (WCDMA)
GPS	Trimble
Computer	Shuttle XPC, P4 3.06GHz, 1GB ram, 160GB hd
Display	7" VGA
OS:	Windows XP
Application software	Nitro (Agilent)

Table 4. List of instrumentation used in the Swedish measurements

Fig. 3. Test receivers and the operator display are shown

5.2 Measurement antenna

For measurements of mobile radio coverage, it is important to have an omni-directional pattern, especially if different networks with different base station locations are to be compared.

Fig. 4. Photograph of the measurement antenna solution with an extra disc added on top of the car

Whereas a dipole provides an omni-directional pattern with maximum radiation in the horizontal plane, this is not the case for a monopole placed on a finite ground plane such as a car roof (Kronberger et al, 1997). To minimize antenna induced errors in the measurement one may consider an alternative ground plane.

Fig. 5. The monopole antenna during verification measurements at Allgon's (today Powerwave) compact range in Stockholm Sweden. The monopole and the disc ground plane are mounted on the roof from a chassi of the same car as in the actualSwdish 3G verification measurements

In order to improve the pattern for the Swedish test, Ribbenfjärd et al (2004) presented a solution with a monopole on a circular dish placed on the car roof. The design was made using simulations in CST Microwave Studio. A photograph of the antenna is seen in Fig. 4. It is very difficult to make accurate measurement or simulation of an antenna mounted on a complete vehicle, and the validation of the design was therefore made on the part of the roof seen in Fig. 5. The result is shown in Fig. 6 where it is compared to the pattern of a monopole mounted directly on the car roof.

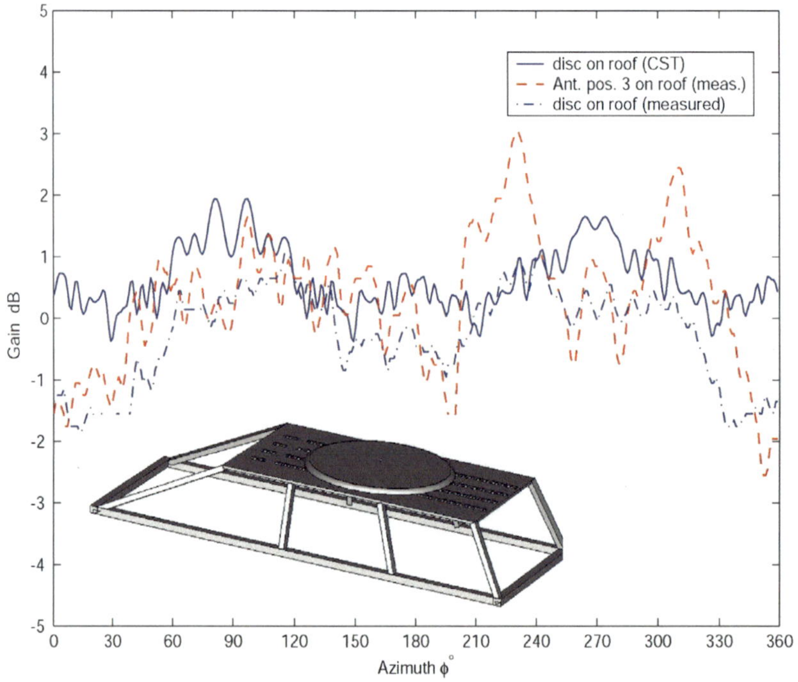

Fig. 6. Simulated and measured azimuth pattern of a monopole antenna with a ground plane disc on the roof at 2GHz The co-polar pattern of a monopole antenna directly on the roof is also shown. The inserted drawing shows the CST-model of the antenna with disc on the car roof. (from Ribbenfjärd et al 2004)

The results show that a substantial improvement in both azimuth and elevation is possible if an additional ground plane in the form of a disc or a cone with corrugated edge is added. The peak-to-peak gain variation in the horizontal plane is reduced from ~5 dB to ~2.5 dB.

6. Compensation for polarization mismatch

For practical reasons, regulators want to measure the vertical field component only, i.e. using a vertically polarized antenna. However, the operators may very well use polarization diversity on reception at the base station. In this case, the base station typically transmits on a 45 degrees slant linear polarization. Therefore, we can expect that the field at the measurement location will have a non-zero horizontally polarized component.

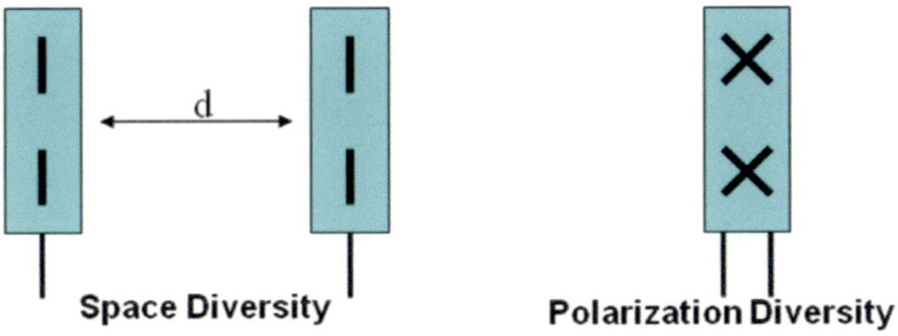

Fig. 7. Illustration of antenna systems for space (left) and polarization (right) diversity reception. In order to receive uncorrelated signals with a space diversity system the distance between the antennas (d) must typically be in the order of 10 wavelength

6.1 Cross polar discrimination

An important parameter related to polarization diversity is the cross polar discrimination or XPD of the radio channel caused by the environment. With this we mean the power ratio of vertically and horizontally polarized wave components incident on our receive base station antenna when a mobile is transmitting.

Environment and source	Mobile orientation	XPD (dB)	Frequency	Correlation between vertical and horizontal field ρ_{env}
Urban (Kozono et al 1984)	Vertical car antenna	4-7	920 MHz	0.02 median
Urban (Vaughan 1990) Sub-urban	30° on large groundplane	7 12	463 MHz	-0.003 0.019
Urban & Sub-urban (Turkmani 1995)	0° 45°	10 4.6-6.3	1790 MHz	<0.7 for 95% <0.7 for 95%
Urban (Lotse et al 1996) Sub-urban	70 ±15 deg in- an outdoor	1-4 2-7	1821 MHz	<0.2 for 90% <0.1 for 90%
Urban & Sub-urban (Eggers et al 1983)	0° 45°	4-7 0	1848 MHz	<0.5 for 93% <0.5 for 93%
Urban (Eggers et al, 1998)	Car mounted monopole	7.6±2.1	970 MHz	0.09±0.09
Urban & Sub-urban (Lempiainen and Laiho-Steffens, 1998)	Random, in- and outdoor	<5	1739 MHz	-0.25 to 0.24
Sub-urban (Wahlberg et al, 1997)	Vertical mobile LOS & NLOS	7	1800 MHz	< 0.2 average
Urban & Sub-urban (Joyce et al, 1999)	Vertical dipole	8-11	900 & 1800 MHz	< 0.7 all values
Urban (Wahlberg et al, 1997) Sub-urban	Vertical monopole	5 10	1800 MHz	-

Table 5. References from measurements of cross polar discrimination in different environments

In Table 5, a number of known references from measurements of cross polar discrimination in different environments are summarized. For example, we find that for an sub-urban environment we typically have XPD = 6dB, between 2 and 5dB in urban environments, and about 5 to 10dB in sub-urban.

6.2 Polarization mismatch

For 3G network verification purposes, we are interested in the amount of power lost if we transmit on slant 45degrees antenna and receive using a vertically polarized antenna only, i.e. the polarization mismatch loss. By virtue of reciprocity, we can use the results for a vertically polarized mobile station transmitting to determine this power. A relevant measurement is then Turkmani et al (1995) where a value of XPD=10dB is reported for a vertically polarized monopole in a suburban environment.

Since this environment discriminates the horizontal polarization a concern in using a 45 degrees slanted polarized antenna for transmission is the increased path loss compared to a vertically polarized. Due to reciprocity the transmission analysis is the same as for the reception, and the maximum average path loss using a slanted 45 degrees antenna vs. a vertically polarized antenna is limited to 3 dB if XPD is infinite.

What is then the difference between transmitting on slanted 45 degrees (P_{45°) antenna compared to vertical (P_{vp}) in a typical sub-urban environment? This can be calculated if one knows the XPD as:

$$P_{45^\circ} = \frac{P_{vp}}{2}(1 + 1/\mathrm{XPD}) \qquad (2)$$

If we use the example mentioned in Turkmani et al (1995) where a value of XPD = 10dB is reported, we find that the mean power difference between transmitting on 45 degrees slant compared antenna compared to a vertical is: ~2.60dB

6.3 Compensation for the polarization mismatch

So, how large should this compensation be? The theory above only shows the expected differences in measured field strength when transmitting on 45 degrees slant polarization compared to vertical in a radio channel which discriminates the horizontal polarization. However, it could also be argued that in some environments the propagation of the horizontal polarization is more favourable than vertical [10]. When that is true then no compensation should really be applied at all.

In conclusion we then find that the polarization mismatch adds an error to the measurement spanning from 0 to 3dB depending on environment. Hence, a general compensation factor of 3dB would be most fair to the operators.

7. Statistical model

The purpose of the test method is to establish whether or not an operator has fulfilled the coverage requirements set out in the license [3]. The method must ensure that the license requirement is fulfilled with sufficient statistical significance and that the sampled data is uncorrelated.

The way to ensure that the data is uncorrelated is to make sure it is received over a large enough geographical area. This is done by dividing the geographical area to be verified for 3G coverage into "test squares". If 95% of the tested squares have a sampled field strength exceeding the one set out in the license requirement, the area is considered to be covered.

7.1 Test squares

The size of the test squares is dependent on the environment and population (Table 6 and Figure 1). The denser the population the smaller the size. In the evaluation only one sample for each test square is used.

Environment	Population (per square)	Size (m)
Rural	0 <= x < 20	500
Suburban	20<= x < 80	250
Urban	80 <= x < 200	125
Dense Urban	x >= 200	50

Table 6. Size and population of test squares

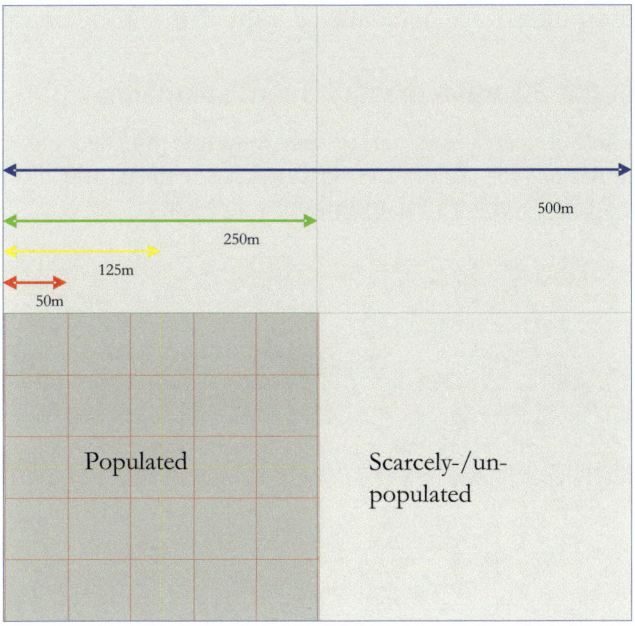

Fig. 8. Size relation between test squares

For practical reasons the measurements are conducted in a drive test using a car. Hence the sampling rate is dependent on the speed of the car and the size of the test squares. Each tested square is allocated the value of "1" if the license requirement is fulfilled and "0" if not. At least 500 test squares are measured in order to assume that the number of "1" are binomially distributed (n; p) where p >= 0.95 in the event that the license requirements are fulfilled.

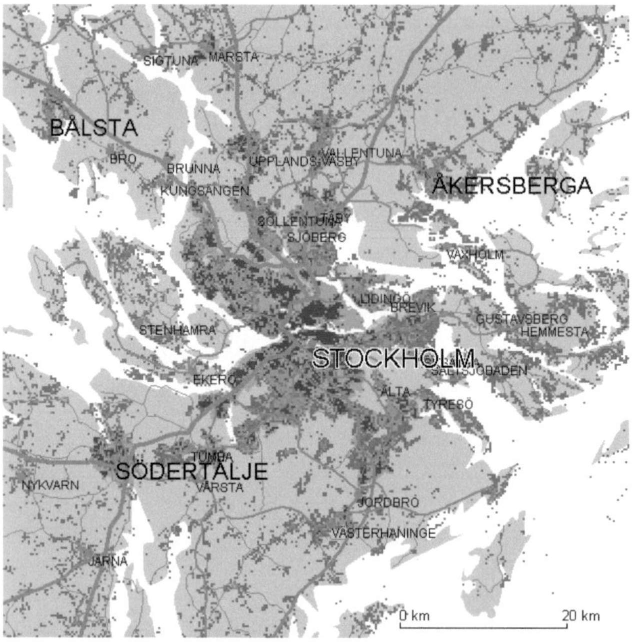

Fig. 9. Graphic illustration of the population density in the Stockholm area

8. Results from the Swedish measurement campaign

In 2007 all Swedish 3G licensees reported that they had fulfilled the modified (see Table 3) coverage requirements. In order to verify these claims the Swedish regulator PTS subsequently conducted some initial and preliminary tests.

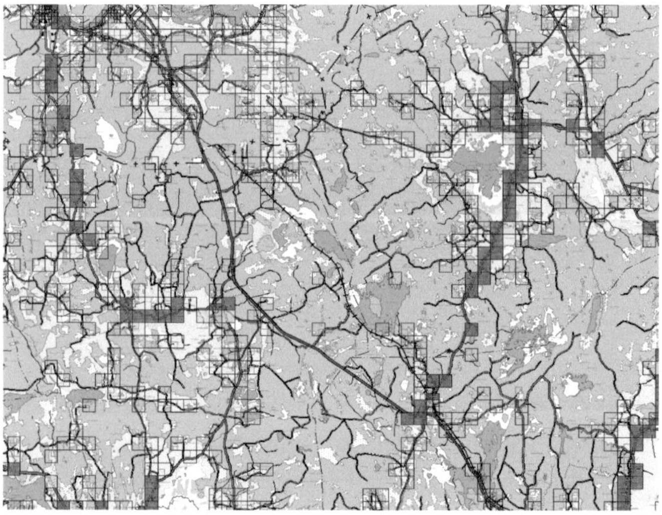

Fig. 10. Graphic illustration of coverage in the Fagersta region at the 52dBµV/m CPICH level. Green Squares indicate Test squares passed, yellow are at the boarder line, and red square are failed

8.1 Suburban environment: test case Fagersta

The first test case was conducted in a typical Swedish suburban environment in an area of and around the city of Fagersta. The field strength requirement was set to 52dBµV/m. In total 535 test squares were measured and in order to pass the test not more than 39 were allowed to fail for the operator to comply with the license requirement.

As shown in Table IV, the result from the measurements show that the operator passes the test easily. Even if the CPICH field strength requirement would be increased to 53dBµV/m would the operator still pass the test indicating that the planning is fairly robust against fading.

Field strength (dBµV/m)	No. Failed Squares
53	31
52	23
51	19
50	17
49	16
48	9
47	6

Table 7. Test results from Fagersta

8.2 Urban environment: test case Sundbyberg

The second test was conducted in a typical Swedish urban environment in the city of Sundbyberg some 10km north of Stockholm. In total 602 test squares were measured and in order to pass the test not more than 43 could fail for the operator to comply with the license requirement. In this environment the required field strength on the CPICH is 58dBµV/m.

Field strength requirement (dBµV/m)	No. Failed Squares
64	11
63	9
62	5
61	3
60	1
59	0
58	0
57	0

Table 8. Test results from Sundbyberg

As is evident from Table 8, the coverage planning is even more robust and the field strength on the CPICH higher in urban areas. Even if the requirement is increased with 6dB the result for the examined operator is still clearly above the limit of 95% area coverage.

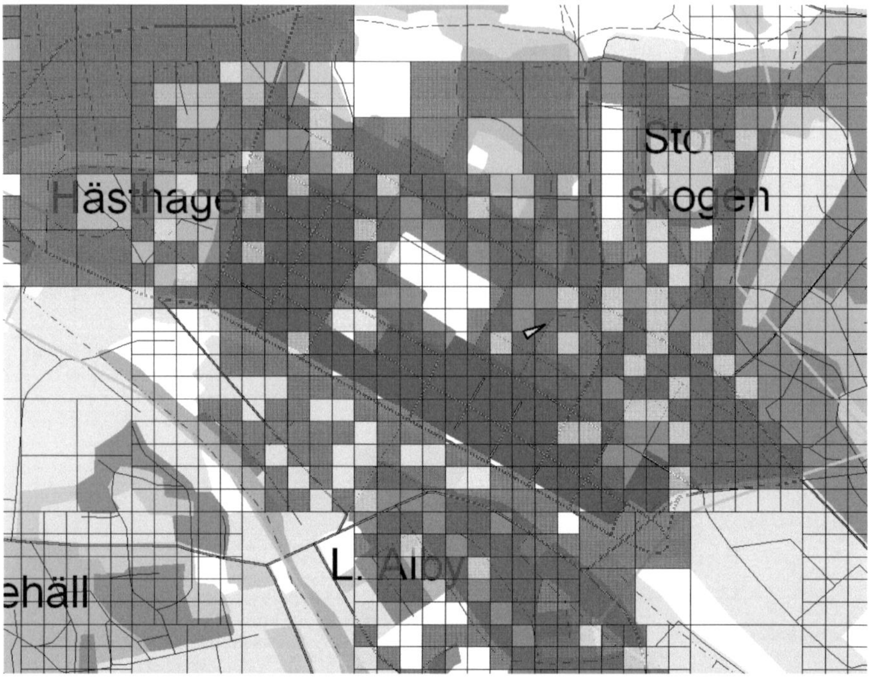

Fig. 11. Graphic illustration of coverage in the Sundyberg region at the 58dBµV/m CPICH level. Green Squares indicate Test squares passed, yellow are at the boarder line, and red square are failed

9. Conclusions

In the beginning of the 21st century, 3G was introduced and most countries in the western world allocated spectrum for this technology. In Europe, the prevailing approach was to allocate spectrum through auctions. However, in Sweden the 3G licenses were awarded after a beauty contest, in which the winners committed themselves to cover a population of 8.886.000 which at the time corresponded to 99.98% of the country's population. The coverage requirements were concrete and measurable and in 2007 it was verified that all Swedish operators complied with the requirements The development of an accepted test method was an important part of this succesfull licensing.

10. Acknowledgment

The Author would like to thank the participants of the 3G test method working group who all contributed in the development of the test. However, I would like to particularly acknowledge Per Wirdemark of Canayma International AB, who has been the principle engineer behind the design of the measurement method, Björn Lindmark at Laird Technologies who was the driving force behind the antenna development and, Lars Eklund and Urban Landmark at the Swedish regulator PTS, who initiated the work and contributed to this book chapter with several of its illustrations and results.

11. References

3GPP (2002), BS radio transmission and reception (FDD) - TS 25.104 V3.10.0 (Release 1999). http://www.3gpp.org, March 2002.

Beckman C., Lindmark B., Karlsson B., Eklund L., Ribbenfjärd D. and Wirdemark P. Verifying 3G licence requirements when every dB is worth a bilion, *European Conference on Antennas & Propagation*: EuCAP 2006

ECC Report 103 (2007). UMTS Coverage Measurements. Nice May 2007. http://www.erodocdb.dk/Docs/doc98/official/pdf/ECCRep103.pdf

Eggers P, Kovacs I., and Olsen K. (1998) Penetration effects on XPD with GSM 1800 handset antennas, relevant for BS polarization diversity for indoor coverage, *in Proc. 48th IEEE Veh. Technol. Conf. Ottawa*, Canada, May 1998, pp. 1959-1963.

Eggers P., Toftgaard J. and Oprea A. (1983) Antenna systems for base station diversity in urban small and micro cells, *IEEE J. Select. Areas Commun.*, vol. 11, pp. 1046-1057.

Holma H. and Toskala A., eds. (2002), WCDMA for UMTS Radio Access for Third Generation Mobile Communications. Chichester, New York,Weinheim, Brisbane, Singapore, Toronto: John Wiley & Sons, Ltd, 2 ed., 2002.

Joyce R., Barker D., McCarthy M. And Feeney M., (1999) A study into the use of polarisation diversity in a dual band 900/1800 MHz GSM network in urban and suburban environments, *IEE National Conference on Antennas and Propagation*. Page(s):316 - 319

Kozono S., Tsuruhara T., and Sakamoto M. (1984) Base station polarization diversity reception for mobile radio, *IEEE Trans. Veh. Technol.*, vol. 33, pp. 301-306, Nov.

Lempiainen J. and Laiho-Steffens K. (1998) The performance of polarization diversity schemes at a base station in small/micro cells at 1800 MHz., *IEEE Trans. Veh. Technol.*, vol. 3, pp. 1087-1092, Aug. 1998.

Lotse F., Berg J.-E., Forssen U., and Idahl P. (1996) Base station polarization diversity reception in macrocellular systems at 1900 MHz, in *Proc. 46th IEEE Veh. Technol. Conf.*, Apr. 1996, pp. 1643-1646.

Northstream AB (2002). 3G rollout status. ISSN 1650-9862, PTSER- 2002:22, available at http://www.pts.se.

PTS (2001) Meddelande av tillståndsvilkor för nätkapacitet för mobila teletjänster av UMTS/IMT-2000 standard enligt 15 § telelagen (1993:597), HK 01-7950, The Swedish National Post and Telecom Agency, PTS March 2001

PTS (2004 II), Coverage Requirements for UMTS, The Swedish National Post and Telecom Agency, PTS, Report Number PTS-ER-2004:32. September 2004

PTS (2004) Method för uppföljning av tillståndsvilkoren för UMTS-näten, The Swedish National Post and Telecom Agency, PTS, Report Number PTS-ER-2004:23. June 2004.

PTS (2008) Dimensionering och kostnad för utbyggnad av UMTS, The Swedish National Post and Telecom Agency, PTS, September 2008.

R. Kronberger, H. Lindenmeier, J. Hopf, and L. Reiter, (1997). Design method for antenna arrays on cars with electrically short elements under incorporation of the radiation properties of the car body, in *IEEE APS Symposium*, Montreal, Canada, pp. 418–421.

Ribbenfjärd D., Lindmark B., Karlsson B., and Eklund L., (2004) Omnidirectional Vehicle Antenna for Measurementof Radio Coverage at 2 GHz, *IEEE Antennas and Wireless Propagat. Letter*, VOL. 3, 269-272, 2004

Turkmani A., Arowojolu A., Jefford P., and Kellett C. (1995) An experimental evaluation of the performance of two branch space and polarization diversity schemes at 1800 MHz, *IEEE Trans. Veh. Technol.*, vol. 44, pp. 318-326, May 1995.

Wahlberg U., Widell S., and Beckman C. (1997) Polarization diversity antennas, in *Proc. Antenna, Nordic Antenna Symp. Göteborg*, Sweden, May 1997, pp. 59-65.

Vaughan R. (1990) Polarization diversity in mobile communications, *IEEE Trans. Veh. Technol.*, vol. 39, pp. 177-186, Aug. 1990.

20

Inter-cell Interference Mitigation for Mobile Communication System

Xiaodong Xu[1], Hui Zhang[2] and Qiang Wang[1]
[1]*Wireless Technology Innovation Institute; Key Laboratory of Universal Wireless Comm., Ministry of Education; Beijing University of Posts and Telecommunications,*
[2]*Nankai University*
China

1. Introduction

With the commercialization of 3G mobile communication systems, the ability to provide diversiform data services, high mobility vehicle communication experiences and asymmetrical services are enhanced further than 2G systems. But at the same time, users still have higher requirement for high-rate and high-QoS mobile services. Many international standardization organizations have launched the research and standardization of 3G evolution system, such as 3GPP Long Term Evolution (LTE) and LTE Advanced project. The primary three standards of 3G are all based on Code Division Multiple Access (CDMA), but with the in-depth research of Orthogonal Frequency Division Multiplexing (OFDM) techniques, OFDM has been emphasized by the mobile communication industry and used as the basic multiple access technique in the Enhanced 3G (E3G) systems for its merit of high spectrum efficiency.

OFDM becomes a key technology in the next cellular mobile communication system. As the sub-carriers in the intra-cell are orthogonal with each other, the intra-cell interference can be avoided efficiently. However, the inter-cell interference problems may become serious since many co-frequency sub-carriers are reused among different cells. Under this background, how to mitigate inter-cell interference and improve the performance for cellular users for vehicular environments become more urgent.

In this chapter, the research outcomes about Intel-cell Interference Mitigation technologies and corresponding performance evaluation results will be provided. The Intel-cell Interference Mitigation strategies introduced here will include three categories, which are interference coordination, interference prediction and interference cancellation respectively.

2. Inter-cell interference coordination

Frequency coordination plays important roles in the Inter-cell Interference Coordination scheme. For frequency coordination, one frequency reuse based Interference Coordination scheme will be introduced, called as Soft Fractional Frequency Reuse (SFFR). Its frequency reuse factor will be derived. Simulation results will be provided to show the throughputs in cell-edge are efficiently improved compared with soft frequency reuse (SFR) scheme.

Especially, for Coordinated Multi-point (CoMP) transmission technology, which is the promising technique in LTE-Advanced, a novel frequency reuse scheme - Coordinated Frequency Reuse (CFR) will be introduced, which can support coordination transmission in CoMP system. Simulation results are also provided to show that this scheme enables to improve the throughputs in cell-edge.

2.1 Soft fractional frequency reuse

In order to improve the performance in cell-edge, the SFFR scheme is introduced, which is based on soft frequency reuse. As shown in Fig.1, the characteristics of such reuse schemes are given as follows: the whole cell is divided into two parts, cell-centre and cell-edge. In cell-centre, the frequency reuse factor (FRF) is set as 1, while in cell-edge, FRF is dynamic and the frequency allocation is orthogonal with the edge of other cells, which can avoid partial inter-cell interference in cell-edge.
Specially, users in each cell are divided into two major groups according to their geometry factors. In cell-edge group, users are interference-limited due to the neighbouring cells, whereas in cell-centre group users are mainly noise-limited. The available frequency resources in cell-edge are divided into non-crossing subsets in SFFR.

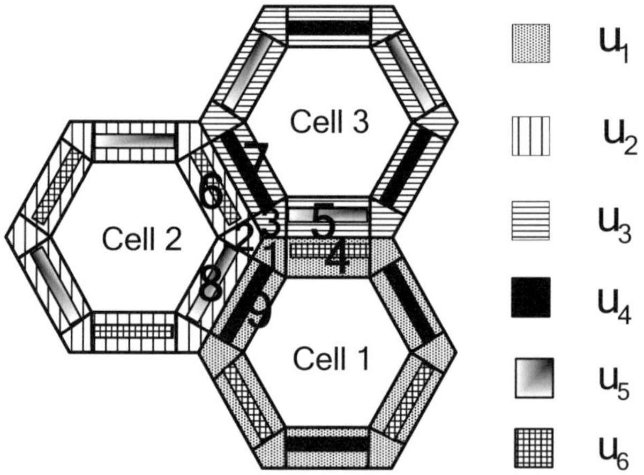

Fig. 1. Concept of Soft Fractional Frequency Reuse

The set of available frequency resources in the cell is allocated as follows: the whole frequency band is divided into two disjoint sub-bands, G and F, where G is allocated to the cell-centre users and F to the cell-edge users. Considering a cluster of 3 cells, as the one shown in Fig. 1, let $F = F_1 \cup F_2 \cup F_3$, where F_i denotes the subset of frequencies allocated to cell i, $(i = 1,2,3)$, and the subsets F_i may be overlapped with each other.
Since the cell-edge users are easily subject to co-frequency interference, the frequency assignments to the cell-edge users greatly rely on radio link performance and system throughput. Generally, the cell-edge can be divided into 12 regions, as the ones marked by 1, 4, and 9 in Cell 1 (see Fig. 1). Therefore, in a cluster of 3 adjacent cells, there are 9 parts in the cell-edge corner, which are in the shaded area. Moreover, we take this SFFR model as an example to deduce the design of the available frequency band assignment for the fields marked by 1, 2, .., 9.

In SFFR, all the available frequencies in cell-edge are divided into 6 non-overlapping subsets. Such subsets are respectively u_1, u_2, u_3, u_4, u_5 and u_6, while the subset in cell-centre is u_0. Firstly, we select frequency from the subsets u_1, u_2, u_3. If it's not enough, choose frequency from u_4, u_5, u_6. If the inter-cell interference increases, we need to add frequency into u_4, u_5, u_6, and decrease the cover area in cell-edge. If such interference is controlled in a low extension, we can decrease the frequency in subsets of u_4, u_5, u_6, and increase the cover area in cell-edge, which enables to improve the frequency utilization. Moreover, we assume $A_{1/3} = \{u_1, u_2, u_3\}$, $A_{2/3} = \{u_4, u_5, u_6\}$ and $A_{3/3} = \{u_0\}$, where $A_{1/3}$ denotes the frequency set with 1/3 reuse, $A_{2/3}$ denotes the frequency set with 2/3 reuse and $A_{3/3}$ denotes the frequency set with FRF equals to 1.
According to the definition of FRF in references, the FRF of SFFR scheme can be obtained as follows:

$$\eta = \frac{\frac{1}{3}\left|A_{1/3}\right| + \frac{2}{3}\left|A_{2/3}\right| + \frac{3}{3}\left|A_{3/3}\right|}{\left|A_{1/3}\right| + \left|A_{2/3}\right| + \left|A_{3/3}\right|} \tag{1}$$

where the symbol $|\cdot|$ stands for the cardinality of frequency set. Taking into account that $|A| = \left|A_{1/3}\right| + \left|A_{2/3}\right| + \left|A_{3/3}\right|$, the following relation is obtained:

$$|A| = |u_0| + 3|u_1| + 3|u_4| \tag{2}$$

Combining Eq.(1) and Eq.(2), the FRF is computed as:

$$\eta = \frac{|u_0|}{|A|} + \frac{1}{3} \times \frac{|u_1|}{|A|} \times 3 + \frac{2}{3} \times \frac{|u_2|}{|A|} \times 3 \tag{3}$$

From Eq.(2), we can get the equation about u_1 as follows:

$$|u_1| = \frac{|A| - 3 \times |u_4| - |u_0|}{3} \tag{4}$$

Following the example of Cell 1, the number of available frequencies in cell-centre is $|u_0|$, whereas in the cell-edge is $|u_1| + 2|u_4|$. Assuming that $|u_0| = k(|u_1| + 2|u_4|)$, where k is a constant parameter, so $|u_4|$ can be got from Eq.(4):

$$|u_4| = \frac{|u_0|}{3k} - \frac{|u_0|}{9} - \frac{|A|}{4} \tag{5}$$

Finally, taking into account Eq.(4) and Eq.(5), Eq.(3) can be expressed in terms of $|u_0|$:

$$\eta = \frac{1}{12} + \left(\frac{1}{3k} + \frac{5}{9}\right)\frac{|u_0|}{|A|} \tag{6}$$

It can be seen from Eq.(6) that as FRF grows, the available frequency resources in cell-centre increase, while those in cell-edge decrease. Moreover, the performance of the SFR scheme is compared with 3GPP LTE simulation parameters and the SFFR scheme.

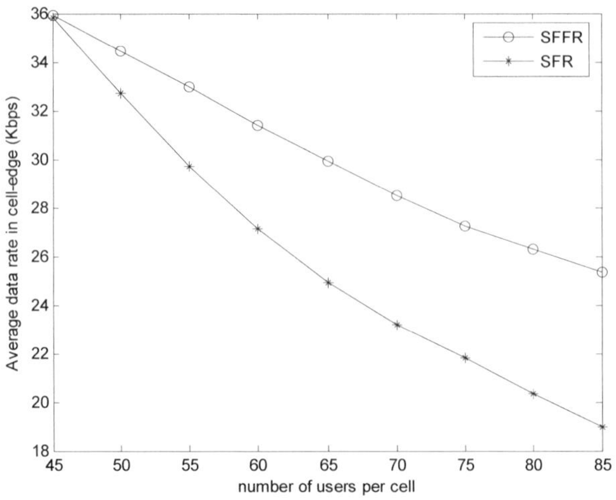

Fig. 2. Comparison of average data rate in cell-edge

Fig. 2 compares the average data rate in cell-edge for SFR and SFFR, where the FRF is set as 8/9. It can be seen that the average data rate in cell-edge decreases as the number of users per cell increases. However, the SFFR scheme outperforms the SFR scheme for a given number of users per cell. Specially, as the increase of users, the improvement by the SFFR scheme is more than that of the SFR scheme, which shows it's more effective when the number of users is large.

In order to mitigate inter-cell interference, a novel inter-cell interference coordination scheme called SFFR is introduced in this part, which can effectively improve the data rate in cell-edge. The numerical results show that compared with the SFR scheme, the SFFR scheme improves the performance in cell-edge.

2.2 Cooperative frequency reuse

In 3GPP LTE-Advanced systems, Coordinated Multi-Point (CoMP) transmission is proposed as a key technique to further improve the cell-edge performance in May 2008. CoMP technique implies dynamic coordination among multiple geographically separated transmission points, which involves two schemes.

a. Coordinated scheduling and/or beamforming, where data to a single UE is instantaneously transmitted from one of the transmission points, and scheduling decisions are coordinated to control.
b. Joint processing/transmission, where data to a single UE is simultaneously transmitted from multiple transmission points.

With these CoMP schemes, especially for CoMP joint transmission scheme, efficient frequency reuse schemes need to be designed to support joint radio resource management among coordinate cells. However, based on the above analysis, most of the existing frequency reuse schemes can not incorporate well with CoMP system due to not considerate multi-cell joint transmission scenario in their frequency plan rule.

In order to support CoMP joint transmission, a novel frequency reuse scheme named cooperative frequency reuse (CFR) will be introduced in this part. The cell-edge areas of each cell in CFR scheme is divided into two types of zones. Moreover, a frequency plan rule is defined, so as to support CoMP joint transmission among neighbouring cells with the same frequency resources. Compared with the SFR scheme, the simulation results demonstrate that the CFR scheme yields higher average throughput in both cell-edge and cell-average points of view with lower blocking probability.

2.2.1 System model

A typical system model for downlink CoMP joint transmission is described in Fig. 3. In the system, cell users are divided into two classes, namely cell-centre users (CCUs) and cell-edge users (CEUs). We assume only CEUs can be configured to work under CoMP mode. Each CEU has a CoMP Cooperating Set (CCS) formed by the cells that provide data transmission service to this CEU, and the serving cell of each CUE is always included in its CCS. The CEU with more than one cell in its CCS is regarded as a CoMP CEU, which can be served by the cells contained in its CCS simultaneously with the same frequency resources. It is assumed that each cell is configured with one transmitting antenna with one receiving antenna for each user.

As shown in Fig. 3, Cell 1, Cell 2 and Cell3 are formed a CCS for user 1. So user 1 is regarded as a CoMP CEU, and can be served by all these three cells simultaneously with the same frequency resources. Since user 2 is not work under CoMP mode, it can only communicate with its serving cell, i.e. Cell 1.

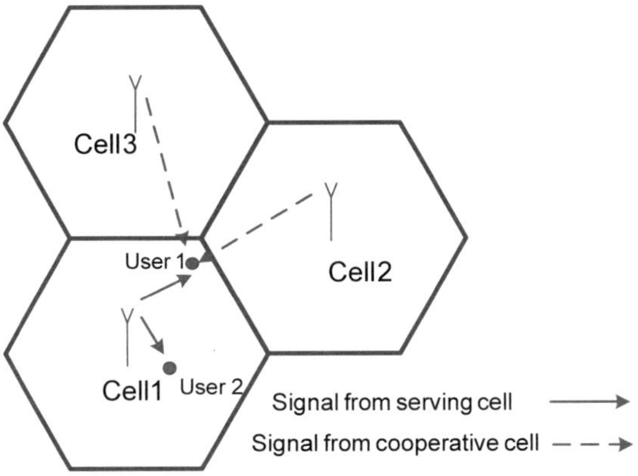

Fig. 3. System Model for downlink CoMP joint transmission

Let Ψ_k denote the CCS of the kth CoMP CEU, Ω denote the overall cells in the system, and $\{\Omega \cap \bar{\Psi}_k\}$ denote the cells in set Ω while not in set Ψ_k . Therefore, the signal to interference plus noise ratio (SINR) on l^{th} physical resource block (PRB) for k^{th} active CoMP CEU connected to i^{th} cell is determined as follows:

$$\gamma_{i,l}^{k} = \frac{\sum_{s \in \Psi} P_{s,l} G_s^k \left| h_{s,l}^k \right|^2}{N_0 + \sum_{n \in \{\Omega \cap \bar{\Psi}_k\}} x_{n,l} P_{n,l} G_n^k} \tag{7}$$

Where $P_{s,l}$ is the transmission power from s^{th} cell on l^{th} PRB. For simplicity, $P_{s,l}$ is constant assuming no power control. G_s^k is the long term gain between s^{th} cell and the kth CoMP CEU, consisting of propagation path loss and the shadow fading. $h_{s,l}^k$ denotes the fast fad gain on

l^{th} PRB for the channel between s^{th} cell and k^{th} CoMP UE. N_0 is the noise power received within each PRB. And $x_{n,l}$ is the allocation indicator of lth PRB, which can be given by:

$$x_{n,l} = \begin{cases} 1, & \text{if } l^{th} \text{ PRB is used in } n^{th} \text{ cell} \\ 0, & \text{otherwise} \end{cases} \tag{8}$$

In 3GPP LTE standards, it was pointed out that interference coordination is handled by the system once every 100ms. The information reported by the users and used by the system is the average SINR value. Thus, $\left|h_{s,l}^k\right|^2$ is replaced by its mean value $E\left(\left|h_{s,l}^k\right|^2\right)=1$, and Eq. (7) can be expressed as

$$\gamma_{i,l}^k = \frac{\sum\limits_{s\in\Psi_k} P_{s,l} G_s^k}{N_0 + \sum\limits_{n\in\{\Omega\cap\bar{\Psi}_k\}} x_{n,l} P_{n,l} G_n^k} \tag{9}$$

For the users who don't work under CoMP mode, they only communicate with their serving cells. The average SINR on l^{th} PRB for k^{th} user of ith cell is then given by:

$$\gamma_{i,l}^k = \frac{P_{i,l} G_i^k}{N_0 + \sum\limits_{n\in\Omega, n\neq i} x_{n,l} P_{n,l} G_n^k} \tag{10}$$

Finally, according to Shannon theorem, the corresponding capacity to the user average SINR on l^{th} PRB can be expressed as:

$$C_{i,l}^k = B\log_2\left(1 + \frac{\gamma_{i,l}^k}{\Gamma}\right) \tag{11}$$

Where B is the bandwidth of each PRB, and Γ called SINR gap is a constant related to the target BER, with $\Gamma = -\ln(5BER)/1.5$.

2.2.2 Cooperative frequency reuse scheme

The principle of the CFR scheme that can support CoMP joint transmission will be introduced here. Each three neighbouring cells are formed as a cell cluster and respectively marked with cell 1, cell 2 and cell 3. The cell-edge area of each cell is then divided into six cell-edge zones according to the six different neighbouring cells. Given the marker of each neighbouring cell, the six cell-edge zones in a cell are then categorized into two types. Hence, there are total six types of cell-edge zones in a cell cluster. As illustrated in Fig.4, each cell-edge zone is marked with A_i^j, where i denotes the cell to which the zone belongs, j is the marker of the dominant interference cell of this zone, note that $i,j=\{1,2,3\}$ and $i \neq j$. For simplifying expression, we just take the cell-edge zones in cell 1 into count:

Zone A_1^2: It is the cell-edge zone of the cells marked with cell 1. Moreover, the dominant interferer of the users in this zone is the nearest neighbouring cell marked with cell 2.

Zone A_1^3: It belongs to the cells marked with cell 1. And the dominant interferer is the nearest neighbouring cell marked with cell 3.

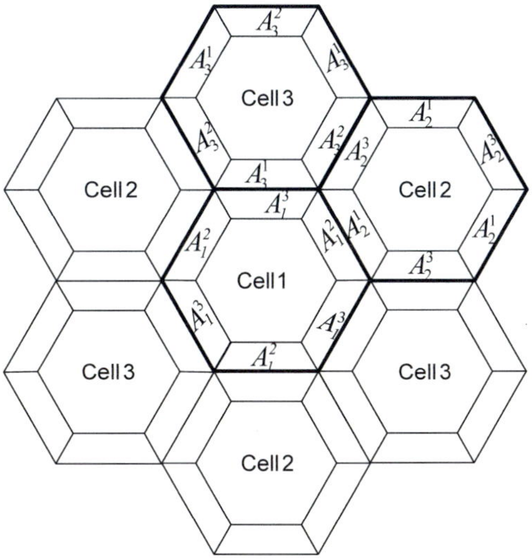

Fig. 4. Cell-edge areas partition for each cell

In order to support multi-cell joint transmission with neighbouring cells, a cooperative frequency subset is defined for each cell in CFR scheme. Then the resources are allocated to users in each cell cluster according to the following frequency reuse rule:

Step1. In each cell, the whole resources are divided into two sets, G and F, where $G \cap F = \varnothing$. Resources in set G are used for CCUs in each cell. While resources in set F are used for CEUs.

Step2. Set F is further divided into three subsets, marked by F_1, F_2, F_3, with $F_i \cap F_j = \varnothing (i \neq j)$.

Step3. For each cell cluster, F_i is assigned for cell i as a cooperative frequency subset, which is used for providing cooperative data transmission for the CEUs in neighbouring cells.

Step4. F_j is assigned for the CEUs in cell-edge zones marked with A_i^j.

Based on the above mentioned frequency reuse rule, the frequency allocation for a cell cluster is shown in Fig. 5.

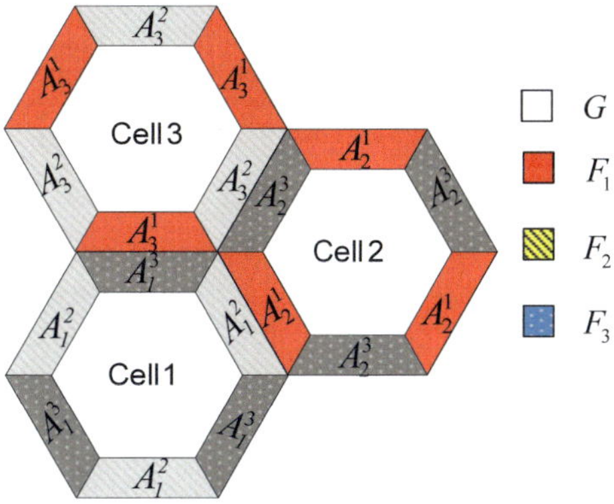

Fig. 5. Frequency assignment for the boundary areas of each cell cluster

On the one hand, orthogonal frequency subsets are allocated to the adjacent cell-edge zones that belong to different cells. Hence, the ICI can be reduced by using different frequency resources in adjacent areas of neighbouring cells. On the other hand, according to the frequency reuse rule, F_j is allocated for cell-edge zone A_i^j. Besides, it is the cooperative frequency subset for cell j, which is the dominant interference cell of zone A_i^j. Hence, for a CoMP CEU located in zone A_i^j, cell i and cell j can form a CCS. And then provide CoMP joint transmission for this CEU simultaneously with the same frequency resources selected from F_j.

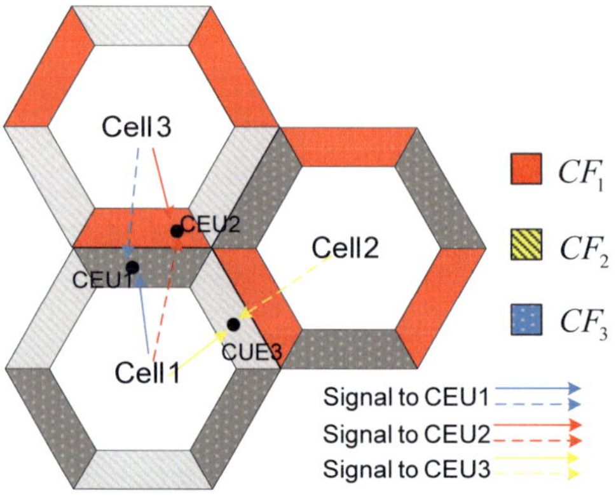

Fig. 6. CoMP joint transmission in CFR system

As shown in Fig.6, when CEU 1 in zone A_1^3 is regarded as a CoMP CEU, its dominant interference cell marked with cell 3 and the serving cell marked with cell 1 can form a CCS. Then CEU 1 can be served by these two cells with the same frequency resources selected from set F_3. What's more, we can see that the whole frequency resources could be reused in all cells. Hence, the frequency reuse factor in CFR scheme can achieve to 1. In CCS selection, we introduce an algorithm for the CCS selection. Let N denote the total number of cells in the system, M denote the maximum number of cells in a CCS of a CEU. The kth CEU's CCS, denoted as Ψ_k, can then be selected according to the user's long term gain G_k^i as follows:

Algorithm: CCS Selection

① $\Psi_k \leftarrow \varnothing$, $count \leftarrow 0$.

② Calculate the long term gain G_k^i between k^{th} CEU and i^{th} cell, for $i = 0,\ldots,N-1$.

$G \leftarrow \{G_k^0, G_k^1, \ldots, G_k^{N-1}\}$

③ Find serving cell for k^{th} CEU

$i \leftarrow \arg\max\left(G_k^i\right), G_k^i \in G$

$s \leftarrow i$

④ Update

$\Psi_k \leftarrow \Psi_k \cup \{i^{th}\ cell\}$

$count \leftarrow count + 1$

⑤ If $count < M$,

$G \leftarrow G - \{G_k^i\}$

$i \leftarrow \arg\max(G_k^i), G_k^i \in G$

Else stop.

⑥ If $G_k^s - G_k^i \leq thr$, go to ④

Else stop.

It has been proved that the maximum size of UE-specific CoMP cooperating set equal to 2 is enough to achieve CoMP gain for 3GPP case 1 in references. Hence, the value of M is set to 2 in this paper. CEUs with two cells in their CCS are regarded as CoMP CEUs, whose SINR can be improved by CoMP joint transmission with the same frequency resources according to the introduced frequency reuse rule.

2.2.3 Performance analysis

System level simulations are performed to evaluate the performance of the introduced CFR scheme. As performance metrics, we used the blocking probability and the average throughput in both the cell-edge and cell-average points of view. The universal frequency reuse (UFR) where PRBs are randomly assigned to the different users in each cell irrespective of their category (CEU or CCU) is taken as a reference scheme. Another reference scheme is SFR scheme, which assigns a fixed non-overlapping cell edge bandwidth to a cluster of three adjacent cells. For the introduced CFR scheme, two cases are studied, where Thr is 0 dB and 5 dB respectively.

We focus on an OFDMA-based downlink cellular system. A number of UEs are uniformly dropped within each cell. The basic resource element considered in the system is the PRB, which consist of 12 contiguous subcarriers. It is assumed that all the available PRBs are transmitted with equivalent power. Only one PRB can be assigned to each active UE. The main simulation parameters listed in Table.1 are based on 3GPP standards.

Parameters	Values
Carrier Frequency	2 GHz
Bandwidth	10 MHz
Subcarrier spacing	15 kHz
Number of subcarriers	600
Number of PRBs	50
The number of cells	21
Cell radius	500m
Maximum power in BS	46 dBm
Distance-dependent path loss	L=128.1+37.6log10 d (dB), d in km
Shadowing factor variance	8dB
Shadowing correlation distance	50m
Inter cell shadow correlation	0.5

Table 1. Simulation Parameters

Fig. 7 shows the blocking probability of the introduced CFR scheme and the conventional SFR scheme as a function of the loading factor. We can see that CFR scheme performs quite better than the SFR scheme. Specially, the blocking probability reduced by SFR scheme is 50% more than SFR scheme. For example, if it is required that the blocking probability must not exceed 5%, Fig.7 indicates that the admissible loading factor of the SFR scheme is only 30%, while the admissible loading factor of the introduced CFR scheme is more than 60% of the total frequency resource. This improvement in the CFR scheme results from the frequency reuse rule designed for each cell cluster. According to the frequency reuse rule, the number of available frequency resources for the cell-edge areas of each cell is twice as great as the conventional SFR scheme.

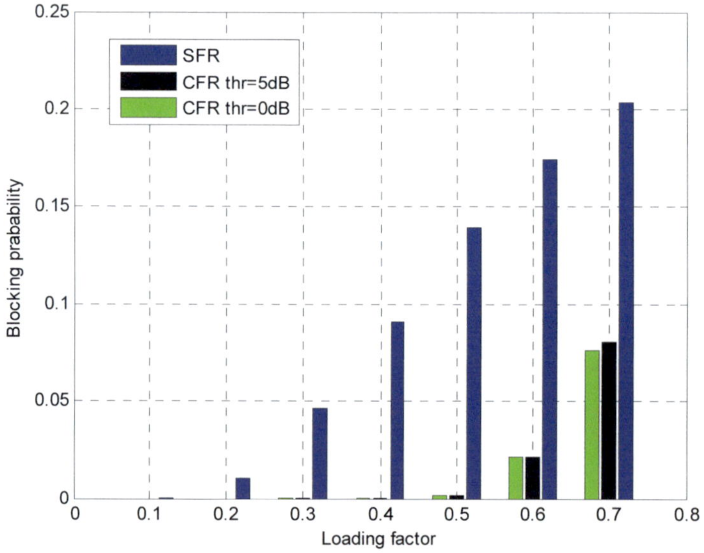

Fig. 7. Blocking probability as a function of the loading factor

Fig. 8 shows the cell-edge average throughput per user for the three different frequency reuse schemes considered in this paper. It can be seen that the average throughput per CEU decreases as the number of users increases in all the three schemes. That is because the probability of PRBs collision increases as the number of users grows. In other words, the ICI increases when the average number of users per cell grows. Moreover, compared with UFR scheme, both CFR scheme and SFR scheme yield a significant improvement in terms of cell-edge average throughput owing to the frequency reuse plans for cell-edge areas.

We can also observe that the introduced CFR scheme achieves higher cell-edge average throughput than SFR scheme. When *Thr* is 0 dB, no user works under CoMP mode. Compared with SFR scheme, the cell-edge average throughput is improved by 4 to 8%, which is achieved mainly owing to the frequency reuse rule designed in CFR scheme. When *Thr* is set to 5dB, the throughput raised by the introduced CFR scheme is 30 to 40% more than the SFR scheme, that is because part of the CEUs are regarded as CoMP users whose throughput can be further improved by CoMP joint transmission.

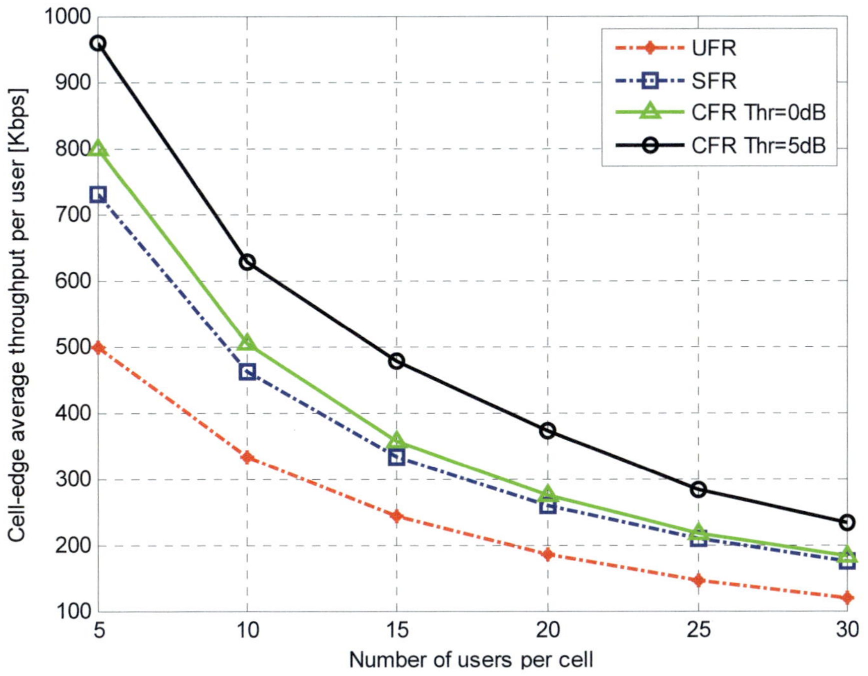

Fig. 8. Cell-edge average throughput per user

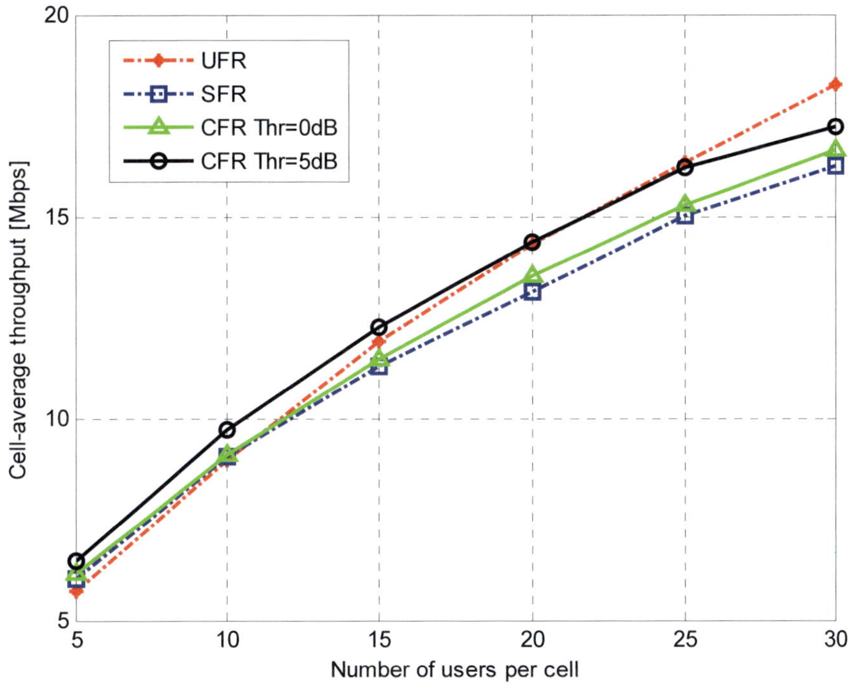

Fig. 9. Cell-average throughput as a function of the number of users per cell

Fig. 9 shows the cell-average throughput of the introduced CFR scheme and the two conventional frequency reuse schemes as a function of the number of users per cell. From the graph, we can see that the cell-average throughput of the introduced CFR scheme outperforms that of the SFR scheme due to better cell-edge performance and lower blocking probability. When *Thr* is set to 0dB, the cell-average throughput is improved by 1 to 3%. While when *Thr* is set to 5dB, the cell-average throughput is improved by 5 to 9%.
It is also can be seen that when the number of users per cell is small, i.e. less than 15, CFR scheme with *Thr* equals to 5dB achieves the best results among the three schemes under consideration. When the number of users is large, UFR scheme achieves the better results than the introduced CFR scheme. However, the payoff for this higher average cell throughput of UFR scheme is a huge decrease in cell-edge average throughput which can be observed from Fig. 8.

2.2.4 Summary

In this part, a novel frequency reuse scheme named CFR is introduced to support CoMP joint transmission and further improve cell-edge performance. First, the method for cell-edge areas partition is introduced, which divides the cell-edge areas of each cell into two types of zones. Then, the frequency plan rule is defined for each cell cluster, which assigns a cooperative frequency subset for each cell and makes CoMP users in cell-edge zones can be served by multi-cell joint transmission with the same frequency resources. In addition, the algorithm is given for the CEUs to select cells in their CCS. The simulation results demonstrate that the introduced CFR scheme significantly outperforms the conventional SFR scheme in terms of blocking probability, cell-edge average throughput and cell-average throughputs.

3. Inter-cell interference prediction

In order to mitigate the inter-cell interference in OFDMA systems, three schemes are given in 3GPP organization, which respectively are interference coordination, interference cancellation and interference randomization. However, the traditional inter-cell interference mitigation schemes belong to passive interference suppression measures, and its effectiveness is still limited. Considering this situation, an active interference mitigation strategy will be introduced in this part, named as interference prediction. By means of the immediate interference prediction in cell, it enables to efficiently avoid and eliminate inter-cell interference, which is a novel type of active interference mitigation strategy.
For interference prediction, this part takes use of the optimal estimation theory. Generally, the problems about optimal estimation theory can be classified into three categories: The first is the model parameter estimation problem, such as the least squares method. The second is time series and optimal filtering estimates problem (the optimal estimation of signal or state). The third is the optimal information fusion estimation. According to the actual situation in inter-cell interference prediction, the second problem of optimal estimation is focused.
The inter-cell interference prediction principle is based on optimal estimation theory, forecasting the co-frequency interference in the next timeslot by means of the former or current channel state, and making the mean square error to be the smallest. The optimal estimation theory includes time-series estimation, optimal filtering estimation method, etc.

Especially, the optimal filtering estimation aims to estimate the signal state, including several filtering estimation algorithms, such as Wiener filter, Kalman filter, and so on.

3.1 Time series

In time series analysis, it aims to establish the time series model, predict and control signal change state based on such model. Moreover, we define the observation sequence as $\{z_t | z_1, z_2, \cdots, z_n, \cdots\}$, the linear mixed coefficient as $\{a_t | a_1, a_2, \cdots, a_n, \cdots\}$, then the future value $\{z_{t+k} | k > 0\}$ can be predicted by means of current and past time series records $\{z_t, z_{t-1}, z_{t-2}, \cdots\}$. The predicted value is written as $\hat{z}_{t+k|t}$, which meets following condition:

$$\hat{z}_{t+k|t} = \sum_{i=0}^{\infty} a_i z_{t-i} \tag{12}$$

The optimal predicted value $\hat{z}_{t+k|t}$ should make the mean square error be minimum, which should obey

$$Min\left\{E\left[\left(\hat{z}_{t+k|t} - z_{t+k}\right)^2\right]\right\} \tag{13}$$

In the above theoretical derivation, the present and past observed records $\{z_t, z_{t-1}, z_{t-2}, \cdots\}$ belong to be infinite series, which is difficult to achieve in practice. Considering this situation, the finite time series of recursive predictor are introduced, such as Box-Jenkins method, Astrom method, etc. Besides, the steps of Box-Jenkins method are as follows:

a. For the observed sequence $\{z_t | t = 1, 2, \cdots, N\}$, calculate its correlation coefficient and partial autocorrelation coefficient, then test whether the sequence is non-stationary white noise sequence. If such sequence is white noise series, go to the end. If such sequence is non-stationary series, take model according to non-stationary time series principle. Else if the sequence is stationary series, take zero for the mean of such sequence and then make model by the Box-Jenkins method.
b. Test the type of zero mean stationary series. Illustrately, determine the series $\{z_t | t = 1, 2, \cdots, N\}$ belong to which model, such as autoregressive (AR) model, moving average (MA) model and autoregressive moving average (ARMA) model.
c. After the model is identified, judge the highest level of such model, and make fitted test from low level to high level. For example, if $\{z_t | t = 1, 2, \cdots, N\}$ belong to AR model, make use of $AR(n, n-1)$, and then the fitted test.
d. Compare with different models and find the right model. On this basis, respectively take adaptive test and error test for the initial model, and select the optimal model.
e. Make prediction by the established model.

3.2 Optimal filter estimation

The nature of filtering is the statistical estimation problem. For example, linear minimum variance estimation methods try to make the variance of estimated value and the actual value minimum. Moreover, such filter is also known as the optimal filter, such as Wiener filter and Kalman filter. The interference prediction process by Kalman filter is shown respectively in Fig. 10 and Fig. 11.

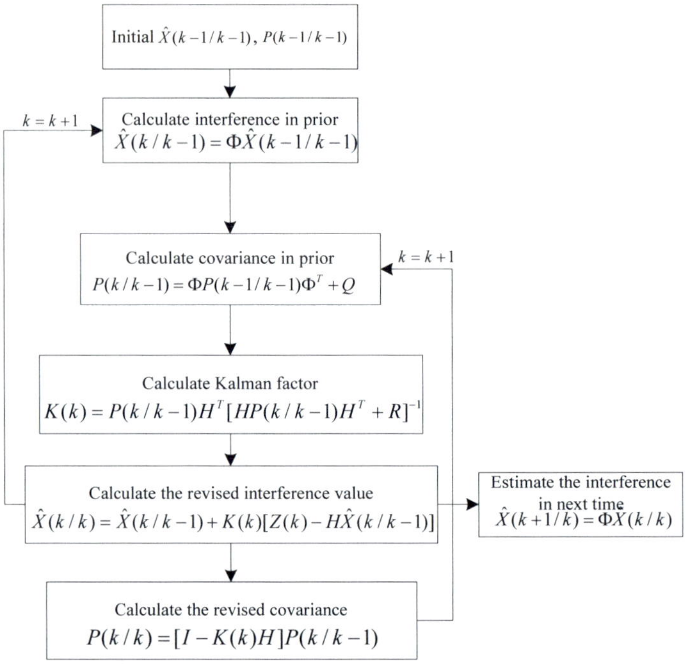

Fig. 10. Interference prediction by Kalman filter

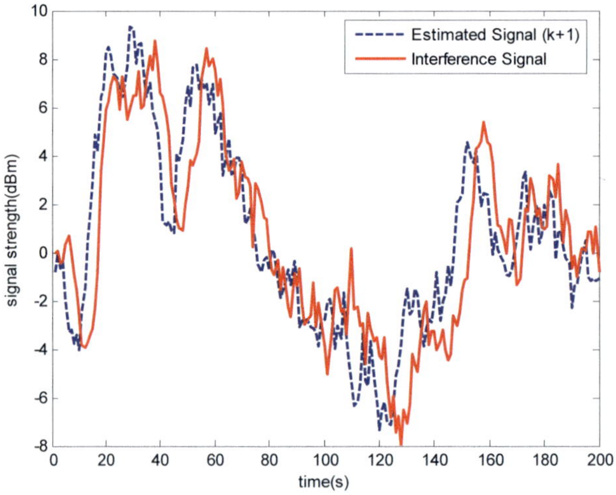

Fig. 11. Interference prediction results by Kalman filter

3.3 Effectiveness principles

The effectiveness of channel prediction criteria was analyzed and the relationship with the predicted time delay and the prediction accuracy of SINR are described in the reference. According to the wireless signal propagation, the mobility rate is closely related with the signal coherence time. If user's mobility rate increases, the coherence time becomes shorter.

Else user's mobility rate decreases, the coherence time becomes longer. The relationship of coherence time and the predicted delay time are divided into three categories, respectively $\tau \gg \Delta t$, $\tau \cong \Delta t$ and $\tau \ll \Delta t$.

Fig.12 shows the prediction results when the coherence time is much greater than the time delay, which is that $\tau \gg \Delta t$. At this time, user is in a slow moving state, and the channel state information (CSI) can be easily obtained. From Fig.12, we can see the predicted SINR in delay time approximates to the actual value.

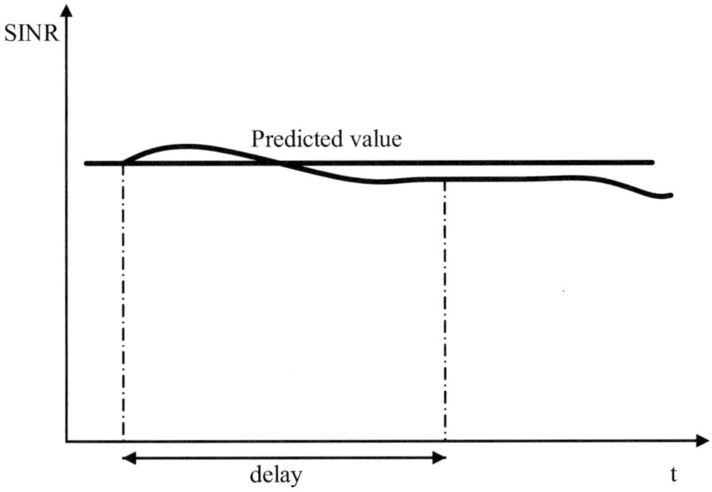

Fig. 12. SINR prediction ($\tau \gg \Delta t$)

Fig.13 shows the prediction results when the coherence time approaches to the time delay, which is that $\tau \cong \Delta t$. At this time, user's moving speed is in a medium state. In order to ensure the continuity of information transmission, the SINR should obey the outage criteria and keep a conservative prediction, which is the threshold SINR value.

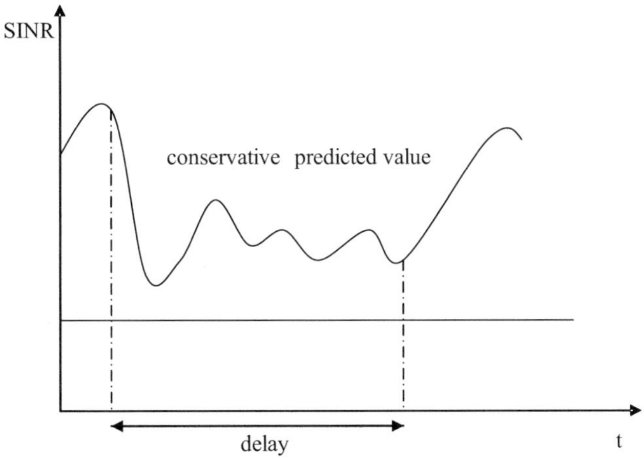

Fig. 13. SINR prediction ($\tau \cong \Delta t$)

Fig. 14 shows the prediction results when the coherence time is far less than the time delay, which is that $\tau \ll \Delta t$. At this time, user's rate is in a high speed state, the coherence time is shorter and the CSI is hard to be obtained. In this situation, we only need to predict the average SINR.

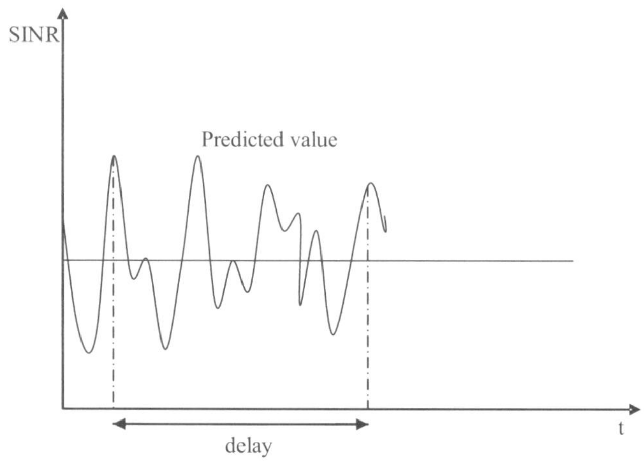

Fig. 14. SINR prediction ($\tau \ll \Delta t$)

3.4 Summary

In this part, the inter-cell interference prediction is introduced, which is an active interference mitigation method. The theoretical basis, which is the optimal estimation theory, is provided with including of two parts: time series and the optimal filter estimation. Besides, the reliability is also analyzed by means of prediction accuracy, which is based on the relationship of the coherent time and the time delay. In addition, the trend for the actual measured radio signals is analyzed with AR model, MA model and ARIMA model. The analytical results are provided to show time series model can efficiently predict the radio signals change and then mitigate the interference effectively.

4. Inter-cell interference cancellation

Inter-cell interference cancellation strategy aims at interference suppression at the user equipments by improving the processing gain. In order to solve this problem, two basic schemes have been discussed in 3GPP proposals. One is to take spatial suppression at the UE side by means of multiple antennas; the other is to directly detect and subtract the inter-cell interference in order to enable inter-cell-interference cancellation. Usually, the inter-cell interference cancellation strategy is used to get the processing gain through suppress strong interference. According to the degree of knowledge available about interferers, interference cancellation methods can be distinguished as three categories, which are blind, semi-blind, and full-knowledge.

Many inter-cell interference cancellation methods are based on generalized spatial diversity. Beam forming is introduced in inter-cell interference cancellation in references. By distinguish different users in space, it effectively reduces interference among users. But on the other hand, it brings with extra interference from main lobe and strong side lobe. A method of subcarrier-based virtual MIMO in inter-cell interference cancellation was proposed, which is introduced in OFDM-based systems with a frequency reuse factor equal to 1. But when UE is located between sectors, inter-sector interference cannot be reduced by the subcarrier-based virtual MIMO (SV-MIMO) due to loss of channel separability. Inter-cell interference cancellation by virtual smart antennas was also studied, which proposes a method for estimating inter-cell symbol timing offsets using multiple signal classification (MUSIC) algorithm. For the use of MUSIC algorithm, the premise is to know the number of source. But in practice, the number of source can not be accurately obtained, which may make MUSIC algorithm not work. Moreover, in most case, many similar algorithm needs to know current channel state information, but at the same time, the complexity of system may be increased if acquire it in downlink. As a result, how to mitigate inter-cell interference in no precise channel is an important problem.

In order to effectively mitigate inter-cell interference in OFDM-based systems, this part focuses on the inter-cell interference cancellation strategy. A novel inter-cell interference mitigation method for OFDM-based cellular systems will be introduced. Compared to the existing methods, the independent component analysis based on blind source separation is presented in inter-cell interference, and the signal to interference plus noise (SINR) is set up as the objective function. This scheme can adapt to the no precise channel conditions, and can mitigate inter-cell interference in a semi-blind state of source signal and channel information.

4.1 Inter-cell interference model

Considering the downlink in cell-edge, assume this MIMO system with q transmission antennas in the serving eNodeB, and p receiving antennas in UE. In such scenario, UE not only receives useful signal from current communicating base station, but also receives noise and interference from other adjacent base stations. The example is shown in Fig. 15.

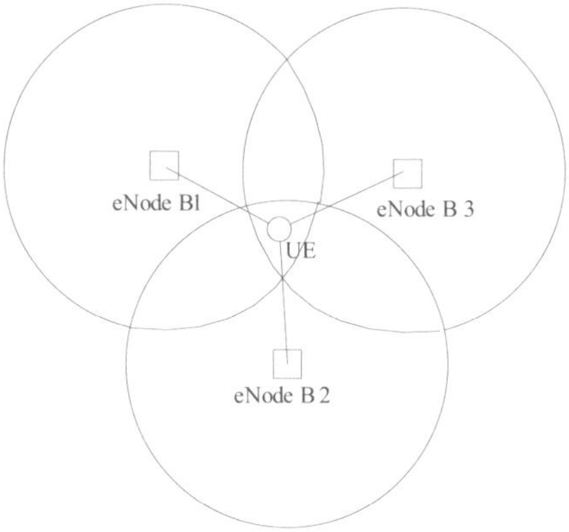

Fig. 15. Inter-cell interference in cell-edge

For many OFDM-based systems, the original signal is transmitted from OFDM transmitter and through MIMO antenna array. The process of inter-cell interference mitigation is shown in Fig.16. Further, we assume the original signal interfered by inter-cell interference and thermal noise, and the channel information is unknown.

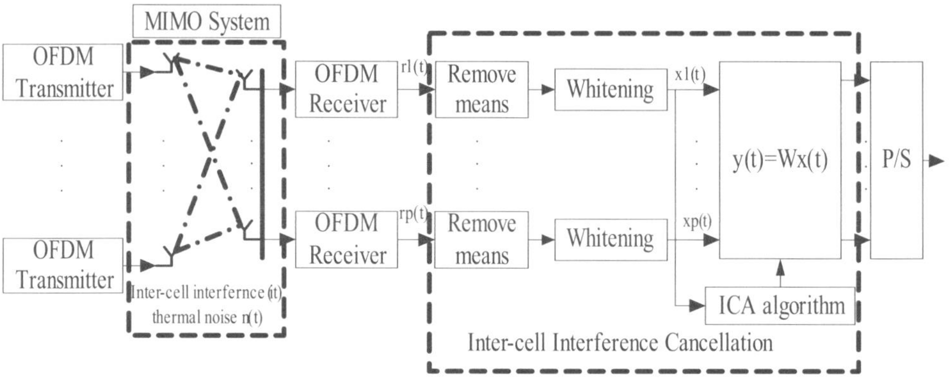

Fig. 16. Inter-cell interference mitigation process

According to the principle in radio signal propagation, the thermal noise can be seen as independent with transmission signals from eNodeB. Compared to the inter-cell interference from other cells, we assume the useful signal is statistically independent with the co-frequency interference from other different cells. So it can be thought that useful signal, unknown inter-cell interference and thermal noises are statistically independent and irrelevant with each other. Moreover, some parameters are defined as follows:

$s(t)=\{u(t),i_1(t),...,i_k(t),...,i_{n-2}(t),n(t)\}$ denotes as the source signal, which is constructed by the useful signal, inter-cell interference and the thermal noise, also written as $(s_1,s_2,...,s_n)^T$. Specifically, $u(t)$ denotes as the useful signal. $i_k(t)$ denotes as the kth unknown additive inter-cell interference with the same frequency, which is in the range of 1 to $n-2$. The dimension for the number of eNodeB reused with the co-frequency subcarriers is $n-2$. $n(t)$ denotes as the additive zero mean thermal noise, also the Gaussian noise.

At the receiving end, we denote $r(t)$ as the received signal, mixed with the useful signal, unknown additive interference and noise. α,β_k and γ are respectively the mixing vectors. So $r(t)$ can be written as the following equation:

$$r(t)=\alpha u(t)+\sum_{k=1}^{n-2}\beta_k i_k(t)+\gamma n(t) \tag{14}$$

Furthermore, Let A denote as linear mixing matrix, which reflects temporal radio signals transmission process and all interference from other adjacent cells are linear mixture. Then the inter-cell interference model can be written as:

$$r(t)=As(t) \tag{15}$$

From Eq.(15), it can been seen that the dimension of $r(t)$ is the same to $s(t)$, which is equal to n. In order to separate the useful signal from inter-cell interference and thermal noise, we take interference cancellation by independent component analysis (ICA), which is an important method in blind source separation (BSS). As shown in Fig.16, we deal with the received signal by remove means and whitening, and such process is set as the transition matrix V. So we can get:

$$x(t)=Vr(t) \tag{16}$$

On the other side, we set up a separation matrix W, and make $WVA = I$ in theory. Moreover, assume $y(t)$ is the separated signal after removing interference and noise, so $y(t)$ can be got by following equation:

$$y(t) = Wx(t) = WVAs(t) = Gs(t) \tag{17}$$

But in fact, because exist with errors and uncertainties, we need to get an optimal approximate solution that can make such a separation matrix W approach to the condition $G = I$. By means of ICA methods, an objective function is established, which takes W as a variable function. When W takes some value, the objective function can achieve to minimum or maximum. At this moment, the variable W is the optimal approximate solution.

4.2 Independent component analysis

In many fields, it needs to separate all the source signals from the mixed signals with no precise knowledge of the source signals and the channel information, whose processes are usually called as blind source separation (BSS). In order to solve such problem, many schemes have been researched. When the source signals are not independent with each other in BSS, some separating schemes are introduced, such as sparse component analysis (SCA), smooth component analysis (SMOCA), non-negative matrix factorization (NMF), and so on. However, the complexity of such algorithms is still high and hard to realize in application.

On the other hand, when the source signals are independent with each other in BSS, the independent component analysis (ICA) schemes are proposed. By principle of independence, the separating complexity is reduced and the results are also improved. Specially, some methods exist in ICA, such as Informax, Fast ICA, generalized eigenvalue decomposition, etc. Informax algorithm is proposed by Bell, whose characteristic is searching for the maximum mutual information between the received signal and the output signal, but its convergence is always slowly. Fast ICA is a fast and fixed-point algorithm proposed by A. Hyvrinen, whose characteristic is computing the maximum kurtosis by iterations. Although its convergence is improved compared with Informax, the effects of thermal noise are always not included in iterations. The ICA based on generalized eigenvalue decomposition is proposed by L. Parra, whose characteristic is decomposing generalized eigenvalue for the received signal. However, this method is limited by the type of source signals.

The critical step in ICA process is to make the estimated independent component gradually approach to the source signal by means of establishing objective function and finding its optimal solution.

According to the classical formula dealing with ICA problems, some requirements must be made in the known conditions in order to get definite solution, as follows:

a. The source signals are all real random signals, and the respective mean is zero. Moreover, these signals are statistically independent with each other.
b. There is at most one source signal whose probability density characteristic is the Gaussian distribution, while the other source signals obey non-Gaussian probability distribution.
c. For the source signals, the approximate probability distribution functions (PDF) need to be acquired.

d. The number of source signals and received signals is equal, and the mixing matrix A is a square matrix.

But in inter-cell interference cancellation, we only need to separate the useful transmission signals from the inter-cell interference and noise, which means we don't need to separate all source signals from the mixed signals. Specially, we can take use of some identity of the useful signals in this situation, such as training sequence.

Based on the above analysis, and in order to verify the inter-cell interference cancellation scheme by ICA methods, some conditions need to be illustrated as follows:

a. Considering the offset of carrier frequency and phase, it may bring with some intra-cell interference actually. In analysis, we neglect such interference, but only consider the inter-cell interference and thermal noise. Moreover, the inter-cell inferences from different cells are seen as statistically independent with each other, and all these signals are linear mixed.
b. By the principle of BSS, the means of the received signal from different eNodeB can be respectively conversion into zero by the process of removing means, and such signals can also be changed into real random signals by whitening, as shown in Fig.16.
c. According to the experienced formulas, the voice signals are usually with Super-Gaussian distribution, for the kurtosis of such signals is positive. While the image signals are usually with Sub-Gaussian distribution, for the kurtosis of such signals is negative. Among the mixed signals, only thermal noises are with Gaussian distribution, whose kurtosis is zero.
d. Specially, if the strength of inter-cell interference from adjacent eNodeB is much stronger than the useful signal in cell-edge, it may reach the handover threshold and the inter-cell handover process may be triggered. Moreover, we set the handover threshold as -10dB in this paper, without considering ping-pong effects.
e. Because only need to separate partial signals from the mixed signals, not all the signals, we don't require all signals received by UE are included in the mixed signals, which means the number of the source signals and the received signals may be unequal. By generalized matrix theory, generalized eigenvalue and generalized inverse matrix can be computed, so it doesn't restrict that A is a square matrix.

Based on the compared analysis, ICA method is taken into inter-cell interference cancellation for OFDM-based systems. Different from traditional ICA methods in BSS, we only need to separate partial useful signals from the mixed signals, not all the signals. Moreover, the training sequence can be used to identify the useful signal in order to avoid the ambiguity of separated signal, and this process may be in a semi-blind state.

By means of the existing ICA methods, a Max-SINR ICA algorithm is constructed, which considers the effects of both interference and thermal noise, raises the speed of convergence and improves the output SNR.

4.2.1 Establish objective function

Considering the signal to interference plus noise ratio (SINR) is one of important measured parameters in communication systems, we set up SINR as the objective function.

Assume the following system parameters are given when the system is set up: the source signal $s(t)$, the received signal $x(t)$, the separated signal $y(t)$, the interference and noise $e(t) = y(t) - s(t)$, and the weighted coefficient in the ith antenna is defined as ω_i, so the SINR in user equipment can be defined as follows:

$$F(y) = SINR = 10\lg\frac{s(t)\cdot s^T(t)}{e(t)\cdot e^T(t)}$$
$$= 10\lg\frac{s(t)\cdot s^T(t)}{[y(t)-s(t)]\cdot[y(t)-s(t)]^T} \tag{18}$$

Assume the separated signal in the ith antenna is $y_i(t)$, the number of receiving antenna is p. For $s(t)$ is unknown in the conditions, we take the weighted arithmetic mean $\overline{y}(t)$ in place of $s(t)$ as an unbiased estimation, that is

$$s(t) = \overline{y}(t) = \frac{1}{p}\sum_{i=1}^{p}\omega_i y_i(t) \tag{19}$$

$$F(y) = 10\lg\frac{\overline{y}(t)\cdot\overline{y}^T(t)}{[y(t)-\overline{y}(t)]\cdot[y(t)-\overline{y}(t)]^T} \tag{20}$$

Also as $y(t) = W\cdot x(t)$, $\overline{y}(t) = W\cdot\overline{x}(t)$, where W is the separation matrix, so Eq. (20) can be equal to

$$F(W) = 10\lg\frac{Wx(t)x^T(t)W^T}{W[x(t)-\overline{x}(t)]\cdot[x(t)-\overline{x}(t)]^T W^T} \tag{21}$$

Then, take expectations on $F(W)$, which is

$$G(W) = E\{F(W)\} \tag{22}$$

For $x(t)$, let $C = E\{x(t)x^T(t)\}$ represent the variance matrix, and let $\tilde{C} = E\{[x(t)-\overline{x}(t)]\cdot[x(t)-\overline{x}(t)]^T\}$ represent the covariance matrix, so we can get

$$G(W) = 10\lg\frac{WCW^T}{W\tilde{C}W^T} \tag{23}$$

4.2.2 Optimize initial separation matrix

In order to get the maximum of $G(W)$ about W, take partial derivative in Eq. (10), and let it equate to zero, that is:

$$\frac{\partial G(W)}{\partial W} = 0 \tag{24}$$

Through Eq. (24), we can get an equation about W. By solve this equation, the solution W can be got. According to the proof in Theorem 1, W is the eigenvector of generalized matrix $\tilde{C}^{-1}C$.

Theorem 1: The separation matrix W is the eigenvector of generalized matrix $\tilde{C}^{-1}C$, where $\tilde{C}^{-1}$ is the general inverse matrix of covariance matrix $\tilde{C}$, and C is the variance matrix.

Proof:

By simplify Eq. (24), we can get

$$\frac{2CW}{WCW^T} = \frac{2\tilde{C}W}{W\tilde{C}W^T} \tag{25}$$

Then combined with the known conditions $C = AC_sA^T$, $\tilde{C} = A\tilde{C}_sA^T$ and $WA = I$, Eq. (25) can be simplified as:

$$WCW^T = WAC_sA^TW^T = C_s$$

$$W\tilde{C}W^T = WA\tilde{C}_sA^TW^T = \tilde{C}_s$$

Put the above expressions into Eq. (25), so

$$\frac{CW}{C_s} = \frac{\tilde{C}W}{\tilde{C}_s} \tag{26}$$

$$\Rightarrow CW = (\frac{C_s}{\tilde{C}_s})\tilde{C}W \tag{27}$$

For the source signals $s(t) = (s_1, s_2, ..., s_n)^T$, for $i = 1,...,n$, and $j = 1,...,n$, when $i \neq j$, because s_i and s_j are independent and irrelevant, so $E\{s_is_j^T\} = 0$; only when $i = j$, it's nonzero, and let $b_i = E\{s_is_i^T\}$, which means the variance. Further, set the covariance as $\tilde{b}_i = E\{[s_i - \overline{s}_i]\cdot[s_i - \overline{s}_i]^T\}$. On this basis, the matrix constructed by variance and covariance is respectively as follows:

$$C_s = E\{s(t)\cdot s^T(t)\} = diag[b_1, ..., b_i, ..., b_n]$$

$$\tilde{C}_s = E\{[s(t) - \overline{s}(t)]\cdot[s(t) - \overline{s}(t)]^T\} = diag[\tilde{b}_1, ..., \tilde{b}_i, ..., \tilde{b}_n]$$

Assume $\lambda_i = \frac{b_i}{\tilde{b}_i}$, by matrix calculating formula, we can get $\frac{C_s}{\tilde{C}_s} = diag[\lambda_1, ..., \lambda_i, ..., \lambda_n]$. Then simplify Eq. (27), as following:

$$\tilde{C}^{-1}C\cdot W = diag[\lambda_1, ..., \lambda_i, ..., \lambda_n]\cdot W \tag{28}$$

As is shown in Eq. (28), W is the eigenvector of generalized matrix $\tilde{C}^{-1}C$. So we can get the separation matrix W by compute the eigenvector of $\tilde{C}^{-1}C$. The proof is the end.

On the other side, the sizes of eigenvalues reflect different signal strength. By Eigenvalue separation, the information about the useful signal can be found among the mixed signals, which enables to optimize the separating process. On this basis, we need to separate partial useful signals from the mixed signals, while remove the inter-cell interference and thermal noise represented by some other eigenvalues.

4.2.3 Max-SINR ICA algorithm

In order to mitigate the inter-cell interference by cancellation, we take SINR as the objective function, making it to the maximum. Based on Fast ICA algorithm in BSS, the steps of this algorithm are introduced as follows:

a. Step1: Remove the mean of the original signals $r(t)$ by filtering.

b. Step2: Perform the data whitening through the equation $x(t) = D^{-1/2}E^T r(t)$, where D is the eigenvalue matrix and E is the eigenvector matrix. Moreover, D and E can be got by the generalized eigenvalue decomposition of the matrix $R = E\{r(t)r^T(t)\}$.

c. Step3: According to the mixed signal $x(t)$, construct the variance matrix and the covariance matrix as: $C = E\{x(t)x^T(t)\}$, $\tilde{C} = E\{[x(t) - \overline{x}(t)] \cdot [x(t) - \overline{x}(t)]^T\}$

d. Step4: Calculate the generalized eigenvalues and generalized eigenvectors by Eq. (29), and find the eigenvalue and its eigenvector $\hat{W}$, which is the initial separation matrix.

$$CW = (\frac{C_s}{\tilde{C}_s})\tilde{C}W \tag{29}$$

e. Step5: Set the initial value $k = 0$, and normalize $\hat{W}$ by the equation $W_0 = \hat{W} / \left\|\hat{W}\right\|$.

f. Step6: Judge whether $\left\| W_k^T W_k \right| - 1 \right| \le \varepsilon$, in which ε is arbitrary small. If set up, output W_k; else go to following iterations:

while $k \le k_{\max}$

while $\left\| W_k^T W_k \right| - 1 \right| > \varepsilon$

do $W_k = E\{(W_{k-1}^T x)^3 x\} - 3W_{k-1}$

$W_k = W_k - \sum_{i=0}^{k-1} (W_k^T W_i) W_i$

$W_k = W_k / \left\| W_k \right\|$

end

$k = k + 1$

end

g. Step7: Separate the useful transmitting signals from the mixed signals.

$$y(t) = Wx(t) \tag{30}$$

4.2.4 Select useful signals

Considering the inter-cell interference from different cells must be with different training sequences, so the useful signals can be selected according to following equation:

$$d_i = \sum_{i=1}^{N_S} \left(\hat{a}_i(t) - a_0^T(t) \right)^2 \tag{31}$$

Where N_s is the length of training sequence constructed by the Gold sequence, $\hat{a}_i(t)$ is the estimated value of training sequence symbol, and $a_0(t)$ is the known value of training sequence symbol.

On the other hand, the estimated training sequence of useful signal and the known training sequence must be with the minimum distance, so it can get

$$d = Min\{d_1, d_2, \ldots, d_n\} \tag{32}$$

By means of Eq.(32), we can select the useful signal $u(t)$ from the separated signals.

4.3 Performance evaluation

In order to verify the results of such inter-cell interference cancellation method introduced in this paper, the static simulation is performed in the OFDM-based environment. The simulation parameters are described in Table 2, and the process of inter-cell interference cancellation by Max-SINR ICA is shown in Fig.16. In static simulation, we assume the position of each UE is fixed, and neglect the Doppler frequency effect.

Parameters	*Values*
Channel environment	Rayleigh fading
Carrier Frequency	2 GHz
Bandwidth	10 MHz
Distance of sub-carrier	15 kHz
FFT size	1024
No. of subcarriers	512
No.of cells	3
Cell radius	1 km
Modulation	BPSK
Convolutional codes rate	1/2
Symbols per frame	200
Length of symbol	1/20 ms

Table 2. Simulation Parameters

In performance evaluation, some parameters are taken as measurement, such as input SINR, output SNR, average number of iterations, etc. Moreover, we assume the number of signals in the received signal subspace equal to the source signal, and there are only four source signals mixed in linear that respectively from three different eNodeB and the thermal noise. One is the useful transmission signal, while the others are seen as the interference.

By the analysis, we compare the Max-SINR ICA algorithm with the classical Fast ICA algorithm, and the results illustrate that the advantage of the introduced algorithm is obvious. Specifically, the convergence of such two ICA algorithms is compared. Two situations are considered in simulation, which respectively the length of processing frame is fixed and the strength of thermal noise is fixed.

4.3.1 Convergence comparison

In order to compare the convergence of Max-SINR ICA algorithm with other classical ICA algorithms, we take the Fast ICA algorithm as an example, and simulation is performed. The result is shown in Fig. 17, where the average number of iterations is as a function of the input SINR ($SINR_{in}$). The length of processing frame is set as 50 symbols, while the input SNR (SNR_{in}) is set as 40 dB.

From Fig. 17, it can be seen that the convergence speed for the introduced Max-SINR ICA is faster than Fast ICA. With the same value of $SINR_{in}$, the average number of iterations for Max-SINR ICA is less than Fast ICA. Generally, the iteration number is determined by the search speed of the separation matrix W, which should satisfy the condition that $\left\| \left| W^T W \right| - 1 \right| \leq \varepsilon$. Moreover, ε is a positive real variable and arbitrarily small.

For Fast ICA algorithm, the Newton iteration is taken into search for the optimized solution, with a superlinear convergence speed. On the other side, it takes random vectors of unity length for the initial separation matrix $\hat{W}$, which averagely needs more iteration to search separation matrix compared with Max-SINR ICA.

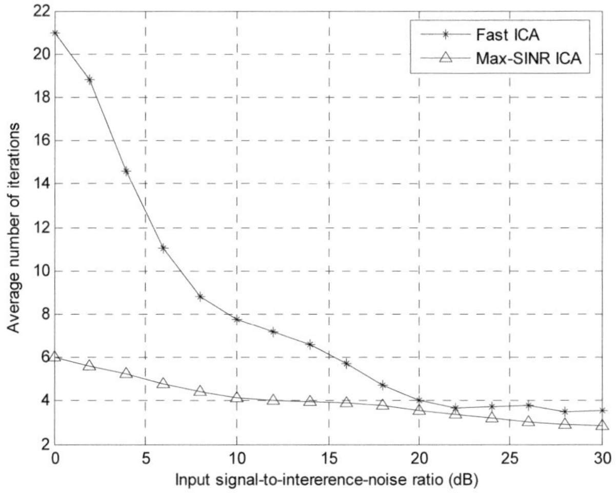

Fig. 17. Iterations of ICA algorithms

But for Max-SINR ICA algorithm, the analytical method is taken in searching for the initial separation matrix, which directly acquires a closed form solution by generalized eigenvalue decomposition. Then the Newton iteration is taken, which is based on the optimized initial separation matrix. Because there isn't an iterative process that search for the optimized initial separation matrix, its iterative steps are less than Fast ICA algorithm. Moreover, we can know from Theorem 1 that the initial $\hat{W}$ is an eigenvector matrix when SINR takes the maximum value in the objective function.

4.3.2 Fix the length of processing frame

In order to compare the performance of the introduced algorithm under different thermal noise, we fix the length of processing frame N, and let $N = 50$ symbols. In such situation, we only need to consider the effects of thermal noise in the mixed signals.

Furthermore, we give a definition about the output SNR (SNR_{out}) and the length of processing frame N, which is as the following equation.

$$SNR_{out} = 10\log\left\{\frac{1}{N}\sum_{k=1}^{N}\frac{u^2(k)}{[u(k)-\hat{u}(k)]^2}\right\} \tag{33}$$

In Eq. (33), $u(k)$ is the kth sample of useful signal $u(t)$, while $\hat{u}(k)$ is the estimated value of $u(k)$. On this basis, simulation is taken. The relationship of SNR_{out} and $SINR_{in}$ is illustrated in Fig. 17, where SNR_{out} is as a function of $SINR_{in}$ with different strength of thermal noise. Moreover, we select SNR_{in} to reflect the fluctuation of such strength. From Fig.17, it can be seen that SNR_{out} with both Max-SINR ICA algorithm and Fast ICA algorithm is robust with the increase of $SINR_{in}$.

As shown in Fig.18, we make $SINR_{in}$ vary from -10dB to 30dB, and SNR_{out} grows slowly with the increase of $SINR_{in}$. One reason is that the output SNR by ICA algorithm is affected by the mutual information among the source signals and the probability distribution of each signal. For such characteristics are determined, the limited change of $SINR_{in}$ plays a little effect in SNR_{out}. When SNR_{in} is equal to 40dB, SNR_{out} is around from 18dB to 22dB. But when SNR_{in} is equal to 10dB, SNR_{out} is around from 9dB to 14dB. Based on this analysis, it can be found that by means of ICA algorithm, the higher SNR_{in} is, the higher SNR_{out} is.

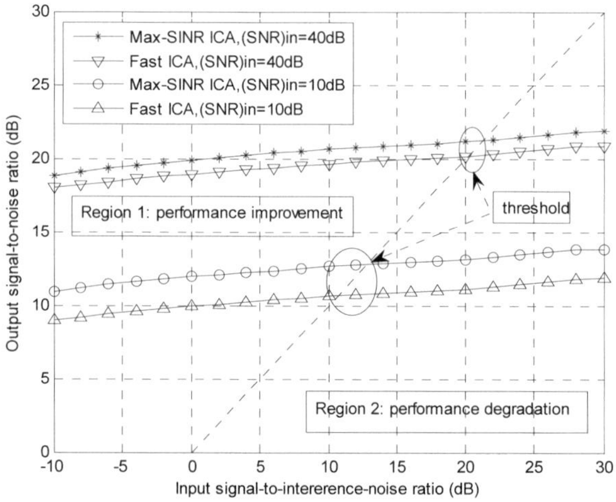

Fig. 18. SNR_{out} and $SINR_{in}$ (fix the length of processing frame)

On the other side, with the same condition, such as SNR_{in} is equal, the Max-SINR ICA algorithm shows a better performance than the Fast ICA algorithm. Especially, SNR_{out} improved by the Max-SINR ICA algorithm is a little more than SNR_{out} improved by the Fast ICA algorithm.

However, with the increase of $SINR_{in}$, the increase of SNR_{out} is still limited, whose growth rate is slower than $SINR_{in}$. As a result, when $SINR_{in}$ increases into some value, it reaches to balance: $SNR_{out} = SINR_{in}$. Moreover, this state is shown as the slash through the origin in Fig.18, which divides the graph into two regions: Region 1 and Region 2.

In Region 1, $SNR_{out} > SINR_{in}$, which means that the interference mitigation by ICA algorithm is effective. But in Region 2, $SNR_{out} < SINR_{in}$, which means that the interference mitigation by ICA algorithm is not only ineffective, but also degrades the performance worse as the growth of $SINR_{in}$.

Compared with the Fast ICA algorithm, the Max-SINR ICA algorithm raises the threshold $SINR_{in}$ of Region 1 and Region 2. It can be seen in Fig.18 that the threshold $SINR_{in}$ for the Max-SINR algorithm is a little larger, which means if $SINR_{in}$ is in this area, the performance is improved by the Max-SINR ICA algorithm, but degraded by the Fast ICA algorithm.

Fig. 19 shows the processing gain for such two ICA algorithms, when the length of processing frame is fixed. It can be found that the processing gain decreases with the increase of $SINR_{in}$. Besides, as $SINR_{in}$ continuously increases, we can set the area with the positive processing gain as Region 1, while the area with the negative processing gain as Region 2. Among Region 1 and Region 2 is the threshold line.

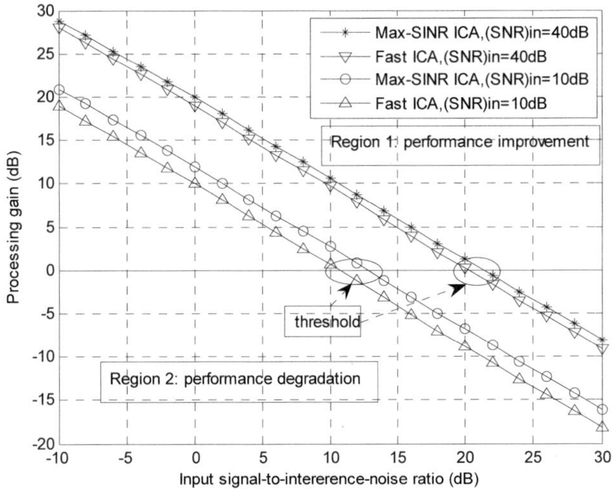

Fig. 19. Processing gain (fix the length of processing frame)

Specially, when $SINR_{in}$ is lower than the threshold, the processing gain is positive, which enables to improve the performance. What's important, the lower the $SINR_{in}$ is, the higher the processing gain is, which is useful to the users in cell-edge. But when $SINR_{in}$ is higher than the threshold, the processing gain is negative, which degrades the performance.

Compared with the performance brought by such two algorithms, the processing gain brought by the Max-SINR ICA algorithm is larger with the same SNR_{in}. Moreover, the introduced algorithm also raises the threshold $SINR_{in}$. When $SINR_{in}$ is among this area, the processing gain can be improved by the Max-SINR ICA algorithm, but degraded by the Fast ICA algorithm.

4.3.3 Fix the strength of thermal noise

In order to measure the effects brought by the length of processing frame, we fix the strength of thermal noise in the mixed signals, which is in a form of fixed signal to noise ratio, $SNR_{in} = 40dB$. Moreover, the simulation result is shown in Fig.20, and SNR_{out} is also set as a function of $SINR_{in}$ with different lengths of the processing frame.

In static simulation, we respectively take the length of the processing frame as 50 and 100, and the performance brought by such two ICA algorithms is compared. Further, it can be seen that the performance can be divided into two regions:

In Region 1, the performance is improved, where $SNR_{out} > SINR_{in}$. With the increase of $SINR_{in}$, it shows that for the same ICA algorithm, the longer the length of the processing frame is, the higher the SNR_{out} is. The reason is that the independence among source signals is easier to be established with longer processing frames. But in Region 2, the performance is degraded, where $SNR_{out} < SINR_{in}$, and it is degraded worse as $SINR_{in}$ increases gradually. Moreover, when the length of the processing frame is longer, the threshold $SINR_{in}$ between Region 1 and Region 2 also becomes a little higher.

The reason why Region 1 and Region 2 exist in Fig. 18 and Fig. 20 is that: The output SNR by ICA algorithm is mainly affected by the mutual information among the source signals and the probability distribution of each signal. Once such characteristics are determined in the mixed signals, the limited change of $SINR_{in}$ plays a little effect in SNR_{out}. At this time, as the growth of $SINR_{in}$, SNR_{out} increases slowly, such curve may gradually reach to the threshold. Before this threshold, it's Region 1. Else, it's Region 2.

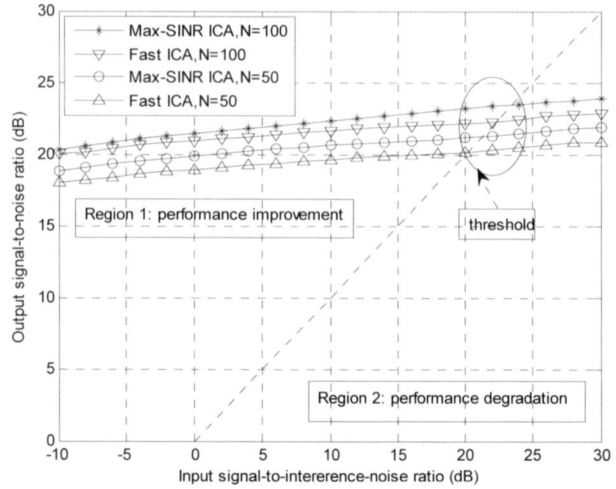

Fig. 20. SNR_{out} and $SINR_{in}$ (fix the strength of thermal noise)

Compared with the Fast ICA algorithm, both SNR_{out} and the threshold $SINR_{in}$ are raised by the Max-SINR ICA algorithm, with the same processing frame. From Fig. 20, it can be seen that in the threshold area, the performance is improved by the Max-SINR ICA algorithm, but degraded by the Fast ICA algorithm.

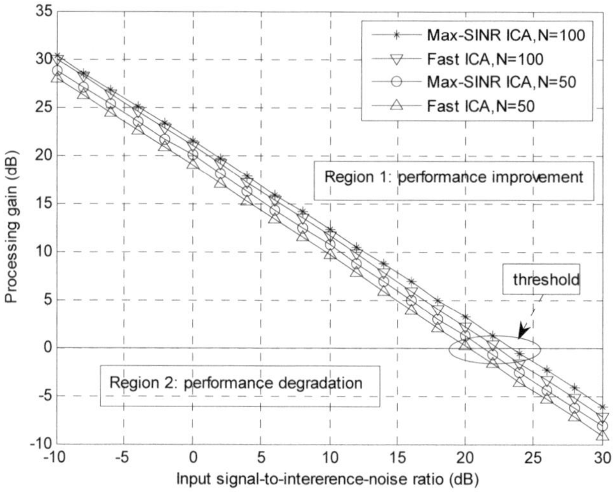

Fig. 21. Processing gain (fix the strength of thermal noise)

Fig.21 shows the processing gain for such two ICA algorithms, when the strength of thermal noise is fixed. It can be found that the processing gain decreases with the increase of $SINR_{in}$. In Region 1, the processing gain is positive, and enables to improve the performance. While in Region 2, the processing gain is negative, and degrades the performance. Similar to Fig.19, it also can be found from Fig.21 that the longer the length of the processing frame is, the higher the processing gain is.

Compared with the Fast ICA algorithm, both the processing gain and the threshold are raised by the Max-SINR ICA algorithm with the same processing frame. The conventional Fast ICA has forced the interference to zero, not considering the effect of the additive thermal noise. Meanwhile, the introduced algorithm minimizes both the interference and noise in order to maximize SINR. Thus the effect of the noise enhancement can be suppressed by the introduced algorithm, which gives the performance improvement.

Based on the above analysis, it's proper to use ICA algorithm under lower $SINR_{in}$, higher SNR_{in} and with longer lengths of the processing frame, which enables to mitigate the inter-cell interference, and improve the performance. Specially, it had better employ such inter-cell interference algorithm in practical application when the range of $SINR_{in}$ is below 10dB, but SNR_{in} is above 10dB.

On the other side, it is worth noting that the effect of user mobility isn't considered because of static simulation. Actually, when the length of processing frame is too large, such mobility can't be tracked for the Doppler frequency effect and time varying channel. In practice, the length of processing frame should be limited by the maximum speed of UE, which need to be researched by dynamic simulation in the future.

4.4 Summary

In order to cancel inter-cell interference, one inter-cell interference mitigation method is introduced, which is based on ICA algorithm. Compared to finding the maximum kurtosis in classical ICA algorithms, such as Fast ICA, Max-SINR ICA algorithm is introduced, which sets SINR as the objective function in this algorithm. As an important measured factor in interference mitigation, it need try to make such function get the maximum value. By optimize the initial separation matrix in iterations, the convergence speed of this introduced algorithm is faster than Fast ICA algorithm. Furthermore, two situations are divided in simulation, which respectively fix the length of processing frame and fix the strength of thermal noise.

By means of ICA algorithm, the output SNR increases as the growth of the input SINR, but the processing gain gradually decreases as the growth of the input SINR. Moreover, the lower the SINR is, the higher the output SNR and the processing gain are.

On the other side, as the growth of the input SINR, there are two regions for the performance. When the input SINR is lower than the threshold, the performance is improved. But when the input SINR is higher than the threshold, the performance is degraded.

Besides, the effects brought by the thermal noise and the length of the processing frame are considered. When the input SNR is higher in the mixed signals, the output SNR is higher. When the length of the processing frame is longer, the output SNR is also higher. What's more, compared with the Fast ICA algorithm, the Max-SINR algorithm raises the output SNR and the processing gain in the same conditions.

According to the above comparison, it can be found that this inter-cell interference cancellation method is performed well with lower SINR. So it's good to improve the quality of service for users in cell-edge where is always in the state of lower SINR. Another advantage is that this algorithm can be performed in a semi-blind state, with no precise knowledge of source signal and channel information. Moreover, it may not bring with extra interference, which is much better than many existing inter-cell interference cancellation algorithms.

5. Conclusion

In this chapter, the inter-cell interference mitigation for mobile communication system is analyzed and three kinds of solutions with inter-cell interference coordination, inter-cell interference prediction and inter-cell interference cancellation are introduced with system models, theoretical analyses and simulation results.

For interference coordination, Soft Fractional Frequency Reuse and Coordination Frequency Reuse schemes are introduced. Their frequency reuse factors are derived. Simulation results are provided to show the throughputs in cell-edge are efficiently improved compared with soft frequency reuse scheme.

The inter-cell interference prediction is an active interference mitigation method. The theoretical basis, which is the optimal estimation theory, is provided with including of two parts: time series and the optimal filter estimation. Besides, the steps of Box-Jenkins method are introduced in addition. The reliability is also analyzed by means of prediction accuracy, which is based on the relationship of the coherent time and the time delay.

For inter-cell interference cancellation, two major technologies are described in this chapter, which are space interference suppression and interference reconstruction/subtraction respectively. Based on the independent component analysis (ICA) technology in blind source separation, a semi-blind interference cancellation algorithm is introduced, named as Max-SINR ICA, which aims to improve the output SNR and optimize the initial iterative separation matrix. Simulation results show that the iterative convergence speed for Max-SINR ICA algorithm is faster than the traditional Fast-ICA algorithm. By the Max-SINR ICA algorithm, the inter-cell interference can be efficiently cancelled in a semi-blind state, especially with lower input SINR, higher input SNR and longer processing frame.

6. References

3GPP. (2005). R1-050507, Soft frequency reuse scheme for UTRAN LTE, Huawei. *3GPP TSG RAN WG1 Meeting* #41, Athens, Greece.

3GPP. (2005). R1-051396. Comparison of bit repetition and symbol repetition for inter-cell interference mitigation.

3GPP. (2006). R1-060416. Combining inter-cell-interference co-ordination/avoidance with cancellation in downlink and TP.

3GPP. (2006). R1-060518. TP for combining beam-forming with other inter-cell interference mitigation approaches.

3GPP. (2006). TR 25.814 v7.1.0, Physical layer aspects for evolved UTRA (Release 7).

3GPP. (2006). TR 25.913, Requirements for Evolved UTRA (E-UTRA) and Evolved UTRAN (E-UTRAN).

3GPP. (2007). TR 25.912. Feasibility Study for Evolved UTRA and UTRAN.

3GPP. (2008). R1-082024, A discussion on some technology components for LTE-Advanced, Ericsson. 3GPP TSGRAN WG1 #53, Kansas City, MO, USA.

3GPP. (2008). R1-083569, Further discussion on Inter-Cell Interference Mitigation through Limited Coordination, Samsung. 3GPP TSGRAN WG1 #54bits, Prague, Czech Republic.

3GPP. (2009). R1-091688, Potential gain of DL CoMP with joint transmission, NEC Group. 3GPP TSGRAN WG1 #57, San Francisco, USA.

3GPP. (2009).TR 36.814 v1.0.1, Further Advancements for E-UTRA Physical Layer Aspects (Release 9).

A. Hyvarienen, J. Karhunen, E. Oja. (2001). Independent Component Analysis, John Wiley and Sons.

Haipeng Le, Lei Zhang, Xin Zhang, and Dacheng Yang. (2007).A Novel Multi-Cell OFDMA System Structure using Fractional Frequency Reuse, In *Proc. of IEEE PIMRC* 2007, pp.1-5.

Hanbyul Seo and Byeong Gi Lee. (2004). A proportional-fair power allocation scheme for fair and efficient multiuser OFDM systems, In *Proc. of IEEE Globecom* '04 ,vol.6, pp.3737-3741.

H.L. Bertoni. (2000). Radio propagation for modern wireless systems. Prentice Hall, Inc.

Huiling Jia, Zhaoyang Zhang, Guanding Yu, Peng Cheng, and Shiju Li. (2007).On the Performance of IEEE 802.16 OFDMA System under Different Frequency Reuse and Subcarrier Permutation Patterns, In *Proc. of IEEE International conference on communications*, ICC 07', pp.5720-5725.

Hui Zhang, Xiaodong Xu, Xiaofeng Tao, Ping Zhang. (2009). An Inter-Cell Interference Mitigation Method for OFDM-Based Cellular Systems Using Independent Component Analysis. *IEICE Transactions on Communications*, Vol.E92-B, No.10.

Hui Zhang, Xiaodong Xu, Jingya Li, Xiaofeng Tao. (2009). Multicell Power Allocation Method based on Game Theory for Inter-Cell Interference Coordination. *Science in China, Series F: Information Sciences*, Vol.52, No.12, pp: 2378-2384.

Hui Zhang, Jingya Li, Xiaodong Xu, Shuang Wang, Ping Zhang. (2009). Multi-cell Subcarrier Allocation based on Interference Forecast by Kalman Filter. *Journal of Beijing University of Posts and Telecommunications*, Vol.32, No.3, pp.86-90.

Hui Zhang, Xiaodong Xu, Jingya Li, and Xiaofeng Tao.(2010) Subcarrier Resource Optimization for Cooperated Multipoint Transmission. *International Journal of Distributed Sensor Networks*, vol. 2010.

Hui Zhang, Jingya Li, Xiaodong Xu, Tommy Svensson. (2009). Channel Allocation based on Kalman Filter Prediction in Downlink OFDMA Systems. *IEEE VTC 2009*-Fall.

Hyvarinen, A. (1999). Fast and robust fixed-point algorithms for independent component analysis. *IEEE Trans. on Neural Networks*, vol.10, no. 3, pp. 626 - 634.

I. Kostanic, W. Mikhael. (2002).Rejection of the co-channel interference using non-coherent independent component analysis based receiver, *Proc. IEEE Midwest Symposium on Circuits and Systems*, vol. 2, Aug.2002.

I.Kostanic, W.Mikhael. (2004).Blind source separation technique for reduction of co-channel interference. *IEE Electronics Letters*, Vol. 38, No. 20, pp.1210 – 1211.

J.G.Andrews. (2005). Interference cancellation for cellular systems: a contemporary overview. *IEEE Wireless Commun.* Magazine, vol.12, no. 2, pp. 19 - 29.

J. Tomcik. (2006). Qualcomm, MBFDD and MB TDD wideband mode, *IEEE* 802.20-05/68r1.

K.N. Lau, K. Yu, K. Ricky.(2006). Channel adaptive technologies and cross layer designs for wireless systems with multiple antennas theory and applications. Canada: John Wiley & Sons, Inc., Publication, pp. 1-503.

K.I. Lee, Y.H. Ko. (2006). An inter-cell interference cancellation method for OFDM cellular systems using a subcarrier-based virtual MIMO. *Proc. IEEE VTC*, pp. 1-5.

Ki Tae Kim, Seong Keun Oh. (2007).A Universal Frequency Reuse System in a Mobile Cellular Environment, In *Proc. of IEEE VTC* 2007-Spring, pp.2855-2859.

Ki Tae Kim, Seong Keun Oh. (2008).An Incremental Frequency Reuse Scheme for an OFDMA Cellular System and Its Performance, In *Proc. of IEEE VTC* 2008-Spring, pp.1504-1508.

K.W. Park, K.I. Lee, Y.S. Cho. (2006). An inter-cell interference cancellation method for OFDM-Based cellular systems using a virtual smart antenna. *IEICE Trans. on Commun.*, vol. E89-B, no.1, pp. 217-219.

L. Prarra, P. Sajda.(2003). Blind source separation via generalized eigenvalue decomposition. *Journal of Machine Learning Research*, no.4, pp. 1261-1269.

M. Barkat. (2005). Signal Detection and Estimation. Artech House Publishers.

Q.H. Spencer, C.B. Peel. (2004). An introduction to the multi-user MIMO downlink. *IEEE Commun. Magazine*, vol.42, pp. 60-67.

S.E. Elayoubi, O. B. Haddada, and B. Fourestie. (2008). Performance Evaluation of Frequency Planning Schemes in OFDMA-based Networks, *IEEE Trans. Wireless Commun.*, vol. 7, no.5, pp. 1623–1633.

S.R. Curnew, J. How. (2007). Blind signal separation in MIMO OFDM systems using ICA and fractional sampling. *Proc. International Symposium on Signals, Systems and Electronics*, pp. 67-70.

T. Ristaniemi, J. Joutsensalo. (1999). Nonlinear algorithm for blind interference cancellation, *Proc. IEEE Signal Processing Workshop on Higher-Order Statistics*, pp.43-47.

T. Yang. (2004). Diversity wireless receivers with efficient co-channel interference suppression. *Proc. IEEE Advances in Wired and Wireless Communication*, pp.145-147.

Xu Fangmin, Tao Xiaofeng, Zhang Ping. (2009). A Frequency Reuse Scheme for OFDMA Systems, *Journal of Electronics&Information Technology*,vol. 3, no.4, pp.903-906.

Xu Xiaodong, Zhang Hui, Li Jingya, Tao Xiaofeng, Zhang Ping. (2009). An Improved Exponential Distributed Power Control Algorithm for MIMO Cellular Systems. *IEEE WiCOM* 2009.

21

Novel Co-Channel Interference Signalling for User Scheduling in Cellular SDMA-TDD Networks

Rami Abu-alhiga[1] and Harald Haas[2]
The University of Edinburgh
United Kingdom

1. Introduction

The latest advancements of the 3rd generation (3G) universal mobile telecommunications system (UMTS) have led to the long term evolution (LTE) standard release (referred to as 3.9G) within the 3rd generation partnership project (3GPP). LTE does not meet the requirements for the fourth generation (4G) systems defined by the international telecommunication union (ITU). Therefore, work on LTE-Advanced within 3GPP has recently started. LTE-Advanced can be seen as the continuous evolution of wireless service provision beyond voice calls towards a true ubiquitous air-interface capable of supporting multimedia services (Sesia et al., 2009).

LTE-Advanced systems face a number of essential requirements and challenges which include coping with limited radio resources, increased user demand for higher data rates, asymmetric traffic, interference-limited transmission, while at the same time the the energy consumption of wireless systems should be reduced. Driven by the ever-growing demand for higher data rates to effectively use the mobile Internet, future applications are expected to generate a significant amount of both downlink (DL) and uplink (UL) traffic which requires continuous connectivity with quite diverse quality of service requirements. Given limited radio resources and various propagation environments, voice over IP applications, such as Skype, and self-generated multimedia content platforms, such as YouTube, and Facebook, are popular examples that impose a major challenge on the design of LTE-Advanced wireless systems. One of the latest studies from ABI Research, a market intelligence company specializing in global connectivity and emerging technology, shows that in 2008 the mobile data traffic around the world reached 1.3 Exabytes (10^{18}). By 2014, the study expected the amount to reach 19.2 Exabytes. Furthermore, it has been shown that video streaming is one of the dominating application areas which will grow significantly (Gallen, 2009).

In order to meet such diverse requirements, especially, the ever-growing demand for mobile data, a number of different technologies have been adopted within the LTE-Advanced framework. These include smart antenna (SA)-based (also known as directional antennas or antenna arrays) multiple-input multiple-output (MIMO) systems (Bauch & Dietl, 2008a;b; Foschini & Gans, 1998; Kusume et al., 2007) and efficient multiuser transmission techniques such as multiuser MIMO using precoding to achieve, for example, space division multiple access (SDMA) (Fuchs, et al., 2007), and networked MIMO, *i.e.* coordinated multipoint (CoMP) systems. Therefore, there is a broad agreement recently among LTE standardization groups that MIMO will be the key to achieve the promised data rates of 1 Gbps and more (Seidel, 2008).

It is well known that co-channel interference (CCI), caused by frequency reuse, is considered as one of the major impairments that limits the performance of current and 4G wireless systems (Haas & McLaughlin, 2008). To outmaneuver such obstacle, various techniques such as joint detection, interference cancelation, and interference management have been proposed. One of the most promising technology is to utilize the adaptability of SAs. Spatial signal pre-processing along with SAs can provide much more efficient reuse of the available spectrum and, hence, an improvement in the overall system capacity. This gain is achievable by adaptively utilizing directional transmission and reception at the base station (BS) in order to enhance coverage and mitigate CCI. One of the key challenges to overcome, however, is the signalling overhead which increases drastically in MIMO systems.

Unlike the traditional resource allocation in single-input single-output (SISO) fading channels, which is performed in time and frequency domains, the resources in MIMO systems are usually allocated among the antennas (the spatial domain). From closed-loop MIMO point of view, channel aware adaptive resource allocation has been shown to maintain higher system capacity compared to fixed resource allocation (Ali et al., 2007; Gesbert et al., 2007; Koutsimanis & Fodor, 2008). In particular, adaptive resource allocation is becoming more critical with scarce resources and ever-increased demand for high data rates.

It is shown that for closed-loop MIMO the optimal power allocation among multiple transmit antennas is achieved through the water-filling algorithm (Telatar, 1999). However, to enable optimal power allocation, perfect channel state information (CSI) at the transmitter is required. Some other work focused on transmit beamforming and precoding with limited feedback (Love, et al., 2005; 2003; Mukkavilli et al., 2002; 2003; Zhou et al., 2005), where the transmitter uses a quantized CSI feedback to adjust the power and phases of the transmitted signals. To further reduce the amount of feedback and complexity, different strategies such as per-antenna rate (an adaptive modulation and coding approach that controls each antenna separately) and power control algorithms have been proposed (Catreux et al., 2002; Chung et al., 2001a;b; Zhou & Vucetic, 2004; Zhuang et al., 2003). By adapting the rate and power for each antenna separately, the performance (error probability (Gorokhov et al., 2003) or throughput (Gore et al., 2002; Gore & Paulraj, 2002; Molisch et al., 2001; Zhou et al., 2004)) can be improved greatly at the cost of slightly increased complexity. Additionally, antenna selection is proposed to reduce the number of the spatial streams and the receiver complexity as well. Various criteria for receive antenna selection or transmit antenna selection are presented, aiming at minimizing the error probability (Bahceci et al., 2003; Ghrayeb & Duman, 2002; Gore et al., 2002; Gore & Paulraj, 2002; Heath & Paulraj, 2001; Molisch et al., 2003) or maximizing the capacity bounds (Molisch et al., 2003; Zhou & Vucetic, 2004). It is shown that only a small performance loss is experienced when the transmitter/receiver selects a good subset of the available antennas based on the instantaneous CSI (Zhou et al., 2004). However, it is found that in spatially correlated scenarios, proper transmit antenna selection cannot just be used to decrease the number of spatial streams, but can also be used as an effective means to achieve multiple antenna diversity (Heath & Paulraj, 2001). When the channel links exhibit spatial correlation (due to the lack of spacing between antennas or the existence of small angular spread), the degrees of freedom (DoF) of the channel are usually less than the number of transmit antennas. Therefore, using transmit antenna selection, the resources are allocated only to the uncorrelated spatial streams so that an enhanced capacity gain can be achieved.

Most of the above work focused on the point-to-point (P2P) link in single user scenario. In a multiuser MIMO (MU-MIMO) context, MIMO communication can offer significant capacity growth by exploiting spatial multiplexing and multiuser scheduling. Therefore, opportunistic approaches have recently attracted considerable attention (Choi et al., 2006; Viswanath et al., 2002). So far, opportunistic resource allocation in a MU-MIMO scenario is still an open issue. Wong *et al.* and Dai *et al.* (Dai et al., 2004; Wong et al., 2003) consider a multiuser MIMO system and focused on multiuser precoding and turbo space-time multiuser detection, respectively. More recent work has addressed the issue of cross-layer resource allocation in DL MU-MIMO systems (Wang & Murch, 2005). In broadcast MU-MIMO channels, dirty-paper coding (DPC) (Costa, 1983) can achieve the maximum throughput (Goldsmith et al., 2003; Vishwanath et al., 2003; Weingarten et al., 2004). In particular, DPC can accomplish this by using successive interference precancelation through employing complex encoding and decoding. Unfortunately, DPC is classified as a nonlinear technique that has very high complexity and is impractical. Due to the fact that DPC is computationally expensive for practical implementations, its contribution is primarily to determine the achievable capacity region of MU-MIMO channel under a per-cell equal power constraint. Therefore, many alternative practical precoding approaches are proposed to offer a trade-off complexity for performance (Airy et al., 2006; Chae et al., 2006; Hochwald et al., 2005; Pan et al., 2004; Shen et al., 2005; Windpassinger et al., 2004). These alternatives considered different criteria and methods such as minimum mean squared error (MMSE) (Schubert & Boche, 2004; Shi et al., 2008), channel decomposition, and zero forcing (ZF) (Chen et al., 2007; Choi & Murch, 2004; Spencer et al., 2004; Wong et al., 2003).

One of the most attractive approaches is the block diagonalization (BD) algorithm which supports orthogonal multiple spatial stream transmission. In BD algorithm, the precoding matrix of each user is designed to lie in the null space of all remaining channels of other in-cell users, and hence the intracell multiuser CCI is pre-eliminated (Chen et al., 2007; Shen et al., 2005; Spencer et al., 2004). In particular, SA-based SDMA, implementing BD algorithm, can multiplex users in the same radio frequency spectrum (*i.e.* same time-frequency resource) within a cell by allocating the channel to spatially separable users. This can be done while maintaining tolerable, almost negligible, intracell CCI enabled by BD signal pre-processing capabilities. Moreover, channel aware adaptive SDMA scheme can be achieved through joint exploitation of the spatial DoF represented by the excess number of SAs at the BS along with multiuser diversity. Generally, the radio channel encountered by an array of antenna elements is referred to as beam. In other words, SA technology along with BD algorithm can enable the BS to adaptively steer multiple orthogonal beams to a group of spatially dispersed mobile stations (MSs) (Choi et al., 2006), as depicted in Fig. 1.

The joint beam selection and user scheduling for orthogonal SDMA-TDD (time division duplex) system is a key problem addressed in this chapter. From precoding point of view, the availability of CSI of all in-cell users at the BS is crucial in multiuser (MU)-MIMO communication scenario to optimally incorporate different precoding techniques such as BD, adaptive beamforming, or antenna selection, in order to increase the overall system spectral efficiency. Basically, there are two methods for providing a BS with CSI of all associated MSs, namely limited (quantized) feedback and analog feedback. Limited feedback (also know as direct feedback) involves the MS to measure the DL channel and to transmit a feedback messages of quantized CSI reports to the BS during the UL transmission. Alternatively, the second method, referred to as UL channel sounding according to LTE terminology, involves the BS to estimate the DL channel based on channel response estimates obtained

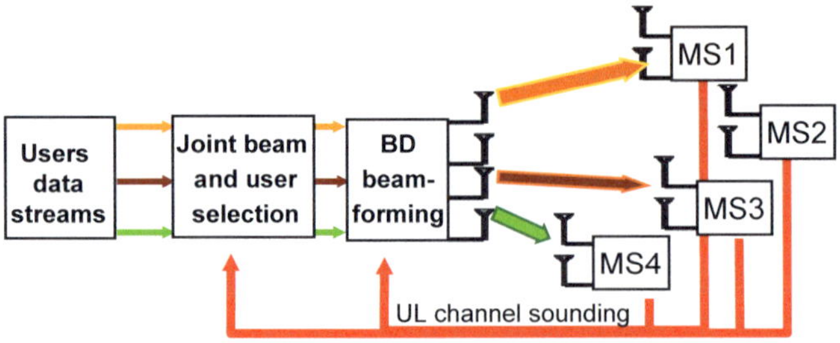

Fig. 1. A block diagram of SA-based MU-MIMO transmission implementing BD beamforming

from reference signals (pilots) received from the MS during UL transmission. Channel sounding offers advantages in terms of overhead, complexity, estimation reliability, and delay. Closed-loop SDMA-TDD networks can benefit from these advantages to avoid outdated feedback scenarios, enhance the network throughput, and reduce the computational cost at the user side. Clearly, TDD systems offer a straightforward way for the BS to acquire the CSI enabled through channel reciprocity (Love, et al., 2004). The advantages of UL channel sounding are discussed later in this chapter and a more detailed treatment can be found, for instance, in the technical documents of the evolved universal terrestrial radio access (EUTRA) study item launched in the LTE concept (Sesia et al., 2009).

In summary, UL channel sounding method is considered as one of the most promising feedback methods for SA-based SDMA-TDD systems due to its bandwidth and delay efficiency. In particular, UL channel sounding avoids the usage of dedicated feedback physical channels which results in utilizing the available bandwidth for data transmission much more effectively. In addition, UL channel sounding requires a shorter duration of time to convey the feedback information to the BS compared to the direct feedback method. This feature reduces the probability of having outdated feedback especially in fast varying channel conditions.

In interference-limited scenarios and according to Shannon capacity formula, the system performance is limited by the CCI from adjacent cells. Meanwhile, conventional channel sounding (CCS) only conveys the channel state information (CSI) of each active user to the BS. Therefore, CSI is only a suboptimal metric for multiuser spatial multiplexing optimization in interference-limited scenarios.

In light of the above, the benchmark system considered in this chapter for the system level analysis of the feedback methodology is a closed-loop SDMA TDD system. In the benchmark system, a BD technique is utilized to optimize the MU-MIMO spatial resources allocation problem based on perfect instantaneous CSI of each in-cell active user obtained from UL channel sounding pilots. The main goal of the benchmark system is to adaptively communicate with a group of users over disjoint spatial streams while optimizing the gains of the MU-MIMO channels. The optimization aims at enhancing the overall system capacity using fixed and uniformly distributed transmit power.

Most of the ZF-based precoding algorithms (e.g. BD) have been designed to *only* mitigate intracell CCI from different users in the same cell without considering CCI coming from transmitters in neighbouring cells. In a cellular environment, especially when full frequency reuse is considered, intercell (also known as other-cell) CCI becomes a key challenge which cannot be eliminated by BD-like algorithms. Moreover, it is shown that intercell CCI can significantly degrade the performance of SDMA systems (Blum, 2003). More specifically, if the

BS schedules a group of users only based on the available CSI, the scheduling decision may be *optimum* for a noise limited system, but high intercell CCI at the respective MSs might render the scheduling decision greatly suboptimum. Therefore, the signal-to-interference-plus-noise ratio (SINR) would be a more appropriate metric in multicell interference limited scenarios, but this metric cannot directly be obtained from CCS. Thus, the key challenge here is to provide knowledge of intercell CCI observed by each user to the BS in addition to CSI. If, furthermore, intercell CCI observed by each SA at the BS itself is taken into account, the beam selection and user scheduling process can be jointly improved for both DL and UL (Abualhiga & Haas, 2008).

2. Contributions and assumptions

The contribution associated with the feedback-based interference management for SDMA presented in this chapter can be split into three main parts:

- A novel interference feedback mechanism is developed. Specifically, it is proposed to weight the UL channel sounding pilots by the level of the received intercell CCI at each MS. The weighted uplink channel sounding pilots act as a bandwidth-efficient and delay-efficient means for providing the BS with both CSI *and* intercell CCI experienced at each active user. Such modification will compensate for the missing interference knowledge at the BS when traditional UL channel sounding is used. In addition, through exploitation of channel reciprocity the technique will act as implicit inter-cell interference coordination (ICIC) avoiding any additional signalling between cells.
- A novel procedure is developed to make the interference-weighted channel sounding (IWCS) pilots usable for the scheduler to optimize the spatial resource allocation during the UL slot. It is proposed to divide the metric obtained from the IWCS pilots by the intercell CCI experienced at the BS. The resulting new metric, which is implicitly dependent on DL and UL intercell CCI, provides link-protection awareness and it is used to jointly improve spectral efficiency in UL as well in DL.
- Finally, in order to facilitate a practical implementation, a heuristic algorithm (HA) is proposed to reduce the computational complexity to solve user scheduling problem.

The key assumptions for the system level analysis of the IWCS pilots performance can be summarized as follows:

- The considered closed-loop SDMA system enjoys perfect knowledge of the MIMO channel coefficients of each active user. Hence, this channel knowledge at both BS and MS is exploited to decompose the channel matrix into a collection of uncoupled parallel SISO channels.
- The considered problem of jointly adapting the MU-MIMO link parameters for a set of flat fading co-channel interfering MIMO links exploits two DoF: transmit antenna selection, and user selection. Since these two DoF are associated with two different layers (the physical (PHY) layer and medium access control (MAC) layer) the problem is considered to be optimized in a cross-layer fashion.
- The time and frequency DoF (*e.g.* frequency channel dependent scheduling and dynamic frequency resource allocation) are not considered in this study.
- This chapter assumes that appropriate methods are in place that completely eliminate or avoid intracell CCI. Therefore, the system is only limited by intercell CCI. However,

the level of intercell CCI usually outweighs thermal noise and the system is, therefore, interference limited.

- According to the definition of the UL channel sounding mechanism adopted by LTE, channel sounding pilots are different from the demodulation pilots dedicated to the process of coherent data detection. This implies that modifying the UL channel sounding pilots does not hamper the channel estimation processes required for coherent data detection. In particular, the only purpose of the proposed modification on the UL channel sounding pilots is to add interference awareness to the channel sounding technique. In addition, according to the LTE technical documents related to the UL channel sounding pilots, the predetermined sounding waveforms are transmitted using orthogonal signals among all active users in all cells using the same frequency band. The sounding pilot sequences are chosen to be orthogonal in frequency domain among all of the users' antennas (Sesia et al., 2009). In summary, the above properties enable the BS to estimate the UL wideband channel for each antenna of each active user without any intracell or intercell CCI between the channel sounding pilots. Moreover, errors in the channel estimation due to the presence of noise is beyond the scope of this chapter. As a consequence, perfect channel estimation is considered as outlined in the first assumption.

3. Overview of feedback methods

Basically, there are two methods for providing a BS with the CSI of all MSs, namely direct feedback and UL channel sounding.

1. **Direct feedback**: According to LTE terminology this feedback method is termed codebook-based feedback (Abe & Bauch, 2007). The MS determines the best entry in a predefined codebook of precoding (beamforming) vectors/matrices and transmits a feedback indication to the BS conveying the index value. In codebook feedback, the MS uses downlink channel estimates to determine the best codebook weight or weights for the BS to use as a precoding vector/matrix. The MS creates a feedback indication that includes the codebook index and then sends the feedback indication to BS. This method can be considered as a candidate option for frequency division duplex (FDD) systems which require an explicit transfer of the DL CSI during the UL transmission due to the absence of channel reciprocity.

   It is worth mentioning that the physical feedback channel needs to have some reference signals to facilitate the coherent detection of the feedback information at the BS.

2. **UL channel sounding**: The MS transmits a sounding waveform on the UL and the BS estimates the UL channel of the MS from the received sounding waveform. The sounding pilot sequences are chosen to be orthogonal between all of the users' antennas and also are designed to have a low peak to average power ratio (PAPR) in the time domain, (Fragouli, et al., 2003; Popovic, 1992). The details of UL channel sounding are given in 3GPP technical documents (Sesia et al., 2009). However, a brief treatment of the uplink channel sounding signal model is given below.

   According to LTE technical documents related to uplink channel sounding, the BS instructs the MS where and how to sound (*i.e.*, send a known pilot sequence) on the uplink. The information obtained from uplink channel sounding at the BS is used to determine DL beamforming weights for MIMO channel dependent scheduling on the uplink, as well as for MIMO channel dependent scheduling on the DL. According to the structure discussed in 3GPP technical documents, channel sounding pilots enable the BS to estimate the UL wideband channel for each antenna of each active user without any intracell or intercell CCI between the channel sounding pilots.

Any codebook-based feedback scheme must account for the number of SAs at the BS. In a codebook-based feedback scheme, the MS must be able to estimate the DL channel no matter how many SAs are available at the BS. Thus, the computational complexity at the MS, and the information that is required to be fed back increase with the number of antennas at BS. In contrast, channel sounding schemes are independent of the number of BS antennas. In other words, the problem of channel estimation is much more difficult in a codebook-based feedback scheme than in a channel sounding scheme. More specifically, in codebook-based feedback, the air-interface must enable the MS to estimate the channel between its antennas and a relatively larger number of SAs at the BS. Such estimation imposes a heavy processing load on a MS in a direct feedback scheme, while in a channel sounding scheme the estimation process takes place at the BS side (Hassibi & Hochwald, 2003).

For instant, consider the case where the BS has eight transmit antennas and the MS has a two receive antennas. In a channel sounding scheme, the BS must estimate the channel between its eight antennas and the two transmit antennas. In contrast, in a codebook-based feedback scheme, the air interface must enable the MS to estimate the channel between its two antennas and the eight transmit antennas (an eight-source channel estimation problem, which is much more difficult).

In TDD systems codebook-based feedback schemes tend to have much higher latency between the time of the channel estimation and the time of the subsequent DL transmission. The resulting outdated CSI can have detrimental effects on the performance of closed-loop transmission schemes, especially in fast fading channels. In contrast, channel sounding reference signals can be transmitted at the end of the UL slot. They can directly be exploited for the subsequent DL transmission. For these reasons, in a TDD system, UL channel sounding is preferred over codebook-based feedback.

4. SDMA with block diagonalization adaptive beamforming

4.1 Overview of SA technology

Originally, SA pre-processing techniques were proposed for military communications. Due to the significant technological advancements over the past two decades, SA-based technologies have become a cost-efficient solution for commercial communication systems to overcome some of the major challenges such as multipath fading, CCI, and capacity limitations especially for the cell-edge users. By exploiting the spatial diversity and the spatial processing capabilities of SA, an efficient utilization of available bandwidth and, hence, an increased system spectral efficiency is facilitated.

This section highlights the major features of SA-based SDMA systems relevant for the main contributions in this chapter. More specifically, the review is aimed at the benchmark SDMA system considered in this chapter. Also, this section briefly describes the generalized BD beamforming method for multiuser SDMA system (Pan et al., 2004), where the BS transmits multiple spatially multiplexed independent data streams to a group of users selected according to a scheduling criterion. Due to physical size constraints at the user side, the MSs are assumed to be equipped with limited number of multiple omnidirectional antennas (OAs) (two throughout this chapter). This assumption is also convenient in order to maintain affordable cost and reduced complexity at the mobile. As depicted in Fig. 2, each SA

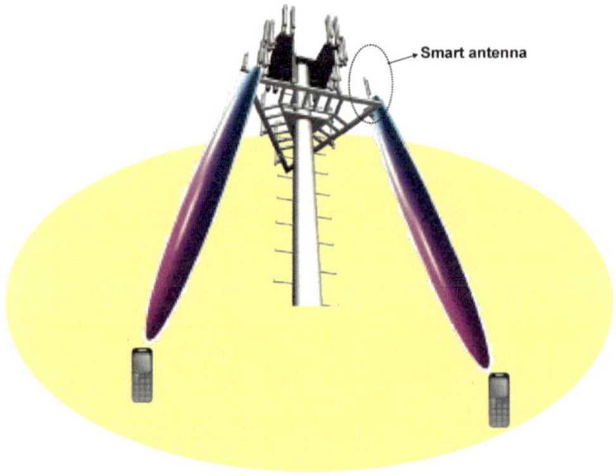

Fig. 2. A schematic illustration of a SA-based beamforming SDMA system

consists of an array of antenna elements and is dedicated to directionally transmit/recieve a single independent spatial stream. Such spatial stream is referred to as an effective antenna according to LTE terminology.

4.2 Multiuser MIMO BD system model

Consider a single-cell downlink MU-MIMO system where the BS is equipped with N_T transmit SAs, and communicates over multiple MIMO channels with K users. It is assumed that all users are equipped with the same number of receiving OAs denoted as N_R. To simplify the following analysis, it is assumed that $N_R < N_T$ and N_T is an integer multiple of N_R in order to serve all users. Fig. 3 illustrates an example of the considered SDMA system where $N_R = 2$.

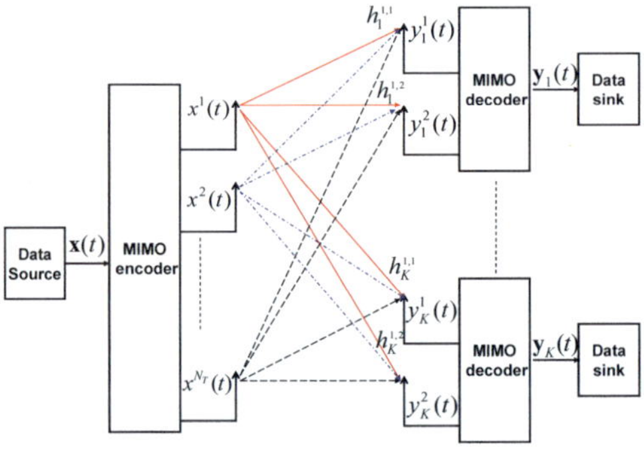

Fig. 3. SDMA system model with multiple MIMO users each equipped with 2 OAs

The flat fading MIMO channel matrix for user k is denoted as $\mathbf{H}_k$ where $h_k^{(j,i)}$ is the fading coefficient between the jth transmit antenna and the ith receive antenna of user k. For analytical simplicity, the rank r_k of $\mathbf{H}_k$ is assumed to be equal to $\min(N_R, N_T)$ for all users. Again, channel estimation errors caused by various reasons such as feedback delay, and feedback quantization error, *etc.* are beyond the scope of this chapter. Hence, it is assumed that the BS has perfect CSI for all users. By assuming that the number of data streams s_k to user k is equal to N_R, the transmitted data streams to user k can be denoted as a N_R-dimensional vector $\mathbf{x}_k$ where $\sum_{k=1}^{K} s_k \leq N_T$. Since CSI is available at both sides of the MIMO link, it is assumed that the MIMO transmission includes linear pre-processing and post-processing performed at both BS and MS, respectively. Prior to transmission, the data vector of user k, $\mathbf{x}_k$, is multiplied by a $N_T \times N_R$ precoding matrix $\mathbf{T}_k$. In this chapter, it is assumed that $\mathbf{T}_k$ is generated using the BD beamforming algorithm which is a member of the zero-forcing (ZF) type multiuser precoding algorithms (Spencer et al., 2004). At the BS, the data vectors of the K users are linearly superimposed and propagated over the channel from all N_T antennas simultaneously. It is also assumed that the elements of $\mathbf{x}_k$ are independent and identically distributed (i.i.d.) with zero mean and unit variance. The signal vector received at user k is

$$\mathbf{y}_k = \mathbf{H}_k \sum_{\iota=1}^{K} \mathbf{T}_\iota \mathbf{x}_\iota + \mathbf{n}_k \tag{1}$$

Equation (1) can be rewritten in the form of summation of the desired signal, the interference signal, and the noise signal as follows

$$\mathbf{y}_k = \mathbf{H}_k \mathbf{T}_k \mathbf{x}_k + \mathbf{H}_k \sum_{\iota \neq k}^{K} \mathbf{T}_\iota \mathbf{x}_\iota + \mathbf{n}_k \tag{2}$$

where the first term on the right-hand-side of (2) is the desired signal, the second term denotes the intracell CCI experienced by user k, and the last term $\mathbf{n}_k$ is the additive white Gaussian noise vector received by user k. According to key principles of BD algorithms, $\mathbf{T}_\iota$ is designed such that $\mathbf{H}_k \mathbf{T}_\iota = 0$ for $\forall \iota \neq k$ and , hence, the intracell CCI is completely eliminated (nulled). The block matrices $\mathbf{H}_S$ and $\mathbf{T}_S$ can be defined as the system channel matrix and the system precoding matrix, respectively, as follows:

$$\mathbf{H}_S = \left[\mathbf{H}_1^H \mathbf{H}_2^H \ldots \mathbf{H}_K^H\right]^H \tag{3}$$

$$\mathbf{T}_S = \left[\mathbf{T}_1 \mathbf{T}_2 \ldots \mathbf{T}_K\right] \tag{4}$$

Under the constraint of zero intracell CCI among users within the same cell, the optimal solution can be obtained by diagonalizing the product of (3) and (4) $\mathbf{H}_S \mathbf{T}_S$. Now, $\tilde{\mathbf{H}}_S$ can be defined as the block matrix of the MIMO links interfering with user k as follows

$$\tilde{\mathbf{H}}_S = \left[\tilde{\mathbf{H}}_1^H \ldots \tilde{\mathbf{H}}_{k-1}^H \tilde{\mathbf{H}}_{k+1}^H \ldots \tilde{\mathbf{H}}_K^H\right]^H \tag{5}$$

In light of the above, the intracell CCI free constraint requires that $\mathbf{T}_k$ is selected to lie in the null space of $\tilde{\mathbf{H}}_S$. The details of designing the BD precoding matrices and the associated constraints can be found in (Chen et al., 2007; Pan et al., 2004; Spencer et al., 2004).

5. Problem statement

To fully exploit multiuser diversity the following questions have to be addressed. In a spatial multiplexing opportunistic SDMA system with an excess number of SAs at the BS, how should the optimal set of spatially separable users be chosen? What is the appropriate allocation of the transmit/receive antennas (spatial beams) targeting the selected users? Since CCS pilots only provide a sub-optimal metric (*i.e.* CSI), how can a better metric be provided (*i.e.* instantaneous SINR) for such optimization problem while maintaining the same inherent feedback bandwidth and delay efficiency? For practical reasons such as cost and physical size, the number of SAs at the BS is greater than the number of OAs at the MS, as is the case

in EUTRA. Algorithms that achieve spatial multiplexing gains such as V-BLAST receivers (Foschini, 1996), however, require that the number of antennas at the receiver is greater than or equal to the number of transmit antennas. Consequently, the number of DL spatial streams is limited by the number of OAs at the MS side. This situation results in a spatial DoF for the selection of the subset of transmit SA for per user DL transmission.
Note that according to the fourth assumption in section 2, it is assumed that the BD algorithm can only eliminate intracell CCI. However, the system is still limited by intercell CCI. Therefore, throughout this chapter, whenever the term interference is mentioned, it refers to CCI originated from transmitters in the adjacent cells.

6. Interference-aware metric for downlink optimization

In the following we make use of channel reciprocity which is best available in the TDD mode. The new interference-weighted feedback concept proposed in this chapter is applied to the UL CCS pilots. The first major contribution of this chapter is the use of modified pilots, termed UL IWCS pilots, which are proposed to replace the UL CCS pilots. In particular, the CCS pilots are modified to become IWCS pilots by weighting (dividing) the UL CCS pilots by the magnitude of the intercell CCI received at each MS. The UL IWCS pilots are then used at the BS to extract the CSI plus the level of the intercell CCI experienced by the respective MS. Thereby, the SINR at each active MS is conveyed to the respective BS without any additional signalling overhead. The key idea of applying the interference-weighting concept to the CCS pilots is depicted in Fig. 4 for a SISO case.

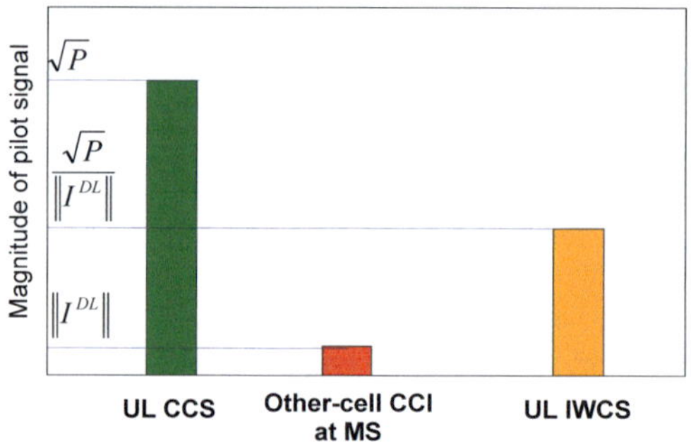

Fig. 4. Interference-weighting concept applied to the pilot signal in a SISO case

Consider an UL CCS transmission of a SDMA cellular interference-limited scenario with a BS equipped with N_{BS} SAs and K active users, each equipped with N_{MS} OAs. A narrowband flat fading channel is assumed, *i.e.*, a frequency subcarrier if orthogonal frequency division multiple access (OFDMA) or single carrier frequency division multiple access is used. We assume that both the BS and MSs experience sufficient local scattering. Therefore, both real and imaginary parts of the entries of $\mathbf{H}_k$ are samples of a zero-mean Gaussian distribution. Hence, the conventional channel sounding MS-to-BS pilots transmission can be modeled as follows:

$$\mathbf{y}_u(t) = \mathbf{H}_k(t)\mathbf{z}_k(t) + \mathbf{n}_k(t) \tag{6}$$

where t is the time index. The predetermined pilot signal $\mathbf{z}_k(t)$ is a N_{MS}-dimensional vector; the received signal $\mathbf{y}_k(t)$ is a N_{BS}-dimensional vector. Conventional channel sounding pilots can be used to estimate two metrics: distance-dependent link gain (link budget), and the multipath fading channel coefficients (small-scale fading). By weighting (6) by the intercell CCI experienced by user k in the downlink, the IWCS transmission can be written as follows:

$$\mathbf{y}_k(t) = \frac{\mathbf{H}_k(t)\mathbf{z}_k(t)}{\|I_k^{\mathrm{DL}}(t-1)\|} + \mathbf{n}(t) \tag{7}$$

where $\|I_k^{\mathrm{DL}}(t-1)\|$ is the amplitude of the intercell CCI experienced by k^{th} MS at $(t-1)^{\mathrm{th}}$ time interval. Consequently, the interference-weighting concept enables the sounding pilots to be used by the BS to obtain the DL interference-aware-metric $\mathbf{O}_k^{\mathrm{DL-IAM}}(t) \in \mathbb{C}^{N_{\mathrm{BS}} \times N_{\mathrm{MS}}}$ which can be formulated as follows

$$\mathbf{O}_k^{\mathrm{DL-IAM}}(t) = \frac{\mathbf{H}_k(t)}{\|I_k^{\mathrm{DL}}(t-1)\|} \tag{8}$$

From an information theoretic point of view, this metric (which can be considered as the square root of the instantaneous MIMO DL signal-to-interference ratio (SIR)) provides better feedback information compared to the CCS case. Consequently, the DL joint user scheduling and transmit SA selection can be improved since the DL interference-aware-metric can be used to obtain the SINR at the user side. Alternatively, quantized SINR can be fed back via the direct feedback method, but this requires transmission resources and longer time (potentially resulting in outdated feedback).

6.1 Interference estimation

From a practical point of view, the CCI experienced by a MS or a BS can be estimated by allowing the respective entity to sense the channel when no intended transmission is taking place. For instance, the CCI sensing period can be set for an active MS to be between the end of the DL period and the beginning of the switching time (DL to UL). Then, the MS can quantize the sensed CCI during the switching time period. Afterwards, a pilot weighted by the quantized magnitude of CCI can be transmitted during the subsequent UL period. This means that strong CCI will cause an 'artificial' attenuation of the pilot signal. This will pretend a bad channel at the BS, *i.e* the probability that the BS will schedule this particular resource block for this MS is reduced.

6.2 Optimization methodology

The general approach used in this chapter to maximize the sum capacity is a brute-force search. As in the example illustrated in Fig. 5, each BS, equipped with 4 SAs, will select 2 MSs each equipped with 2 OAs to access its spatial resources. Clearly, each MS experiences independent levels of CCI, and is subject to independent channel conditions on the desired link. For instance, the BS in cell 1 can schedule 2 users out of 3 possible candidates. For any active user, a BS can assign 2 out of 4 SAs to establish communication links. Using combinatorial basics, the BS has $\binom{3}{2}\binom{4}{2} = 18$ options to select antennas and user pairs in the given example.

The procedure followed to extract the optimization metrics provided by IWCS pilots is illustrated in the example depicted in Fig. 6 which is based on the example of Fig. 5. Basically, each way in which the BS can distribute the 4 SAs among 2 out of 3 users forms a possible solution. From Fig. 6, two arbitrary solutions are highlighted by shading them with squares of different colors and different styles for the borderline. By considering the solution shaded by blue squares of solid borderline, it can be seen that MS1 is allocated the first two SAs, while

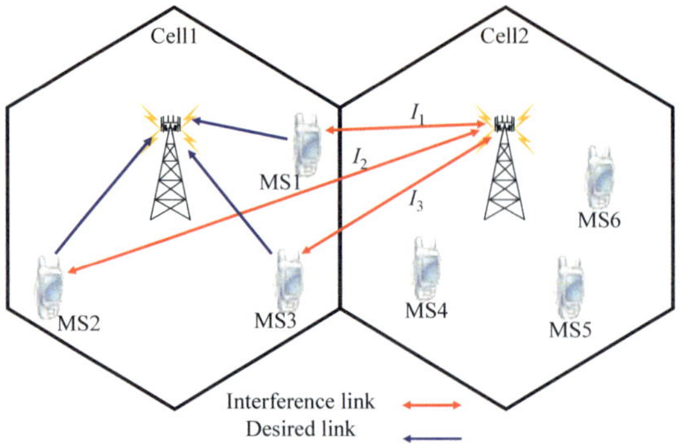

Fig. 5. Interference-limited multiuser MIMO system where each BS is equipped with 4 antennas, each MS is equipped with 2 antennas.

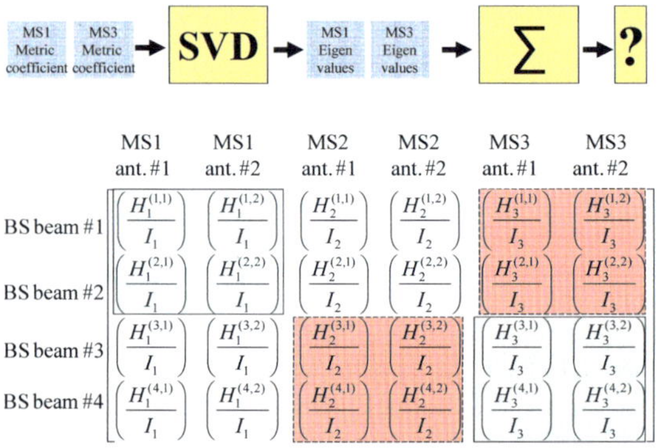

Fig. 6. Interference-limited virtual MU-MIMO matrix for the example illustrated in Fig. 5.

MS3 is allocated the last two antennas. Clearly, it can be seen that each SA allocation forms a (2×2) square matrix containing the coefficients of the optimization metric.

The next step is to obtain the eigenvalues of each square matrix via singular value decomposition (SVD). Afterwards, the eigenvalues of each group of selected users are summed. Finally, the summation of eigenvalues will be used to find the optimum solution among all possible solutions according to the different scheduling criteria. To examine all possible solutions, two approaches are considered: exhaustive search (ES) and HA. The details of the two approaches are discussed in section 8.

7. Link-protection-aware metric for uplink and downlink optimization

The main purpose of this section is to propose a method to accommodate the IWCS pilots to suit UL optimization. Since the amplitude of CCI experienced at the BS $\|I_j^{\mathrm{UL}}(t-1)\|$ (referred to as UL interference) where $(j \in 1, \ldots, N_{\mathrm{BS}})$ is different from the CCI experienced at the MS $\|I_k^{\mathrm{DL}}(t-1)\|$, the DL interference-aware-metric in (8) is not suitable for UL optimization. In order to use the IWCS pilots for UL optimization, the BS weights each row of the metric defined in (8) by the received interference at the associated SA $\|I_j^{\mathrm{UL}}(t-1)\|$ at the BS, which is assumed to be known at the BS side. Thus, the new optimization metric, referred to as link-protection-aware-metric, $\mathbf{O}_k^{\text{LPAM}}(t) \in \mathbb{C}^{1\times N_{\text{MS}}}$ can be formulated as follows:

$$\mathbf{O}_k^{\text{LPAM}}(t) = \frac{\mathbf{H}_k^{(j)}(t)}{\|I_k^{\text{DL}}(t-1)\|\|I_j^{\text{UL}}(t-1)\|} \tag{9}$$

where $\mathbf{H}_k^{(j)}(t)$ is the jth row in $\mathbf{H}_k(t)$.
Basically, this metric allows the BS to decide which subset of receive antennas it should use for user k, and which users should be grouped together. Since the link-protection-aware-metric is inversely proportional to $\|I_k^{\text{DL}}(t-1)\|$, a MS experiencing low interference level has higher chances to be selected. Due to channel reciprocity, a user that receives little interference from a set of users (in Tx mode) in a particular time slot, this user only causes little interference to the same set of users (now in Rx mode) in a different time slot. Therefore, the link-protection-aware-metric decreases the probability of scheduling users that are potential strong interferers. Consequently, this forms an inherent link-protection for the already established links in the neighboring cells. Similarly, this metric can be used for DL optimization to jointly select a subset of users experiencing favorable channel conditions and low intercell CCI to receive from a subset of SAs causing low intercell CCI to the neighboring cells. Note that the link-protection-aware-metric is simultaneously used to improve both UL and DL performance. Hence, the cross-layer scheduling for UL has to be the same as for DL, which reduces the scheduling time.
According to the MIMO literature (Foschini & Gans, 1998; Telatar, 1999), if the channel matrix $\mathbf{H}_k$ is known at the BS, then the instantaneous DL MIMO channel capacity of user k using fixed transmit power $\frac{P_t}{N_{\text{BS}}}$ per antenna can be expressed as the sum of capacities of r SISO channels each weighted with power gain λ_{k_i} where $(i \in 1,...,r)$, r is the rank of the channel, and λ_{k_i} are the eigenvalues of $\mathbf{H}_k\mathbf{H}_k^H$. P_t is the total transmit power at the BS. Assuming an interference-limited system, the instantaneous MIMO capacity of user k can be expressed as follows:

$$C_k = \sum_{i=1}^{r} \log_2(1 + \frac{P_t}{N_{\text{BS}} \times \|I_k^{\text{DL}}\|^2}\lambda_{k_i}) \tag{10}$$

By using the system model introduced above and (10), the primary objective is to find the optimum way, according to the scheduling criterion in use, in which the BS distribute N_{BS} antennas among $\frac{N_{\text{BS}}}{N_{\text{MS}}}$ spatially separable users out of user population of size K, where a selected user communicates with exactly N_{MS} SAs at the BS. For instance, in the case of a maximum sum rate scheduling criterion, the optimum solution maximizes the capacity of the multiuser MIMO channel at the expense of fairness.
The sample-space population (SP) (the size of the pool containing all possible solutions) of such problem is formulated later in this chapter. The resulting optimization problem can be written as follows:

$$C_{\max} = \arg\max_{v\in\text{SP}} \sum_{k=1}^{\frac{N_{\text{BS}}}{N_{\text{MS}}}} C_k^v \tag{11}$$

where SP are all possible choices of allocating the beams to the members of v, while v represents a possible choice of grouping the scheduled users.

7.1 Summary of the optimization metrics

In the DL, the following four metrics are used in conjunction with the different scheduling criteria (section IV summarizes them):

1 The scheduler only uses perfect instantaneous measurements of the MIMO multipath fading channel coefficients (ignoring the distance-dependent link-gain). This metric is fair by nature due to the uniform distribution of the users and the i.i.d. distributed fading, and it is referred to as DL blind-metric $\mathbf{O}_k^{\mathrm{DL-BM}}(t)$.

2 The scheduler uses instantaneous measurements of the distance-dependent link-gain. In contrast, this metric is greedy by nature because it tends, in most of the cases, to select those users which are closer to the BS, and it is referred to as DL link-gain-aware-metric $\mathbf{O}_k^{\mathrm{DL-LGAM}}(t)$.

Note, the first two metrics above are supported by the conventional channel sounding pilots and they do not provide the BS with information about interference at the mobile, $\|I_k^{\mathrm{DL}}(t-1)\|$.

3 The DL interference-aware-metric $\mathbf{O}_k^{\mathrm{DL-IAM}}(t)$ defined in (8) is used. This metric is supported by the IWCS pilots or by means of explicit signalling which, however, requires additional resources for signalling traffic.

4 To foster link-protection awareness the metric developed in (9), $\mathbf{O}_k^{\mathrm{LPAM}}(t)$, is used.

Two cases are examined for the UL:

1 The scheduler uses instantaneous measurements of the CSI of each spatial stream (each row of $\mathbf{H}_k$) divided by the interference level experienced by the associated SA at the BS. This metric, referred as UL interference-aware-metric, $\mathbf{O}_k^{\mathrm{UL-IAM}}(t) \in \mathbb{C}^{1\times N_{\mathrm{MS}}}$ can be expressed as follows:

$$\mathbf{O}_k^{\mathrm{UL-IAM}}(t) = \frac{\mathbf{H}_k^{(j)}(t)}{\|I_j^{\mathrm{UL}}(t-1)\|} \tag{12}$$

It is important to mention that this metric can be obtained using UL CCS pilots since UL interference $\|I_j^{\mathrm{UL}}(t-1)\|$ is available at the BS without feedback.

2 This is similar to the fourth case for the DL metrics which is defined as the link-protection-aware-metric in (9). This metric jointly considers UL and DL performance.

In summary, the results shown in section 10 are associated with four different metrics for DL transmission, and two different metrics for UL transmission.

7.2 Numerical example

To show the link-protection feature of the new metric defined in (9), a simple example is presented in Fig. 7. In this example, the arbitrary numbers quantifying the gain of each link and the interference experienced by each entity are used to estimate the achievable capacity using (10). It is assumed that cell 2 has an established DL transmission with MS3 and the argument for the achievable DL capacity is $\frac{\mathrm{H}_{\mathrm{MS3}}}{\mathrm{I}_{\mathrm{MS3}}} = \frac{9}{3}$. The neighboring BS, BS1, has got assigned the same time slot for UL transmission and it attempts to select between MS1 and MS2. If BS1 uses the UL interference-aware-metric, MS1 is scheduled for UL; $\frac{\mathrm{H}_{\mathrm{MS1}}}{\mathrm{I}_{\mathrm{BS1}}} = \frac{6}{2} > \frac{\mathrm{H}_{\mathrm{MS2}}}{\mathrm{I}_{\mathrm{BS1}}} = \frac{5}{2}$. Hence, the argument for the achievable UL capacity is $\frac{\mathrm{H}_{\mathrm{MS1}}}{\mathrm{I}_{\mathrm{BS1}}} = \frac{6}{2}$. As a result, it is assumed that the interference at MS3 increases to $\mathrm{I}_{\mathrm{MS3}} = 4.5$ due to the low shadowing

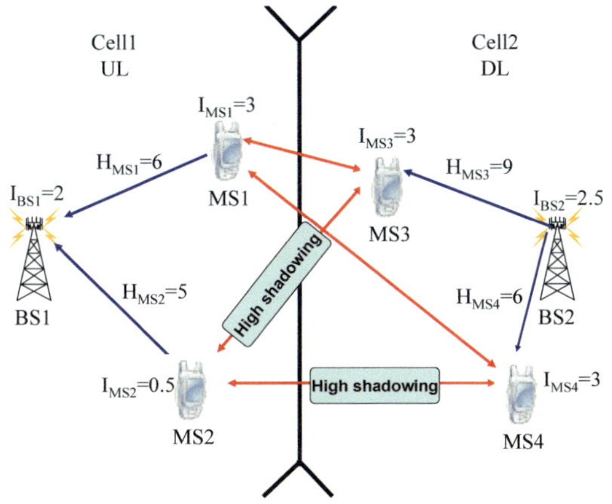

Fig. 7. Example of interference-limited 2-cell scenario with which the basic working principle of the link-protection-aware-metric is illustrated.

conditions between MS3 and MS1. This reduces the argument of the current achievable DL capacity in cell 2 to $\frac{9}{4.5}$. Thus, the arguments of the achievable system capacity are $\frac{6}{2}$ and $\frac{9}{4.5}$. In contrast, if BS1 uses the link-protection-aware-metric, MS2 is scheduled; $\frac{H_{MS1}}{I_{MS1} \times I_{BS1}} = \frac{6}{3 \times 2} < \frac{H_{MS2}}{I_{MS2} \times I_{BS1}} = \frac{5}{0.5 \times 2}$. Hence, the argument of the achievable UL capacity is $\frac{H_{MS1}}{I_{BS1}} = \frac{5}{2}$. Consequently, the interference at MS3 does not change (due to the high shadowing between MS2 and MS3), and therefore, the arguments of the achievable cell capacity are $\frac{5}{2}$ and $\frac{9}{3}$ for cell 1 and cell 2, respectively. In summary, in the first case the sum of the arguments of the cell capacity is 5 whereas in the second case the sum is 5.5. Clearly, it can be seen from this example that the link-protection-aware-metric used in the second case improves the overall spectral efficiency.

8. Heuristic algorithm for reduced computational complexity

8.1 Introduction

Generally the ES is not practical due to its computational complexity. Therefore, in this section a heuristic algorithm is proposed to reduce the involved complexity.

8.2 Exhaustive search mathematical model

The complexity of exhaustive search approach for scheduling optimization of SDMA system depends on the total number of users K, the number of SAs at the BS N_{BS}, and the number of antennas at each MS N_{MS}. The search burden for the scheduler is equivalent to the SP size. Using combinatorial fundamentals, (13) and (14) can be obtained. By comparing (13) with (14), it can be noticed that the multiuser diversity plays a significant role when $K > \frac{N_{BS}}{N_{MS}}$; due the fact that not all users can be scheduled.

If $K \leq \frac{N_{BS}}{N_{MS}}$

$$\text{SP} = \underbrace{\frac{N_{BS}!}{\left((N_{MS}!)^K \, (N_{BS} - K\, N_{MS})!\right)}}_{\text{SDMA DoF}}; \qquad (13)$$

if $K > \frac{N_{BS}}{N_{MS}}$

$$\mathrm{SP} = \underbrace{\frac{K!}{\left(\left(\frac{N_{BS}}{N_{MS}}\right)!\left(K-\frac{N_{BS}}{N_{MS}}\right)!\right)}}_{\text{Multiuser DoF}}\underbrace{\frac{N_{BS}!}{(N_{MS}!)^{\left(\frac{N_{BS}}{N_{MS}}\right)}}}_{\text{SDMA DoF}}. \tag{14}$$

It is important to mention that (13) and (14) are applicable only for scenarios in which N_{BS} is an integer multiple of N_{MS}. To gain insight into the influence of each dimension of the DoF on the SP, Fig. 8, Fig. 9(a), and Fig. 9(b) are obtained assuming N_{MS} to be 2. By comparing

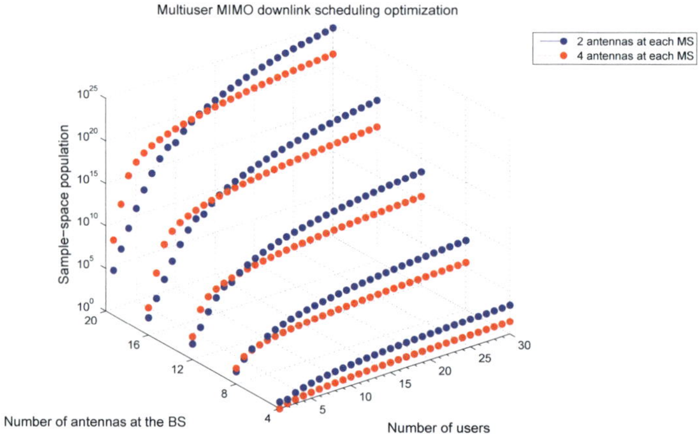

Fig. 8. The search complexity versus the number of SAs at the BS and the number of users, assuming 2 and 4 antennas at each MS

Fig. 9(a) with Fig. 9(b), it can be seen that SP grows polynomially with the number of users, and exponentially with the number of antennas at the BS (Learned et al., 1997). Clearly, the ES approach is computationally expensive for large number of users and antennas.

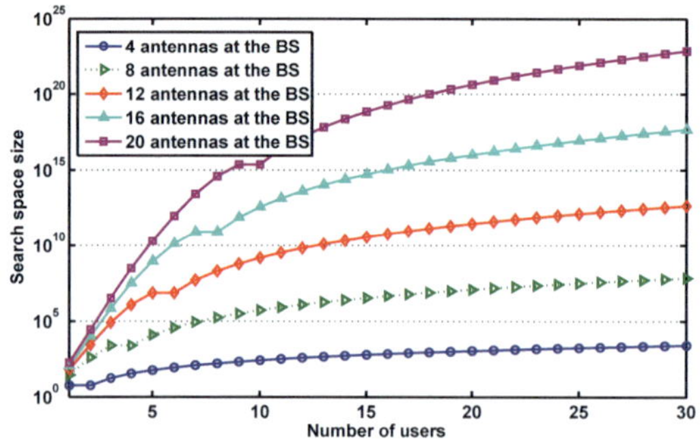

(a) Search space size versus the number of users

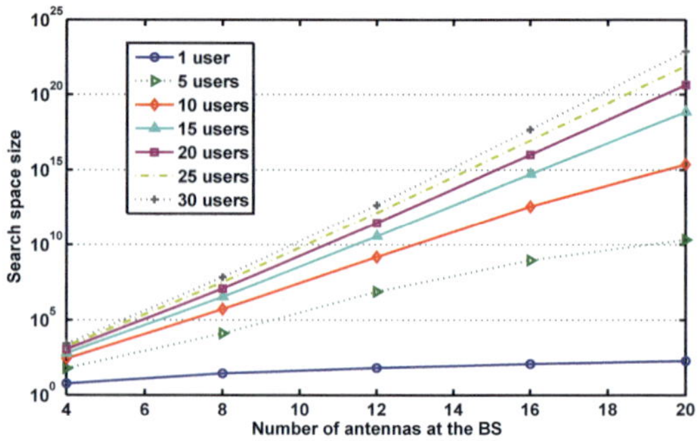

(b) Search size versus the number of SAs at the BS

Fig. 9. Search space size assuming 2 antennas at each MS

8.3 Heuristic to reduce complexity

A HA is developed to significantly reduce the computational complexity of finding a close to optimum solution. The algorithm exploits two basic principles. The first one is based on the elimination of users that experience excessive interference while the second is based on an angle of arrival (AoA) sectorization approach. The former aims to reduce the SP by suppressing those users exposed to interference levels greater than a predetermined interference threshold. The latter converts the cell of interest (CoI) optimization problem into smaller optimization problems, which can be solved simultaneously. This can be done by sectorizing the cell based on the AoA of the uplink channel sounding signal. Subsequently, the antennas at the BS are distributed among these sectors according to the user density of each sector. As a consequence, the original SP of the CoI will be reduced significantly as the complexity is split into smaller parallel search spaces.

8.4 Numerical examples

To demonstrate the working mechanism of the HA, let us consider a cell with 28 users, each equipped with 4 antennas, and the BS equipped with 16 antennas. According to (14), the scheduler has to search through approximately 3.027×10^9 possibilities to find the optimum solution. By applying the interference based user elimination approach, let us assume that 8 users have been found to be exposed to severe interference levels. Similarly, assume that the cell can be equally divided into four sectors with the same user density. According to the proposed HA, the original optimization problem becomes four identical sector optimization problems. Hence, each sector has 4 associated antennas at the BS, and $\frac{28-8}{4} = 5$ admissible users, each equipped with 4 antennas. Since only one user can be supported per sector, the scheduler has to search through 5 possibilities per sector, which can be done in parallel for all sectors.

To show the effect of the sectorization and the number of antennas at the MS, another example is used. In this example, a cell with 6 users, each equipped with 2 antennas, and the BS equipped with 12 antennas, is considered. Following a similar procedure as in the previous example, assume that all users are admissible, and that the cell can be equally sectorized into only 2 sectors with the same user density. Thus, each sector has 6 associated antennas at the BS, and 3 users, each equipped with 2 antennas. According to (13), the optimization problem with a search space of 3.992×10^7 is converted into 2 identical optimization problems, each with a search space of 120. It is worth noting in this example that multiuser diversity has not been exploited, and a simple sectorization technique is applied. However, the scheduler search burden is significantly reduced, and thus the practicability of the proposed interference-aware antenna selection and scheduling algorithm is greatly enhanced.

9. Scheduling criteria

Scheduling criteria can be generally classified into two groups: greedy and fair. In this chapter, the considered optimization metrics are tested using three scheduling criteria. Specifically, a greedy criterion referred to as maximum capacity (MC) (Borst & Whiting, 2003; Gesbert et al., 2007), and two fair criteria referred to as proportional-fair (PF) (Chaponniere et al., 2002; Viswanath et al., 2002) and score-based (SB) (Bonald, 2004) are considered.

9.1 Maximum capacity scheduling criterion

MC criteria always assigns the antennas to those users with the highest eigenvalue sum, hence, it is efficient in terms of overall throughput but may look oppressive for low SINR users, typically located far from the BS (cell-edge users).
The achievable capacity according to MC scheduling criterion can be formulated as follows:

$$C^{\mathrm{MC}} = \arg\max_{v \subset \mathrm{SP}} \{C_v\}, \tag{15}$$

where C^{MC} is the achievable capacity using MC criterion, C_v is the instantaneous capacity of the vth option.

9.2 Proportional fair scheduling criterion

The PF criterion seeks to assign the antennas to those users with the highest eigenvalue sum relative to their mean eigenvalue sum. This causes the scheduler to realize a reasonable trade-off between throughput efficiency and fairness.
The achievable capacity according to PF scheduling criterion can be formulated as follows:

$$C^{\mathrm{PF}} = \arg\max_{v \subset \mathrm{SP}} \left\{ \frac{C_v}{\overline{C}_v} \right\}, \tag{16}$$

where C^{PF} is the achievable capacity using PF criterion, C_v and $\overline{C}_v$ are the instantaneous and average capacity of the vth option, respectively.

9.3 Score-based scheduling criterion

The SB scheduler assigns the antennas to those users with the best scores. The score corresponds to the rank of the current eigenvalue sum among the past values observed over a time window of a particular length, W (Bonald, 2004).
The achievable capacity according to SB scheduling criterion can be formulated as follows:

$$C^{\mathrm{SB}} = \arg\min_{v \subset \mathrm{SP}} \{S_v\}, \tag{17}$$

where C^{SB} is the achievable capacity using the SB criterion, and S_v is the score of the instantaneous capacity of the vth option among the past values observed over the time window W. The throughput-fairness trade-off is customizable by changing the window size W. More specifically, setting the window size equal to one, this will result in a similar behavior as observed with the MC criterion. A large window size converges to the behavior of the PF scheduler.
Since the score, S_v, is an integer number several scheduling scenarios could result in the same best rank (the first position). In this case, the solution which achieves the highest throughput is selected. Therefore, in this chapter the SB criterion effectively results in a hybrid version of both the SB scheduler and the MC scheduler.

10. Results and discussion

In this section, the performance of the different optimization metrics is assessed using the cumulative distribution function (cdf) of per-user capacity and cell capacity for both UL and DL. The per-user capacity in this work is the 2×2 spatial multiplexing MIMO capacity which can be calculated using (10). The cell capacity is the sum capacity of the group of scheduled users. In order to shed some light on fairness of the considered optimization metrics, the cdf of the scheduled (served) user distance to the BS is obtained for all considered scheduling criteria.

A two-tier cellular platform consisting of hexagonal cells is used in the simulation. In each cell, the BS and MSs are equipped with (6 or 8) and 2 antennas, respectively. Using the parameters given in Table 1, the 19-cell cellular TDD system with uniform user distribution, as shown in Fig. 10, is simulated using the Monte Carlo method. The channel matrix of kth user $\mathbf{H}_k$ is a zero-mean i.i.d. complex Gaussian random matrix. The system level performance evaluation is based on both central cell and wrap-around techniques. The DL and UL performances are

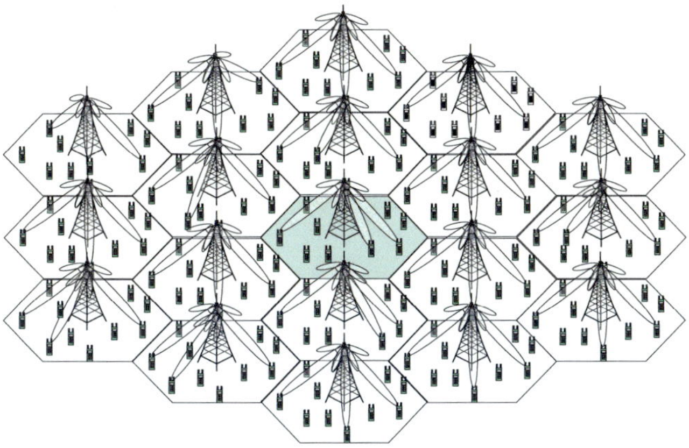

Fig. 10. Interference-limited multiuser MIMO system implementing orthogonal SDMA

analyzed for asynchronous UL/DL transmissions, where the two possible link-directions (UL or DL) occur with equal probability and independently from each other. All entities (BSs and MSs) in the system use fixed transmit power of 30 dBm.

In the wrap-around technique, the multicell layout is folded such that cells on the right side of the network are connected with cells on the left side and, similarly, cells in the upper part of the network get connected to cells in the lower part. The created area may be seen as borderless, but with a finite surface, and it may be visualized as a torus. The wrap-around technique is suitable for both downlink and uplink simulations. One of the main advantages of wrap-around technique is that the decision taken by a scheduler in a particular cell influences the scheduling behavior in the adjacent cells. Such an observation cannot be collected using the central cell technique. Another advantage compared to the central cell technique is that simulation data can be collected from all cells, which may reduce the required simulation time to collect sufficient statistics.

10.1 Downlink performance with interference awareness

In order to avoid a huge SP size, the BS is assumed to be equipped with 6 SAs. Therefore, the same time-frequency resource is simultaneously used, at maximum, three times within the same cell for three different users. The results in Fig. 11(a) show that the cell capacity can be significantly enhanced if knowledge of interference is taken into account in the antenna selection and MIMO user scheduling process. For example it can be seen from Fig. 11(a), for the MC criterion (unmarked curves), that the median of the DL capacity for the

Channel propagation environment	Suburban micro cell
BS-to-BS distance	300-500 m
Number of interfering tiers	2
Pilot Tx power	30 dBm
Number of users	300-1000
Carrier frequency	2 GHz
Minimum distance between MS and BS	35 m
Number of antennas at BS	6-8
Number of antennas at MS	2
log-normal shadowing standard deviation, σ_S	8 dB
Correlation distance of shadowing	50 m (as in UMTS 30.03)
Path loss model (dB)	$31.5 + 35 \log 10(d)$

Table 1. Simulation parameters

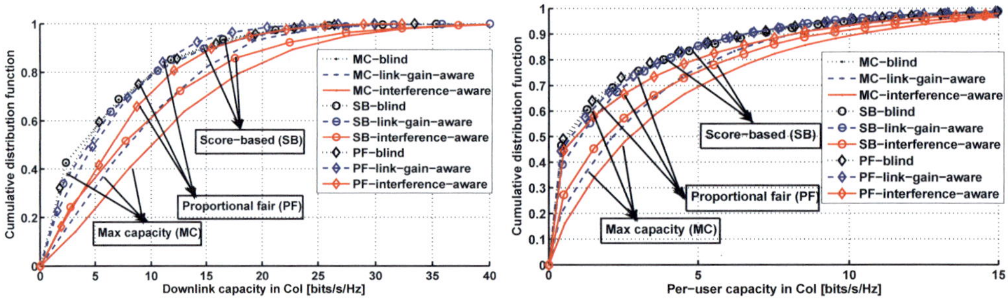

(a) Cell performance (b) User performance

Fig. 11. Downlink performance comparison among the DL interference-aware and both blind and link-gain-aware metrics using ES approach

DL interference-aware-metric (using the novel concept of interference-weighted pilots) is 10 bps/Hz while it is 7.5 and 4 bps/Hz for the link-gain-aware-metric and the blind-metric, respectively. As expected from the greedy characteristics of the link-gain-aware-metric, it can be seen that it outperforms the blind-metric. The greediness of these metrics will be revisited later when the issue of fairness is discussed. Similarly, the per-user capacity in the case of the DL interference-aware-metric is substantially improved for all scheduling criteria. From Fig. 11(b), for the SB criterion (circle-marked curves), it can be seen that the median of the per-user capacity resulting from the DL interference-aware-metric is 2 bps/Hz while it is 1 and 0.5 bps/Hz for link-gain-aware-metric and the blind-metric, respectively.

The results in Fig. 12 are obtained for the case of a cell radius of 300 m. It is shown that both DL interference-aware-metric and link-gain-aware-metric prioritize the users closer to the BS when the MC criterion is used. For instance, Fig. 12, assuming the MC criterion (unmarked curves), shows that the median of the served user distance to the

BS of the DL interference-aware-metric and the link-gain-aware-metric are 160 m and 140 m, respectively, while it is 190 m for the blind-metric case. Alternatively, when the PF criterion (diamond-marked curves) is used the DL interference-aware-metric shows high level of fairness (median is 220 m) compared to both blind-metric (median is 175 m) and link-gain-aware-metric (median is 210 m).

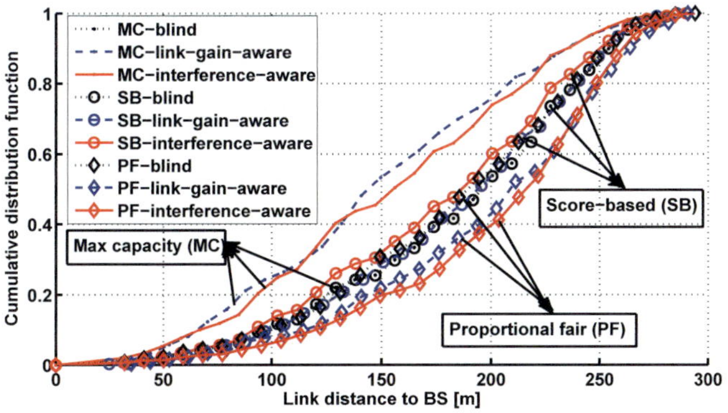

Fig. 12. Fairness comparison among the three considered scheduling criteria in terms of the distance between the BS and the scheduled user.

It is worth noting that the distance distribution of the served users associated with the blind-metric is only slightly affected by using the SB and the PF criteria. This is expected due to the inherent fairness of this metric which is a consequence of the i.i.d. distributed small-scale fading and the random distribution of the users.

10.2 Performance of heuristic algorithm

In order to study the efficiency of the HA, a DL scenario is simulated under the following assumptions: 8 SAs at the BS, 2 antennas at each MS, fixed sectorization is used to equally divide the cell into two sectors, only the best 8 users in terms of the experienced intercell CCI are considered in each cell, and the MC criterion is used. It has been found using simulations that the search space has been reduced from 176400 to 80 on average (depending on the user density per sector). This is achieved at the following cost: The median of the achievable throughut using DL interference-aware-metric in combination with the HA approach only reaches approximately 90% of the median of the optimum capacity using the exhaustive search approach. This can be seen in Fig. 13.
Since the ES approach is computationally too expensive, simulation results for larger SP sizes cannot be obtained within reasonable time. Therefore, all the subsequent results are obtained using the HA approach due to its low complexity and close to optimum performance as demonstrated in Fig. 13. Furthermore, in order to increase the spatial SDMA DoF, the BS is assumed to be equipped with 8 SAs, hence 4 users can be scheduled.

10.3 Uplink performance with link-protection awareness

From Fig. 14(a), for the MC criterion (square-marked curves), it can be seen that the UL median cell capacity using the link-protection-aware-metric, (9), is 27 bps/Hz which corresponds to an enhancement of 12.5% of the UL median cell capacity using the UL interference-aware-metric, (12), which is 24 bps/Hz.

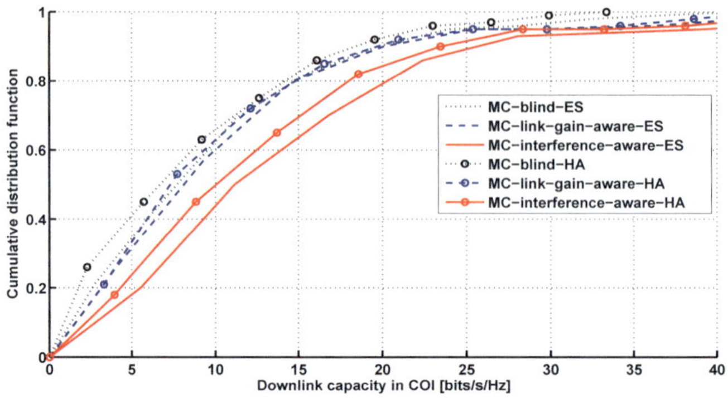

Fig. 13. Throughput comparison for the metrics considered in the previous section for the MC criterion using both ES and HA approaches

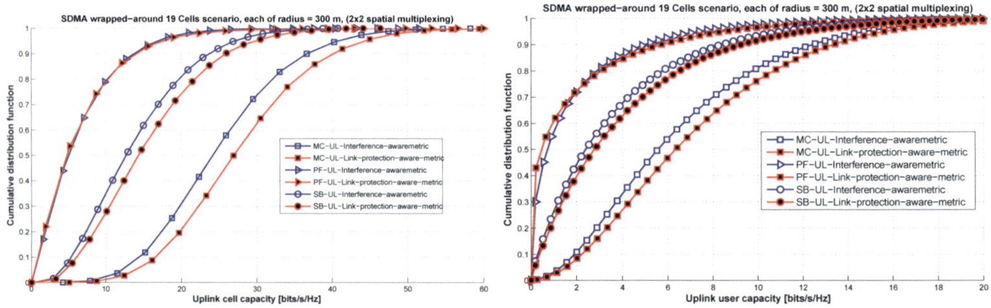

(a) Cell performance (b) User performance

Fig. 14. Uplink performance comparison between the UL interference-aware-metric and the link-protection-aware-metric

Notice that the link-protection-aware-metric, (9), jointly considers UL and DL and it implicitly provides the BS with CCI interference observed at the MSs. Since channel reciprocity is assumed, a user that observes high interference in the DL will in turn cause high interference to the corresponding users in the other cell when they are receiving data from their BS. Therefore, not scheduling a user that observes high interference is beneficial in two ways: (a) other users might observe less interference at these particular transmission resources, and scheduling these users would result in higher user capacity or reduced transmission power; and (b) interference caused to other users in the neighboring cells is automatically kept a low level. This effectively results in interference-aware radio resource management. These observations are affirmed by noting the results of the per-user capacity of the link-protection-aware-metric, which outperforms the UL interference-aware-metric for all scheduling criteria considered. For instance, from Fig. 14(b), for the SB criterion (circle-marked curves), it can be seen that the median of the per-user capacity resulting from the link-protection-aware-metric is 2.9 bps/Hz while it is 2.5 bps/Hz for the UL interference-aware-metric. However, the PF criterion (triangle-marked curves) does not provide a noticeable improvement due to the inherent fairness. The fairness constraint prioritizes a fair allocation of the spatial resources among the users over preventing strong interferers from being scheduled.

10.4 Downlink performance with link-protection awareness

The same analysis carried out in the previous subsection for the UL is now applied to the DL. From Fig. 15(a), for the MC criterion (square-marked curves), it can be seen that the median

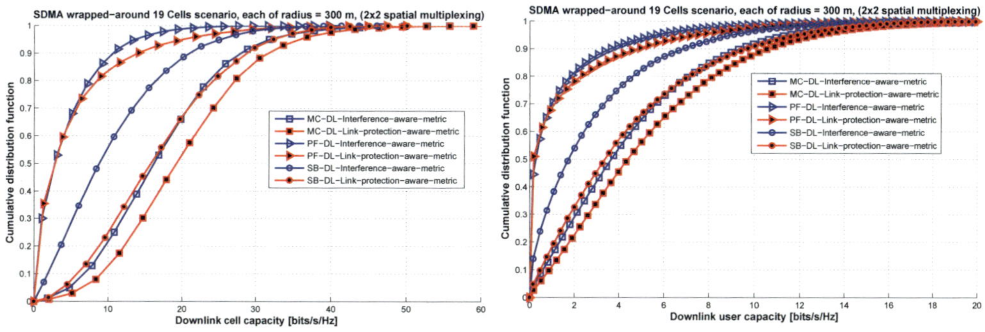

(a) Cell performance (b) User performance

Fig. 15. Downlink performance comparison between the DL interference-aware-metric and the link-protection-aware-metric.

of the DL capacity using the link-protection-aware-metric, (9), is 19 bps/Hz while it is 15 bps/Hz for the DL interference-aware-metric, (8). Similarly, the per-user capacity in the case of the link-protection-aware-metric is improved for all scheduling criteria. From Fig. 15(b), for the SB criterion (circle-marked curves), it can be seen that the median of the per-user capacity resulting from the link-protection-aware-metric is 3.5 bps/Hz while it is 1.9 bps/Hz for the DL interference-aware-metric.

Irrespective of the nature of the employed scheduling criterion, the results show that the gain in the UL is lower compared to the DL. This primarily is due to the fact that BS-to-BS interference degrades the link-protection selectivity in UL. Given that all antenna groups at the BS site are in close proximity, the large-scale fading (log-normal shadowing) on the interference links can, thus, be considered the same for this case. Therefore, the variations among the interference links depend only on the small-scale fading.

On the other hand, the tendency of greediness of the SB criterion, explained in section 9, allows for the exploitation of multiuser diversity in the DL given the high user density. Furthermore, due to the relatively high variance of the statistics of DL interference, a marginal improvement is achieved using the PF criterion, as it can be seen in Fig. 15. The gains are only materialized at the high interference percentiles due to the emphasis on fairness when using the PF criterion (triangle-marked curves).

11. Conclusions

Weighting the UL channel sounding pilots by the instantaneous CCI observed locally (referred to as interference-weighted channel sounding pilots) enables the users to implicitly convey the level of CCI to its BS along with channel state information. In addition, ICIC is achieved without any explicit inter-cell signalling as follows: A user that experiences high CCI in the DL would equally cause high interference to the same users that caused this interference when the transmission directions are reversed (due to channel reciprocity). The pilot-weighting mechanism "artificially" deteriorates the channel in this case so that this particular user will most probably be scheduled on a different channel which is beneficial for this user as well as for the interfering users in the neighboring cells (resulting in interference coordination). If additionally the modified pilots are weighted by the UL interference (observed at the BS), this provides combined knowledge of the interference at both ends. It has been shown that this additional information, for example, enables interference-aware user scheduling to improve the capacity compared to systems which only utilize the conventional channel sounding pilots.

It has been found that compared to both blind-metric and link-gain-aware-metric, a capacity gain of 150% and 35%, respectively, at the 10th percentile can be achieved when the novel downlink interference-aware-metric is used assuming the maximum capacity criterion. By considering the score-based policy, simulations show that using the link-protection-aware-metric results in a capacity gain of 230% and 15% at the 10th percentile compared to both downlink and uplink interference-aware-metric, respectively. Marginal capacity gains have been obtained for the PF policy which ensures fairness at the expense of capacity efficiency. However, please notice that for the sake of conciseness only a single channel has bee assumed in this study. Higher gains are envisaged for the PF policy if a broadband OFDMA system with multiple resource blocks would have been considered.

Utilizing the proposed heuristic algorithm significantly reduces the computational complexity to approximately 0.05% of the complexity of the exhaustive search approach. This reduction in complexity is achieved at the cost of 8% loss at the 10th percentile cell capacity.

12. References

Abe, T. & Bauch, G. (2007). Differential codebook mimo precoding technique, *Global Telecommunications Conference, 2007. GLOBECOM '07. IEEE*, pp. 3963 –3968.

Abualhiga, R. & Haas, H. (2008). Implicit Pilot-Borne Interference Feedback for Multiuser MIMO TDD Systems, *Proc. of the International Symposium on Spread Spectrum Techniques and Applications (ISSSTA)*, IEEE, Bologna, Italy, pp. 334–338.

Airy, M., Bhadra, S., Heath, R. & Shakkottai, S. (2006). Transmit Precoding for the Multiple Antenna Broadcast Channel, *Proc. of the 63rd Vehicular Technology Conference (VTC 06)*, Vol. 3, IEEE, Melbourne, Australia, pp. 1396–1400.

Ali, S. H., Lee, K.-D. & Leung, V. C. M. (2007). Dynamic Resource Allocation in OFDMA Wireless Metropolitan Area Networks, *IEEE Wireless Communications* 14(1): 6–13.

Bahceci, I., Duman, T. & Altunbasak, Y. (2003). Antenna Selection for Multiple-Antenna Transmission Systems: Performance Analysis and Code Construction, *IEEE Transactions on Information Theory* 49(10): 2669–2681.

Bauch, G. & Dietl, G. (2008a). Enhanced mimo for imt-advanced wireless systems, *2008 IET Seminar on Wideband and Ultrawideband Systems and Technologies: Evaluating current Research and Development*, pp. 1 –21.

Bauch, G. & Dietl, G. (2008b). Multi-user mimo for achieving imt-advanced requirements, *International Conference on Telecommunications (ICT 2008)*, pp. 1 –7.

Blum, R. (2003). MIMO Capacity with Interference, *IEEE Journal on Selected Areas in Communications* 21(5): 793–801.

Bonald, T. (2004). A Score-Based Opportunistic Scheduler for Fading Radio Channels, *Proc. of the European Wireless Conference (EWC)*, Barcelona, Spain.

Borst, S. & Whiting, P. (2003). Dynamic channel-sensitive scheduling algorithms for wireless data throughput optimization, *IEEE Transactions on Vehicular Technology* 52(3): 569–586.

Catreux, S., Driessen, P. & Greenstein, L. (2002). Data Throughputs Using Multiple-Input Multiple-Output (MIMO) Techniques in a Noise-Limited Cellular Environment, *IEEE Transactions on Wireless Communications* 1(2): 226–235.

Chae, C.-B., Mazzarese, D. & Heath, R. W. (2006). Coordinated Beamforming for Multiuser MIMO Systems with Limited Feedforward, *Fortieth Asilomar Conference on Signals, Systems and Computers (ACSSC)* pp. 1511–1515.

Chaponniere, E. F. Black, P. J., Holtzman, J. M. & Tse, D. N. C. (2002). Transmitter Directed Code Division Multiple Access System Using Path Diversity to Equitably Maximize Throughput, *US Patent 6449490* .

Chen, R., Heath, R. W. & Andrews, J. G. (2007). Transmit Selection Diversity for Unitary Precoded Multiuser Spatial Multiplexing Systems With Linear Receivers, *IEEE Transactions on Signal Processing* 55(3): 1159–1171.

Choi, L.-U. & Murch, R. (2004). A Transmit Preprocessing Technique for Multiuser MIMO Systems Using a Decomposition Approach, *IEEE Transactions on Wireless Communications* 3(1): 20–24.

Choi, W., Forenza, A., Andrews, J. G. & Heath Jr., R. W. (2006). Capacity of Opportunistic Space Division Multiple Access with Beam Selection, *Proc. of the Global Telecommunications Conference (GLOBECOM 06)*, IEEE, San Francisco, USA, pp. 1–5.

Chung, S. T., Lozano, A. & Huang, H. (2001a). Approaching Eigenmode BLAST Channel Capacity Using V-BLAST With Rate and Power Feedback, *Proc. of the 54th Vehicular Technology Conference (VTC 01)*, Vol. 2, Atlantic City, New Jersey, pp. 915–919.

Chung, S. T., Lozano, A. & Huang, H. (2001b). Low Complexity Algorithm for Rate and Power Quantization in Extended V-BLAST, *Proc. of the 2001 IEEE 53rd Vehicular Technology Conference*, Vol. 2, Atlantic City, New Jersey, pp. 910–914.

Costa, M. (1983). Writing on Dirty Paper, *IEEE Transactions on Information Theory* 29(3): 439–441.

Dai, H., Molisch, A. & Poor, V. H. (2004). Downlink Capacity of Interference-Limited MIMO Systems with Joint Detection, *IEEE Transactions on Wireless Communications* 3(2): 442–453.

Foschini, G. J. (1996). Layered Space-Time Architecture for Wireless Communication in a Fading Environment when Using Multi-Element Antennas, *Bell Labs Technical Journal* 1(2): 41–59.

Foschini, G. J. & Gans, M. J. (1998). On Limits of Wireless Communications in a Fading Environment when Using Multiple Antennas, *Wireless Personal Communications* 6(6): 311–335.

Fragouli, C., Al-Dhahir, N. & Turin, W. (2003). Training-Based Channel Estimation for Multiple-Antenna Broadband Transmissions, *IEEE Transactions on Wireless Communications* 2(2): 384–391.

Fuchs, M. & Del Galdo, G. & Haardt, M. (2007). Low-Complexity Space–Time–Frequency Scheduling for MIMO Systems With SDMA, *IEEE Transactions on Vehicular Technology* 56(5): 2775–2784.

Gallen, C. (2009). In 2014 Monthly Mobile Data Traffic Will Exceed 2008 Total, *ABI Research*, Retrieved January 3, 2011, from www.abiresearch.com/press/

Gesbert, D., Kiani, S. G., Gjendemsjø, A. & Øien, G. E. (2007). Adaptation, Coordination, and Distributed Resource Allocation in Interference-Limited Wireless Networks, *Proc. of the 7th IEEE International Symposium on Wireless Communication Systems* 95(12): 2393–2409.

Ghrayeb, A. & Duman, T. (2002). Performance analysis of MIMO Systems with Antenna Selection Over Quasi-Static Fading Channels, pp. 333–337.

Goldsmith, A., Jafar, S., Jindal, N. & Vishwanath, S. (2003). Capacity Limits of MIMO Channels, *IEEE Journal on Selected Areas in Communication* 21(5): 684–702.

Gore, D., Heath, R. & Paulraj, A. (2002). Statistical Antenna Selection for Spatial Multiplexing Systems, *Proc. of the International Conference on Communications (ICC 02)*, Vol. 1, New York, USA, pp. 450–454.

Gore, D. & Paulraj, A. (2002). MIMO Antenna Subset Selection with Space-Time Coding, *IEEE Transactions on Signal Processing* 50(10): 2580–2588.

Gorokhov, A., Gore, D. & Paulraj, A. (2003). Receive Antenna Selection for MIMO Flat-Fading Channels: Theory and Algorithms, *IEEE Transactions on Information Theory* 49(10): 2687–2696.

Haas, H. & McLaughlin, S. (eds) (2008). *Next Generation Mobile Access Technologies: Implementing TDD*, Cambridge University Press, ISBN: 13:9780521826228.

Hassibi, B. & Hochwald, B. M. (2003). How Much Training is Needed in Multiple-Antenna Wireless Links?, *IEEE Transactions on Information Theory* 49: 951–963.

Heath, R. & Paulraj, A. (2001). Antenna Selection for Spatial Multiplexing Systems Based on Minimum Error Rate, *Proc. of the International Conference on Communications (ICC 01)*, Vol. 7, Helsinki, Finland, pp. 2276–2280.

Hochwald, B., Peel, C. & Swindlehurst, A. (2005). A Vector-Perturbation Technique for Near-Capacity Multiantenna Multiuser Communication. part II: Perturbation, *IEEE Transactions on Communications* 53(3): 537–544.

Koutsimanis, C. & Fodor, G. (2008). A Dynamic Resource Allocation Scheme for Guaranteed Bit Rate Services in OFDMA Networks, *Proc. of the IEEE International Conference on Communications (ICC 08)*, pp. 2524 – 2530.

Kusume, K., Joham, M., Utschick, W. & Bauch, G. (2007). Cholesky factorization with symmetric permutation applied to detecting and precoding spatially multiplexed data streams, *IEEE Transactions on Signal Processing* 55(6): 3089 –3103.

Learned, R., Willsky, A. & Boroson, D. (1997). Low complexity optimal joint detection for oversaturated multiple access communications, *IEEE Transactions on Signal Processing* 45(1): 113–123.

Love, D. J. & Heath, R. (2005). Limited Feedback Unitary Precoding for Spatial Multiplexing Systems, *IEEE Transactions on Information Theory* 51(8): 2967–2976.

Love, D. J., Heath, R. & Strohmer, T. (2003). Grassmannian Beamforming for Multiple-Input Multiple-Output Wireless Systems, *Proc. of the International Conference on Communications (ICC 03)*, Vol. 4, IEEE, pp. 2618–2622.

Love, D. J., Heath, R., Santipach, W. & Honig, M. L. (2004). What is the Value of Limited Feedback for MIMO Channels, *IEEE Communications Magazine* .

Molisch, A., Win, M. & Winters, J. (2001). Capacity of MIMO Systems with Antenna Selection, *Proc. of the International Conference on Communications (ICC 01)*, Vol. 2, pp. 570–574.

Molisch, A., Win, M. & Winters, J. (2003). Reduced-Complexity Transmit/Receive-Diversity Systems, *IEEE Transactions on Signal Processing* 51(11): 2729–2738.

Mukkavilli, K., Sabharwal, A., Aazhang, B. & Erkip, E. (2002). Performance Limits on Beamforming with Finite Rate Feedback for Multiple Antenna Systems, *Proc. of the 36th Asilomar Conference on Signals, Systems and Computers*, Vol. 1, pp. 536–540.

Mukkavilli, K., Sabharwal, A., Erkip, E. & Aazhang, B. (2003). On Beamforming with Finite Rate Feedback in Multiple-Antenna Systems, *IEEE Transactions on Information Theory* 49(10): 2562–2579.

Pan, Z., Wong, K.-K. & Ng, T.-S. (2004). Generalized Multiuser Orthogonal Space-Division Multiplexing, *IEEE Transactions on Wireless Communications* 3(6): 1969–1973.

Popovic, B. (1992). Generalized chirp-like polyphase sequences with optimal correlation properties, *IEEE Transactions on Information Theory* 38: 1406–1409.

Schubert, M. & Boche, H. (2004). Solution of the Multiuser Downlink Beamforming Problem with Individual SINR Constraints, *IEEE Transactions on Vehicular Technology* 53(1): 18–28.

Seidel, E. (2008). Progress on "LTE Advanced" – the New 4G Standard, *Newsletter*, NOMOR, Munich, Germany.

Sesia, S., Toufik, I. & Baker, M. (2009). *LTE - The UMTS Long Term Evolution: From Theory to Practice*, Wiley.

Shen, Z., Chen, R., Andrews, J., Heath, R. & Evans, B. (2005). Low Complexity User Selection Algorithms for Multiuser MIMO Systems with Block Diagonalization, *Proc. of the 39th Asilomar Conference on Signals, Systems and Computers.*, pp. 628–632.

Shi, S., Schubert, M. & Boche, H. (2008). Downlink MMSE Transceiver Optimization for Multiuser MIMO Systems: MMSE Balancing, *IEEE Transactions on Signal Processing* 56(8): 3702–3712.

Spencer, Q., Swindlehurst, A. & Haardt, M. (2004). Zero-Forcing Methods for Downlink Spatial Multiplexing in Multiuser MIMO Channels, *IEEE Transactions on Signal Processing* 52(2): 461–471.

Telatar, E. (1999). Capacity of Multi-Antenna Gaussian Channels, *European Transaction on Telecommunication* 10(6): 585–595.

Vishwanath, S., Jindal, N. & Goldsmith, A. (2003). Duality, Achievable Rates, and Sum-Rate Capacity of Gaussian MIMO Broadcast Channels, *IEEE Transactions on Information Theory* 49(10): 2658–2668.

Viswanath, P., Tse, D. & Laroia, R. (2002). Opportunistic Beamforming Using Dumb Antennas, *Proc. of the International Symposium on Information Theory*, IEEE, p. 449.

Wang, C. & Murch, R. (2005). Adaptive Cross-Layer Resource Allocation for Downlink Multi-User MIMO Wireless System, *Proc. of the 61st Vehicular Technology Conference (VTC 05)*, Vol. 3, pp. 1628–1632.

Weingarten, H., Steinberg, Y. & Shamai, S. (2004). The Capacity Region of the Gaussian MIMO Broadcast Channel, *Proc. of the International Symposium on Information Theory (ISIT 04)*, Chicago, USA, pp. 174–182.

Windpassinger, C., Fischer, R., Vencel, T. & Huber, J. (2004). Precoding in Multiantenna and Multiuser Communications, *IEEE Transactions on Wireless Communications* 3(4): 1305–1316.

Wong, K.-K., Murch, R. & Letaief, K. (2003). A Joint-Channel Diagonalization for Multiuser MIMO Antenna Systems, *IEEE Transactions on Wireless Communications* 2(4): 773–786.

Zhou, S., Wang, Z. & Giannakis, G. (2005). Quantifying the Power Loss When Transmit Beamforming Relies on Finite-Rate Feedback, *IEEE Transactions on Wireless Communications* 4(4): 1948–1957.

Zhou, Z., Dong, Y., Zhang, X., Wang, W. & Zhang, Y. (2004). A Novel Antenna Selection Scheme in MIMO Systems, *International Conference on Communications, Circuits and Systems (ICCCAS 04)*, Vol. 1, pp. 190–194.

Zhou, Z. & Vucetic, B. (2004). MIMO Systems with Adaptive Modulation, *Proc. of the 59th Vehicular Technology Conference (VTC 04)*, Vol. 2, pp. 765–769.

Zhuang, H., Dai, L., Zhou, S. & Yao, Y. (2003). Low Complexity Per-Antenna Rate and Power Control Approach for Closed-Loop V-BLAST, *IEEE Transactions on Communications* 51(11): 1783–1787.

22

Demodulation Reference Signal Design and Channel Estimation for LTE-Advanced Uplink

Xiaolin Hou and Hidetoshi Kayama
DOCOMO Beijing Communications Laboratories Co., Ltd.
China

1. Introduction

The merits of 3GPP long term evolution (LTE), such as high spectral efficiency, very low latency, support of variable bandwidth, simple architecture, etc, make it the most competitive candidate for the next generation mobile communications standard. In the first release of LTE, only single transmit antenna is supported in the uplink due to its simplicity and acceptable performance. However, in order to keep its current leading position, LTE needs further evolvement (known as LTE-Advanced (LTE-A)) to provide better performance, including a higher uplink spectrum efficiency. Therefore, multiple transmit antennas must be supported in the LTE-A uplink and one important issue is the uplink demodulation reference signal (DMRS) design, which will influence uplink channel estimation accuracy and eventually determine uplink reliability and throughput.

In this study we first briefly review the current status of DMRS in LTE uplink and then different DMRS enhancement schemes are investigated for LTE-A uplink multiple-input multiple-output (MIMO) transmission. Also, two-dimensional channel estimation algorithms are provided to realize accurate uplink channel estimation. With computer simulations, the performances of several candidate LTE-A uplink DMRS design schemes are evaluated and compared. Finally some basic conclusions are provided together with the latest standardization progress.

2. DMRS in LTE uplink

LTE uplink is based on single-carrier frequency division multiple access (SC-FDMA) due to its low peak-to-average power ratio (PAPR). There are two types of reference signal in LTE uplink: DMRS used for data reception and sounding reference signal (SRS) used for scheduling and link adaptation. In this study we only focus on DMRS design and related channel estimation for the physical uplink shared channel (PUSCH).

Take frame structure type 1 for example, each LTE radio frame is 10ms long and consists of 20 slots of length 0.5ms. A subframe is defined as two consecutive slots. For the normal cyclic prefix (CP) case, each slot contains 7 symbols. The two-dimensional time-frequency resource can be partitioned as resource blocks (RBs) and each RB corresponds to one slot in the time domain and 180 kHz in the frequency domain. In LTE uplink, the DMRS for PUSCH in the frequency domain will be mapped to the same set of physical resource blocks (PRBs) used for the corresponding PUSCH transmission with the same length expressed by the number of subcarriers, while in the time domain DMRS will occupy the $4th$ SC-FDMA symbol in each slot for the normal CP case, as shown in Fig. 1.

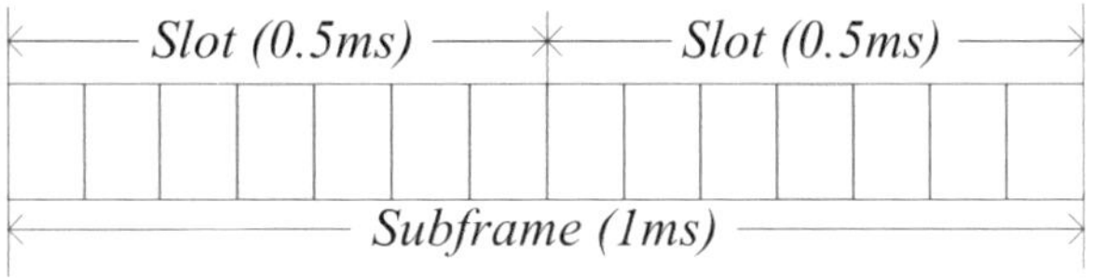

Fig. 1. DMRS in LTE uplink

In order to support a large number of user equipments (UEs) in multiple cells, a large number of different DMRS sequences are needed. A DMRS sequence $r_{u,v}^{(\alpha)}(n)$ is defined by a cyclic shift (CS) α of a base sequence $\bar{r}_{u,v}(n)$ according to

$$r_{u,v}^{(\alpha)}(n) = e^{j\alpha n} \cdot \bar{r}_{u,v}(n), 0 \leq n < M_{sc}^{RS} \tag{1}$$

where $M_{sc}^{RS} = mN_{sc}^{RB}$ is the length of DMRS sequence, m is the RB number and N_{sc}^{RB} is the subcarrier number within each RB. When the subcarrier bandwidth is set as 15kHz, each RB will contain 12 subcarriers, i.e., $N_{sc}^{RB} = 12$. Multiple DMRS sequences can be derived from a single base sequence through different values of α.

The definition of the base sequence depends on the sequence length. For $M_{sc}^{RS} \geq 3N_{sc}^{RB}$, the base sequence is defined as the cyclic extension of the Zadoff-Chu sequence (Chu, 1972)

$$\bar{r}_{u,v}(n) = x_q(n \bmod N_{ZC}^{RS}), 0 \leq n < M_{sc}^{RS} \tag{2}$$

$$x_q(m) = e^{-j\frac{\pi q m(m+1)}{N_{ZC}^{RS}}}, 0 \leq m < N_{ZC}^{RS} - 1 \tag{3}$$

where $x_q(m)$ is the q_{th} root Zadoff-Chu sequence and N_{ZC}^{RS} is the length of Zadoff-Chu sequence that is given by the largest prime number such that $N_{ZC}^{RS} < M_{sc}^{RS}$. For $M_{sc}^{RS} < 3N_{sc}^{RB}$, the base sequence is defined as the computer generated constant amplitude zero autocorrelation (CG-CAZAC) sequence

$$\bar{r}_{u,v}(n) = e^{j\varphi(n)\pi/4}, 0 \leq n < M_{sc}^{RS} \tag{4}$$

where the values of $\varphi(n)$ are given in (3GPP, TS 36.211).

Base sequences $\bar{r}_{u,v}(n)$ are divided into 30 groups with $u \in \{0, 1, ..., 29\}$. Each group contains one base sequence ($v = 0$) with $1 \leq m \leq 5$ and two base sequences ($v = 0, 1$) with $6 \leq m \leq N_{RB}^{max,UL}$, where $N_{RB}^{max,UL}$ is the maximum RB number in the uplink. In order to reduce inter-cell interference (ICI), neighboring cells should select DMRS sequences from different base sequence groups. Furthermore, there are 3 kinds of hopping defined for the DMRS in LTE uplink, i.e., group hopping, sequence hopping and CS hopping, where CS hopping should always be enabled in each slot.

The CS value α in a slot is given by $\alpha = 2\pi n_{cs}/12$ with

$$n_{cs} = (n_{DMRS}^{(1)} + n_{DMRS}^{(2)} + n_{PRS})/12 \tag{5}$$

where $n_{DMRS}^{(1)}$ is a broadcast value, $n_{DMRS}^{(2)}$ is included in the uplink scheduling assignment and n_{PRS} is given by a cell-specific pseudo-random sequence. Obviously, there are 12 usable CS values in total for DMRS in LTE uplink.

3. DMRS design and channel estimation for LTE-A uplink

3.1 DMRS enhancement

Current LTE uplink DMRS only considers UE with single transmit antenna. However, in order to boost the uplink spectrum efficiency, multiple transmit antennas must be supported in LTE-A uplink. Therefore, the uplink DMRS must be enhanced for MIMO transmission and each UE now may have multiple DMRS sequences, depending on its transmit antenna number (without precoding) or spatial layer number (with precoding).

There are several possible solutions, including CS extension, orthogonal cover code (OCC), interleaved frequency division multiplexing (IFDM) and their combinations. Considering the backwards compatibility with LTE and the low PAPR requirement for uplink transmission, IFDM should be excluded first. Then CS, OCC and their combinations are promising candidates for DMRS enhancement and will be discussed in more details in the following text.

3.1.1 Baseline: CS extension

Considering the backwards compatibility, it is agreed that cyclic shift separation is the baseline for the LTE-A uplink DMRS enhancement (3GPP, TR 36.814). Without loss of generality, uplink precoding is not considered in the following text, therefore, transmit antenna and spatial layer are equivalent and interchangeable. For single-user MIMO (SU-MIMO) transmission with $n_T \geq 2$ spatial layers, it is natural to assign multiple CS values to separate the multiple spatial layers. Then the questions remained to be answered are how to assign different CS values to different spatial layers and how to ensure the backwards compatibility to LTE.

If we assign multiple CS values with the following constraint

$$n_{cs,i} = (n_{cs,0} + \frac{C}{n_T} \cdot i)\mathrm{mod}(C), i = 0, 1, ..., n_T - 1 \tag{6}$$

where $n_{cs,i}$ corresponds to the CS value of DMRS for the ith spatial layer and C is the constant value 12 for PUSCH. Then the CS value of DMRS for the first spatial layer $\alpha_0 = 2\pi n_{cs,0}/12$ is exactly the same as that for the single transmit antenna case in LTE. Therefore, all the original CS signaling and hopping designs for the single transmit antenna UE in LTE can be kept unchanged for the multiple transmit antennas UE in LTE-A, once the constraint in Eq. (6) is satisfied.

Because this DMRS design can be viewed as binding together the CS values of DMRS as well as the channel impulse response (CIR) positions of different spatial layers with the maximum distance constraint, as illustrated in Fig. 2 (Note that the relationship between α_i and α_0 will keep unchanged during CS hopping), we simply call it maximum distance binding (MDB).Its benefits include:

- First, the distance between CIRs of different spatial layers in the time domain can be always maximized, thus the interference between DMRS of different spatial layers can be minimized;
- Second, no additional signaling is required for CS notification and hopping when support uplink MIMO transmission, therefore, it is completely backwards compatible to LTE;
- Third, it can support time-domain inter-slot interpolation that is necessary for moderate to high mobility cases.

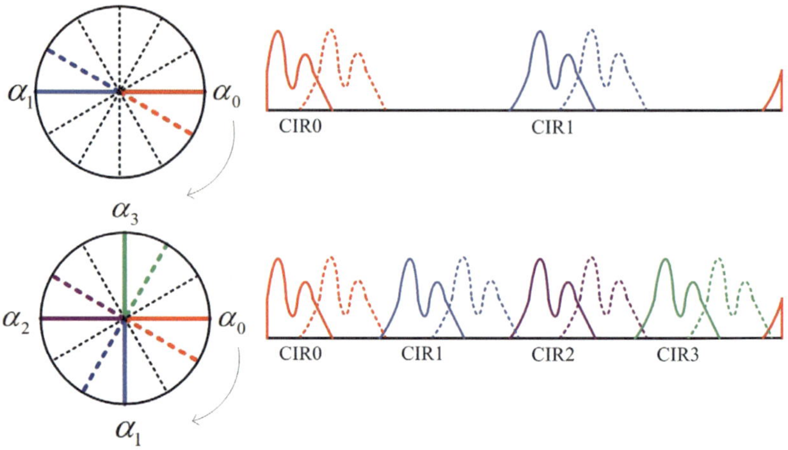

Fig. 2. CS extension with MDB

Actually, the same DMRS design principle can also be applied to the uplink multi-user MIMO (MU-MIMO) transmission with single transmit antenna UEs. Now it only requires some constraint in the uplink scheduling assignment for the CS values of multiple DMRS (because $n_{DMRS}^{(1)}$ and n_{PRS} are the same for all the UEs in the same cell, respectively) as follows

$$n_{DMRS,i}^{(2)} = (n_{DMRS,0}^{(2)} + \frac{C}{n_T} \cdot i)\text{mod}(C), i = 0, 1, ..., n_T - 1 \quad (7)$$

where $n_{DMRS,0}^{(2)}$ is the scheduled value for the first UE.
In order to support the above CS scheduling constraint for MU-MIMO transmission, we have two possible options:

- Option 1: No signaling modification

  Because the current LTE specification only supports 8 possible values for $n_{DMRS}^{(2)}$ (3GPP, TS 36.211), a limited number of combinations can be chosen in the uplink scheduling with the MDB constraint (7) satisfied. Therefore, for the 2-user case, $n_{DMRS,i}^{(2)} \in \{(0,6),(2,8),(3,9),(4,10)\}$; while for the 4-user case, $n_{DMRS,i}^{(2)} \in \{(0,3,6,9)\}$.

- Option 2: Slight signaling modification

  If the specific field in downlink control information (DCI) format 0 for the CS of DMRS can be increased from 3 bits to 4 bits, all the possible combinations in the CS scheduling for MU-MIMO transmission can be supported with the MDB constraint (7) satisfied.

3.1.2 Further enhancement: CS + OCC

For high-order SU-MIMO, MU-MIMO and coordinated multi-point (CoMP) reception that will be supported in the further evolvement of LTE, the number of superposed spatial layers will increase to four or even eight. In order to reduce the interference between multiple spatial layers, OCC, such as $[+1,+1]$ and $[+1,-1]$, can be further introduced across the two DMRS symbols within the same subframe.

For MU-MIMO and CoMP reception, CS + OCC can provide some special advantage compared to CS only scheme, such as capability to multiplex UEs with different transmit bandwidths and robustness to timing difference of multiple UEs. For SU-MIMO, CS + OCC may also be attractive for high-order MIMO transmission and/or high-order modulation. The combination of CS and OCC could have two variations, i.e., CS + OCC with identical CS and CS + OCC with offset CS (TI, 2009), as illustrated in Fig. 3 (a) and Fig. 3 (b), respectively, taking four spatial layers for example.

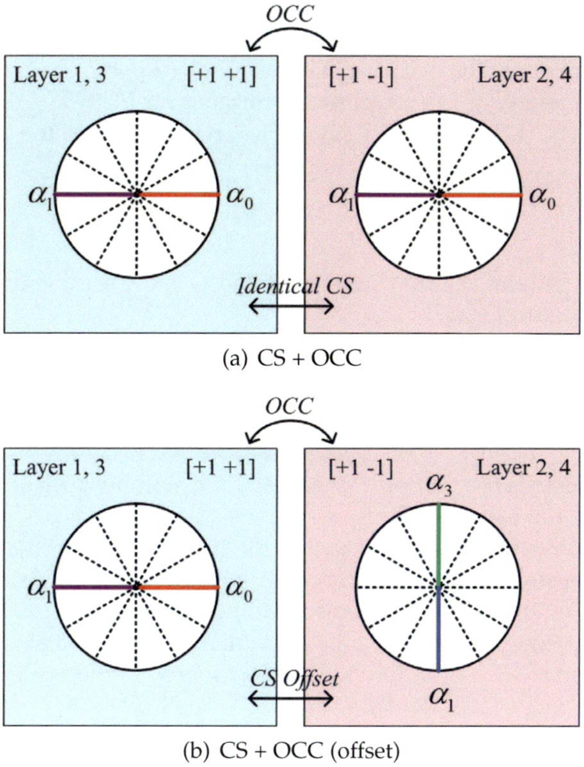

Fig. 3. Combination of CS and OCC

However, OCC will lose its effectiveness in some cases, such as when the mobility increases from low to moderate or PUSCH hopping happens within one subframe. In the aforementioned situations, CS + OCC with identical CS, abbr. as CS + OCC, cannot work at all; while CS + OCC with offset CS , abbr. as CS + OCC (offset), still can work, but in essence only CS takes effect now. Obviously, CS + OCC (offset) occupies twice CS resources compared to CS + OCC. Meanwhile, to introduce OCC into LTE-A uplink DMRS design, some additional control signaling may be needed. Otherwise, the linkage between OCC and CS must be defined to avoid increasing control signaling, i.e., the notification of OCC could be realized in an explicit way.

3.2 Two-dimensional channel estimation

In order to obtain the time-frequency two-dimensional channel state information (CSI) in the SC-FDMA uplink, two-dimensional channel estimation is needed for each subframe. Without loss of generality, assume that the inter-symbol interference (ISI) and the inter-carrier interference (ICI) are small and neglectable. Therefore, for PDSCH and corresponding DMRS within one subframe, the received signal at the k-th subcarrier in the l-th SC-FDMA symbol can be written as

$$Y(k,l) = H(k,l) \cdot X(k,l) + N(k,l), k_0 \leq k < k_0 + 12 \cdot N_{RB}^{UL} - 1, 0 \leq l < 14 \tag{8}$$

where $X(k,l)$, $H(k,l)$ and $N(k,l)$ denote the transmitted signal, the channel frequency response (CFR) and the additive white Gaussian noise (AWGN) with zero mean and variance σ^2 for the k-th subcarrier in the l-th SC-FDMA symbol, respectively. k_0 is the frequency starting position of PUSCH and N_{RB}^{UL} is the uplink RB number for PUSCH. For the multipath wireless channel within one SC-FDMA symbol, the CFR can be related to the CIR as

$$H(k,l) = \sum_{g=0}^{G-1} h(g,l) \cdot e^{-j2\pi kg/K} \tag{9}$$

where $h(g,l)$ is the g-th multipath component and G is the sample number corresponding to the maximum multipath delay.

The first step of two-dimensional channel estimation is to obtain the initial estimated superposed channels within the two DMRS symbols, i.e., $\hat{H}(k,3)$ and $\hat{H}(k,10)$, as follows

$$\hat{H}(k,l) = Y(k,l) \cdot conj(r_{u,v}^{(\alpha_0)}(k)), l = 3, 10 \tag{10}$$

where $conj(\cdot)$ represents the complex conjugate and without loss of generality, PUSCH and DMRS hopping are not considered.

To facilitate the following description, define the final estimated channel as $\tilde{H}(k,l)$. Then the target of two-dimensional channel estimation is to derive each data symbol's $\tilde{H}(k,l)$ from $\hat{H}(k,3)$ and $\hat{H}(k,10)$. Taking the implementation complexity into account, two concatenated one-dimensional channel estimation, i.e., frequency-dimensional channel estimation and time-dimensional channel estimation, will be considered in this study.

3.2.1 Frequency-dimensional channel estimation

For frequency-dimensional channel estimation, discrete-time Fourier transform (DFT) based channel estimation (Edfors et al., 2000) could be utilized. However, because the RB allocation to a given UE is generally only a small portion of the overall uplink bandwidth, the CIR energy leakage will be observed in practice, as shown in Fig. 4, where the first and the second rows are for two-antenna and four-antenna cases, while the left and the right columns are for RB# = 1 and RB# = 10 cases, respectively. It's obvious that the smaller the RB number, the more severe the CIR energy leakage. This phenomenon will make the CIRs from different transmit antennas superposed together and difficult to be separated with each other, especially when the transmit antenna number becomes larger. Furthermore, the frequency domain Gibbs phenomenon (Oppenheim et al., 1999) will appear at the edges of assigned consecutive RBs for a given UE due to the signal discontinuities. Therefore, the estimation accuracy of traditional DFT-based channel estimation will deteriorate significantly in practice.

In order to mitigate the aforementioned problems, an improved DFT-based channel estimation was proposed for LTE(-A) uplink (Hou et al., 2009), which is illustrated in Fig. 5 for each receive antenna of eNB. After serial-to-parallel (S/P) conversion and K-point fast Fourier transform (FFT), the received signal is transformed into the frequency domain. Because each UE (except for UEs in the same MU-MIMO transmission) occupies different RBs in the uplink, we can first separate different UEs by way of frequency division multiplexing (FDM). Then taking channel estimation for UE1 for example, multiply the separated received DMRS by

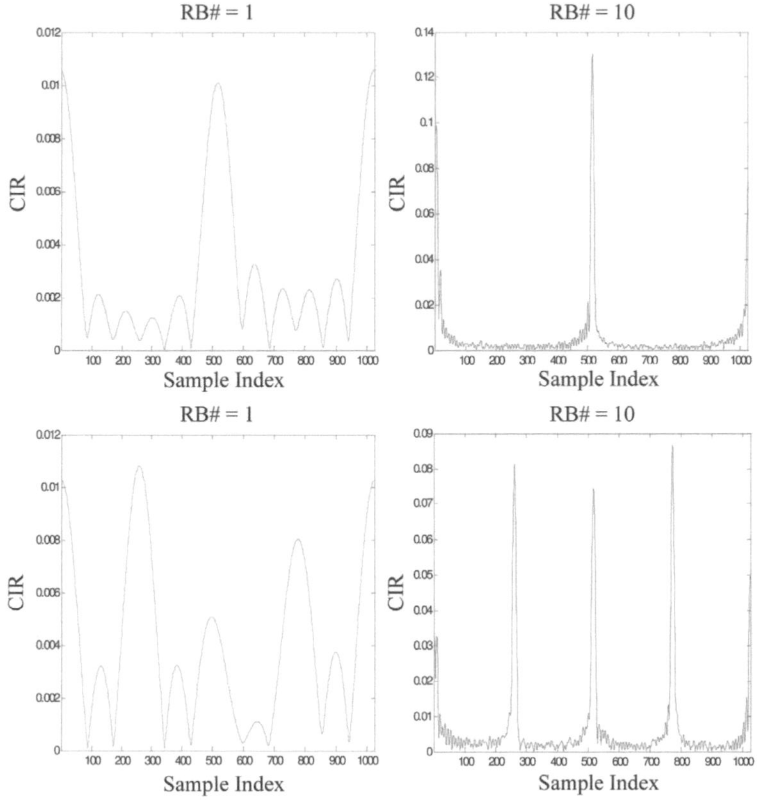

Fig. 4. CIR energy leakage

the complex conjugate of the DMRS sequence assigned for the $1st$ spatial layer and perform K-point inverse FFT (IFFT) to get the superposed CIRs in the time domain, i.e.,

$$\hat{h}(g,l) = ifft(\hat{H}(k,l)), 0 \leq k < K, l = 3,10 \tag{11}$$

After the operation of dynamic CIR reservation (DCIR^2), we can seperate the CIRs for different spatial layers by way of different CS values. As for the operation of DCIR^2, the dynamically reserved CIR for each spatial layer consists of 2 parts with respect to the timing positions, i.e., $(\frac{C}{n_T} \cdot i) \cdot K, i = 0,1,...,n_T - 1$:

- Right part
  There are $\lambda \cdot CP$ samples preserved with the following right boundary

$$(\frac{C}{n_T} \cdot i) \cdot K + \lambda \cdot CP - 1, i = 0,1,...,n_T - 1 \tag{12}$$

  where CP is the cyclic prefix length of the SC-FDMA symbol and λ is an adjustable parameter $(0 \leq \lambda < 1)$ that can be optimized in practical implementations.

- Left part
  There are $\mu \cdot \Delta$ samples preserved with the following left boundary

$$[(\frac{C}{n_T} \cdot i) \cdot K - \mu \cdot \Delta + K]\text{mod}(K), i = 0, 1, ..., n_T - 1 \quad (13)$$

where Δ is the main lobe width of CIR energy leakage ($\Delta = \frac{K}{12 \cdot RB\#}$) and μ is an adjustable parameter ($0 \leq \mu < \frac{K/n_T - CP}{\Delta}$) that can be optimized in practical implementations. In order to simply the adjustment, we can define $\tilde{\Delta} = \frac{K}{12}$ and $\tilde{\mu} = \frac{\mu}{RB\#}$, therefore, $\tilde{\Delta}$ becomes a constant and only $\tilde{\mu}$ should be adjusted.

The proper choices of λ and $\tilde{\mu}$ are mainly determined by the noise level, the multipath delay profile and the assigned RB number for a given UE. And afer DCIR^2, we can obtain the CIR for the i-th spatial layer as $\tilde{h}_i(g, l)$.

Finally, the frequency-dimensional channel estimation result of DMRS symbols for the i-th spatial layer can be achieved by K-point FFT and provided to the following time-dimentional channel estimation block.

$$\tilde{H}_i(k, l) = ifft(\tilde{h}_i(g, l)), 0 \leq g < K, l = 3, 10 \quad (14)$$

Another point should be emphasized is the operation of frequency domain windowing/dewindowing . Due to the frequency domain Gibbs phenomenon caused by the discontinuities at the edges of assigned consecutive RBs for a given UE, the overall channel estimation accuracy will be degraded, especially at the edges of assigned consecutive RBs. Therefore, some frequency domain window, such as Hanning window, Hamming window, Blackman window, etc. (Oppenheim et al., 1999), can be further added (see the dashed-line blocks in Fig. 5) to improve the channel estimation accuracy with some additional complexity. For example, Blackman window will be adopted in our following computer simulations.

$$w(n) = 0.42 - 0.5cos(2\pi n/M) + 0.08cos(4\pi n/M) \quad (15)$$

where M is the window length and $0 \leq n \leq M$. In order not to eliminate the useful signals within the assigned RBs, the window length should be larger than the assigned bandwidth ($12 \cdot RB\#$) for the corresponding UE.

Note that the improved DFT-based channel estimation can be applied to not only LTE-A MIMO uplink, but also LTE single-input single-output (SISO) or single-input multiple-output (SIMO) uplink.

3.2.2 Time-dimensional channel estimation

After frequency-dimensional channel estimation, we only obtain channel estimation results for two DMRS symbols within each subframe. In order to further acquire channel estimation result for each data symbol, time-dimensional channel estimation is needed, i.e., inter-slot interpolation via two DMRS symbols within each subframe. Two practical schemes are time-dimensional linear interpolation (TD-LI) and time-dimensional average or despreading (TD-Average/Despreading), i.e.,

$$\tilde{H}(k, l) = c_l \cdot \tilde{H}(k, 3) + (1 - c_l) \cdot \tilde{H}(k, 10), 0 \leq l < 14 \quad (16)$$

$$\begin{aligned} c_l = (10 - l)/7 \quad &, \quad TD - LI \\ c_l = 1/2 \quad &, \quad TD - Average \end{aligned} \quad (17)$$

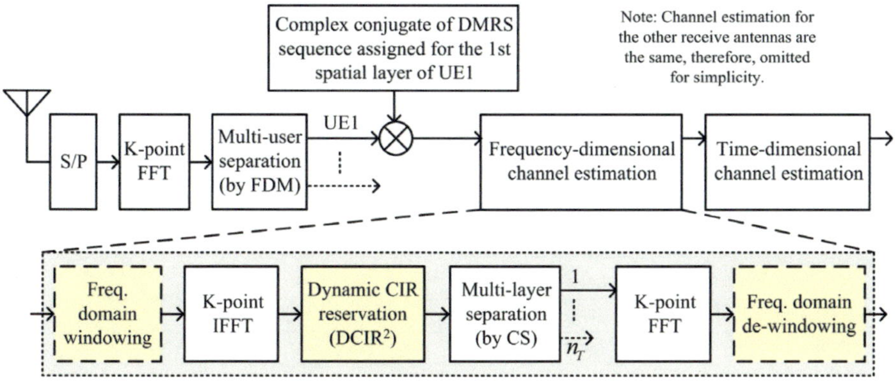

Fig. 5. The improved DFT-based channel estimation

It should be noted that for the case of CS + OCC with identical CS, TD-Despreading must be carried out before frequency-dimensional channel estimation.

4. Performance evaluation

Computer simulation results, including both block error rate (BLER) and throughput performances, will be provided in this section to compare different DMRS design schemes, i.e., CS only, CS + OCC and CS + OCC (offset).

The simulation parameters are listed in Table 1. Notice that the FFT size is larger than the usable subcarrier number because of the existence of guard band. There are totally 50 RBs in the uplink and we consider two RB# allocation cases with RB# $=$ 4,10, respectively. Furthermore, 2 typical MIMO configurations, i.e., 2×2 and 4×4, are both simulated. The MIMO transmission scheme is spatial multiplexing without precoding and the MIMO detection scheme is minimum mean square error (MMSE) detection. Without loss of generality, synchronization error and PUSCH hopping are not considered. For the improved DFT-based frequency-dimensional channel estimation, we simply chose $\lambda = 0.5$ and $\tilde{\mu} = 0.2$ and the frequency domain window length is set to be $M = 1.1 \cdot RB\# \cdot 12$. The channel model is selected as typical urban (TU) with mobility of 3km/h or 30km/h.

First, BLER performances of different DMRS design schemes will be compared. The same frequency-dimensional channel estimation is utilized for different DMRS design schemes and time-dimensional channel estimation could be different, i.e., CS + OCC can only use TD-Despreading, while CS only and CS + OCC (offset) can use TD-LI or TD-Average/Despreading. Furthermore, the curve with perfect CSI is also provided in each figure for comparison. The BLER performances are evaluated with two representative configurations, i.e., 10RB with 16QAM and 4RB with 64QAM (the coding rate is 1/2), for 2×2 and 4×4 MIMO, respectively.

When the mobile speed is as low as 3km/h, Fig. 6 shows that for the 2×2 MIMO case different DMRS design schemes have almost the same BLER performance. The only difference comes from time-directional channel estimation, i.e., TD-Average/Despreading can achieve slightly better performance than TD-LI due to the noise averaging effect in low mobility cases. And from Fig. 7, it can be observed that for the 4×4 MIMO case the introduction of OCC is helpful to improve the BLER performance in low mobility cases, especially when the RB number is small and/or the modulation order is high.

Parameters	Values
Carrier frequency	2GHz
Bandwidth	10MHz
FFT size	1024
Usable subcarrier #	600
Cyclic prefix	72
Assigned RB #	4, 10
MIMO configuration (Spatial multiplexing)	2×2 4×4
MIMO detection	MMSE
Modulation	QPSK,16QAM,64QAM
Channel coding	Turbo (Coding rate 1/2, 2/3, 3/4)
Synchronization	Perfect
PUSCH hopping	Disabled
DMRS design	CS Only CS + OCC CS + OCC (offset)
Frequency-dimensional channel estimation	Improved DFT-based
Time-dimensional channel estimation	TD-LI TD-Average/TD-Despreading
Channel model	Typical Urban (TU)
Mobile speed	3km/h, 30km/h

Table 1. Simulation parameters

However, if the mobile speed increases from low to moderate, i.e., from 3km/h to 30km/h, the aforementioned conclusion has to be revised. As shown in Fig. 8, now CS only with TD-LI can achieve the best performance and OCC will lose its effectiveness. The reason behind is that when the mobile speed is as high as 30km/h, the wireless channels between two consecutive slots are relatively fast time-varying, which makes TD-Average/Despreading cannot work well. On the other hand, TD-LI can still track the time-varying channel effectively. Therefore, from the mobility point of view, OCC has its apparent limitation, i.e., OCC will mainly work in the low mobility cases. However, considering the major application scenario of MIMO is the low mobility environment, OCC is still attractive for DMRS enhancement. And in the following simulations only 3km/h is considered.

In order to provide a more comprehensive comparison between different DMRS design schemes, the throughput performances with different modulation level and coding rate are provided in Fig. 9 and Fig. 10 for 2×2 and 4×4 MIMO, respectively. Three different modulation schemes (QPSK, 16QAM, 64QAM) and three different coding rates (1/2, 2/3, 3/4) are simulated, so in total there are nine combinations of modulation and coding. Considering the mobile speed is low, TD-Average/Despreading will be adopted instead of TD-LI. Also under this situation, because CS + OCC and CS + OCC (Offset) have neglectable performance difference, only CS + OCC is simulated, together with CS only and perfect CSI. Therefore, in each figure there are 27 curves, shown by different line styles and markers. For each DMRS design scheme, only the envelop of nine curves (each with one specific combination of modulation and coding) is highlighted to show the highest achievable throughput, which

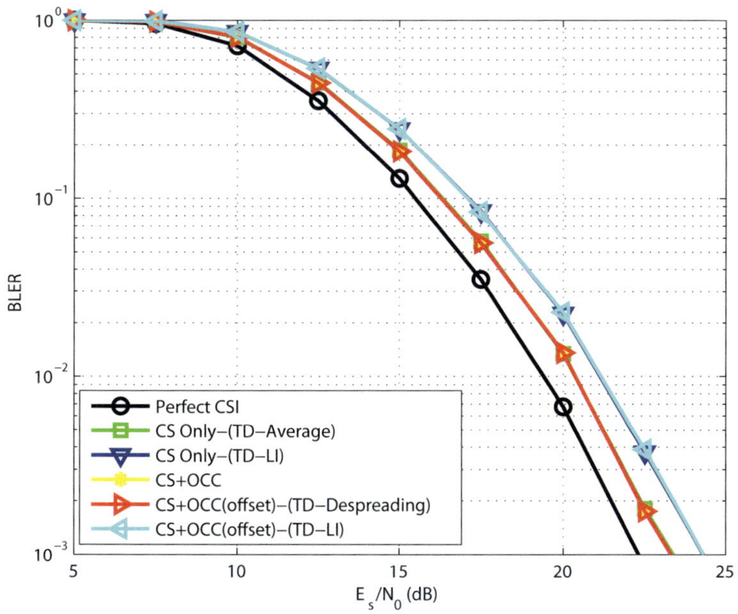

(a) 10RB, 16QAM

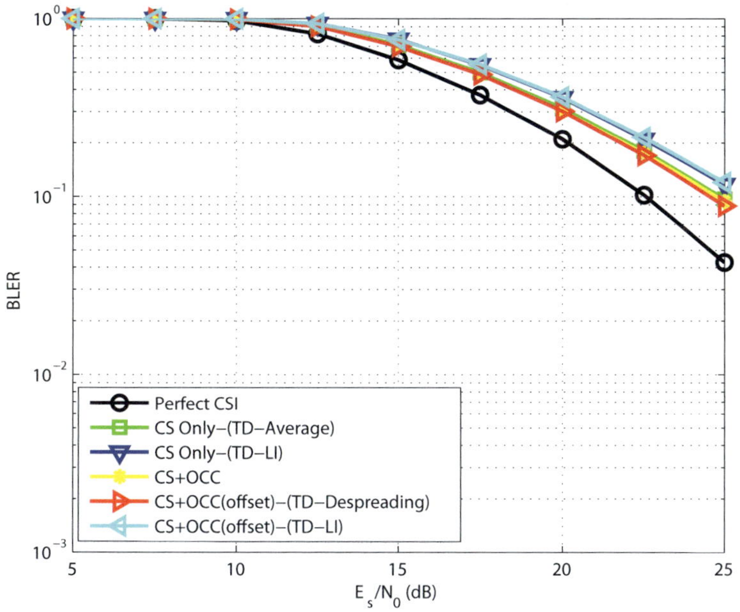

(b) 4RB, 64QAM

Fig. 6. BLER performance (2×2 MIMO, TU, 3km/h)

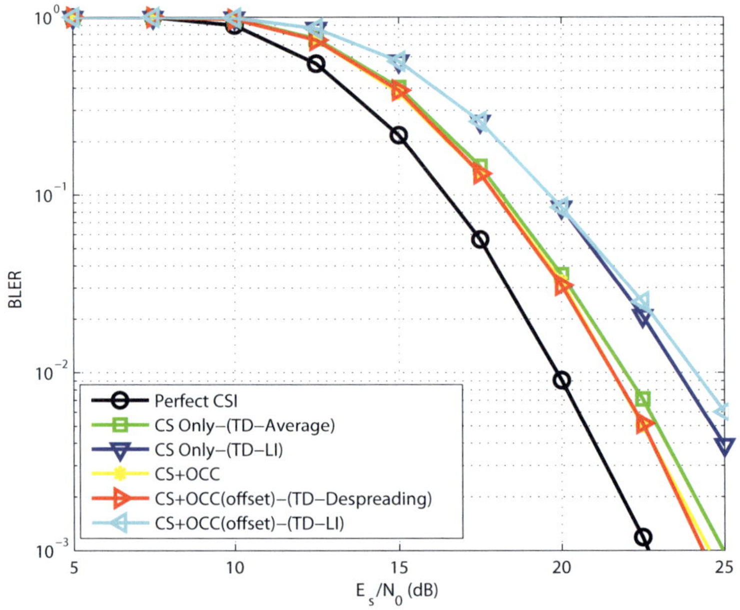

(a) 10RB, 16QAM

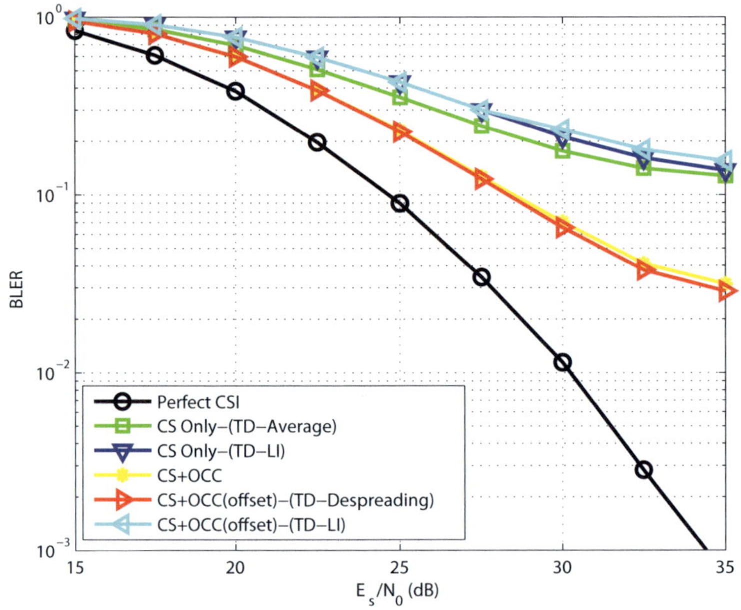

(b) 4RB, 64QAM

Fig. 7. BLER performance (4×4 MIMO, TU, 3km/h)

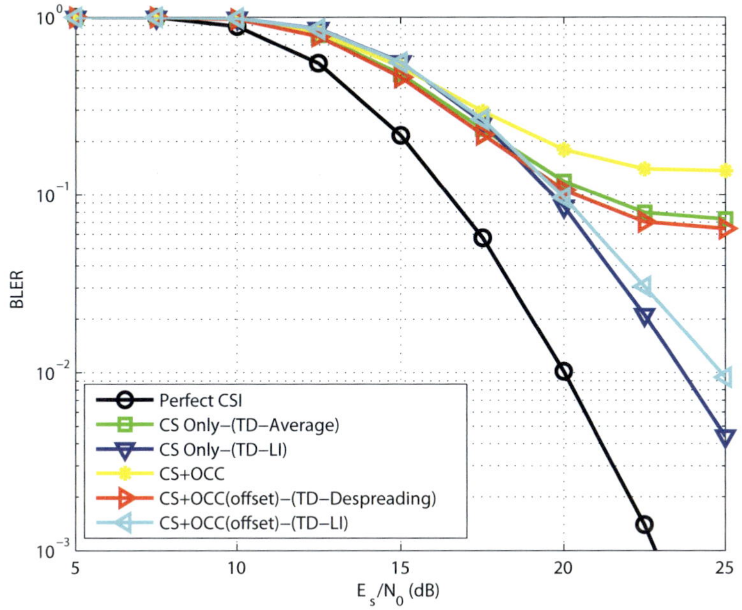

Fig. 8. BLER performance (4 × 4 MIMO, 10RB, 16QAM, TU, 30km/h)

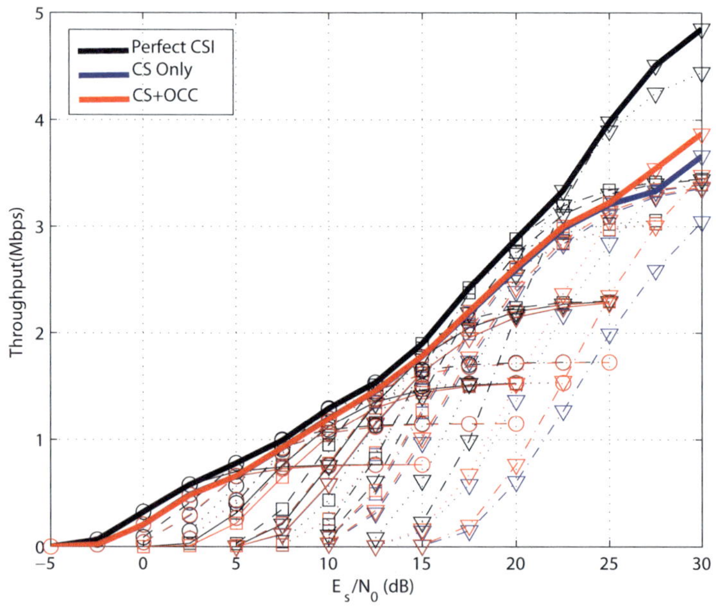

(a) 4RB

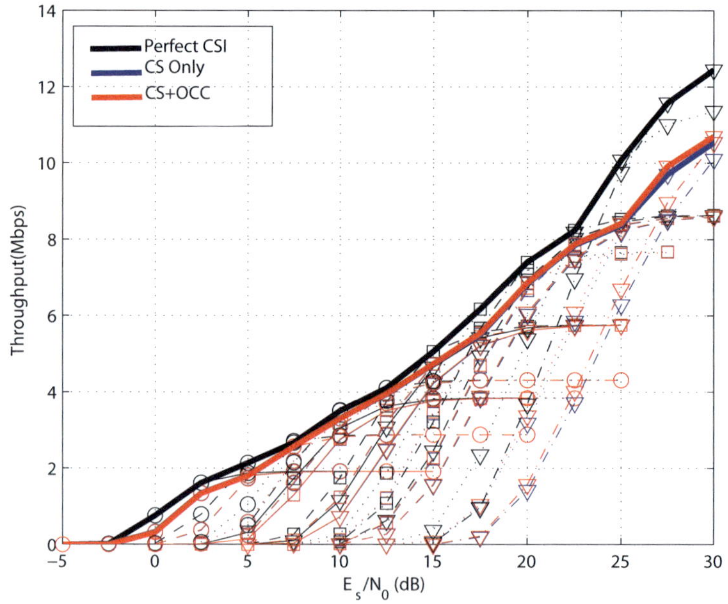

(b) 10RB

Fig. 9. Throughput performance (2×2 MIMO, TU, 3km/h)

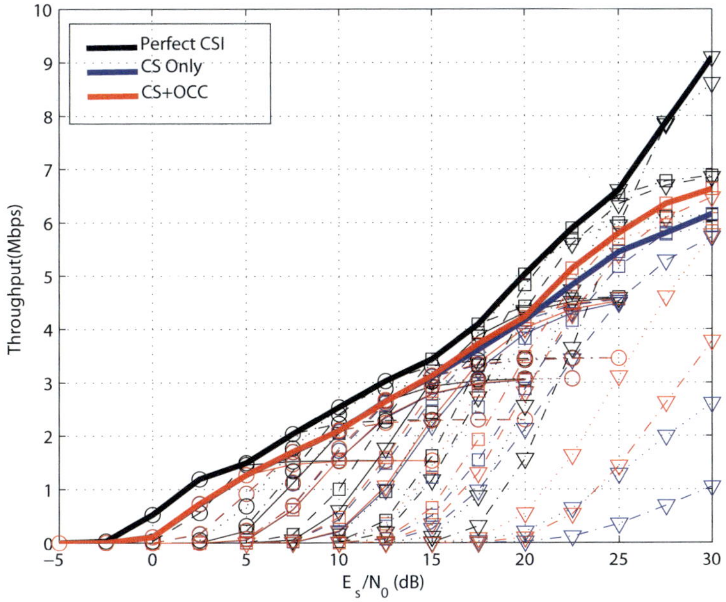

(a) 4RB

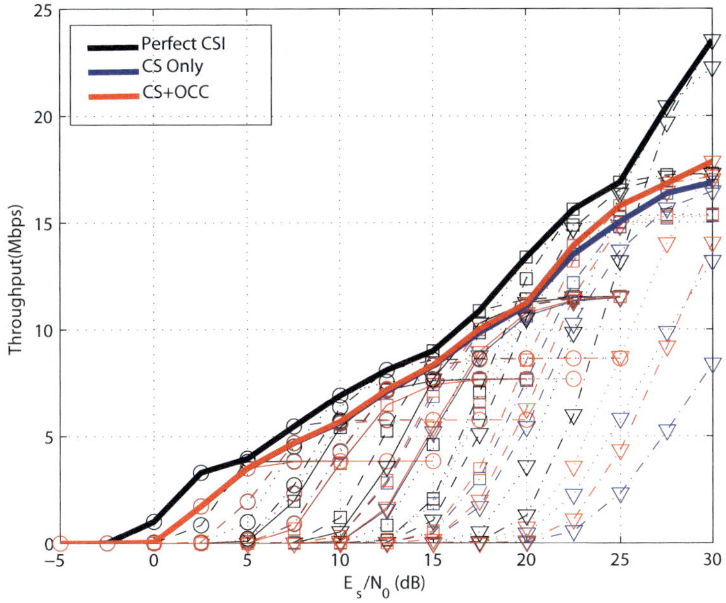

(b) 10RB

Fig. 10. Throughput performance (4×4 MIMO, TU, 3km/h) could be viewed as the throughput performance with the ideal adaptive modulation and coding. From Fig. 9 and Fig. 10, we can observe that the introduction of OCC can improve throughput to some extent when some of the following situations are satisfied, i.e., the antenna number is large, the RB number is small, and the sigal-to-noise ratio is high.

5. Some basic conclusions and standardization progress

DMRS enhancement is a key step to support MIMO transmission, including SU-MIMO, MU-MIMO and CoMP, for LTE-A uplink. In this study different DMRS design schemes as well as channel estimation are investigated. In addition to the baseline of CS extension, the further combination of CS with OCC is also discussed. In addition to the special advantage for MU-MIMO and CoMP, CS + OCC is also attractive for high-order SU-MIMO to further suppress the interferences among the increasing multiple spatial layers. With the enhanced DMRS design and improved channel estimation, a higher spectrum efficiency can be realized in LTE-A uplink. Meanwhile, considering the backwards compatibility, as less as possible modification to the current LTE specification is preferred.
In the recent 3GPP RAN1 meetings, it was agreed that (3GPP, R1-102601)

- Introduce the OCC in Rel-10 without increasing uplink grant signaling overhead
- OCC can be used for both SU-MIMO and MU-MIMO

More design details about DMRS enhancement, such as CS and OCC linkage, DMRS hopping, etc., are still under discussions and hopefully the uplink DMRS enhancement for LTE-A will be finalized by the end of 2010.

6. References

3GPP. R1-102601: Final report of 3GPP TSG RAN WG1 #60bis, *3GPP TSG RAN WG1 Meeting #61*, Montreal, Canada, May 10-14, 2010.

3GPP TR 36.814 *http://www.3gpp.org/ftp/specs/html-info/36814.htm*

3GPP TS 36.211 *http://www.3gpp.org/ftp/specs/html-info/36211.htm*

Chu, D. C. (1972). Polyphase codes with good periodic correlation properties. *IEEE Trans. Info. Theory*, Vol. 18, No. 4, July 1972, pp. 531-532, ISSN 0018-9448

Edfors, O.; Sandell, M.; van de Beek, J. J.; Wilson, S. K. & Borjesson, P. O. (2000). Analysis of DFT-based channel estimators for OFDM. *Wireless Pers. Commun.*, Vol. 12, No. 1, Jan. 2000, pp. 55-70, ISSN 0929-6212

Hou, X.; Zhang, Z. & Kayama, H. (2009). DMRS design and channel estimation for LTE-Advanced MIMO uplink, *Proc. IEEE VTC09-Fall*, Alaska, USA, Sept. 20-23, 2009.

Oppenheim, A. V.; Schafer, R. W. & Buck, J. R. (1999). *Discrete-Time Signal Processing, 2nd ed.*, Prentice Hall, ISBN 013216292X, New Jersey

Texas Instruments. R1-091843: Discussion on UL DM RS for SU-MIMO, *3GPP TSG RAN WG1 Meeting #57*, San Francisco, USA, May 4-8, 2009.

Made in the USA
Columbia, SC
01 July 2026

f6816350-74ba-4ceb-ae96-f34d9f56c9b3R01